Lecture Notes in Computer Science 13207

More information about this series at https://link.springer.com/bookseries/558

Jan Mazal · Adriano Fagiolini ·
Petr Vasik · Michele Turi ·
Agostino Bruzzone · Stefan Pickl ·
Vlastimil Neumann · Petr Stodola (Eds.)

Modelling and Simulation for Autonomous Systems

8th International Conference, MESAS 2021
Virtual Event, October 13–14, 2021
Revised Selected Papers

 Springer

Editors
Jan Mazal 🆔
NATO M&S COE
Rome, Italy

Adriano Fagiolini 🆔
University of Palermo
Palermo, Italy

Petr Vasik 🆔
Brno University of Technology
Brno, Czech Republic

Michele Turi
NATO M&S COE
Rome, Italy

Agostino Bruzzone
Savona Campus
University of Genoa
Savona, Italy

Stefan Pickl
Bundeswehr University Munich
Neubiberg, München, Germany

Vlastimil Neumann
University of Defence
Brno, Czech Republic

Petr Stodola
University of Defence
Brno, Czech Republic

ISSN 0302-9743   ISSN 1611-3349   (electronic)
Lecture Notes in Computer Science
ISBN 978-3-030-98259-1   ISBN 978-3-030-98260-7   (eBook)
https://doi.org/10.1007/978-3-030-98260-7

This Springer imprint is published by the registered company Springer Nature Switzerland AG
The registered company address is: Gewerbestrasse 11, 6330 Cham, Switzerland

Preface

This volume contains selected papers presented at the Modelling and Simulation for Autonomous Systems (MESAS) Conference, held during on October 13–14, 2021. MESAS 2021 was planned to take place in Rome COVID-19 pandemic but was instead held virtually due to the ongoing COVID-19 pandemic.

The initial idea to launch the MESAS project was introduced by the NATO Modelling and Simulation Centre of Excellence in 2013, with the intent to bring together the Modelling and Simulation and the Autonomous Systems/Robotic communities and to collect new ideas for concept development and experimentation in this domain. From that time, the event has gathered (in regular, poster, and way ahead sessions) fully recognized experts from different technical communities in the military, academia, and industry.

The main topical parts of the 2021 edition of MESAS were "Future Challenges of Advanced M&S Technology", "M&S of Intelligent Systems", and "AxS in Context of Future Warfare and Security Environment". The community of interest submitted 52 papers for consideration. Each submission was reviewed by three Technical Committee members or selected independent reviewers. The committee, in the context of the review process outcome, decided to accept 34 papers to be presented and 32 of these papers were accepted to be included in the conference proceedings.

December 2021

Jan Mazal
Adriano Fagiolini
Petr Vasik
Michele Turi
Agostino Bruzzone
Stefan Pickl
Vlastimil Neumann
Petr Stodola

Preface

MESAS 2021 Organizer

NATO Modelling and Simulation Centre of Excellence
(NATO M&S COE)

The NATO M&S COE is a recognized international military organization activated by the North Atlantic Council in 2012 and does not fall under the NATO Command Structure. Partnering nations provide funding and personnel for the centre through a memorandum of understanding. The Czech Republic, Italy, the USA, and Germany are the contributing nations, as of this publication. The NATO M&S COE supports NATO transformation by improving the networking of NATO and nationally owned M&S systems, promoting cooperation between nations and organizations through the sharing of M&S information, and serving as an international source of expertise.

The NATO M&S COE seeks to be a leading world-class organization, providing the best military expertise in modelling and simulation technology, methodologies, and the development of M&S professionals. Its state-of-the-art facilities can support a wide range of M&S activities including, but not limited to, education and training of NATO M&S professionals on M&S concepts and technology with hands-on courses that expose students to the latest simulation software currently used across the alliance; concept development and experimentation using a wide array of software capability and network connections to test and evaluate military doctrinal concepts as well as new simulation interoperability verification; and the same network connectivity that enables the COE to become the focal point for NATO's future distributed simulation environment and services. Further details can be found at https://www.mscoe.org/.

Organization

General Chairs

Michele Turi	NATO M&S COE, Italy
Adriano Fagiolini	University of Palermo, Italy
Agostino Bruzzone	University of Genoa, Italy
Stefan Pickl	Universität der Bundeswehr München, Germany
Vlastimil Neumann	University of Defence, Czech Republic

Organization Committee Chair

Jan Mazal	NATO M&S COE, Italy

Organization Committee

Petr Vasik	Brno University of Technology, Czech Republic
Petr Stodola	University of Defence, Czech Republic
Dana Kristalova	University of Defence, Czech Republic
Marco Biagini	NATO M&S COE, Italy
Jiri Novotny	NATO M&S COE, Italy
Luca Palombi	NATO M&S COE, Italy

Proceedings Committee Chair

Petr Vasik	Brno University of Technology, Czech Republic

Proceedings Committee

Roman Byrtus	Brno University of Technology, Czech Republic
Anna Derevianko	Brno University of Technology, Czech Republic
Ivan Eryganov	Brno University of Technology, Czech Republic
Jiri Novotny	NATO M&S COE, Italy

Technical Committee Chair

Petr Stodola	University of Defence, Czech Republic

Technical Committee

Ronald C. Arkin	Georgia Institute of Technology, USA
Özkan Atan	Van Yüzüncü Yil University, Turkey
Richard Balogh	Slovak University of Technology in Bratislava, Slovakia
Marco Biagini	NATO M&S COE, Rome, Italy
Antonio Bicchi	University of Pisa, Italy
Marcin Bielewicz	NATO Joint Force Training Centre, Poland
Dalibor Biolek	University of Defence, Czech Republic
Erdal Cayirci	University of Stavanger, Norway
Andrea D'Ambrogio	University of Rome Tor Vergata, Italy
Frederic Dalorso	French Air Force, France
Walter David	Ronin Institute, USA
Jan Faigl	Czech Technical University in Prague, Czech Republic
Jan Farlik	University of Defence, Czech Republic
Pavel Foltin	University of Defence, Czech Republic
Petr Františ	University of Defence, Czech Republic
Małgorzata Gawlik-Kobylińska	War Studies University, Poland
Kamila Hasilová	University of Defence, Czech Republic
Václav Hlaváč	Czech Technical University in Prague, Czech Republic
Jan Hodický	University of Defence, Czech Republic
Jan Holub	Czech Technical University in Prague, Czech Republic
Jaroslav Hrdina	Brno University of Technology, Czech Republic
Martin Hubáček	University of Defence, Czech Republic
Karel Hájek	University of Defence, Czech Republic
Thomas C. Irwin	US Department of Defense, USA
Piotr Kosiuczenko	Military University of Technology, Poland
Patrik Kutílek	Czech Technical University in Prague, Czech Republic
Václav Křivánek	University of Defence, Czech Republic
Dana Kristalova	University of Defence, Czech Republic
Jan Leuchter	University of Defence, Czech Republic
Paweł Maciejewski	Czech Technical University in Prague, Czech Republic
Pavel Maňas	University of Defence, Czech Republic
Vladimír Mostyn	Technical University of Ostrava, Czech Republic
Pierpaolo Murrieri	Leonardo S.p.A., Italy
Andrzej Najgebauer	Military University of Technology, Poland
Jan Nohel	University of Defense, Czech Republic
Petr Novák	Technical University of Ostrava, Czech Republic
Lucia Pallotino	University of Pisa, Italy
Stefan Pickl	Universität der Bundeswehr München, Germany
David Place	Naval Postgraduate School, USA
Josef Procházka	University of Defence, Czech Republic
Dalibor Procházka	University of Defence, Czech Republic
Paolo Proietti	MIMOS, Italy

Contents

AxS/AI in Context of Future Warfare and Security Environment

M&S of Intelligent Systems – R&D and Application

UAV Based Vehicle Detection on Real and Synthetic Image Pairs: Performance Differences and Influence Analysis of Context and Simulation Parameters

Michael Krump$^{(\boxtimes)}$ and Peter Stütz

Bundeswehr University Munich, Institute of Flight Systems, Neubiberg, Germany
{michael.krump,peter.stuetz}@unibw.de

Abstract. The automated evaluation of the airborne acquired sensor data plays a decisive role in the use of UAV systems (Unmanned Aerial Vehicles). To generate or augment training data for respective algorithms, virtual environments are often used to replace complex and cost-intensive real flights. However deviation in data characteristics between these two domains can cause performance differences, leading to the so called *"Reality Gap"*. Our current research focuses on the comparison of detection performance between different real and synthetic training and test datasets and the analysis of the underlying influencing factors and parameters. For a targeted evaluation, dedicated UAV flights are performed and a dataset with over 3300 aerial images of vehicles under different conditions is recorded. Based on the telemetry data and a remodeling of the test area in the virtual environment, coupled real and synthetic image pairs are generated. Using the deep learning based detector network YOLOv3, the performance differences between this image pairs are evaluated for different training set compositions (real, synthetic-only and mixed training data). In addition, the impact of specific object, context, environmental, or simulation parameters on detection performance will be investigated to identify design guidelines that improve training data generation and lead to more robust detection models.

Keywords: UAV · Vehicle detection · Virtual environment · YOLO · Deep learning · CNN · Detection analysis · Performance factor · Image pairs · Synthetic duplicates · Reality gap · DJI M210

1 Introduction

Continuous developments and a wide variety of applications have led to a significant increase in the use of UAVs for the acquisition of airborne image data. Automated and real-time sensor data processing is thereby necessary to achieve the desired level of autonomy. Vehicle detection on such UAV-based aerial imagery is a frequently considered use case in this context [1–6]. Since objects vary in size, shape, orientation, and background depending on the flight situation, this poses special challenges to the underlying deep-learning based detector network [2, 7]. To reduce the number of time-consuming and

© Springer Nature Switzerland AG 2022
J. Mazal et al. (Eds.): MESAS 2021, LNCS 13207, pp. 3–25, 2022.
https://doi.org/10.1007/978-3-030-98260-7_1

costly flight missions while still capturing the required variance, virtual environments are often used to generate synthetic sensor data that augment or completely replace real training datasets [8–12]. This raises the question of how such virtual environments must be designed to minimize the performance difference, called the *Reality Gap*, that occurs when using sensor data from different domains (reality and simulation). Analysis of coupled real and synthetic image pairs is used in the following investigations to identify influential context, object, environmental, sensor and simulation parameters.

2 Object of Research

In [12], using different training configurations with aerial imagery from both domains, we investigated the influence of synthetic training data on the detection performance of the common object detector YOLOv3 [13] in UAV-based vehicle detection. The UAVDT dataset [7] served as a benchmark and provided the real image data, while the synthetic images were generated using the Presagis M&S Suite [14]. An evaluation on the real UAVDT test set, when trained with the associated UAVDT training data, resulted in an Average Precision (AP) of 69.9% at an intersection over union threshold of 0.3. A hybrid training dataset with 20% admixed synthetic data improved the detection performance respectively the AP by more than 2% point. Pure synthetic training data alone was not sufficient and resulted in relatively low detection performances with AP values around 15% due to the *Reality Gap*.

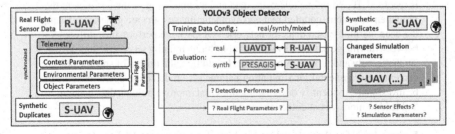

Fig. 1. Visualization and grouping of the different objects of investigation in this paper. Red corresponds to real datasets, blue corresponds to synthetic datasets and green describes the evaluation in terms of detection performance and the respective parameters analyzed. (Color figure online)

The models trained in [12] with real, synthetic and mixed data serve as a starting point for the investigations presented in this paper. We will now use coupled real and synthetic image pairs to investigate in more detail which correlations play a role in the composition of the *Reality Gap* and which of the following parameters influence the detection performance. An extended annotation of the image pairs enables the decoupled influence analysis of object, context and environment parameters, a subsequent overlay of the data sets the analysis of sensor effects and an adjustment of the render settings the analysis of different simulation parameters. Overall, conclusions are to be drawn about the stability of the black-box detector models and about possible improvements of the real and synthetic training data generation.

Figure 1 shows the different steps that are investigated in this paper to answer the following research questions:

(a) Which experimental setup is useful for **real UAV flights for dataset generation** and which parameter variations have to be recorded? A suitable storage of the telemetry data and the parameter states is necessary for the later decoupled analysis and for the generation of the synthetic duplicates.

(b) How do the detector models trained in [12] with different training data compositions behave on the real and synthetic image pairs and what are the differences between these two domains? Which **context, environment and object parameters** have a significant impact on the detection performance and therefore influence the stability of the detector model?

(c) How does the change of individual **sensor and simulation parameters** affect the *Reality Gap* and the detection performance and which of these should therefore be more varied and considered in future training data generation?

3 Real and Synthetic Datasets

The following is an overview of the training and test datasets used for UAV based vehicle detection. Both real datasets and data from the synthetic domain are considered. The selection is downsized to datasets containing annotations in the form of bounding boxes (BB) around the occurring vehicles. In order to be able to train stable detector models that also have the necessary generalization capability, the data must fulfill various requirements [15]: Scale diversity due to different flight altitudes, perspective variation due to different viewpoints, different object orientations and high background complexity. Special attention is given to the generation of coupled real and synthetic sensor data that meet these requirements and have nearly identical image content due to the direct replication, thus allowing a reliable analysis of the detector performance differences on both domains.

3.1 R-UAV/S-UAV: Real Flight Setup and Synthetic Re-modeling

Special attention is given to the generation of coupled real and synthetic sensor data that meet these requirements and have nearly identical image content due to the direct replication, thus allowing a reliable analysis of the detector performance differences on both domains.

Hardware Setup. To record suitable real UAV sensor data, we conducted our own flight missions at the test area of the University of the Bundeswehr Munich. The flight system to be used had to meet several specifications. It should have an integration possibility for a gimbal with a corresponding camera system. The payload must be sufficiently dimensioned to carry an additional computer board that handles reproducible mission planning and data storage via a programming interface. A built-in GPS receiver with Real Time Kinetics (RTK) is required for highly accurate acquisition of current position and telemetry data. A DJI Matrice M210 RTK V2 multicopter with Zenmuse XT2 sensor system was used for the flights, enhanced with an Nvidia Jetson TX2 computer board. This had access to the onboard SDK (Software Development Kit) of the multicopter via ROS (Robot Operating System) based inter-process communication and was connected to the control station on the ground via WLAN. Software programmed in C++ enabled control of the gimbal and camera and waypoint-based specification of the flight path, and thus fully automated and reproducible execution of the entire flight mission under various parameter variations. This setup is graphically illustrated in Fig. 2.

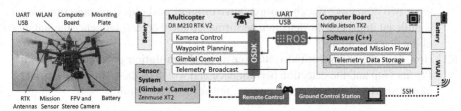

Fig. 2. *Left:* Multicopter system used with the respective components. *Right:* Schematic visualization of the underlying hardware and software setup and the associated interfaces. RTK: Real Time Kinetics; FPV: Front Person View; OSDK: Onboard SDK; SSH: Secure Shell

Experimental Setup. The goal of this automated execution is to capture reproducible scenarios with as many decoupled parameter variations as possible. The question arises which test setup and which gradation of the different parameters is reasonable. Since the image-based vehicle detection is in the focus and no tracking algorithms are investigated, the test vehicles were positioned statically.

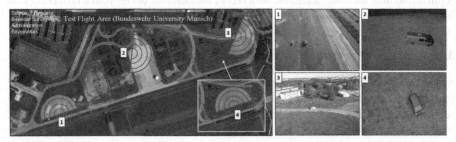

Fig. 3. Overview map of the test flight area and the selected test locations together with a visualization of the flight path. Sample images for each position are shown on the right.

In Fig. 3 the four selected locations are shown with associated sample images. Care was taken in the selection process to consider different substrates and road shapes as well as different background scenarios. There are three test vehicles with different characteristics in terms of color, size and vehicle type included in the dataset: Mercedes Sprinter (van, YOM 2004), Honda HR-V (SUV, YOM 2018) and Honda Jazz (small car,

YOM 2010). These form the point of interest (POI) of the waypoint-based flight path. A semicircular arrangement of the waypoints and the acquisition points ensures that all object orientations are acquired in equal numbers without duplicates. The resulting flight path for each location is also visualized to scale in the overview map in Fig. 3.

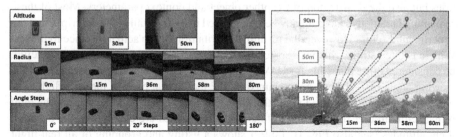

Fig. 4. Composition of the selected step sizes for the parameters altitude, radius and angle to the object. On the right side, the resulting variation in terms of the viewing angles is plotted, which together with the semicircular flight path cover the entire parameter space.

In the next step, the corresponding horizontal and vertical distances to the object have to be defined, at which a sensor image is acquired (see Fig. 4). This is based on the investigations in [12], which used synthetic data to show that despite discrete values for flight altitude, distance and object orientation in the training data, it is possible to generalize to arbitrary values during later evaluation. The step sizes used in that earlier work serve as the basis for the selection made here. The distance to the object is increased stepwise in a relatively uniform manner from 0 to 80 m, resulting in five semicircular flight patterns, where the first corresponds to a rotation of the multicopter above the test vehicle. To obtain uniformly distributed object sizes, the distances between flight altitudes are increased as the flight altitude increases. At each of the four selected altitudes (15 m, 30 m, 50 m, 90 m), the semicircular flight pattern is flown again. An image is taken in angular steps of 20°, with a short hover sequence beforehand to be able to exclude influences of motion blur and temporal delays in the telemetry data. Besides capturing the different object orientations, this also leads to a variation in the background, especially at low viewing angles.

Figure 4 illustrates that the selected step sizes cover all viewing angles and distances approximately uniformly. In addition, the test setup is adapted to the flight duration and generates 200 images per scenery.

Dataset Description. A total of 3329 UAV aerial images of vehicles were generated using the presented real flight setup. In order to suppress the influence of camera distortions especially in the outer image areas, the acquired sensor data are calibrated in advance. The original camera resolution of 4000 × 3000 pixels is reduced to 1000 × 750 pixels due to better comparability with other datasets and compatibility with the detector network used. To be able to investigate the detection performance, the test vehicles have to be manually annotated with appropriate bounding boxes. An *Ignore Area* around randomly located additional vehicles in the image prevents an influence of these on the evaluation. Figure 5 shows that this *Ignore Area* covers only a very small range in percentage for almost all images (<1%). In addition, the environmental conditions are classified manually for each image. The labels used describe characteristics such as season, shadows cast, wet or spotty road surface, reflections in the area of the vehicle, fog, car light, night or images for which a human observer would also have problems with recognition. Figure 5 shows not only the percentage of these environmental conditions but also the distribution of the other object and context parameters. Most of these remaining parameters have already been discussed in more detail in the description of the experimental setup. However, the resulting distribution of the actual distance to the test vehicle is interesting, which covers a range between 15 m and 120 m and can therefore be considered quite plausible for UAV applications. The accumulation of images with a viewing angle of −90° can be explained by the fact that at each of the four flight altitudes, the recording is started vertically above the test vehicle and then, due to the radii of the semi-circular flight paths, different viewing angles are captured at the different flight altitudes (see also Fig. 4). Table 1 shows a listing and grouping of all real flight parameters examined and annotated.

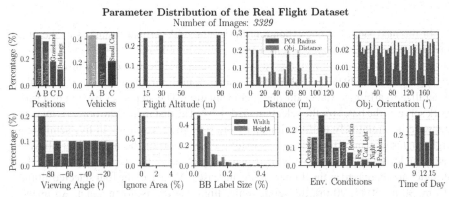

Fig. 5. Distribution of the different environmental, context and object parameters in the UAV dataset generated with the described real flight setup.

It should be mentioned that the real dataset generated in this way is not primarily intended to replace existing training datasets. The goal is rather the decoupled analysis of the parameter variations that were captured and annotated and, based on this, a generation of synthetic duplicates for the targeted investigation of the *Reality Gap*.

Synthetic Duplication. The physics-based sensor simulation of the Presagis Modeling and Simulation Suite [14] was used for the re-modeling of the test flight site in the virtual world in CDB (Common Database) format. The basis of the modeling is a georeferenced aerial image with a ground resolution of 20 cm per pixel, which is overlaid on the elevation data. A material classification of the subsurface is used to enhance fine structures and details on roads or grass areas by multiplying the aerial image with matching texture patterns, so-called *HyperTextures*. True-to-scale grid models of the existing buildings are textured accordingly and finally represent the occurring objects together with 3D volume trees and adapted 3D duplicates of the test vehicles. Synchronized telemetry data in the form of GPS coordinates, flight altitude of the multicopter and gimbal orientations are stored simultaneously with each sensor image acquired during the flight. Based on these, virtual poses are simulated in the virtual world and the resulting rendered synthetic duplicates are stored. Properties related to the sensor, such as resolution or *Field of View* (FOV), are taken into account as well as the corresponding environmental influences are simulated with the integrated atmospheric attenuation simulation *MOSART* [16]. The simulation environment also offers the possibility to adjust various sensor or simulation parameters such as noise, quality of shadow cast or antialiasing techniques, which is mainly used for the investigations of point (c) of the research questions. The annotations of the synthetic images in the form of BBs around the test vehicles are automatically generated following the procedure described in [12].

Table 1. Listing of the varied and annotated real flight parameters, which are divided into the subgroups context, object and environmental parameters. In Sect. 5.3 their influence on the detection performance is examined in more detail.

	Context Parameters	Obj. Parameters	Env. Parameters
Real Flight Params. (RF)	Position (A-D)	Flight Altitude	Season (Summer; Autumn; Winter)
	Test Vehicle (A-C)	POI Radius	Shadow
	Ignore Proportion	Object Distance	Wet Surface
	BB (%): x, y, all	Orientation	Spotty Surface
	Occlusion	Viewing Angle	Reflection
	Time of Day		Fog
			Car Light
			Night
			Problem (Difficult Detection)

3.2 Training Datasets

In the following, we now present the training datasets that were used in [12] to train the deep-learning based object detector YOLOv3. The resulting models serve as a basis for the further investigations presented here.

Synthetic Training Dataset. In [12], we also described a method for automated generation of an annotated training dataset in virtual environments using the same synthetic database and the Presagis programming interface. Nested loops were used to iterate over 38 different 3D vehicle models (extended to 80 models by re-coloring them according to the worldwide car color distribution), six locations and other parameters like object orientation, flight altitude or camera radius. The six positions cover different scenarios and backgrounds, the vehicle orientation varied in 30° increments, and discrete flight altitudes of 15 m, 30 m, 50 m, and 90 m were considered. By permanently pointing the camera at the vehicle object and varying the camera distances between 0 m, 20 m, 40 m and 80 m, different view angles were also generated. In this way, a training dataset was generated with over 93000 automatically annotated images and 86000 occurring vehicles. Random values for time of day, visibility and noise fraction should additionally increase the variation. Figure 6 shows some examples from the dataset, which serves as a basis for the analysis of relevant image properties and parameters when using synthetic training data. To compare the performance differences with real training data and to be able to validate the evaluation of the acquired real and synthetic UAV image pairs (R-UAV/S-UAV) against real benchmark data from another geographic location, an established real dataset is also needed.

Fig. 6. Annotated sample images from the synthetic training data set also generated using the re-modeled virtual environment.

UAVDT. The UAVDT [7] dataset is an open-access benchmark dataset for various UAV applications such as object detection, single- and multi-object tracking. It includes the required variation in flight altitude, viewing angle, and environmental conditions, and unlike our previously presented datasets, it captures more urban environments with multi-lane roads and intersections. The training dataset is used to generate a real trained detector model and the test dataset as a separately identified subset serves as a real benchmark for comparison with our UAV datasets described in Sect. 3.1.

Fig. 7. Annotated sample images from the real UAVDT benchmark dataset [7].

4   Variations of Sensor and Simulation Parameters

In order to better analyze and understand the influencing factors and susceptibilities of the trained detection models, the R-UAV and the S-UAV dataset are overlaid with different sensor effects and in this way several test sets are generated, which allow a decoupled analysis of the properties. Figure 8 shows an overview of the parameters considered, which are described in more detail below.

Color-Size: This group changes various color properties and the resolution. A color cast is simulated by randomly changing the V-value in the HSV color space. To investigate the general importance of colors, the monochrome condition is considered. Also included is a reduction in JPEG compression quality to 25/100 and a reduction in resolution to 400 × 300 pixels, whereas the input size used into the YOLOv3 network was 608 × 608.

Lighting: This includes an increase in contrast and gamma. In addition, a method of data augmentation presented in [17] is used, which simulates different lighting conditions by randomly overlaying them with parallel brightness gradients or spots. In white balance, the pixel values are divided by the average intensity value and then normalized, which in this form serves to selectively increase the brightness and thus mimic reflections.

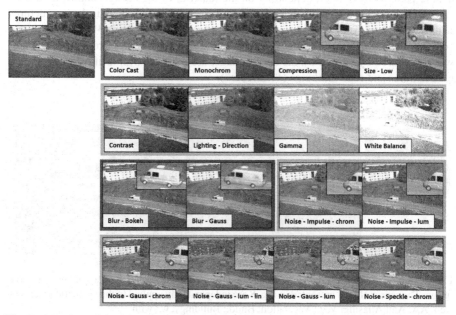

Fig. 8. Overview of the investigated sensor effects based on a real example image. The groups Color-Size (orange), Lighting (gray), Blur (blue) and Noise (yellow) are distinguished. chrom.: Chrominance; lum.: Luminance (Color figure online)

Blur: Here, on the one hand, a Gaussian blur with a kernel size of 5 × 5 pixels is considered, as it is also used in the pre-processing steps of various image processing algorithms, as well as a simulation of the Bokeh effect [18]. This describes a background blur as it is often used in photography and also simulates slight color errors as they can also occur with real lenses.

Noise: This group includes various forms of noise. A distinction is made between luminance noise, which only distorts the brightness values, and chrominance noise, which affects all color channels. Impulse noise, which is also often referred to as salt-and-pepper noise, corresponds to a manipulation of individual pixel values in the image. Gaussian noise results in an additive superposition of image and noise pattern, which in one case also has linear structures. Speckle noise, in contrast, corresponds to a multiplicative superposition and therefore affects brighter areas to a greater extent.

Simulation Parameters: Figure 9 also shows an overview of various simulation effects, for each of which a separate dataset was generated in addition to the sensor effects when creating the synthetic duplicates (S-UAV). These include the FXAA (*Fast Approximate Anti-Aliasing*) anti-aliasing method and the complete omission of anti-aliasing reduction methods. The *"Shadow-low"* group reduces the quality of the synthetic shadow generation by using a lower resolution and number of shadow maps, while in *"Shadow-no"* no shadows are simulated at all. *HyperTextures* overlay the aerial image with fine texture patterns by default, depending on the material of the background. It is investigated which effects and dependencies the detector has learned towards these structures with the *"HyperTextures-no"* group. Furthermore, datasets without 3D vegetation and additionally without 3D buildings are generated. Finally, in the last group, the road networks, textured by default by the aerial image, are overlaid with artificially generated road textures and road markings (*"wStreets"*).

Fig. 9. Overview of the additionally investigated simulation effects based on a synthetic example image. The sensor effects presented in Fig. 8 are independently applied to the synthetic images as well. AA: Anti Aliasing; Veg.: Vegetation; Build.: Buildings; w.: with

5   Evaluation and Results

In a first step, we analyzed how the performance of the real and synthetically trained detector models behaves on the generated real and synthetic image pairs compared to the benchmark datasets. We then investigated how the detection performance depends on the annotated real flight parameters and which sensor effects and simulation parameters play an influencing role in the generation of synthetic training data.

5.1 Experimental Setup

In the following, we briefly discuss the setting we used in [12] to train those models that will now be used for the analysis. The YOLOv3 [13] object detector pre-trained on ImageNet was adapted to our use case of UAV-based vehicle detection by transfer learning. Three types of training datasets were used: the real UAVDT dataset [7] (see Sect. 3.2), our synthetically generated dataset (see Sect. 3.2) and a mixed dataset where the real UAVDT dataset is augmented with a 20% synthetic data portion. The associated test sets (UAVDT and our synthetic test set) and the real and synthetic image pairs independent of them (see Sect. 3.1) are not included in the training data. The built-in methods of data augmentation were used, suitable anchor boxes were calculated for each training set and an input image size of 608 × 608 was used. The initial learning rate was 0.004 and was adjusted during training using exponential learning rate scheduling. By looking at the shape of the loss function and detection performance, the number of iterations was determined for each training configuration to avoid overfitting and underfitting. The performance metric used is *Average Precision* (AP) (as defined by the Pascal VOC Challenge since 2010 [19]), which corresponds to the area under the interpolated *Precision-Recall* (PR) curve at a specific *Intersection over Union* (IoU) threshold. IoU describes the overlap ratio between detected and ground truth BB and was set to 0.3 in our case to minimize the influence of inaccuracies and differences between manual hand-labeled real annotations and automatically generated annotations of the synthetic data.

5.2 Comparison of Performance Differences for Real and Synthetic Training Data

Figure 10 compares the results for the different test sets both in terms of the *Reality Gap* between different domains and in terms of the differences between datasets from the same domain. The real UAVDT training set yields a general detector that performs similarly in both domains and on all datasets. This means that the real data contains all the features that the detector network needs for detection even on synthetic data and that the *Reality Gap* in this direction is small. In general, better results are obtained on the synthetic data because they contain more pronounced features and lower noise components. It is worth mentioning that the precision of the detections on the real UAV images (R-UAV) drops comparatively at high recall values, even though the dataset is from the same domain. This indicates that the detection of unknown real data with partially different sceneries and objects is the most complex form of evaluation. This should be taken into account when evaluating performance.

The effects of purely synthetic training are shown in Fig. 10 right. The performance on the unknown synthetic test data is almost ideal (AP: 99%), where in contrast the learned model is not suitable for the real UAVDT test data (AP: 15%). In [12] we listed multiple reasons for this *Reality Gap*, such as different scenery between synthetic and UAVDT dataset, general dataset composition or object size and number. It stands to reason that overfitting to the synthetic data and features occurred during training.

There are numerous studies in the literature that have also found similar behavior and insufficient generalization ability with purely virtual training data without additional adaptations such as domain adaptation [8, 20–25]. Johnson-Roberson et al. [9] attribute this, among other things, to the fact that the variation and training benefit of a synthetic image is smaller than that of a real image, since the former has less complexity in terms of illumination, colors, textures, and patterns, which in turn has to be compensated by a larger number of synthetic images.

In this context, it is now interesting to consider the real and synthetic image pairs. On the synthetic duplicates, an AP of over 93% is achieved. Although these were also generated with the same simulation environment, similar parameters and similar geographical positions, a drop of six percentage points can thus be observed compared to the synthetic test set. This is most likely due to the fact that the test vehicles in the S-UAV dataset did not appear in the synthetic training dataset in terms of color and model. For the real R-UAV images, in contrast to the real UAVDT test images, a quite plausible detection performance was achieved despite the purely synthetic training (AP: 50%). This is probably because the synthetic training images capture very similar sceneries and context and object parameters (e.g., road types, vehicle positions, object sizes, building types, vegetation, etc.), since they were also generated on the re-modeled test flight area. Conversely, this means that the observed very large *Reality Gap* between UAVDT and synthetic test set (AP: 15% ↔ 99%) is due to two causes. One proportion (UAVDT, AP: 15% ↔ R-UAV, AP: 50%) is, as we conjectured in [12], solely due to differences in context and object parameters and is also called *Content Gap* [26]. Only the second fraction (R-UAV, AP: 50% ↔ S-UAV, AP: 93%), also called *Appearance Gap* [26], is actually due to the influence of synthetic image generation, since the image duplicates (R-UAV/S-UAV) are identical in terms of image content as a result of exact re-modeling. Kar et al. [26] used this fact to minimize the content based proportion of the *Reality Gap* by optimizing the attributes of the scene graph for synthetic dataset generation. In agreement with [27], it is also shown that the differences between different datasets of the same domain can be of a similar order of magnitude as the differences between domains.

In the last step, the detection performance for a mixed training set is now investigated. This is composed of all UAVDT training images augmented with 20% synthetic data. The thereby increased generalization ability of the learned model leads to a partly very significant increase of the detection performance for all test data compared to pure real training, as shown in Table 2. The higher recall values of the curves in Fig. 10 below underline the increased generalization ability, as this means that a larger fraction of the vehicles occurring in the test set is detected. In general, the mixed training achieves the best performance of all training configurations. Several publications [8, 10, 22, 28] also concluded, for both the object detection and semantic segmentation use cases, that augmenting the real training with synthetic data increases performance. Whether the real training images are extended with synthetic images or methods of domain adaptation are used or finetuning and pretraining of the models with synthetic data takes place, does not play a role for the effect in the first step.

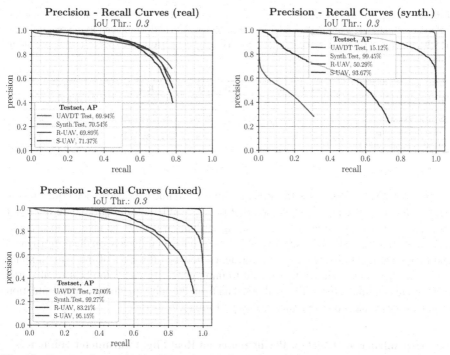

Fig. 10. Comparative analysis of detection performances for different training configurations (real, pure synthetic, mixed) on real and synthetic benchmark datasets (UAVDT/Synth. Test) and the presented real and synthetic image duplicates (R-/S-UAV). PR curves are used for evaluation. The AP values derived from this are shown in the legend in percent to the right of the associated test set.

Detailed conclusions allow again the consideration of the real and synthetic image pairs. As before, the precision-recall curves and thus the detection performances of the R-UAV and S-UAV dataset are between those of UAVDT and synthetic test set. This again argues for a split of the Reality Gap into Content and Appearance Gap, albeit in a diminished form due to the mixed training, but still present. Since the UAVDT training and test data are very similar in terms of scenery, object and context parameters, the improvement in the UAVDT test set due to the synthetic training portion is only 2.1% points. In contrast, despite the same performance with real training (UAVDT, AP =: 69.9% ↔ R-UAV, AP: 69.9%), the performance on the R-UAV dataset improved very significantly by 13.3% points with mixed training. This result is noteworthy in several respects. It again shows that performance differences between test datasets of the same domain can turn out to be quite significant. The added synthetic training data was generated at the re-modeled university and test flight area where the real flights for the R-UAV dataset generation took place, and thus contains similar conditions and scenarios. Overall, this comparison shows that selective augmentation of general real-world benchmark training data with synthetic imagery of later operational areas can lead to significantly better adaptation of the detector to the prevailing operational conditions there.

Table 2. Comparison of the detection performances of the different training configurations on the different test datasets using AP. In addition, the change of the AP compared to the real training, which serves as baseline, is given in each case. The highest performance per test set is printed in bold.

	UAVDT Test	Synth. Test	R-UAV	S-UAV
Real Training	69.9 %	70.5 %	69.9 %	71.4 %
Synth. Training	15.1 % -54.8	**99.5 %** +29.0	50.3 % -19.6	93.7 % +22.3
Mixed Training (Real + 20 % Synth.)	**72.0 %** +2.1	99.3 % +28.8	**83.2 %** +13.3	**95.2 %** +23.8

In summary, it can be stated that real benchmark training data lead to detection models with relatively stable performance and that purely synthetic training data are often not sufficient to train a general detector. During evaluation it must be taken into account that the domain as well as the scenery and context have an influence on the performance and thus significant differences in performance between test data of the same domain can occur. It has been shown that mixed training generally yields the best results, and that targeted augmentation with synthetic data from the subsequent operational situation promises the greatest possible increase in performance.

5.3 Dependence of Detection Performance on Real Flight Parameter Influences

The YOLOv3 detector, like most deep-learning based neural networks, provides a black-box model that generally does not allow any conclusions about the obtained detections without special analysis tools. In order to be able to determine the relevant influencing factors and to obtain starting points for an improved compilation of the training data and for a targeted extension with synthetic data material, it will be shown in the following how the detection performance depends on certain parameters that are considered as decoupled as possible. The R-UAV dataset serves as the basis for the evaluation, since, in contrast to publicly available real benchmark datasets, it contains the associated object, context and environmental conditions for each image in addition to the common BB annotations. Only by this special dataset generation described in Sect. 3.1 conclusions on the decoupled parameters and thus an indirect analysis of the Content Gap is possible.

Figure 11 shows an overview of the considered parameters and analyzes the changes in terms of detection performance for certain subsets of images compared to all images of the R-UAV dataset. In the first step, the synthetically trained detector model is analyzed. The achieved AP over all images is 50.3% as already discussed in Sect. 5.2. If we look at the detection performances averaged in it for the four occurring positions individually, clear differences become visible. While at the positions "Grassland" and "Parking Lot" above average detection results are achieved, the performance at the positions "Street" and "Buildings" is very low. This can have several reasons. One reason is most likely the ground or the surrounding of the vehicle. At the positions with the high AP values, the vehicle is standing on a larger homogeneous surface (asphalt or grass, see Fig. 3). This has the effect that even at oblique viewing angles the contours of the test vehicle can

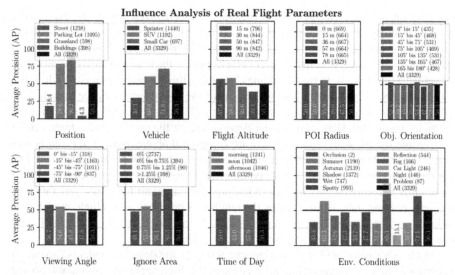

Fig. 11. Overview plot of the influences of various annotated parameters on detection performance measured in AP. The results refer to the real R-UAV dataset as the test set and the synthetically trained detector model.

be distinguished relatively well from the surroundings, thus enabling easier and more reliable detection and thus a higher TP percentage.

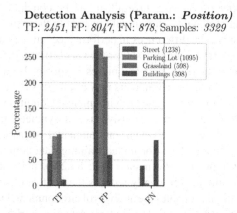

Fig. 12. Analysis of TP/FP/FN detections in relation to the number of images in each subcategory for the "Position" parameter. Each image contains only one test vehicle at a time. The basis is the evaluation of the synthetically trained detector model on the real R-UAV dataset.

Figure 12 describes the percentage of TP/FP/FN proportions for the individual subgroups for the parameter "Position" in relation to the number of occurring images in the respective subgroup. For the positions "Parking Lot" and "Grassland" the percentage is very close to 100%. Since only one car occurs in the image at a time due to the test setup, this means that almost all test vehicles are correctly recognized despite the exclusively synthetic training. The poorer performance for the "Street" position is partly due to the lower TP percentage and partly also to the higher FP percentage. This in turn is most likely due to the scenery, as the concrete blocks next to the road can easily be mistaken for vehicles (see Fig. 3). In the "Buildings" position, this FP proportion is very low, but the TP proportion is also very low, which leads to very poor performance overall. Due to the lower number of images in this position (398), this may well interact with the "Vehicle" parameter. Here, too, a clear dependence on the type of test vehicle can be seen. While the SUV and the small car are recognized above average, the van is recognized worse despite its more pronounced contour and its size. This is due to the fact that the proportion of similar vehicle types in the 38 different 3D vehicle models used for training is comparatively low. Both findings are important for future generation of synthetic training data. By increasing the number of positions and vehicle models, the variation with respect to these parameters should be increased, thereby achieving higher stability in the range of these parameters during evaluation.

As already shown in [12] using synthetic test data, the detector model is capable of detecting arbitrary continuous gradations despite the discrete parameter gradations used in the synthetic training data generation process for the parameters altitude, radius to the Point of Interest (POI) and object orientation. This can also be confirmed in the case considered here for the evaluation on the real R-UAV data. For the parameters "Flight Altitude", "POI Radius", "Object Orientation" and "Time of Day" only relatively small differences can be observed for the individual subgroups and also the viewing angles generated by them show a similar performance. Only with increasing flight altitude the quality of the detections tends to decrease, which, however, is to be expected due to the associated smaller object size. Overall, it can be concluded that the synthetic generation process was reasonably chosen with respect to these parameters and that no adjustments are necessary. The differences in performance with regard to the proportion of the "Ignore Area" are less meaningful, since only very little image material is available for high proportions as desired (90 and 180 pieces, respectively) and thus side effects from the "Position" and "Vehicle" parameters play a role. It is to be expected that these side effects also distort the results of the "Environmental Conditions".

In general, overfitting due to the exclusive use of synthetic training data makes the model highly susceptible to individual environmental or parameter variations. In contrast, Fig. 13 shows the analysis for mixed training data composition. In addition to a significantly higher mean performance (AP: 83.2%), it is clear that the higher generalization ability of the model greatly reduces the fluctuations in single parameter variations in general. This also enables the analysis of environmental influences. Changes in vegetation due to different seasons do not affect detection performance, nor does the presence of a shadow cast. It is worth mentioning that due to a high variation in this regard, even a wet or spotty surface after rain does not cause a difference in performance. On the other hand, images taken at night or with the vehicle lights switched on, images with test vehicles that are difficult for the human observer to recognize (*"Problem"*) or images with reflections in the area of the test vehicle lead to a lower detection performance. Surprisingly, the condition *"Fog"* results in an above-average AP. In part, this is because

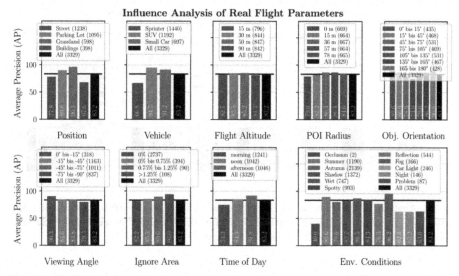

Fig. 13. Similar plot of parameter influences on detection performance as in Fig. 11 for the R-UAV dataset, but for a mixed trained detector model.

the recordings in *"Fog"* exclusively include the positions *"Grassland"* and *"Parking Lot"* and the test vehicle *"Small Car"*. Nevertheless, it can be stated that fog does not lead to a significant degradation of detection performance when it was included in the training data (UAVDT + synthetic training set).

Overall, an analysis of the *Content Gap* was possible by specifically varying and annotating certain real flight parameters. It turned out that the parameters *"Position"* and *"Vehicle"* play a decisive role and have to be considered in considered in more detail in the synthetic data generation, while object parameters like orientation or viewing angle have little influence under the conditions examined here. Images taken at dusk or reflections from the vehicle degrade performance, while other environmental factors such as season, shadows cast, wet surfaces or fog have no effect. A model trained with mixed data is generally more stable to such changing parameters.

5.4 Dependence of Detection Performance on Sensor and Simulation Parameters

The overall objective of this paper is to investigate relevant factors influencing the detection performance when comparing real and synthetic image data. Since the black-box detector model does not allow any conclusions to be drawn about this, an analysis of the performance differences when certain sensor and simulation parameters are changed can also provide clues to the causes of the observed *Reality Gap* and any vulnerabilities of the trained model.

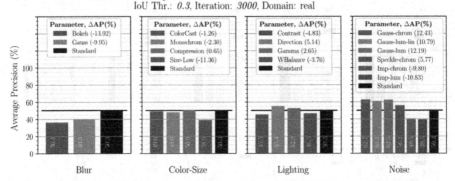

Fig. 14. Analysis of the influence of different sensor effects on the AP for the real R-UAV dataset using the purely synthetically trained detector model. WBalance: White Balance; chrom.: Chrominanz; lum.: Luminance; lin.: linear; Imp.: Impulse

This is particularly interesting in the context of the approach known as domain randomization. It states, that if there is sufficient variation in the synthetically generated training data with respect to sensor effects, environmental conditions and simulation styles, reality is considered by the detector as just another variation [29, 30]. In the following, we therefore analyze the performance differences for the variants of the R-UAV and S-UAV datasets described in Sect. 4 using the synthetically trained detector model.

Figure 14 gives an overview of the performance changes on the real R-UAV dataset with an AP of 50.3% as baseline and Fig. 15 on the synthetic S-UAV dataset with an AP of 93.7% as baseline. The evaluation on the real data is more important, since it is more similar to the later use case, and furthermore the influences are biased in the synthetic S-UAV data due to the very high similarity to the synthetic training data and the resulting very high detection performance.

Both types of blur investigated lead to a significant decrease in detection performance in the real data. An analysis of the PR curves shows that the Gaussian blur only leads to a downward shift of the curve towards lower precision values, while for the simulated Bokeh effect the additional color errors (see Fig. 8) rather cause a reduction of the recall. Overall, the blurring leads to a lower saliency of the relevant features, which results in significant performance degradation especially for the complex real test data. Therefore, different forms and strengths of blur should definitely be considered in future synthetic training data generation. In the case of synthetic data, this effect does not have an impact due to the more pronounced features. It remains to be mentioned that nevertheless, for both real and synthetic data, the number of FP detections has also decreased due to the more unclear features.

Changes in color (color cast or grayscale) and compression artifacts have comparatively small effects on real data, which indicates a good generalization capability of the detector in this respect. A reduction of the resolution decreases the AP by more than 11% points. In the synthetic data (see Fig. 15) on the other hand, color changes do have negative effects. This again speaks for an overfitting to the synthetic training data. A

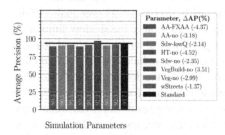

Fig. 15. Analysis of the influence of different sensor effects on the AP for the synthetic S-UAV dataset using the purely synthetically trained detector model, similar to Fig. 14. In addition, the effects of various simulation parameters are also investigated here. AA: Anti-Aliasing; Sdw.: Shadow; lowQ: low Quality; HT: *HyperTextures*; Veg.:Vegetation

reduction of the resolution probably has no negative effect on the synthetic data due to the already low level of detail. It is not possible to derive specific design guidelines from this group.

In both real and synthetic evaluation, increasing contrast and simulating local blinding through white balance has a negative impact on detection performance. Both lead to lower recall values and can therefore be used to increase the variance in the future generation of synthetic training data. This is in agreement with Sect. 5.3, where real images with reflections in the area of the vehicle also lead to a reduced detection performance. An increase of the gamma value or the lighting conditions has no or even a positive influence.

In addition, different forms of luminance as well as chrominance noise were considered. It is noticeable that with synthetic training all types of Gaussian noise and also Speckle noise lead to a significant increase in AP, in some cases by more than 10% points, whereas impulse noise leads to a decrease in performance, also on the order of 10% points. The superposition of all images with random intensities of white Gaussian luminance noise during the generation process of the underlying synthetic training data (see [12]) is probably the cause for this effect. Impulse noise or salt-and-pepper noise, on the other hand, leads to changes in individual pixel values in the image. Several publications [31–34] from the separate research areas "One Pixel Attack" and "Adversarial

Attack" show that a deception of the detector and a decrease of the detection performance can be expected by targeted pixel manipulations or noise. For the synthetic test data the effects are not visible for the reasons mentioned above. Therefore, it can be concluded that for future training data generation, the proportion of images overlaid with Gaussian noise should be reduced and instead a proportion with impulse noise should be added to increase the variance in the training set.

For the S-UAV dataset, Fig. 15 additionally investigated the influence of different simulation parameters presented in Sect. 4. The effects are rather small due to the already very high detection performance, but still provide important clues for image properties that have an influence on the detection result. Changing the antialiasing method or omitting it completely when creating the synthetic duplicates leads to a deterioration of the test result. This means that the network is sensitive to the features introduced or changed by antialiasing. Variation in this respect in the synthetic training data generation should therefore be considered as tending to be positive. In contrast, a lower quality of the synthetic shadow generation and even the complete omission of the shadow simulation has comparatively little effect. The absence of *HyperTextures*, on the other hand, leads to a reduction in AP of over 4% points. From this it can be concluded on the one hand that not only the vehicle but also the background plays a role in the detection and on the other hand that the detector has learned the structures used and therefore a variation should be incorporated in this respect in the future. As expected, the result improves when there are no 3D buildings and no 3D vegetation in the image, because then the focus is on the 3D vehicle object and the FP fraction decreases drastically. The mere absence of 3D vegetation, on the other hand, results in lower performance due to a higher FP fraction, and the addition of artificial road textures has no significant effect.

Overall, it can be stated that the effects are smaller for the synthetic S-UAV dataset, since it comes from the same domain as the training data and contains more pronounced features. The sensor effects of blur, contrast, white balance, and impulse noise should be considered in future synthetic training data generation in favor of higher stability and variation, and in favor of reducing overfitting. From the group of simulation parameters, this concerns various anti-aliasing methods, *HyperTextures*, and vegetation. With respect to coloration, compression artifacts and changing illumination effects, however, the synthetically trained detector is comparatively stable and insensitive.

6   Conclusion and Future Work

In this paper, we described the generation process of coupled real and synthetic image pairs for the use case of UAV-based vehicle detection and investigated the influence parameters and the components of the *Reality Gap* for different training data configurations using these data. In the first part, the hardware setup used to acquire the real UAV aerial images of vehicles is described. Automated execution of the flight missions enabled the generation of reproducible imagery under different environmental conditions. Besides the description of the generated dataset with more than 3300 vehicle images, the annotated and specially selected real flight parameters like flight altitudes, object orientations, positions or viewing angles are also discussed. Based on the recorded telemetry data and with the help of the test flight site remodeled in the virtual environment, the synthetic image duplicates are finally generated. The detector models based on the YOLOv3 network generated in [12] with real, synthetic and mixed training data were used for evaluation. Using the paired images with identical image content, it could be concluded that the *Reality Gap* is composed of two components: the *Appearance Gap* and the *Content Gap*. This also causes that performance differences between datasets originating from the same domain can have a similar magnitude due to content differences as the performance differences between domains. Mixed training data generally leads to a higher generalization capability of the detector and it was further shown that by selectively augmenting general real-world benchmark training datasets with synthetic images of the subsequent application areas, a significantly better fit of the detector model to the prevailing deployment conditions there can be achieved.

The method presented in the second part of the paper for performing an influence analysis of a wide variety of parameters on detection performance can be used for any detector and application. The *Content Gap* analysis performed here with respect to real-flight parameters (context, object and environmental parameters) showed that vehicle model and position in particular had a strong influence on performance and must be considered to a greater extent in future training data generation. In contrast, parameters such as flight altitude, object orientation, viewing angle, or shadowing had little or no effect. A similar analysis of various sensor effects revealed that the trained detector is susceptible to blur, contrast changes, overexposure and impulse noise, whereas high stability to changes in color and illumination conditions was observed. With respect to different simulation parameters, anti-aliasing methods and changes in *HyperTextures* showed the greatest effect.

Future work will investigate how these findings can be used to increase variance in the training data generation process by varying the influential parameters identified here. This should further increase the generalization capability of the trained detector models and optimize the use of synthetic sensor data.

References

1. Benjdira, B., Khursheed, T., Koubaa, A., Ammar, A., Ouni, K.: Car detection using unmanned aerial vehicles: comparison between faster R-CNN and YOLOv3. In: 1st International Conference on Unmanned Vehicle Systems-Oman, UVS 2019, pp. 1–6 (2019)
2. Li, Q., Mou, L., Xu, Q., Zhang, Y., Zhu, X.X.: R^3-Net: a deep network for multi-oriented vehicle detection in aerial images and videos. IEEE Geosci. Remote Sens. Soc. **57**, 5028–5042 (2019). https://doi.org/10.1109/TGRS.2019.2895362
3. Tayara, H., Soo, K.G., Chong, K.T.: Vehicle detection and counting in high-resolution aerial images using convolutional regression neural network. IEEE Access **6**, 2220–2230 (2017)
4. Xu, Y., Yu, G., Wang, Y., Wu, X., Ma, Y.: Car detection from low-altitude UAV imagery with the faster R-CNN. J. Adv. Transp. **2017**, 1–10 (2017)
5. Tang, T., Deng, Z., Zhou, S., Lei, L., Zou, H.: Fast vehicle detection in UAV images. In: RSIP 2017 - International Workshop on Remote Sensing with Intelligent Processing, pp. 1–5 (2017)
6. Lechgar, H., Bekkar, H., Rhinane, H.: Detection of cities vehicle fleet using YOLO V2 and aerial images. Int. Arch. Photogramm. Remote Sens. Spat. Inf. Sci. ISPRS Arch. **42**, 121–126 (2019)
7. Du, D., et al.: The unmanned aerial vehicle benchmark: object detection and tracking. In: Ferrari, V., Hebert, M., Sminchisescu, C., Weiss, Y. (eds.) ECCV 2018. LNCS, vol. 11214, pp. 375–391. Springer, Cham (2018). https://doi.org/10.1007/978-3-030-01249-6_23

8. Gaidon, A., Wang, Q., Cabon, Y., Vig, E.: Virtual worlds as proxy for multi-object tracking analysis. In: Proceedings of the IEEE Computer Society Conference on Computer Vision and Pattern Recognition, pp. 4340–4349 (2016)
9. Johnson-Roberson, M., Barto, C., Mehta, R., Sridhar, S.N., Vasudevan, R.: Driving in the matrix: can virtual worlds replace human-generated annotations for real world tasks? In: IEEE International Conference on Robotics and Automation (ICRA), pp. 746–753 (2017)
10. Shafaei, A., Little, J.J., Schmidt, M.: Play and learn: using video games to train computer vision models. In: British Machine Vision Conference (2016)
11. Hummel, G., Smirnov, D., Kronenberg, A., Stütz, P.: Prototyping and training of computer vision algorithms in a synthetic UAV mission test bed. In: AIAA SciTech 2014, pp. 1–10 (2014)
12. Krump, M., Ruß, M., Stütz, P.: Deep learning algorithms for vehicle detection on UAV platforms: first investigations on the effects of synthetic training. In: Mazal, J., Fagiolini, A., Vasik, P. (eds.) MESAS 2019. LNCS, vol. 11995, pp. 50–70. Springer, Cham (2020). https://doi.org/10.1007/978-3-030-43890-6_5
13. Redmon, J., Farhadi, A.: YOLOv3: An Incremental Improvement (2018)
14. Presagis - COTS Modelling and Simulation Software. https://www.presagis.com/en/. https://www.presagis.com/en/page/academic-programs/
15. Lu, J., et al.: A vehicle detection method for aerial image based on YOLO. J. Comput. Commun. **06**, 98–107 (2018)
16. Cornette, W.M.: MOSART: modeling the radiative environment of earth's atmosphere, terrain, oceans, and space. J. Washingt. Acad. Sci. **98**, 27–46 (2012)
17. Fan, Z.: Adjust Local Brightness for Image Augmentation. Medium. https://medium.com/@fanzongshaoxing/adjust-local-brightness-for-image-augmentation-8111c001059b
18. Ravikumar, R.: Bokehlicious Selfies. https://rahulrav.com/blog/bokehlicious.html
19. Everingham, M., Van Gool, L., Williams, C.K.I., Winn, J., Zisserman, A.: The pascal visual object classes (VOC) challenge. Int. J. Comput. Vis. **88**, 303–338 (2010)
20. Carrillo, J., Davis, J., Osorio, J., Goodin, C., Durst, J.: High-fidelity physics-based modeling and simulation for training and testing convolutional neural networks for UGV systems. In: Modelling and Simulation for Autonomous Systems, MESAS 2019 (2019)
21. Rozantsev, A., Lepetit, V., Fua, P.: On rendering synthetic images for training an object detector. Comput. Vis. Image Underst. **137**, 24–37 (2015)
22. Pepik, B., Stark, M., Gehler, P., Schiele, B.: Teaching 3D geometry to deformable part models. In: 2012 IEEE Conference on Computer Vision and Pattern Recognition, pp. 3362–3369. IEEE (2012)
23. Ros, G., Sellart, L., Materzynska, J., Vazquez, D., Lopez, A.M.: The SYNTHIA dataset: a large collection of synthetic images for semantic segmentation of urban scenes. In: 2016 IEEE Conference on Computer Vision and Pattern Recognition (CVPR), pp. 3234–3243. IEEE (2016)
24. Sun, B., Saenko, K.: From virtual to reality: fast adaptation of virtual object detectors to real domains. In: Proceedings of the British Machine Vision Conference 2014, pp. 82.1–82.12. British Machine Vision Association (2014)
25. Vazquez, D., Lopez, A.M., Marin, J., Ponsa, D., Geronimo, D.: Virtual and real world adaptation for pedestrian detection. IEEE Trans. Pattern Anal. Mach. Intell. **36**, 797–809 (2014)
26. Kar, A., et al.: Meta-sim: learning to generate synthetic datasets. In: Proceedings of IEEE International Conference on Computer Vision, pp. 4550–4559 (2019)
27. Prakash, A., et al.: Structured Domain Randomization: Bridging the Reality Gap by Context-Aware Synthetic Data (2018)

28. Richter, S.R., Vineet, V., Roth, S., Koltun, V.: Playing for data: ground truth from computer games. In: Leibe, B., Matas, J., Sebe, N., Welling, M. (eds.) ECCV 2016. LNCS, vol. 9906, pp. 102–118. Springer, Cham (2016). https://doi.org/10.1007/978-3-319-46475-6_7
29. Tobin, J., Fong, R., Ray, A., Schneider, J., Zaremba, W., Abbeel, P.: Domain randomization for transferring deep neural networks from simulation to the real world. In: IEEE/RSJ International Conference on Intelligent Robots and Systems (2017)
30. Tremblay, J., et al.: Training Deep Networks with Synthetic Data: Bridging the Reality Gap by Domain Randomization (2018)
31. Su, J., Vargas, D.V., Kouichi, S.: One pixel attack for fooling deep neural networks (2017)
32. Qiu, S., Liu, Q., Zhou, S., Wu, C.: Review of artificial intelligence adversarial attack and defense technologies. Appl. Sci. **9**, 909 (2019)
33. Vargas, D.V., Su, J.: Understanding the One-Pixel Attack: Propagation Maps and Locality Analysis (2019)
34. Xu, H., et al.: Adversarial attacks and defenses in images, graphs and text: a review. Int. J. Autom. Comput. **17**, 151–178 (2020)

Obstacle Detection in Real and Synthetic Harbour Scenarios

Nicolò Faggioni[1]([✉]), Nicola Leonardi[2], Filippo Ponzini[1], Luca Sebastiani[2], and Michele Martelli[1]

[1] DITEN – Department of Naval Architecture, Electric, Electronic and Telecommunication Engineering, University of Genoa, 16145 Genoa, Italy
nicolo.faggioni@edu.unige.it
[2] SEASTEMA, Viale Brigate Partigiane 92R, 16129 Genoa, Italy

Abstract. In the last decade, the autonomous vehicle has been investigated by both academia and industry. One of the open research topics is obstacle detection and avoidance in real-time; for such a challenge, the most used approaches are based on deep learning, especially in the automotive sector. Usually, trained neural networks are used to detect the obstacles by receiving the point clouds from LiDAR as input data. However, this approach is currently not feasible in the marine sector as there are no large datasets of LiDAR point clouds and relatively few RGB images available to train networks. For such a reason, this paper aims to present the first step for the design of an alternative approach that integrates unsupervised and supervised learning algorithms for the detection and tracking of both fixed and moving obstacles. A virtual scenario that can be customized according to the users' purpose has been developed and used to collect data by emulating the LiDAR and camera behaviour. Moreover, the preliminary on-field LiDAR recording is presented and processed. The unsupervised clustering algorithms have been tested, and the pros and cons of the different clustering approaches are shown.

Keywords: LiDAR point cloud · Clustering · Autonomous navigation · Marine scenario

1 Introduction

The enabling technologies suitable to obtaining autonomous navigation have been deeply investigated and discussed in the last decade as they represent challenging topics. Such aspects were confirmed by the "Global Marine Technology Trends 2030" [1], which indicates "smart ships" as a development key of the maritime sector. Thus, the guidance, navigation, and control (GNC) systems are profoundly studied and analyzed. In particular, the obstacle detection that represents a subsystem of the collision avoidance logic has been studied with several experimental and numerical approaches, based on one or multi-sensor.

© Springer Nature Switzerland AG 2022
J. Mazal et al. (Eds.): MESAS 2021, LNCS 13207, pp. 26–38, 2022.
https://doi.org/10.1007/978-3-030-98260-7_2

A numeric geometrical approach for collision avoidance, tailored for a ship equipped with AIS (Automatic Identification System), is shown and discussed in [2]. Such an approach is limited since AIS is not installed on all surface vehicles. In order to create a multi-purpose detection logic, it is necessary to sample data via additional sensors in addition to AIS [3, 4] since only cargo ships of 300 gross tonnages or more and all passenger ships are equipped with AIS.

Only recently, research on the detection of collision avoidance in the maritime sector has begun to be investigated and studied in detail, mainly with deterministic geometric approaches. Contrarily, in the automotive field, the developments concerning detection are much broader and are widely based on machine learning [5]. Indeed, sizeable datasets to train the learning algorithms are available in open access; KITTI and ApolloScape are mainly used [6, 7]. These datasets contain the LiDAR (Light Detection and Ranging) point clouds, images, and positions data. These types of data are used in the deep learning approach [8, 9]. In particular, Convolutional Neural Networks (CNNs) are trained through the LiDAR point clouds to recognize detecting obstacles like pedestrians, cars, and trucks [10, 11].

Unfortunately, the available trained neural networks used for this task are not feasible for the marine sector to recognize ships, buoys, yachts, and other floating objects. Specifically, only one open-access dataset is available for the marine sector obtained by motorboat equipped with LiDAR, RADAR, infrared camera, and RGB camera sensors [12]. However, the available marine dataset is not as widespread as the automotive ones; furthermore, no machine learning-based research studies have been found. Moreover, no marine trained network is available for testing. Hence, nowadays, in the marine sector, the only way to use a machine learning approach based on a LiDAR point cloud neural network is to train a new one; the side effect is that the training of a neural network is a profoundly time-consuming activity and a specific experimental campaign is needed to collect data. This critical aspect is widely felt. Indeed, several studies regarding the maritime sector that show different approaches can be found [13–15].

For such reason, an alternative approach has been developed, and it is based on both camera images and LiDAR point clouds data. This approach detects and tracks multiple objects using a data-fusion method using lidar and images data. The proposed workflow is shown in Fig. 1. The following sections present a detailed description and analysis of the data processing regarding RGB images and LiDAR point clouds. The multi-object tracking by means of data fusion activity is still under development.

Moreover, the authors suggest an alternative approach to generate and collect point clouds and images. Such an approach consists of emulating a customized marine environment by means of a synthetic scenario to obtain "virtual" LiDAR point clouds and the rendered images. Using a virtual scenario has several advantages [16]. For instance, it allows generating data when it is not possible to carry out on-field experiments.

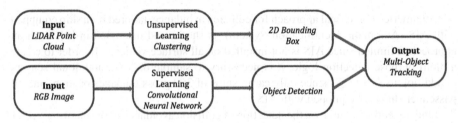

Fig. 1. Obstacle detection workflow.

2   LiDAR Point Cloud Acquisition

In recent years, LiDAR-type instrumentation has become increasingly frequent, mainly when the need arises to know the surrounding environment, particularly the position and shape of surrounding objects. Objects are described using laser beams with a specific wavelength. The data are contained in a point cloud structure in which spatial coordinates (x, y, z) and intensity (as a function of colour, shape and laser encounter angle) are available for each point acquired. The LiDAR model used for the experimental campaign is HESAI PandarXT LiDAR. It provides 32 equally spaced channels of infrared laser light and a variable rotation speed defining different resolutions in a working range of 120 m.

2.1   Synthetic Scenario

The lack of a large available dataset in the marine environment forces the acquisition of multiple point clouds and relative images to develop and test the algorithms and tools necessary to carry out the activity. The possibility of having a tailored dataset of fast construction was one of the main objectives. In particular, this allows users to obtain a wide variety of datasets reducing the efforts significantly. Moreover, the creation of LiDAR data can be carried indoor without misspend, economic and time resources. In addition, LiDAR equipment is not necessary. For these reasons, a procedure designed to extract point clouds compatible with LiDAR technology from a totally virtual environment has been developed.

The desired scenario has been built by means of three-dimensional modelling software. First, the LiDAR acquisition system is virtualized by drawing straight lines from the chosen observation point to mimic the laser beams. Eventually, the points are placed in the intersections between beams and objects.

The position of points is exported as a point cloud, and it represents a single LiDAR acquisition frame. In such a case, the point cloud is composed only of surrounding environment positions; the intensity is fixed to a constant value for all objects. This approach allows users to emulate several LiDARs and the different sensor settings, for instance, the rotor speed and the number of laser beams.

The image realization related to the acquired scenario takes place by importing the 3D tailored scenario in a rendering environment. The desired materials and backgrounds are assigned, and the image is rendered. The possibility of customizing the product is wide. By acting on the parameters available in the rendering environments, it is possible to virtualize any image acquisition tool by imposing its technical specifications. The extreme usefulness of this method lies in the possibility of total customization of the scenario; for example, it is possible to build a series of acquisitions in which one or more boats move by simulating path, manoeuvre or speed, tailored on a specific required activity. Therefore, the proposed procedure results are an image-point cloud pair obtained in a relatively short time. Eventually, it is possible to reproduce any scenario acquired with any instrument for which the technical specifications are available. Despite countless advantages, the results produced by a virtual approach do not consider the information related to the real use of LiDAR in a hostile environment such as the marine one. Several expected results, such as water reflections and data loss, cannot be simulated sufficiently accurately due to the nature of these events.

Furthermore, it is not possible to directly impose the reflectivity of the laser beam inside the point cloud structure since the intensity is not a geometric parameter. However, it can be included in the point cloud structure with a post-processing activity. Nevertheless, it is challenging to accurately emulate the intensity acquired by a LiDAR in a real environment. Eventually, if the intensity must necessarily be used, the proposed approach presents some limits.

2.2 Experimental Marine Environment

Several experimental campaigns have been carried out in order to acquire a dataset composed of image and LiDAR data. In order to acquire both information, a portable layout has been designed. Furthermore, LiDAR support, mounted on the camera (Nikon D7200), has been projected and printed via additive manufacturing, in particular fused deposition modelling (FDM). Hence, it is possible to know in detail, a priori, the distance between the camera and the LiDAR. This information is crucial for the calibration procedure in the data fusion procedure. Hereinafter, the same frames acquired simultaneously via camera and LiDAR sensor have been reported.

Fig. 2. Camera image.

Figure 2 shows an experimental acquisition via camera. It is possible to distinguish two boats: a black yacht and a *gozzo* handle by a person.

Fig. 3. LiDAR point cloud.

Figure 3 shows the same frame through a point cloud. In the centre of the image, it is possible to see three human figures, the researchers, and in the lower part, the points belonging to the quay. The sea surface was not acquired. On the right side, part of the *gozzo* is distinguishable, while the motor yacht is slightly visible on the left side. The authors want to highlight the poor acquisition of the black hull of the motor yacht due to optical phenomena.

3   Clustering Methods

The clustering activity is a task achieved by various algorithms that differ significantly in understanding what constitutes a cluster. Cluster a point cloud means to group points belonging to the same objects that share a certain similarity. In the proposed activity, the clustering algorithms are used to identify, among the detected points, those that belong to a single object and separate them from other points that define different objects.

Clustering on LiDAR points is a widely documented activity. Furthermore, to decrease the computational time of the clustering activities, the point cloud is projected on the XY plane by working on a 2D plane where Z is the vertical component. Such simplification has been done because it is improbable that two different objects are over the other during navigation in a marine scenario.

3.1   DBSCAN Approach

DBSCAN [17] is a clustering algorithm that requires two threshold parameters, distance and a number of points to search within the range, and it is based on the concept of density-reachability. The algorithm presents a particularly refined conception of noise and a very effective cluster definition. However, it has several disadvantages in the analyzed scenario. The rotating multi-beam LiDAR has laser channels that diverge with the distance; the acquired points are therefore more sparse with increasing distance. In addition, the data are less dense in areas further away from the observation point because many laser channels have already encountered an object at shorter distances. A density-based algorithm presents problems when the density is not constant. Indeed, threshold values must be imposed a priori and must be respected throughout its operating range. However, adequate threshold values can accurately perform in a specific interest space region if the range of distance of interest is contained. Some general-purpose solutions to use data with a varying density adapting different solutions are present in literature [18, 19]. Future development could be designed a tailored algorithm based on rotating multi-beam LiDAR data by implementing a threshold density value capable of automatically varying according to the distance from the observation point.

The main problem regarding the DBSCAN clustering is the high computational time, making the algorithm unsuitable for real-time applications. The DBSCAN clustering results in the virtual and real scenarios have been reported. For both scenarios, the time needed by using a standard laptop to process 50k points is about 10 s (Fig. 4).

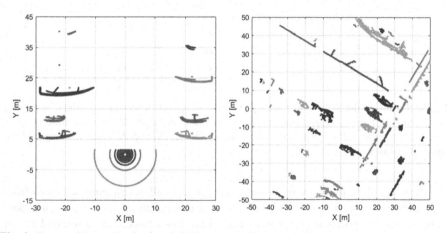

Fig. 4. Reported the result of a virtual (left) and real (right) acquisition processed using DBSCAN clustering. The result has a high quality. Indeed, the boat shapes are clearly distinguished in both figures, but the computational time involved is very high.

3.2 Euclidian Approach

The Euclidean approach could be considered as a DBSCAN approach simplification. Indeed, it is based on a single threshold value, the Euclidean distance. The points that are at a reciprocal distance less than or equal to the threshold distance have been segmented within the same group. Such a simple algorithm gives less accurate results, has no noise cluster, and is subject to group propagation phenomena through bridging objects.

However, it is affected in a non-incisive way by the variation in the density of points if the threshold distance value is carefully chosen. Thanks to its simplicity, it needs extremely low computational times, even more than an order in magnitude compared to DBSCAN. Furthermore, a noise cluster can be easily implemented by imposing a minimum and a maximum number of points for acceptable clusters. Therefore, while not as sophisticated a definition as DBSCAN, it works accurately in the scenarios studied.

Hereinafter are reported the results regarding the virtual scenario and the real one obtained by using the Euclidian algorithm. The clustering was performed with the same condition reported in the previous approach. The computational time is about 1 s, a magnitude order less than DBSCAN, and the result is quite similar (Fig. 5).

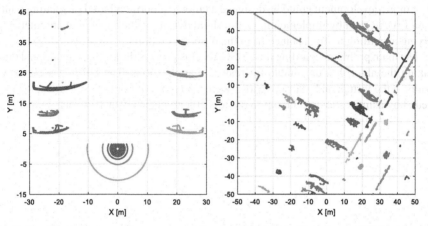

Fig. 5. Results of the same frame seen previously of virtual (left) and real (right) scenarios, processed with the Euclidean clustering algorithm, were reported. The computational time is about a magnitude order less than the taken one by DBSCAN, and the result is comparable in terms of accuracy.

Moreover, the addition of noise and filters concepts allows cleaning scenarios from any interference, such as the wake reflections and acquisition errors has been added. In particular, the points acquired from water have low intensity in correspondence at sea level in the real acquisition due to the previously mentioned phenomena. Therefore, although the point cloud is projected in 2D, a filter that excludes points with a vertical position lower than a specific threshold value that identifies the sea level and, simultaneously, a value of intensity lower than the value that identifies noisy phenomena has been implemented. In this way, the detected sea points are eliminated while preserving the points belonging to the hull that would be eliminated with a single filter on the vertical positioning (Fig. 6).

4   2D Bounding Box

The clustering algorithm labels every point that composes the point cloud, indicating their group membership. Therefore, the clustering output must be processed to obtain a single representative point for each detected group to achieve a tracking algorithm. In particular, this procedure has been carried out via the bounding box approach to

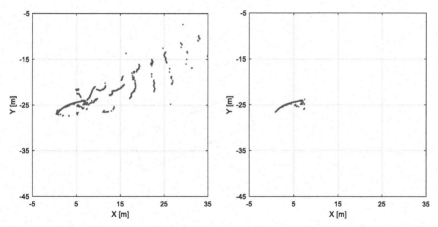

Fig. 6. Reported raw acquired (left) and filtered (right) point cloud data regarding a boat in a real scenario. In the left image, no filter and no noise conception are applied, while in the second one, points compatible with wake characteristics are deleted, the discussed basic definition of noise is applied.

finding its centroid. Since the point cloud is projected in a plane, the bounding box is also expressed by a 2D geometry to reduce the computational load and facilitate the implementation of the process in a possible real-time working system.

In literature, there are several methods to build a bounding box. The developed method is based on PCA (principal components analysis) approach [20]. It guarantees a low computational load, a faithful representation of the size of the group of points and the possibility of obtaining a bounding box oriented according to the principal components of the distribution of points. In particular, the main components of point distribution have been evaluated. The corresponding most significant directions are those for which the data show the greatest variance, identifying the Cartesian system of groups. This 2D system identifies the two main directions representing the object (ships or other floating obstacles). The projection of points in these directions identifies the minimum and maximum limits of the group of points regarding the new reference system. The centroid can be evaluated with two different approaches (*i*) as the centre of gravity of the distribution of points or (*ii*) as the geometric centre of the bounding box figure.

Figure 7 shows LiDAR points that identify a sailing boat clustered and enclosed in the bounding box built adopting the method based on PCA. In blue and red are reported the segments that lie on the main directions and delimited by the minimum and maximum values of the points projected on the aforementioned axes, thus defining the bounding box (in black). The black 'x' marker identifies the centroid defined as the geometric centre of the bounding box. Thus, it approximates accurately enough the centre of the boat.

5 Object Detection

The second branch (as reported in Fig. 1) of the proposed multi-tacking method, regards image processing, has been described hereinafter. The goal of Computer Vision is the

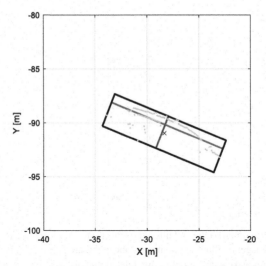

Fig. 7. PCA bounding box.

extraction of meaning from visual signals, in particular from images. The image data is processed with a tensor format; in particular, the dimension is (*hxwxn*) where; *h* and *w* are the numbers of pixels along with height and width, respectively, and *n* is the number of RGB components (equal to 3). Hence, each pixel is described by a value for each RGB channel, as shown in Fig. 8.

Computer vision embraces numerous tasks, including classification, object detection, semantic segmentation, pose estimation, and tracking. The first CNNs were inspired by Hubel & Wiesel's modelling [21] of visual cell types in the brain and the notion of specialized local filters. The last decade had seen CNN dominate overall tasks related to images starting in 2012 when AlexNet [22] won the ImageNet Large Scale Visual Recognition Challenge (ILSVRC) [23]. Four basic building blocks can describe a traditional architecture of a CNN: (*i*) convolutional layer, (*ii*) pooling layer, (*iii*) fully connected layer, (*iv*) non-linear activation layers. In this paper a CNN has been used for the object detection task; indeed, the position within the image, in terms of the bounding box, and semantic category of each subject in the scene have been returned as the output of the neural network. In particular, the wide variation in size and shape regarding boats and ships and semi-submerged objects, which are very difficult to detect, are widespread issues.

5.1 Trained CNN

The CNN typology used for the object detection task is YOLO (You Only Look Once), in particular, the YOLOv4.

YOLO [24] was proposed by R. Joseph et al. in 2015 and was the first one-stage detector network in the deep learning era. YOLO interprets the detection as a single regression problem that outputs the coordinates of the bounding boxes and the probabilities of the classes from the input image. Indeed, it can predict which objects are

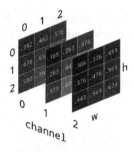

Fig. 8. Visual illustration of the tensor.

present and where they are placed with a single swipe. As the name itself implies, the novelty introduced is to completely abandon the previous two-step *"proposal detection + verification"* paradigm favouring an entirely different solution, namely applying a single neural network to the entire image. This network divides the image into regions and predicts the bounding boxes and probabilities for each region simultaneously. The novelties introduced in the v2 [25], v3 [26], v4 [27] versions have improved the detection accuracy while keeping the inference speed at the real-time level. The CNN was trained with a large number of RGB images of marine scenarios. The neural network is trained to recognize and give a certain degree of fidelity regarding the object detected [40% to 100%]. In particular, it is able to detect and label *"boat"* and *"person"*. Hereinafter, the synthetic marine scenario results and the experimental one have been reported.

Fig. 9. YOLOv4 object detection results on the simulated scenario.

Figure 9 shows a tailored rendered marine scenario, where six boats are present. It has been created to evaluate overlaps between the boats. It is possible to observe that two sailing yachts in the background are detected with a good fidelity level. Also, the motorboat partially visible in the foreground is well caught, but the bounding box created contains another motorboat piece that was not detected. Contrarily, the sailing yacht with sails lowered near the camera is detected with a lower fidelity than 50%. In the scenario, the CNN is able to detect four of the six objects present because the two that are not seen are visible only in a small piece from the image, and the residual bodies do not allow the identification of the objects.

Fig. 10. YOLOv4 object detection results on experimental scenario.

Figure 10 shows a single frame of scenario captured by the camera. It is possible to observe that there are distinguishable two moving boats (motor yacht and *gozzo*) within it. In the background, a semi-submerged mooring buoy and a fueling station platform are present. These objects are labelled with the tag *"boat"* because the CNN has been trained to recognize boats, not specifically buoys or other floating objects. Nevertheless, CNN detects all the objects with good fidelity. Moreover, it adds the bounding box also to the person driving the boat.

6   Conclusions

The approach proposed in the paper was tested with both virtual and real data, performing accurately in both cases. Furthermore, the authors suggest using virtual data during the concept phase when the resources available are limited.

The proposed approach lets to benefit from the advantages offered by the two sensor types.

Thanks to the LiDAR analysis, it is possible to know precisely the position data (speed and route are available performing the tracking activity) of the object, while from the image detection, it is possible to carry out the labelling activity.

Furthermore, by combining the outputs of the sensors, it is possible to avoid missing readings of one of the two, e.g., the black hull not detected by the LiDAR or the lack of detection of an object not observable from the camera due to poor visibility due to adverse weather conditions.

The other main advantage is that no neural network trained on LiDAR point clouds in a maritime scenario is required. With the excellent performance of point clouds clustering, this approach has significant advantages as it allows no time to be spent training any network.

In addition, the performed object detection CNN offers excellent results. CNN has been able to identify vessels with good accuracy even in conditions of partially occluded or out-of-view objects.

The multi-object tracking obtained by means of the data fusion approach is still under development. Future developments will be devoted to setting the sensor hierarchy and adding sensors to improve both the detection capability and safety.

Acknowledgement. Part of research activities reported in the paper is carried out within the project - "MARIN - Naval Integrated Remote Environmental Monitoring", (KATGSO3); fundend by "Programma operativo FESR 2014 – 2020 Obiettivo Convergenza – Regolamento Regionale n. 17/2014 – Titolo II capo 1 – Aiuti ai programmi di investimento delle grandi imprese – Contratti di Programma Regionali" And supported by the COMPASS laboratory of the University of Genova.

References

1. Shenoi, R.A., et al.: Global marine technology trends 2030 (2015)
2. Zaccone, R., Martelli, M.: A collision avoidance algorithm for ship guidance applications. J. Mar. Eng. Technol. **19**(sup1), 62–75 (2020)
3. Son, N.-S., Kim, S.-Y.: On the sea trial test for the validation of an autonomous collision avoidance system of unmanned surface vehicle, ARAGON. In: OCEANS 2018 MTS/IEEE Charleston. IEEE (2018)
4. Thombre, S., et al.: Sensors and AI techniques for situational awareness in autonomous ships: a review. IEEE Trans. Intell. Transp. Syst. (2020)
5. Li, Y., et al.: Deep learning for LiDAR point clouds in autonomous driving: a review. IEEE Trans. Neural Netw. Learn. Syst. **32**(8), 3412–3432 (2020)
6. Geiger, A., et al.: Vision meets robotics: the kitti dataset. Int. J. Robot. Res. **32**(11), 1231–1237 (2013)
7. Huang, X., et al.: The apolloscape open dataset for autonomous driving and its application. IEEE Trans. Pattern Anal. Mach. Intell. **42**(10), 2702–2719 (2020)
8. Gao, H., Cheng, B., Wang, J., Li, K., Zhao, J., Li, D.: Object classification using CNN-based fusion of vision and LIDAR in autonomous vehicle environment. IEEE Trans. Industr. Inf. **14**(9), 4224–4231 (2018)
9. Wu, B., et al.: Squeezeseg: convolutional neural nets with recurrent crf for real-time road-object segmentation from 3D lidar point cloud. In: 2018 IEEE International Conference on Robotics and Automation (ICRA). IEEE (2018)
10. Wu, B., et al.: Squeezesegv2: improved model structure and unsupervised domain adaptation for road-object segmentation from a lidar point cloud. In: 2019 International Conference on Robotics and Automation (ICRA). IEEE (2019)

11. Xu, C., et al.: Squeezesegv3: spatially-adaptive convolution for efficient point-cloud segmentation. In: Vedaldi, A., Bischof, H., Brox, T., Frahm, J.-M. (eds.) ECCV 2020. LNCS, vol. 12373, pp. 1–19. Springer, Cham (2020). https://doi.org/10.1007/978-3-030-58604-1_1

12. AUV Lab, MIT Sea Grant Marine Perception Dataset. https://seagrant.mit.edu/auvlab-datasets-marine-perception-1/

13. Muhovic, J., Mandeljc, R., Bovcon, B., Kristan, M., Pers, J.: Obstacle tracking for unmanned surface vessels using 3-D point cloud. IEEE J. Oceanic Eng. **45**(3), 786–798 (2020)

14. Sorial, M., et al.: Towards a real time obstacle detection system for unmanned surface vehicles. In: OCEANS 2019 MTS/IEEE SEATTLE. IEEE (2019)

15. Villa, J., Aaltonen, J., Koskinen, K.T.: Path-following with lidar-based obstacle avoidance of an unmanned surface vehicle in harbor conditions. IEEE/ASME Trans. Mechatron. **25**(4), 1812–1820 (2020)

16. Martelli, M., Faggioni, N., Zaccone, R.: Development of a navigation support system by means of a synthetic scenario. In: Sustainable Development and Innovations in Marine Technologies, pp. 481–487. CRC Press (2019)

17. Ester, M., et al.: A density-based algorithm for discovering clusters in large spatial databases with noise. In: KDD, vol. 96. no. 34 (1996)

18. Zhu, Y., Ting, K.M., Carman, M.J.: Density-ratio based clustering for discovering clusters with varying densities. Pattern Recogn. **60**, 983–997 (2016)

19. Hu, X., Wang, D., Wu, X.: Varying density spatial clustering based on a hierarchical tree. In: Perner, P. (ed.) MLDM 2007. LNCS (LNAI), vol. 4571, pp. 188–202. Springer, Heidelberg (2007). https://doi.org/10.1007/978-3-540-73499-4_15

20. Jolliffe, I.: Principal component analysis. In: Encyclopedia of Statistics in Behavioral Science (2005)

21. Poggio, T., Serre, T.: Models of visual cortex. Scholarpedia **8**(4), 3516 (2013)

22. ImageNet. https://image-net.org/challenges/LSVRC/. Accessed 2 July 2021

23. Alom, M.Z., et al.: The history began from alexnet: a comprehensive survey on deep learning approaches. arXiv preprint arXiv:1803.01164 (2018)

24. Redmon, J., et al.: You only look once: Unified, real-time object detection. In: Proceedings of the IEEE Conference on Computer Vision and Pattern Recognition (2016)

25. Redmon, J., Farhadi, A.: YOLO9000: better, faster, stronger. In: Proceedings of the IEEE Conference on Computer Vision and Pattern Recognition (2017)

26. Redmon, J., Farhadi, A.: Yolov3: an incremental improvement. arXiv preprint arXiv:1804.02767 (2018)

27. Bochkovskiy, A., Wang, C.Y., Liao, H.Y.M.: Yolov4: optimal speed and accuracy of object detection. arXiv preprint arXiv:2004.10934 (2020)

Fault Detection and Identification
on Pneumatic Production Machine

Barnabás Dobossy(✉) ⓘ, Martin Formánek ⓘ, Petr Stastny ⓘ,
and Tomáš Spáčil

Brno University of Technology, Technická 2896, 616 69 Brno, Czech Republic
barnabas.dobossy@vutbr.cz

Abstract. Pneumatic cylinders have become integral parts of today's production machinery. In the age of just-in-time inventory system and with it the related production process, new, increased requirements were introduced. As a result, even the smallest fault in the system can lead to degradation in the product's quality in addition to this it can cause unplanned downtime leading to delays in production, not to mention higher costs. The availability of cheap sensors, big data, and algorithms from the field of predictive maintenance made the aforementioned problem tractable.

This paper examines whether signal-based condition indicators provide commercially viable and affordable basis for development of a health monitoring system for pneumatic actuator-based production machinery. The experiments and their results presented in this paper served two objectives. The first was to examine if faults on such equipment can be detected. The second was to identify the best combination of sensors, which are able to detect and identify fault with required accuracy. The evaluation of the sensors was not solely based on fault detection capabilities, but other practical aspects (price and durability of the sensors) were also taken into account.

Keywords: Health monitoring · Fault detection and isolation ·
Pneumatic cylinder · Production machinery

1 Introduction

Nowadays in the world of meticulously planned and timed production processes, even the smallest faults in production machinery can have serious consequences, leading to performance degradation, decrease in production, unplanned downtime, delays in production, problems with logistics not to mention safety hazards. Mechanical actuators, e.g. motors, hydraulic actuators, and pneumatic actuators are amongst the most vital parts of production machinery.

This research was funded by the Faculty of Mechanical Engineering, Brno University of Technology under the projects FSI-S-20-6407: "Research and development of methods for simulation, modelling a machine learning in mechatronics", and FV-21-03: "Laboratory model: Inertially driven inverse pendulum".

J. Mazal et al. (Eds.): MESAS 2021, LNCS 13207, pp. 39–60, 2022.
https://doi.org/10.1007/978-3-030-98260-7_3

This paper examines the applicability of state-of-the-art techniques for monitoring the evolution of the health conditions of pneumatic cylinders-based production machinery. The presupposition of these methods was affordability and increased accessibility of sensors which has led to the availability of big data and revolution of the maintenance procedure. A new approach called predictive maintenance has been developed which, contrary to the previously used techniques, introduced improvements in the field of condition monitoring, providing advancement in:

– fault detection and isolation (FDI),
– predicting the remaining useful life (the time interval in which the machine is expected to work as intended).

Condition monitoring is a term defining a group of methods that aim to evaluate the health of a system and its components by analysing and interpreting the data collected from the given system through sensors or transducers. Condition monitoring includes detection, isolation (diagnosis), and prediction of the faults in a system in the earliest possible stage. According to the definition of Prof. Rolf Isermann [1]: "A fault is unpermitted deviation of at least one characteristic property (feature) of the system from the acceptable, usual, standard condition". Eventually, it can result in loss or reduction of the capability to perform the required function and lead to failure and malfunction [1].

Fault detection and isolation deals with the detection of fault occurrence and pinpointing its source (location) in the system. Current applications of fault detection and identification systems are widespread, ranging from wind farms [2,3], electro-mechanical systems [6,12], pneumatic systems [4,5] up to machinery used for oil refinement in the petrochemical industry [13], or fault detection of sensors in a safety-critical control application [8].

Early fault detection methods were based on limit-checking of the monitored quantity, meaning the system is considered healthy if the monitored quantity Y lies between pre-defined upper and lower threshold values Y_{max} respectively Y_{min} [1]. Despite the method's simplicity, its significance was lost to condition indicator-based methods as the system's complexity increased.

Condition indicators are features of the data, that characterise the degradation process, meaning their value changes reliably and in a predictable manner as the system's health condition degrades. Due to their reliability, they are helpful in distinguishing healthy from faulty conditions, and they may even serve as a way to identify the source of the given faulty condition (e.g., excessive amount of vibration measured on a DC motor might suggest worn-out bearings) [19]. Generally, two main types of condition indicator are used:

– model-based condition indicator
– signal-based condition indicator

Of the two above-mentioned methods, the model-based condition indicators were the older approach [20]. The signal-based condition indicators were only developed in the 1980s as a result of technological advancements in the field of digital signal processing. These new methods called signal-based methods were based on real-time processing of the measured signals which, included extraction and identification of properties or patterns within the signals that were correlated with changes in health conditions of the system, hence the name condition indicators. Deviation from the healthy values of the condition indicator is a sign of change in health conditions, which can be used to detect and identify different fault conditions.

The experiments in this paper on signal-based fault detection techniques will provide the foundation for the development of a commercially available health monitoring solution for pneumatic production machines. The planned solution should work as an early warning system by being able to detect faults before they would even occur, and warn the operators about their occurrence, and at the same time pinpoint their source. As a result, targeted maintenance action can be facilitated on the exact location of the fault at the right time, thus avoiding harm to the products or to the machine itself, saving expenses and making the manufacturing process more efficient. The application of the mathematical apparatus, and the algorithms in this paper is not only limited to pneumatic manufacturing equipment, but can be applied to other systems e.g., unmanned autonomous vehicles (UAV) with pneumatic suspensions, pneumatic break systems.

The paper examines, whether faults in pneumatic machines are detectable, identifiable and whether their source can be identified from analysis of sensor readings. The presented experiments will first and foremost concern the two most common sources of faults in real-life applications:

- faults of initial configuration of the machine (e.g. incorrect tuning of pressure level)
- faults emerging with usage of the machine as a result of wear

Several experiments were designed and performed to detect and identify these faults with the highest possible granularity, as well as to complete the following tasks:

- to identify the extent of detectability and identifiability of faults of a pneumatic actuator-based manufacturing machine from the available sensors
- to identify optimal sensor(s) that satisfy not just the precision requirements on condition monitoring but also requirements necessary for their deployment in a real-world industrial application, such as the following ones:
  - durability of the sensor (if possible, no mechanical parts)
  - no maintenance requirements
  - simple mounting
  - cost efficiency

A vital part of the signal-based approach is the extraction of the right condition indicators, from which the most expressive set of indicators can be selected for fault detection and isolation. A very powerful and popular group of condition indicators are features extracted from the frequency domain [3,4,12,15]. In [4] the authors reached the conclusion that faults in pneumatic cylinders can be very well detected in the frequency domain of acoustic emission signal, while the emergence of new peaks in the frequency domain is correlated with faults in the system. In [15] authors successfully used frequency features of vibration signal successfully for health monitoring of bearings of DC motors.

Another popular source of condition indicators is the time-domain signal. In works [2] and [17] time-domain features such as kurtosis, crest factor, peak value, and RMS are used for health monitoring of a wind turbine. In [18] used similar time-domain condition indicators to detect faults in a gearbox. In [16] the author proposed a method, which exploited average time-domain vibration signal to detect health status changes of the gearbox.

The structure of this paper is as follows. Section 2 presents theoretical background with an emphasis on the workflow used and the processing steps carried out in each of its stages. Sections 3 and 4 describe the pneumatic testbench and the data acquisition signal chain used in the experiments. Section 3 takes a detailed look on the hardware, including the used sensors and adjustable parameters of the bench. Section 4 deals with the description of the data acquisition signal chain. The experiments and their results are presented in Sect. 5.

2   Signal-Based Condition Indicators

Keeping manufacturing machinery working effectively and minimising unavoidable downtime are two vital objectives in every production plant which are only achievable through a good maintenance strategy. Therefore maintenance timing is essential for efficient and cost-effective production. Based on the timing of the service intervention the following three maintenance strategies can be distinguished:

- *Reactive maintenance*:
    - the maintenance action takes place after a fault has already occurred
    - it is unexpected therefore the maintenance personnel cannot prepare for it
    - usually results in downtime of the machinery
- *Preventive maintenance*:
    - is carried out in fixed time intervals, regardless of whether the upkeep was needed
    - it was designed to keep parts in good condition
- *Predictive maintenance*:
    - the maintenance action is based on the analysis of information concerning the actual state of the machine
    - the upkeep action is targeted
    - it can be planned in advance
    - rarely results in a total shutdown of the whole factory
    - the parts are replaced right before a fault would occur, therefore their full lifetime is used

A more detailed description and comparison of the advantages and disadvantages of maintenance types is available in [7]. In the following section closer look on predictive maintenance workflow will be taken, while breaking the process up to its stages. In the following sections theoretical background will be explained in relations to the methodology used in our experiments.

3   Data Processing Workflow

The block diagram in Fig. 1 illustrates the main stages taken in the development of our fault detection and identification procedure. Contrary to the general predictive maintenance workflow which ends with implementation and deployment of the developed algorithm, this paper intends to serve only as a proof of concept of application of predictive maintenance for condition monitoring of a pneumatic actuator.

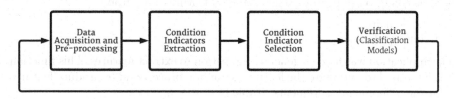

Fig. 1. Data processing workflow

In the following subsections, a detailed look at the individual stages of the above-presented workflow will be taken.

3.1   Data Acquisition and Pre-processing

Successful implementation of condition monitoring algorithms requires a data set of appropriate size and properties. The exact size of the data set is unknowable, therefore it has to be determined empirically, while taking into account the following properties also discussed in [9]:

- complexity of the problem
- complexity of the classification algorithm

It is also required for the used data set to include sufficient representation of the system's behaviour under a range of healthy and faulty conditions. This requirement is necessary in order to train a classifier with good generalisation properties. Further details on the parameters used to acquire data will be discussed in Sect. 5.1.

Pre-processing of the acquired data is the second stage in our workflow and it is an integral part of the process as the quality of the extracted condition indicator and also the accuracy of fault detection is dependent on the information content of the measured signal. Therefore, it is important to improve the quality of the raw data by pre-processing it. Due to the high quality of sensors used and measurement hardware, the level of noise and other unwanted disturbances was acceptable, therefore pre-processing was limited only to the following operations:

- signals from the sensors were converted to physical quantities
- derivation of velocity and acceleration from the measured position data
- acquiring frequency domain representation of the signals through FFT

3.2   Condition Indicator Extraction

For the purpose of this paper MATLAB's Diagnostic Feature Designer was used to extract and select signal-based condition indicators from the measured signals. Signal-based condition indicators are the features of the signal, quantities that describe its behaviour, shape, frequency content and their value changes simultaneously with the system's degradation. They serve as a simple, computationally efficient, powerful, and reliable tools to produce features that can be used as condition indicators. The extracted features were from:

- time domain
- frequency domain

The application was used to generate a function to extract features. This function was later used to automate the feature extraction process and to produce features from the different measurements (Table 1).

Table 1. List of condition indicators extracted from the measured signals

List of condition indicators	
Domain	Condition indicator
Time-domain	Mean
	Standard deviation
	RMS
	Shape factor
	Kurtosis
	Skewness
	Peak value
	Crest factor
	Impulse factor
	Clearance factor
	Total harmonic distortion
	Signal-to-noise ratio
	SINAD
Frequency-domain	Peak amplitude
	Peak frequency
	Band power

3.3   Condition Indicator Selection

Feature selection is the fourth stage of the process. It is used to reduce the dimensionality of the feature space, which ranks and discards features that have bad predictive power. Feature selection is favourable because it improves the quality of the classification model, for the following reasons:

- prevents overfitting
- improves model size
- improves accuracy
- reduces training time

   There is an abundance in various feature selection algorithms, the selection of which was made based on the findings of the paper [11]. The authors of the paper compared the performance of different selection algorithms on synthetic data, and the Relieff algorithm came out on top. In addition to the fact that this method belongs to the category of feature selection algorithms called filter methods (model with lowest computational cost), it performs well in rejecting correlated and redundant features and is not susceptible to non-linearity and noise of the features.

3.4   Verification of the Results

The fifth and final stage in our workflow was the verification of the chosen features, and simultaneously with it, the selection of sensors, i.e., sensor number reduction. To verify the fault detection quality of the sensors and their features, machine learning models were trained (no deep learning models) and the training results were validated using 6-fold cross-validation.

Table 2. Classifiers used during the verification procedure. In the column *Usage*, the letter M stands for multi-class classifier and B for binary classifier

Classification algorithms		
Family	Algorithm	Usage
Tree	Bagged	M
	Boosted	M
	Fine	B
	Medium	B
k-Nearest Neighbour	Fine	B/M
	Medium	B/M
	Weighted	B/M
Support Vector Machine	Cubic	M
	Medium Gaussian	M
	Quadratic	M
Discriminant Analysis	Subspace	M

The performed experiments from the perspective of how the classification problem is posed (i.e., number of classes) are of two kinds, binary and multi-class classification problems. Therefore suitable classifiers had to be chosen - 11 different classification models overall. The selected models are the best performing classifier models based on earlier experiments on the test data set. The selected algorithms are from 4 basic families of classifiers, ranging from Tree and Support Vector Machine to k-Nearest Neighbour and Discriminant analysis. The complete list of classifiers is available in Table 2.

The trained classification models were then tested by a cross-validation algorithm. Eventually, these results served as an indicator of the quality of the sensor and were used to pick the best combination of sensors that will be used during the further stages of our research and development activity.

MATLAB's Classification Learner application was used to train and test the machine learning models on the feature set.

4   Testbench Description

Data for the experiments was collected on a custom-built pneumatic device, which was designed and built to model the behaviour of pneumatic manufacturing equipment. The device has been developed to be able to model different stages of the manufacturing process (e.g., drilling, press-fit, etc.) and to mimic their behaviour not just under healthy conditions but also to simulate fault conditions.

In order to have versatile test equipment that can replicate accurately the behaviour of a pneumatic actuator under different production stages and also simulate fault condition, the testbench was designed to have adjustable parameters. The alterations can be made in:

– pressure of compressed air
– pressure reduction of the reduction valve
– the amount of the load acting on the cylinder (the platform's own weight is 7 kg)
– damping
   • both shock absorbers on two ends disengaged
   • both shock absorbers on both ends deployed
   • only one type of shock absorber on both ends is connected
      * with adjustable damping (9 levels of damping)
      * with constant damping

From the above-mentioned tuneable parameters the air pressure and the adjustments of the reduction valve are parameters, whose faults can occur from the negligence of the operator. While the other two parameters (e.g.: load, and damping) provide an easy way to replicate faults related to the manufacturing process (Fig. 2).

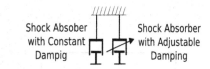

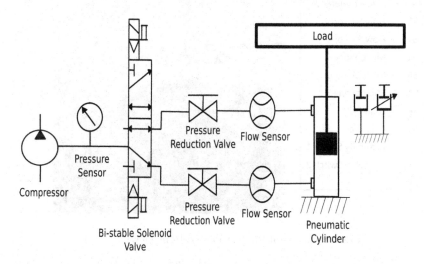

Fig. 2. Schematics of the pneumatic testbench

4.1 Sensors

Information on the motion and the state of the bench was obtained through a wide variety of sensors (8 types, 14 sensors altogether), in order to capture the most possible information, these sensors were mounted on different parts of the test bench (Fig. 3 and Fig. 4). Table 3 contains information regarding their type and the measured quantities:

Table 3. List of sensors mounted on the pneumatic test bench

List of sensors			
Code	Sensor type	Manufacture number	Range
S1	Accelerometer	TE Connectivity 4030-006-120	±6 g
S2	Flow sensor	Festo SFAB-50U-WQ6-2SV-M12	0–50 l/m
S3	Proximity switch	Festo SMT-8M-A-PS-24V-E-0.3-M8D	–
S4	Load cell	Burster 8524	0–2 kN
S5	Pressure sensor	Festo SDEI-D10-G2-MS-L-P1-M12	0–10 bar
S6	Microphone	VMA309	50–50 kHz
S7	Thermocouple	Omega SA2	−50–200 °C

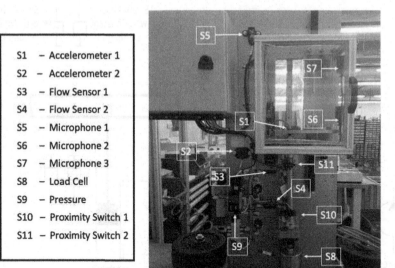

S1	– Accelerometer 1
S2	– Accelerometer 2
S3	– Flow Sensor 1
S4	– Flow Sensor 2
S5	– Microphone 1
S6	– Microphone 2
S7	– Microphone 3
S8	– Load Cell
S9	– Pressure
S10	– Proximity Switch 1
S11	– Proximity Switch 2

Fig. 3. Sensors of the test bench (front view)

S12	– Linear Encoder

Fig. 4. Sensors of the test bench (rear view)

4.2 Reference Settings

Three common processes from automated manufacturing were chosen as the subjects of our experiment, namely:

- drilling
- assembly by press-fitting of parts
- transfer of assembled parts between workstations

Before the experiments could be carried out, the unknown values for the bench's adjustable parameters for the above-mentioned processes had to be identified. Altogether 7 different prescriptions for ideal working conditions were obtained (2x drilling, 2x assembly by press-fitting, 3x transfer). These settings will be used during the experiments as reference states (healthy conditions), since they define how the testbench should behave under normal operating conditions. The following table presents the mentioned healthy conditions:

Table 4. Healthy (reference) states of adjustable parameters of the pneumatic test bench

Operations		Transfer			Assembly		Drilling	
ID		11	12	13	21	22	31	32
Load [kg]		6,25	5	0	0	1,25	0	1,25
Pressure [bar]		6	6	6	6	6	5,5	6
Reduction valve 1		4	4	3	3	3	5	5
Reduction valve 2		2	2	3	3	3	3	3
Bottom shock absorber	Adjustable damping	4	4	3	3	3	5	5
	Constant damping	4	4	3	3	3	5	5

5 Data Acquisition

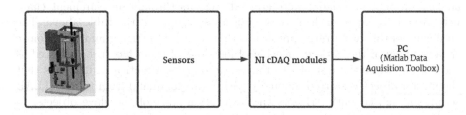

Fig. 5. Block diagram of the data acquisition signal chain

The block diagram in Fig. 5 presents the data acquisition signal chain used in our experiments. The used measurement chain consists of six NI cDAQ modules, the complete list of modules, connection diagram, and sampling information are presented in Table 5. To control the timing and the data transfer from the individual modules, NI's cDAQ-9172 chassis was used.

In addition to the high slot count among other advantages of this chassis is the possibility of multi-rate sampling of the modules (3 objects each with its own sampling rate). This property was highly advantageous due to the fact that the measured quantities have different dynamics. Therefore, the measured signals were grouped into three groups based on their sampling rate. Three microphones were sampled at a rate of 40 kHz, the thermocouple at 0,1 Hz, and the rest of the sensors at 1 kHz.

Table 5. Connection diagram of the used test bench

Sensor type	HW	Signal type	Sampling rate [kHz]
Pressure sensor	NI 9221	Analog	1
Flow sensor	NI 9221	Analog	1
Accelerometer	NI 9221	Analog	1
Linear encoder	NI 9401	Analog	1
Load cell	NI 9211	Analog	1
Proximity sensor	NI 9411	Digital	1
Microphone	NI 9215	Analog	40
Thermocouple	NI 9419	Analog	0,1

5.1   Data Acquisition Firmware

The configuration of the data acquisition hardware and the data measurement session was controlled by MATLAB's Data Acquisition Toolbox, which provided a powerful yet simple-to-use environment to exploit the device specific features of the NI hardware. As it was mentioned above, due to the different nature of the measured quantities, multi-rate sampling was deemed an optimal technique to use. To utilise NI hardware's multi-rate functionality, three separate data acquisition objects ("daq") were created. Sensors through measurement channels were assigned to each of these objects according to the table above. The specificity of this multi-rate approach is that in order to trigger and synchronize the execution of the data acquisition objects one of them has to be selected as the master, while the rest of them are slave objects. Practically, it means that the master object is started manually with a run command from the MATLAB script and subsequently triggers the acquisition start of the slave objects. In addition to scripts that control the data acquisition, further conversion functions were created which convert the electrical quantities from the sensors to the given physical quantity of the measured signals. Lastly, a function converts the individual timetables with measurement data (each sensor has its own) to a large table, containing the measurements from the data acquisition objects to a format suitable for further processing with the feature extractor application called Diagnostic Feature Designer - part of MATLAB's Predictive Maintenance Toolbox.

6   Experimental

Among the goals of this paper was to experimentally identify the extent to which faults can be detected and identified with sufficient detection accuracy. Therefore, experiments with increasing complexity were designed. Starting from detection of fault (distinguishing between healthy and faulty conditions) to identifying the source and the degree by which the faulty configuration is offset from the healthy conditions. The of experiments in execution order is presented below:

– **Single fault condition**
   1. Fault detection in one specific process
   2. Fault detection and isolation in one specific process
      a identification of the source of the fault
      b identification of the source (exact setting of the adjustable parameter)
– **Combination of two fault conditions**
   1. Fault detection and isolation in one specific process, where the combination of two fault conditions occur at the same time

The course of the experiments was identical in every case. With each setting of the adjustable parameters, 20 cycles (in our case a cycle consists of a full extrusion of the cylinder's rod and its return to its original position) worth of data was captured. The number of cycles was determined by taking into account two factors. The first of them was the selection of classification algorithms, which was constrained to supervised learning models not requiring deep learning. As result, there were fewer parameters to learn during the training of the classifier, and at the same time the training data set could be kept reasonable small. The second factor was know-how, collected from previous experiments on the testbench.

Each set of measurements for the given reference setting consisted of data representing both healthy and faulty conditions. The healthy conditions were the reference settings themselves (available in Table 4 while the faulty conditions were defined as any deviation of adjustable parameters from the prescribed reference settings.

7   Results

Table 6 displays the results of the first experiment for process 11 are displayed. As the results point out, the best fault detection accuracy is provided by the sensors such as linear encoder and microphone, where the accuracy reaches 100%. As for the other sensors the fault detection accuracy is very high as well (greater than 90%). The other processes showed similar results to the ones presented in Table 6 (Fig. 6).

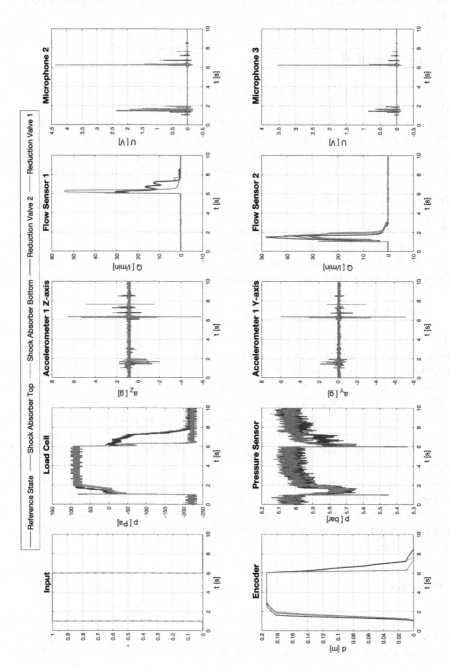

Fig. 6. Illustration of measured signals for each of the examined health conditions

Table 6. Classification results of single-fault fault detection. In column *Acc.* the classification accuracy is presented in percentage and in column *Classifier*, the name of the best performing classifier model's name is displayed.

Transfer between stages		
Sensor	11	
	Acc.	Classifier
Accelerometer 1 Y-axis	99.6	Fine KNN
Accelerometer 1 Z-axis	98.2	Weigh. KNN
Accelerometer 2 Y-axis	95.6	Bagged Tree
Accelerometer 2 Z-axis	93.3	Medium Tree
Encoder	100	Fine KNN
Flow Sensor 1	99.7	Fine KNN
Flow Sensor 2	99.4	Fine KNN
Microphone 1	97.3	Fine KNN
Microphone 2	99.9	Weight. SVN
Microphone 3	100	Fine KNN
Load Cell	99.4	Fine Tree
Pressure	90.8	Boost. Tree

Based on the results, it can be concluded that all of the available sensors passed the first test and can undergo further investigation in order to find the extent to which these sensors are suitable for a more complex tasks.

Table 7. Classification results of single-fault fault detection and isolation for the process transfer between stages. In column *Acc.* the classification accuracy is presented in percentage and in column *Classifier*, the name of the best performing classifier model's name is displayed.

Transfer between stages						
Sensor	11		12		13	
	Acc.	Classifier	Acc.	Classifier	Acc.	Classifier
Accelerometer 1 Y-axis	98.9	Quadr. SVM	93.3	Cubic SVM	98.6	Gauss. SVM
Accelerometer 1 Z-axis	92.9	Weigh. KNN	91.8	Weigh. KNN	94.7	Quadr. SVM
Accelerometer 2 Y-axis	93.9	Quadr. SVM	82.3	Boost. Tree	85.8	Bagged Tree
Accelerometer 2 Z-axis	91.4	Quadr. SVM	87.0	Boosted Tree	95.8	Cubic SVM
Encoder	99,6	Fine KNN	99.9	Fine KNN	100	Fine KNN
Flow Sensor 1	99.4	Quadr. SVM	99.9	Cubic SVM	98.8	Cubic SVM
Flow Sensor 2	98.5	Quadr. SVM	99.1	Cubic SVM	94.6	Bagged Tree
Microphone 1	95.8	Quadr. SVM	82.4	Bagged Tree	88.5	Bagged Tree
Microphone 2	93.6	Quadr. SVM	93.3	Bagged Tree	94.0	Boosted Tree
Microphone 3	90.8	Bagged Tree	91.8	Boosted Tree	92.2	Boost. Tree
Load Cell	99.1	Quadr. SVM	98.8	Quadr. SVM	98.8	Bagged Tree
Pressure	80.6	Weigh. KNN	78.5	Quadr. SVM	87.5	Boosted Tree

In the case of the next experiment (Tables 7, 8 and 9), the fault detection and identification capabilities of the sensors were tested. This time just like in the case of the previous experiment only a single fault condition was modeled (only one adjustable parameter was changed against the reference states) at a time.

Table 8. Classification results of single-fault fault detection and isolation for the process assembly by press-fitting. In column *Acc.* the classification accuracy is presented in percentage and in column *Classifier*, the name of the best performing classifier model's name is displayed.

Assembly by press-fitting				
Sensor	21		22	
	Acc.	Classifier	Acc.	Classifier
Accelerometer 1 Y-axis	81.0	Cubic SVM	81.8	Gauss. SVM
Accelerometer 1 Z-axis	90.6	Quadr. KNN	87.9	Weigh. KNN
Accelerometer 2 Y-axis	73.8	Boost. Tree	82.6	Gauss. SVM
Accelerometer 2 Z-axis	76.8	Quadr. SVM	85.6	Quadr. SVM
Encoder	99,4	Cubic SVN	99.7	Fine KNN
Flow Sensor 1	97.7	Quadr. SVM	100	Cubic SVM
Flow Sensor 2	98.2	Quadr. SVM	99.9	Fine KNN
Microphone 1	84.3	Quadr. SVM	85.3	Boosted Tree
Microphone 2	84.0	Weigh. SVM	81.1	Quadr. SVM
Microphone 3	89.9	Quadr. SVM	88.5	Quadr. SVM
Load Cell	93.2	Bagged Tree	98.0	Cubic SVM
Pressure	63.2	Boost. Tree	88.0	Weigh. KNN

As a result of the higher requirements on the desired output, the classifier had to distinguish more nuanced differences within the data in order to identify the source of the fault.

Table 9. Classification results of single-fault fault detection and isolation for the process drilling. In column *Acc.* the classification accuracy is presented in percentage and in column *Classifier*, the name of the best performing classifier model's name is displayed.

Drilling				
Sensor	31		32	
	Acc.	Classifier	Acc.	Classifier
Accelerometer 1 Y-axis	62.4	Boosted Tree	62.3	Quadr. SVM
Accelerometer 1 Z-axis	84.8	Bagged Tree	85.6	Quadr. SVM
Accelerometer 2 Y-axis	44.1	Med. KNN	46.5	Quadr. SVM
Accelerometer 2 Z-axis	52.1	Quadr. SVM	48.2	Quadr. SVM
Encoder	98.5	Quadr. SVM	99.1	Quadr. SVM
Flow Sensor 1	99.5	Quadr. SVM	95.2	Cubic. SVM
Flow Sensor 2	96.3	Bagged Tree	98.5	Weigh. SVM
Microphone 1	88.4	Boosted Tree	81.8	Cubic SVM
Microphone 2	94.1	Boosted Tree	91.2	Quadr. SVM
Microphone 3	94.8	Bagged Tree	90.3	Bagged Tree
Load Cell	99.3	Quadr. SVM	99.2	Quadr. SVM
Pressure	82.4	Weigh. SVM	85.9	Quadr. SVM

Table 10. Single-fault detection and isolation (with identification of the exact settings) for the process transfer between stages. In column *Acc.* the classification accuracy is presented in percentage and in column *Classifier*, the name of the best performing classifier model's name is displayed.

Transfer between stages						
Sensor	11		12		13	
	Acc.	Classifier	Acc.	Classifier	Acc.	Classifier
Accelerometer 1 Y-axis	83.5	SubSp. Disc	81.7	Cubic SVM	81.0	SubSp. Disc.
Accelerometer 1 Z-axis	80.8	Cubic SVM	78.3	Quadr. SVM	83.5	SubSp. Disc.
Accelerometer 2 Y-axis	70.6	SubSp. Disc	68.5	Bagged. Tree	57.3	Gauss. SVM
Accelerometer 2 Z-axis	63.0	Gauss. SVM	61.8	Bagged Tree	67.7	Gauss. SVM
Encoder	96.5	Cubic SVM	99.1	Fine KNN	100	Fine KNN
Flow Sensor 1	99.6	Cubic SVM	93.2	Quadr. SVM	93.0	SubSp. Disc.
Flow Sensor 2	88.9	Quadr. SVM	92.9	SubSp. Disc	88.8	Quadr. SVM
Microphone 1	80.6	Gauss. SVM	70.0	Bagged Tree	72.2	Quadr. SVM
Microphone 2	80.8	Cubic KNN	86.4	Quadr. SVM	83.3	Quadr. SVM
Microphone 3	80.6	Gauss. SVM	85.8	Quadr. SVM	84.8	Cubic SVM
Load Cell	91.7	Quadr. SVM	92.7	Bagged Tree	92.5	Bagged Tree
Pressure	63.0	Quadr. SVM	64.1	Bagged Tree	66.7	SubSp. Disc

As expected, the more refined differences in data made the classifiers work complex, which resulted in decreased classification performance in case of every used sensor. The biggest decrease in performance (50%), in comparison to the first experiment, occurred in the case of the accelerometer fixed to the frame of the test bench.

Table 11. Single-fault detection and isolation (with identification of the exact settings) for the process assembly by press-fitting. In column *Acc.* the classification accuracy is presented in percentage and in column *Classifier*, the name of the best performing classifier model's name is displayed.

Assembly by press-fitting				
Sensor	21		22	
	Acc.	Classifier	Acc.	Classifier
Accelerometer 1 Y-axis	57.9	SubSp. Disc	49.9	SubSc. Disc.
Accelerometer 1 Z-axis	75.0	SubSc. Disc	67.1	SubSc. Disc.
Accelerometer 2 Y-axis	41.4	SubSc. Disc	46.2	SubSc. Disc.
Accelerometer 2 Z-axis	39.9	SubSc. Disc	45.3	SubSc. Disc.
Encoder	99,4	Cubic SVN	98.9	Quadr. SVM
Flow Sensor 1	92.5	Quadr. SVM	97.0	Cubic SVM
Flow Sensor 2	94.0	SubSp. Disc	95.0	SubSc. Disc.
Microphone 1	62.2	Bagged Tree	58.8	Bagged Tree
Microphone 2	59.12	Quadr. SVM	61.2	Quadr. SVM
Microphone 3	66.8	Quadr. SVM	69.1	Quadr. SVM
Load Cell	87.5	Cubic SVM	98.9	Quadr. SVM
Pressure	61.0	SubSp. Disc	75.0	SubSc. Disc

Table 12. Single-fault detection and isolation (with identification of the exact settings) for the process drilling. In the column *Acc.* the classification accuracy is presented in percentage and in column *Classifier*, the name of the best performing classifier model's name is displayed.

Drilling				
Sensor	31		32	
	Acc	Classifier	Acc	Classifier
Accelerometer 1 Y-axis	43.6	Quadr. SVM	37.2	Quadr. SVM
Accelerometer 1 Z-axis	61.9	Quadr. SVM	55.7	Cubic SVM
Accelerometer 2 Y-axis	26.7	SubSp. Disc	23.8	Gauss. SVM
Accelerometer 2 Z-axis	30.5	Quadr. SVM	27.1	Gauss. SVM
Encoder	98.6	Cubic SVM	94.6	Quadr. SVN
Flow Sensor 1	90.5	SubSp. Disc	92.7	Quadr. SVM
Flow Sensor 2	92.4	SubSp. Disc	87.4	Cubic SVM
Microphone 1	66.7	Quadr. SVM	61.8	Quadr. SVM
Microphone 2	84.7	Quadr. SVM	79.2	Quadr. SVM
Microphone 3	82.2	Quadr. Tree	76.4	Quadr. SVM
Load Cell	92.9	Bagged Tree	94.7	Quadr. SVM
Pressure	77.1	Weigh. SVM	62.3	Quadr. SVM

In the case of the other sensors, the drop in performance was less significant. Among the best performing sensors were the linear encoder, both flow sensors, load cell, and the accelerometer mounted on the moving part that lifts the weights. The accuracy of these sensors was above 90%.

Another interesting fact is that the condition indicators from processes such as drilling and assembly by press-fitting turned out to be harder to detect.

Table 13. Classification accuracy for detection and isolation of combination of two faults. In the column *Acc.* the classification accuracy is presented in percentage and in column *Classifier*, the name of the best performing classifier model's name is displayed.

Assembly by press-fitting		
Sensor	32	
	Acc.	Classifier
Accelerometer 1 Y-axis	84.0	Cubic SVM
Accelerometer 1 Z-axis	80.1	Quadr. SVM
Accelerometer 2 Y-axis	59.4	Bagged Tree
Accelerometer 2 Z-axis	59.3	Bagged Tree
Encoder	99.2	Cubic SVM
Flow Sensor 1	95.5	Quadr. SVM
Flow Sensor 2	94.4	Cubic SVM
Microphone 1	66.7	Bagged Tree
Microphone 2	84.8	Bagged Tree
Microphone 3	84.0	Bagged Tree
Load Cell	94.3	Bagged Tree
Pressure	67.0	Bagged Tree

In the last experiment which dealt with the detection and identification of a single fault, the goal of the experiment was to detect and identify faults to the extent in which the classifier can tell not just the exact source of the fault, but also how much the current setting differs from the reference setting, e.g., in the case of reference state 11, the *Reduction* Valve 1 was set to position 9 instead of 4. In this case, an overall slight worsening of accuracy is noticeable (Tables 10, 11 and 12). Sensors such as the linear encoder, load cell, and both flow sensors are still providing results with over 90% accuracy, however, the rest of the sensors have inferior capabilities to detect faults with such detailedness.

The last experiment was designed to detect fault detection and isolation capabilities of the sensors for the combination of two fault conditions occurring at the same time. In this experiment, the results (Table 13) were the same as in the case of experiment number 3. The best overall performance was obtained from the features from sensors such as the linear encoder, load cell, and both flow sensors. The encoder was the best by significant margin, reaching 99.2% accuracy.

In each case where classification is discussed, it is vital to mention the stability of the results after multiple re-training of the classifier models. The overall fluctuation in the results of classification accuracy was around 2–3%, while in the case of the encoder, the stability of the results was even greater around 0.1 percentage points.

By considering factors, namely cost and practicality related to the use and mounting of the sensor, there is a slight alteration in the end result (Table 14). Sensors such as flow sensor and load cell become impractical. In the case of the load cell, it is due to its cost and problematic mounting. Based on the results above, it is clear that condition indicators from encoders provide incomparable classification performance despite their relatively high cost. Accelerometers are not suited for detection of the degree by which the fault condition of the machinery differs from the ideal state, however, they make up for it in the other two vital perspectives.

Table 14. Comparison of the four best sensors based on classification accuracy (*Accuracy*), price *Cost*, and ease of usage and application Practicality

Sensor	Accuracy	Cost	Practicality
Linear encoder	★ ★ ★ ★ ★	★ ★	★ ★
Flow sensor	★ ★ ★ ★ ★	★ ★ ★	★
Accelerometer	★ ★ ★	★ ★ ★★	★ ★ ★ ★ ★
Load cell	★ ★ ★ ★ ★	★	★

8　Conclusion

The main contribution of our work is proof showing that signal-based condition indicators are a viable approach to monitor the health condition and to detect and identify faults in the case of pneumatic actuator-based production machines. The presented methods can be further applied to a variety of pneumatic systems e.g., suspension of UAVs, break systems.

Based on the presented results from the previous section, it can be concluded that condition indicators from sensors, such as the encoder, flow sensor, or load cell are capable of identifying even the degree by which the fault condition of the machinery differs from the ideal, healthy state of the machine. It was also shown that condition indicators from these sensors are capable of detecting single fault and also the combination of two faults, which occur at the same time. In the case of reduced requirements, which only require identification of the source of the fault, the accelerometer mounted onto the moving part of the machine can be a viable option.

Inclusion of other aspects to our evaluation such as practicality or the cost of the sensors changes the end result. Based on the new evaluation criterion, the encoder will become the single best sensor for all of the examined tasks.

However, if we are satisfied only with the identification of the source of the fault, accelerometer is an obvious choice due to its low cost and ease of usage.

Acknowledgment. We would like to thank Mechatronic Design Solution ltd. for their help in designing and building the pneumatic test bench and for their expertise with pneumatic actuator-based production machinery.

References

1. Isermann, R.: Fault-Diagnosis Systems: An Introduction from Fault Detection to Fault Tolerance. Springer, Berlin (2006)
2. Hu, A., Xiang, L., Zhu, L.: An engineering condition indicator for condition monitoring of wind turbine bearings. Wind Energy (Chichester, England) **23**, 207–219 (2020). https://doi.org/10.1002/we.2423
3. Yang, W., Tavner, P.J., Crabtree, C.J., Wilkinson, M.: Cost-effective condition monitoring for wind turbines. IEEE Trans. Ind. Electron. **57**, 263–271 (2010). https://doi.org/10.1109/TIE.2009.2032202
4. Mahmoud, H., Mazal, P., Vlašic, F.: Detecting pneumatic actuator leakage using acoustic emission monitoring. Insight (Northampton) **62**, 22–26 (2020). https://doi.org/10.1784/insi.2020.62.1.22
5. Dunbar, W.B., de Callafon, R.A., Kosmatka, J.B.: Coulomb and viscous friction fault detection with application to a pneumatic actuator. In: 2001 IEEE/ASME International Conference on Advanced Intelligent Mechatronics. Proceedings (Cat. No. 01TH8556), vol. 2, pp. 1239–1244. IEEE (2001). https://doi.org/10.1109/AIM.2001.936889
6. Pandhare, V., Singh, J., Lee, J.: Convolutional neural network based rolling-element bearing fault diagnosis for naturally occurring and progressing defects using time-frequency domain features. In: 2019 Prognostics and System Health Management Conference (PHM-Paris), pp. 320–326. IEEE (2019). https://doi.org/10.1109/PHM-Paris.2019.00061
7. Lacey, J.: The Role of Vibration Monitoring in Predictive Maintenance. https://www.schaeffler.com/remotemedien/media/_shared_media/08_media_library/01_publications/schaeffler_2/technicalpaper_1/download_1/the_role_of_vibration_monitoring.pdf
8. Grepl, R., Matejasko, M., Bastl, M., Zouhar, F.: Design of a fault tolerant redundant control for electro mechanical drive system. In: 2015 21st International Conference on Automation and Computing (ICAC), pp. 1–6. Chinese Automation and Computing Society in the UK - CACS (2015). https://doi.org/10.1109/IConAC.2015.7313984
9. Brownlee, J.: How Much Training Data is Required for Machine Learning? www.machinelearningmastery.com/much-training-data-required-machine-learning/
10. Data Preprocessing for Condition Monitoring and Predictive Maintenance. www.mathworks.com/help/predmaint/ug/data-preprocessing-for-condition-monitoring-and-predictive-maintenance.html
11. Bolón-Canedo, V., Sánchez-Maroño, N., Alonso-Betanzos, A.: A review of feature selection methods on synthetic data. Knowl. Inf. Syst. **34**, 483–519 (2013)
12. Bouzida, A., Touhami, O., Ibtiouen, R., Belouchrani, A., Fadel, M., Rezzoug, A.: Fault diagnosis in industrial induction machines through discrete wavelet transform. IEEE Trans. Ind. Electron. **58**, 4385–4395 (2011). https://doi.org/10.1109/TIE.2010.2095391

13. Zhang, Y., Jiang, J., Flatley, M., Hill, B.: Condition monitoring and fault detection of a compressor using signal processing techniques. In: Proceedings of the 2001 American Control Conference (Cat. No. 01CH37148), vol. 6, pp. 4460–4465. IEEE (2001). https://doi.org/10.1109/ACC.2001.945681
14. Srividya, A., Verma, A.K., Sreejith, B.: Automated diagnosis of rolling element bearing defects using time-domain features and neural networks. Int. J. Min. Reclama. Environ. **23**, 206–215 (2009). https://doi.org/10.1080/17480930902916437
15. Abbasion, S., Rafsanjani, A., Farshidianfar, A., Irani, N.: Rolling element bearings multi-fault classification based on the wavelet denoising and support vector machine. Mech. Syst. Signal Process. **21**, 2933–2945 (2007). https://doi.org/10.1016/j.ymssp.2007.02.003
16. Mcfadden, P.D.: Examination of a technique for the early detection of failure in gears by signal processing of the time domain average of the meshing vibration. Mech. Syst. Signal Process. **1**, 173–183 (1987). https://doi.org/10.1016/0888-3270(87)90069-0
17. Zhu, J., Nostrand, T., Spiegel, C., Morton, B.: Survey of Condition Indicators for Condition Monitoring Systems (Open Access) (2014)
18. Večeř, P., Kreidl, M., Šmíd, R.: Condition indicators for gearbox condition monitoring systems. Acta polytechnica (Prague, Czech Republic: 1992) **45** (2005). https://doi.org/10.14311/782
19. Condition Indicators for Monitoring, Fault Detection, and Prediction. https://www.mathworks.com/help/predmaint/ug/condition-indicators-for-condition-monitoring-and-prediction.html
20. Dai, X., Gao, Z.: From model, signal to knowledge: a data-driven perspective of fault detection and diagnosis. IEEE Trans. Ind. Inform. **9**, 2226–2238 (2013). https://doi.org/10.1109/TII.2013.2243743

Ground Visibility Analyses, Algorithms and Performance

Dana Kristalova[1]([⊠]), Vlastimil Neumann[1], Pavla Zakova[2], Stepan Konecky[1],
Jiri Kralicek[1], Pavel Foltin[1], Libor Kutej[1], Jan Zezula[1], and Radomir Scurek[2]

[1] University of Defence, Brno, Czech Republic
{dana.kristalova,vlastimil.neumann,stepan.konecky,pavel.foltin,
libor.kutej,jan.zezula}@unob.cz
[2] VŠB, Ostrava, Czech Republic
pavla.zaky@seznam.cz, radomir.scurek@gmail.com

Abstract. The paper deals with the topic of efficient visibility analyses, its algorithms and performance in context of the area shape and its size. Visibility analyses are fundamental component of almost any tactical or environmental estimations and has to be executed in very large numbers, usually in "point to area" or "area to area" version. Thus, its effective solution and convenient algorithm selection is important and has to be solved in wider context and relations.

Keywords: Terrain visibility calculation · Operational decision support · Visibility analyses · Surface analyses · Visibility algorithm

1 Introduction

The development of modern C4ISTAR systems is accompanied with requirement increase for complex operational analyses and optimizations, which are connected with tactical factors and conditions. Majority of these factors fell undress the two groups as a visibility and manoeuvre analyses. These analyses have to be executed for almost any potential or desired position/s (usually combinations) of selected (usually all) tactical entities within a common operations picture (COP) what leads to enormous numbers of cases. This fact is the main reason why these algorithms for those analyses has to be very efficient and parallelized if possible.

In this paper it is introduced the comparison of two visibility algorithms, which were developed by the authors and each of them follow little a bit different approach. We assume, that these algorithms could differ in performance under certain conditions, which are subject of investigation within this paper. We search for the guide to distinguish, based on the known parameters, coming as an input to these algorithms (for this case it was simplified to the angle and range), which algorithm probably performs better and should be used in particular case.

© Springer Nature Switzerland AG 2022
J. Mazal et al. (Eds.): MESAS 2021, LNCS 13207, pp. 61–73, 2022.
https://doi.org/10.1007/978-3-030-98260-7_4

2 Problem Statement

The problem is defined as a search for the decision to distinguish which algorithm should be used for particular area visibility calculation (analyses). Based on the previous practice, the Ray-Cast and Floating Horizon algorithms seems to be the most effective algorithms for altitude matrix defined areas, so we initially assume these two. For operational calculations, the terrain square resolution usually varies between 5–10 m, as a good compromise between data precision, computation time and final operational benefit of particular analyses (in our case we use a 5 m resolution). Each of mentioned algorithms has some performance advantage in specific area shape processing and the question is, if there exists a generic math model or algorithm, which will point to the concrete solution (Ray-Cast vs Floating Horizon) with the best calculation performance (time).

```
for(y = 1; y < Range; y++) // Moves from initial position to the Distance/Range  - cycle
        {
                d = 1.0f/( y+1); //Transformation variable init for the row operation.
        for(x = - (y* Angle / Range); x <=  (y* Angle/ Range); x++) //Process the row-par-
ticular horizon
                {
  Fx = x * d * Rozl_x + Rozl_x;   //Transforms (project)  the land X component to the pro-
jection surface.
tAltitude =     (AltitudeMatrix[StandingPositionX + x, Standing PositionY +y] – Observation-
PointHeight ) * d;
                        if(tAltitude  > VisibitityArray[floor(Fx+0.5)]) //Check visibility
                        { //Terrain point is visible
                                VisibilityMatrix(  StandingPositionX + x , Standing-
                                PositionY + y )  = 1;
                        }
                        else
                        {//Terrain point is NOT visible
                                VisibilityMatrix(  StandingPositionX + x , Standing-
                                PositionY + y )  = 0;
                        }
                }
            UpdateVisibilityArray(); //The function interpolates the contemporary horizon
        and updates the
        VisibilityArray,
                }
```

When we look on both algorithms, there are some significant features that could intuitively lead to a performance dilemma, but without proper statistical evaluation, it is just an empirical estimation. Because the search for the existing appropriate C/C++ algorithmic solution was not successful (some algorithms were found, most of them were performance ineffective, using goniometric functions, for instance those from NVIDIA CUDA SDK, some of them were generally described but no-code available), we develop two algorithms fulfilling the initial conditions. These algorithms are part of broader framework, ensuring its functionality, what is not included, like variable initialization and final results storage and further operations. Presented are only a "south to north" parts, processing the symmetric triangle area. In any case main principles should be apparent from following descriptions. The Floating-Horizon visibility algorithm principles looks like:

The ray-cast algorithm was little a bit simpler, but process some points several times again comparing to the floating point algorithm, which, on the other hand has to contain a visibility resolution array to store the maximum transformed height in all directions (within a range -45 to $45°$), which has to be maintained and the height values between two neighborhood vertices are interpolated. The main body of ray-cast algorithm:

```
for(ty = 1; ty <=  STy ; ty++) // process all Ray-cast vectors from Observer position to spec-
ified range.
        {
            for(tx = 0; tx <=  2*STx; tx++) //Process all Ray-cast vectors within particular Y
                {
                 d= 1/(1+ ty); //Projection-transformation coefficient
                 Rays[tx].y = ty * Rays[tx].x; //Calculation of the X,Y in the Altitude matrix,
Rays[i].x stores X              component of direction vector, Y = 1.
                 Tz = AltitudeMatrix[x+Rays[tx].y,y+ty].Altitude - Vz; //Get the Altitude cell
from the Altitude Matrix.
                 if( Rays[tx].z  <  d*Tz )       //Compare if the new point is visible
                    {
                    VisibilityMatrix(((x+Rays[tx].y), (y+ty)) = 1;//point is visible;
                    Rays[tx].z = d*Tz; //Store the highest value
                    }
                }
        }
```

3   Literature Analyses

Today's battlefield clearly tends to use artificial intelligence tools, whether it is technical data and the robotic automation of the equipment themselves, whether it is the including of software or algorithms for planning, visibility and decision-making processes into these resources or it is the question of obtaining suitable datasets and methods of evaluating key data parameters.

The study of the literature has pointed to a large number of high-quality scientific articles which describe the basic key approaches to the evaluation of visibility, unfortunately the time of publication of them is dated to the end of the 20th century. Nevertheless, they cannot be neglected. Current literary sources focus on very specific issues.

Basic information about the analyses of visibility surfaces are summarized in publication [1] and elementary information and the development of algorithms for determining visibility are drawn from knowledge base in [2–4].

Specific effective algorithms for assessing visibility are recorded in [5]. Previous research on the calculation and evaluation of visibility from different types of terrain data models and the selection of the use of visibility criteria is set out in [6–9].

The issue of determining uncertainty and probability in the question of visibility assessment is addressed in the articles [10, 11].

The other approaches to algorithms evaluating visibility in vector or raster data models on different topographical surfaces are state in the following works [12–17].

A comparison of some various computational algorithms for determining visibility have been published in the sources [18–20].

Algorithms used to evaluate visible or hidden locations are closely correlated with moving planning and calculation of time and economic costs for required activities. Algorithm development for finding optimal routes and evaluating of a terrain traffic ability are part of terrain analyses and are published, for example, in [21–23]. Terrain analyses for Cross-Country Movement is other very significant branch of using similar computing algorithms and in this topic an impact of geographical factors can be crucial. More are reported in [24–29].

Choosing key parameters and data that should be used or not used is an endless question. In the same way, optimization methods for the searching of suitable routes or options of suitable ground observation points can be constantly improved and refined and also these computational algorithms are closely linked to the visibility evaluation [30, 31]. Furthermore, in the literature articles can be found, which deal with the mathematical modelling and use of successful algorithms and their applications for various key operations, especially directly in the military topics [32–39].

The efficient algorithms and their applications can also be used to optimize transport, for example [39, 40], or for use in crisis management [41]. Another area where there are controversies over whether or not to use algorithms is the issue of human resources, namely the possibility of the commander to adopt advanced solutions [42].

In any case, it is good to realize that algorithms for evaluating visibility often work with huge amounts of data, and in proportion to the accuracy and volume of data, these are complex computational processes, where any acceleration of computation can ultimately mean a huge progress, see [43, 44]. Also, the level of visualization of evaluated visibility can be considered within decision-making processes.

4 Approach to the Solution

As the problem was defined in Sect. 2, a search for the decision to distinguish which algorithm should be used for particular area visibility, there were identified following steps:

1. Development and implementation of two competing terrain visibility algorithms within a real terrain model. (See picture Nr. 2: Terrain model 20 × 20 km used for evaluation, resolution 5 × 5 m.)
2. Measuring the performance of both algorithms on a variety of areas.
3. Performance analyses and comparison in context of the area shape and size.

Some of key characteristics of used algorithms was already mentioned, both has a different approach and differs in the potential precision. Both methods are point to area (multipoint) processes and follow the search from front to back (from observer to far horizon) to the floating-point version creates particular horizons and which has an important impact on a solution performance and precision in various cases.

In any case, for operational purposes, both of them are acceptable, because, lot of analyses takes a statistical format and usually large terrain models are about several months old, thus small precision handicap is still acceptable.

Both algorithms could also count with potential visibility or hidden areas, which could be or could not be visible in a real situation, because these areas are close to the theoretical edge of visibility, which distinguish between visible or not.

Unclarity about the real particular horizon point visibility appears when the "Uncertainty coefficient" varies between $(-1,1)$.

$$U_c = \frac{\left(HD_{dif} - M_{prec}\right)}{M_{prec}} \tag{1}$$

Where U_c – *Uncertainty coefficient*
HD_{diff} – *Difference between actual horizon and (minimal) visibility horizon - the last visible line from "eye" at the same direction with minimal angle*
M_{prec} – *3D model precision*

Demonstration of these features are visible on following pictures (See Fig. 1):

Both algorithms are convenient for the fast terrain visibility processing. Due to the slightly different approach, differences in processing performance/efficiency can be identified. Both algorithms were evaluated for the visibility performance of the real territory in two steps.

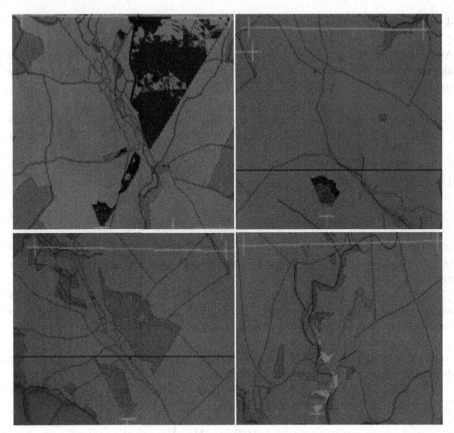

Fig. 1. Example of graphical visualization of potentially hidden areas. The difference of the particular horizon height and height of the terrain point, determines whether the area is "potentially" visible or not. Absolute value of the small difference (this threshold, could be derived from the resolution, precision and other attributes) indicates an uncertain situation that cannot be precisely decided, whether or not is the assessed area visible. The different colour shows the area where the visibility is "tricky", so it implies an unclear solution in practice. The dark (under the horizon) and light (above the horizon) colour indicates the areas close to particular horizon edge and this space may be visible or not.

The first step is the angular calculation range, where each step was about 1° incremented (Graph - Angle Range Axis).

The second step just repeat the first one with incrementing the distance (Graph - Visibility Distance Axis). The interval of this step was set to 100 m and each algorithm configuration (angle, distance) was executed 400 times on the 20 × 20 km digital model as illustrated on following figure (Fig. 2).

Fig. 2. Terrain model 20 × 20 km used for the evaluation, resolution 5 × 5 m (on the left). On the right there is a picture representing the ground observation situation. There are 400 (20 × 20) observation posts (starting with red peaks at the bottom). From each position, all particular visibility calculation was carried out, it ensures comprehensive coverage of the test area and sufficient algorithm performance evaluation.

Finally, the solution time data for each iteration and algorithm was collected and 3D graphs was developed, as presented on Fig. 3 and 4.

From graphs, it is possible to derive following statement: Ray-Cast is more accurate for a closer area - approximately up to 6 km, but from this distance its performance decreases rapidly, and the time required for calculation increases significantly. From this distance, it would be more efficient to use the Floating Horizon Algorithm.

More detailed investigation and math surface regression analyses will be a subject of further research.

5 Discussion of Other Alternative Approaches

As it was mentioned before, there exists other alternative approaches improving the speed or other factors of solution. One of the terrain visibility calculation approach is transformation of the depth map from the 3D rendering process, what could success-fully utilize well established and highly parallel rendering pipeline of common graphic cards. For testing purposes, the sample application was developed, but performance did not reach the previous solution and further optimization and improvement could be a subject of further research. Performance of this approach is more than 100 times slower, comparing the mentioned algorithms, but visibility results are much more detailed and convenient for specific tasks, like urban or complex area analyses, for example search-ing for a potential sniper positions, estimating cover and concealment and so. Results of mentioned approach is illustrated on following figure (Fig. 5).

Next, very perspective approach is exploitation of parallel architecture for generic computation of contemporary graphics cards, most common platforms are OpenCL and

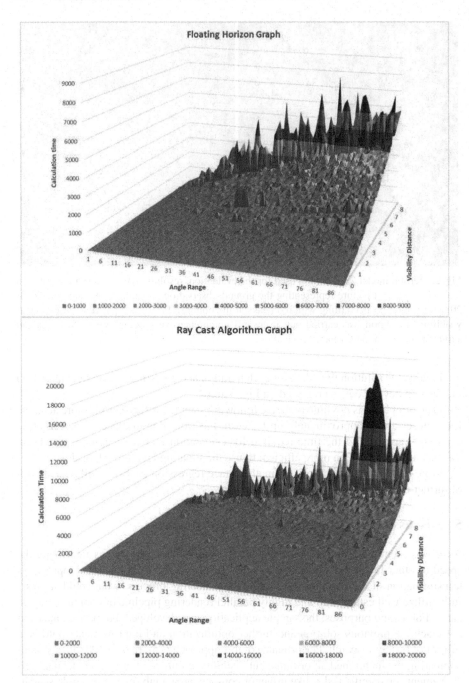

Fig. 3. 3D graphical evaluation performance of the Floating Horizon Algorithm (above) and the Ray Cast Algorithm (below). The space is gradually evaluated in terms of visibility. From the observation point, the angular range first increases, then the distance (indicated in km). The calculation time is given in milliseconds.

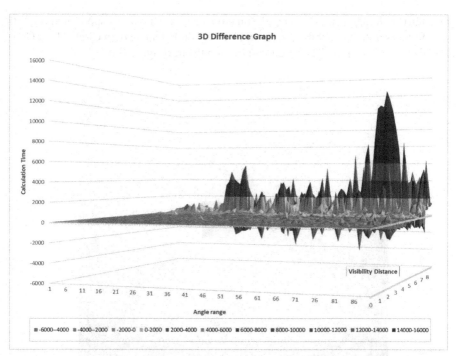

Fig. 4. Graphical expression of comparative analysis. The comparative analysis clearly shows an increase in time calculation values with the increasing angular range of the evaluated area as well as the length of the intentional beam. As mentioned in the article, the decisive limit distance is about 6 km, for a closer area it is advisable to use ray cast algorithm, for more distant and large areas the Floating Horizon Algorithm is significantly faster.

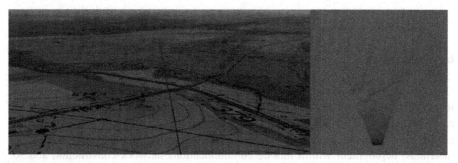

Fig. 5. A depth map is obtained by reverse transformation, where the visual rendering transforms each individual map pixel, even with objects. This approach brings superior results, millions of pixels are calculated.

CUDA. The last mentioned platform was used to test performance of the Floating Horizon Algorithm in Area to Area version, what actually do not require deep modification of the algorithm itself. Demonstration application was developed in this manner and the percentage visibility shadow map of destination area was calculated approximately 150 times faster (solution under the 3 s, NVIDIA RTX 2060) than modern CPU's (Intel Core I7/3 GHz) could achieve. The result is shown on the next figure (Fig. 6).

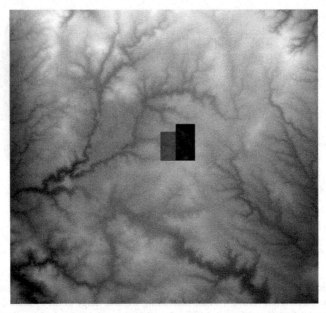

Fig. 6. Parallel execution of Floating Horizon Algorithm on CUDA platform, solution performance is about 150 times faster, comparing to the modern CPU processor (average time 2832 ms, observer source area 1 × 2 km, visibility destination area 1.4 × 2.5 km, and terrain square resolution 5 m)

6   Conclusion

Usually, the problem of calculation efficiency of any algorithm is important, especially in those cases, where these algorithms are executed very often, what is obviously the case of terrain visibility analyses, necessary for operational calculations and other optimization, for instance search for the best distribution of communication technology (antennas) deployment within mobile communications network construction and so. Presented solution show, that, there exists some specifics of selected „state of the art" approaches, which leads to a difference in performance, based on the destination area shape and size. These aspects could be stored in the database and potentially mathematically modelled (subject of further research). This paper shows the solution and results, which could lead to a significant speed up of the particular visibility analyses, when the estimation of selected algorithm performance is known in advance, thus the user or computer could choose the faster version for every particular case.

References

1. Caldwell, D., Mineter, M., Dowers, S., Gittings, B.: Analysis and Visualization of Visibility Surfaces (2003). https://www.researchgate.net/publication/255610771_Analysis_and_Visualization_of_Visibility_Surfaces
2. Ray, C.: Representing visibility for sighting problems. PhD Dissertation, Rensselaer Polytechnic Institute (1994a)
3. Ray, C.: A new way to see terrain. Mil. Rev. **2**, 81–89 (1994)
4. Richburg, R.F., Ray, C.K., Campbell, L.W.: Terrain analysis from visibility metrics. In: Proceedings of the SPIE Conference, April 1995, vol. 2486, pp. 208–219 (1995)
5. De Floriani, L., Magillo, P.: Algorithms for visibility computation on terrains: a survey. Environ. Plann. B. Plann. Des. **30**, 709–728 (2003)
6. De Floriani, L., Falcidieno, B., Nagy, G., Pienovi, C.: Polyhedral terrain description using visibility criteria. TR, Institute for Applied Mathematics, National Research Council, Genova (1989)
7. De Floriani, L., Montani, C., Scopigno, R.: Parallelizing visibility computations on triangulated terrains. Int. J. Geogr. Inf. Syst. **8**(6), 515–531 (1994). https://doi.org/10.1080/02693799408902019
8. Magillo, P., Floriani, L., Bruzzone, E.: Updating visibility information on multiresolut ion terrain models. In: Frank, Andrew U., Kuhn, Werner (eds.) COSIT 1995. LNCS, vol. 988, pp. 279–296. Springer, Heidelberg (1995). https://doi.org/10.1007/3-540-60392-1_18
9. De Floriani, L., Magillo, P.: Multiresolution mesh representation: models and data structures. In: Iske, A., Quak, E., Floater, M.S. (eds.) Tutorials on Multiresolution in Geometric Modelling, pp. 363–417. Springer, Heidelberg (2002). https://doi.org/10.1007/978-3-662-04388-2_13
10. Fisher, P.F.: Algorithms and implementation of uncertainty in viewshed analysis. Int. J. Geogr. Inf. Syst. **7**, 331–347 (1993)
11. Fisher, P.F.: An exploration of probable viewsheds in landscape planning. Environ. Plann. B. Plann. Des. **22**, 527–546 (1995)
12. Blelloch, G.E.: Vector Models for Data-parallel Computing. MIT Press, Cambridge (1990)
13. Shapira, A: Visibility and terrain labelling, master thesis, Department of Electrical, Computer, and System Engineering, Rensselaer Polytechnic Institute, Troy, NY (1990)
14. Lee, J.: Analyses of visibility sites on topographic surfaces. Int. J. Geogr. Inf. Syst. **5**, 413–429 (1991)
15. Mills, K., Fox, G., Heimbach, R.: Implementing an intervisibility analysis model on a parallel computing system. Comput. Geosci. **18**, 1047–1054 (1992)
16. Teng, Y.A., De Menthon, D., Davis, L.S.: Region-to-region visibility analysis using data parallel machines. Concurr. Pract. Exp. **5**, 379–406 (1993)
17. Rallings, P.J., Ware, J.A., Kidner, D.B.: Parallel distributed viewshed analysis, in Proceedings of ACMGIS 1998, pp. 151–156. ACM Press, New York (1998)
18. Sorensen, P., Lanter, D.: Two algorithms for determining partial visibility and reducing data structure induced error in viewshed analysis. Photogramm. Eng. Remote. Sens. **59**, 1129–1132 (1993)
19. Van Kreveld, M.: Variations on sweep algorithms: efficient computation of extended viewsheds and class intervals. In: Kraak, M.J., Molenaar, M (eds.) Proceedings of the Symposium on Spatial Data Handling 1996, Faculty of Geodetic Engineering, University of Technology, Delftpp 13A.15–13A.27 (1997)
20. Alipour, S., Ghodsi, M., Ugur, G., Golkari, M.: Approximation algorithms for visibility computation and testing over a terrain Societa Italiana di Fotogrammetria e Topografia (SIFET). Appl. Geomat. **9**, 53–59 (2017). https://doi.org/10.1007/s12518-016-0180-9

21. Křišťálová, D.: Evaluation of the data applicable for determining the routes of movements of military vehicles in tactical operation. In: The Complex Physiognomy of the International Security Environment, Sibiu, Romania: "Nicolae Balcescu" Land Force Academy Publishing House, pp. 197–203 (2015). ISBN 978-973-153-215-8

22. Kristalova, D., Mazal, J.: The effect of land cover on movement of vehicles in the terrain, world academy of science. Eng. Technol. Int. J. Civil Arch. Eng. **8**(11), 1917–1922 (2014)

23. Křišťálová, D.: A traffic ability of the terrain. Economics and Management (2/2014), pp. 38–47 (2014). ISSN 1802-3975-24. Without the main author. Procedural Guide for Preparation of DMA Cross-Country Movement (CCM) Overlays. Defence Mapping School. Fort Belvoir, Virginia 1993

24. Rybanský, M.: Cross-Country Movement - The Impact and Evaluation of Geographical Factors. The Czech Republic, Brno, p. 114 (2009). ISBN: 978-80-7204-661-4

25. Křišťálová, D., et al.: Geographical data and algorithms usable for decision-making process. In: Modelling and Simulation for Autonomous Systems, Italy, Roma, pp. 226–241. Springer, Heidelberg (2016). https://doi.org/10.1007/978-3-319-47605-6_19, ISSN 0302-9743, ISBN 978-3-319-47604-9

26. Rada, J., Rybanský, M., Dohnal, F.: Influence of quality of remote sensing data on vegetation passability by terrain vehicles. ISPRS Int. J. Geo-Inf. **9**(11), 684 (2020). ISSN 2220–9964, IF 2.899

27. Křišťálová, D.: An effect of sandy soils on the movement in the Terrain. In: MESAS, pp. 262–273. Springer International Publishing, Rome (2014). ISBN 978-3-319-13823-7

28. Mazal, J., Rybanský, M., Bruzzone, A., Kutěj, L., Scurek,R., Foltin, P., Zlatník, D. Modelling of the microrelief impact to the cross country movement. In: Bottani, E., Bruzzone, A.G., Longo, F., Merkuryev, Y., Piera, M.A. (eds.) Proceedings of the 22nd International Conference on Harbor, Maritime and Multimodal Logistic Modeling & Simulation, HMS 2020, vol. 22, pp. 66–70 (2020). ISSN 27240339, ISBN 978-8-885-74146-1

29. Mazal, J., Bruzzone, A., Kutěj, L., Scurek, R., Foltin, P., Zlatník, D.: Optimization of the ground observation (2020). ISSN 2724-0339. ISBN 978-88-85741-46-1

30. Hurtado, F., et al.: Terrain visibility with multiple viewpoints. Int. J. Comput. Geom. Appl. (2014). ISSN (print): 0218-1959IISSN (online): 1793-6357, https://doi.org/10.1142/S02181 95914600085

31. Mazal, J., Stodola, P., Hrabec, D., Kutěj, L., Podhorec, M., Křišťálová, D.. Mathematical Modeling and optimization of the tactical entity defensive engagement. Int. J. Math. Models Methods Appl. Sci. **9**(summer 2015), 600–606 (2015). ISSN 1998-0140

32. Mazal, J., Stodola, P., Procházka, D., Kutěj, L., Ščurek, R., Procházka, J.: Modelling of the UAV safety manoeuvre for the air insertion operations. In: Modelling and Simulation for Autonomous Systems, MESAS 2016, pp. 337–346. Springer, Rome (2016). ISSN 0302-9743. https://doi.org/10.1007/978-3-319-47605-6_27, ISBN 978-3-319-47604-9

33. Nohel, J.: Possibilities of raster mathematical algorithmic models utilization as an information support of military decision making process. In: Mazal, J. (ed.) MESAS 2018. LNCS, vol. 11472, pp. 553–565. Springer, Cham (2019). https://doi.org/10.1007/978-3-030-14984-0_41

34. NATO Modelling and Simulation Centre, pp. 553–565 (2019). ISSN 0302-9743, ISBN 978-3-030-14984-0

35. Časar, J., Farlík, J.: The possibilities and usage of missile path mathematical modelling for the utilization in future autonomous air defense systems simulators. In: Mazal, J., Fagiolini, A., Vasik, P. (eds.) Modelling and Simulation for Autonomous Systems, LNCS, vol. 11995, pp. 253–261 (2020). Springer Nature, Cham (2020). https://doi.org/10.1007/978-3-030-43890-6_20, ISSN 0302-9743, ISBN 978-3-030-43890-6

36. Drozd, J., Stodola, P., Křišťálová, D., Kozůbek, J.: Experiments with the UAS reconnaissance model in the real environment. In: Modelling and Simulation for Autonomous Systems,

pp. 340–349. Springer, Cham (2018). https://doi.org/10.1007/978-3-319-76072-8_24, ISSN 0302 9743, ISBN 978-3-319-76071-1

37. Nohel, J., Flasar, Z.: Maneuver control system CZ. In: Mazal, J., Fagiolini, A., Vasik, P. (eds.) Modeling and Simulation for Autonomous Systems. MESAS 2019. Lecture Notes in Computer Science, vol. 11995, pp. 379–388. Springer, Cham (2020). https://doi.org/10.1007/978-3-030-43890-6_31, ISBN 978-3-030-43889-0

38. Bruzzone, A.G., Massei, M.: Simulation-based military training. In: Guide to Simulation-Based Disciplines, pp. 315–361. Springer, Heidelberg (2017). https://doi.org/10.1007/978-3-319-61264-5_14

39. Mazal, J., Bruzzone, A., Turi, M, Biagini, M., Corona, F., Jones, J.: NATO use of modelling and simulation to evolve autonomous systems. In: Complexity Challenges in Cyber Physical Systems: Using Modeling and Simulation (M&S) to Support Intelligence, Adaptation and Autonomy, pp. 53–80. John Wiley & sons, Hoboken (2019). ISBN 978-1-119-55239-0

40. Rybansky, M.: Modelling of the optimal vehicle route in terrain in emergency situations using GIS data. In: 8th International Symposium of the Digital Earth (ISDE8) 2013, Kuching, Sarawak, Malaysia 2014 IOP Conference Series: Earth Environmental Science, vol. 18, p. 012071 (2014). https://doi.org/10.1088/1755-1315/18/1/012131, ISSN 1755–1307

41. Mokrá, I.: Modelový přístup k rozhodovacím aktivitám velitelů jednotek v bojvých operacích. Disertační práce. Brno: Univerzita obrany v Brně, Fakulta ekonomiky a management, 120 s (2012)

42. Fishman, J., Haverkort, H., Toma, L.: Improved visibility computation on massive grid terrains. In: Proceedings 17th ACM SIGSPATIAL International Conference on Advances in Geographic Information Systems, pp. 121–130. ACM (2009)

43. Haverkort, H., Toma, L.: A comparison of I/O-efficient algorithms for visibility computation on massive grid terrains (2018). [1810.01946v1]

44. Coll, N., Narcís Madern, J., Sellarès, A.: Good-visibility maps visualization. Visual Comput. 26(2), 109–120 (2010). https://doi.org/10.1007/s00371-009-0380-y

High-Frequency Vibration Reduction for Unmanned Ground Vehicles on Unstructured Terrain

Hamza El-Kebir[1]([✉]), Taha Shafa[1,2], Amartya Purushottam[2,3], Melkior Ornik[1], and Ahmet Soylemezoglu[2]

[1] Department of Aerospace Engineering, University of Illinois Urbana-Champaign, Urbana, IL 61801, USA
elkebir2@illinois.edu
[2] Construction Engineering Research Laboratory, US Army Corps of Engineers, Champaign, IL 61822, USA
[3] Department of Electrical and Computer Engineering, University of Illinois Urbana-Champaign, Urbana, IL 61801, USA

Abstract. High-frequency vibrations encountered during land transit of sensitive payloads have long been known to be a possible cause of payload damage and subsequent mission failure. As sensors are also adversely affected by this phenomenon, we aim to provide a solution to minimize high-frequency noise vibrations without reliance on high performance sensing. Naturally, this presents the need for on-board adaptive control capabilities to reduce sensor noise and damage to secured payloads. Thus, we present a novel approach to reducing high-frequency vibration content (HVC) encountered during transit, with the explicit goal of maintaining a desired vehicle speed while keeping high-frequency vibrations below a given threshold regardless of the terrain characteristics. To this end, we present a two-stage solution consisting of a vibration-compensating speed controller and an optimal tracking controller for control command determination. The proposed controller is implemented on a Clearpath Jackal unmanned ground vehicle and subjected to a priori unknown mixed terrain types. Experiments performed on these varying terrains show that the proposed control architecture is able to adjust the desired robot trajectory to remain below the vibration thresholds defined by the mission objective.

Keywords: Adaptive trajectory planning · Navigation on unstructured terrain · Vibration mitigation

1 Introduction

Adverse effects from high-frequency vibrations can adversely affect critical mission tasks beyond the introduction of sensor noise [7,9]. These vibrations can

---

H. El-Kebir and T. Shafa—These authors contributed equally to this work.

J. Mazal et al. (Eds.): MESAS 2021, LNCS 13207, pp. 74–92, 2022.
https://doi.org/10.1007/978-3-030-98260-7_5

cause damage to secured payloads [10,22] and aggravate the conditions of injured personnel during medical evacuations [3,21]. These issues naturally give rise to the problem of vibration suppression.

In this work, we present a novel approach to reducing the high-frequency vibration content encountered during transit with the explicit goal of maintaining a desired vehicle speed, while keeping high-frequency vibrations below a given threshold regardless of the current terrain type. Our theoretical work and presented implementation are motivated by programs like the Squad Multipurpose Equipment Transport (SMET) program, a US Army program providing a robotic "mule" for military personnel; tasks include reducing soldiers' weight burden and medical evacuation of harmed personnel. Sensitive payloads like an injured soldier, or sensitive electronic or chemical equipment, cannot endure high-energy, high-frequency vibrations above a certain threshold without posing a major safety concern during transit. Such settings present a need for ground vehicles to reduce high-frequency vibrations while traversing unknown terrain, without significantly compromising movement speed.

Since many state-of-the-art robotic platforms—Jackal, Argo J8, and Clearpath J5 - come unequipped with vehicle suspension, we are particularly concerned with achieving high-frequency noise mitigation through a novel controller architecture. Moreover, we wish for our approach to be based on portable commercial off-the-shelf components, so as to ensure the widest possibility of proliferation on existing ground vehicle fleets, without the need of a costly retrofitting campaign. To accomplish this objective, we develop a two-step approach to trajectory planning based on a set of mission-dependent waypoints. Given a desired nominal velocity, as well as a maximum permitted velocity, our approach aims to smoothly vary the vehicle's speed to actively suppress high-frequency vibration normal to the terrain. To this end, we introduce an intuitive vehicle- and terrain-agnostic high-frequency vibration measure which is used to command a desired maximum velocity. The frequency content is measured using a low-cost off-the-shelf inertial measurement unit. We use an optimal feedback controller to minimize incurred cost while converging to the desired waypoint.

2   Prior Work

Previous work on noise and vibration mitigation has chiefly focused on terrain determination based on the pseudo-spectral density of the vibrations, allowing for discrete gain or controller switching [1,11,25]. Such approaches only work when the terrain that will be encountered is known a priori, with the classification model depending directly on the vehicle under consideration. Unlike these methods, we provide a means of continuously adapting to changing terrain conditions without the need for pre-classifying the terrain types that may be encountered [14,19] or the need to introduce additional haptic sensors [4,8,15]. In addition, only the tracking controller is dependent on the vehicle properties, with no knowledge assumed about the frequency response properties of the vehicle to vibrations.

Fig. 1. Clearpath Jackal unmanned ground vehicle (Video of the controller applied to the Jackal UGV can be found at https://uofi.box.com/s/9lvx16hrrw7kr4ew7og91w4uspid3n02.)

Beyond these terrain-classifying approaches, other traditional approaches to vibration mitigation can be divided into two categories: approaches that are concerned with mitigating vibration induced by the vehicle's internal components (e.g., the engine, transmission, suspension, etc.), and those that deal with external effects (e.g., speed humps, degraded road surfaces, unstructured terrain). We refer to the first category as internal vibrations, and the latter as external vibrations. Internal vibration mitigation has chiefly been centered around damping the resonance response among internal components [23], whereas external vibration mitigation can be further divided into active and passive approach.

Passive external vibration suppression is governed by the fixed suspension design of the vehicle, whereas active suspension systems make use of controlled actuators to complement the passive springs and dampers [27]. In this work, we do not expect the vehicle to have any type of suspension system, thus limiting our control inputs to wheel torque/velocity, steering angle, etc.

We leverage an off-the-shelf inertial measurement unit to obtain a measure of the severity of the vehicle vibrations. In particular, we focus on vibrations that are perpendicular to the vehicle. Frequency properties of the vibrations are obtained by applying a fast Fourier transform (FFT) to a small time window of sampled data. We use frequency content in a precompensated integral controller (PCIC) to obtain the maximum vehicle speed needed to suppress the vehicle's vibrations. This maximum velocity is then applied to the control input - obtained from a tracking linear quadratic requlator (LQR) - by simply scaling the control input to produce an admissible control input. The LQR controller is obtained from a discretized and periodically linearized kinematic model of the vehicle.

We demonstrate the proposed control law on a Clearpath Jackal unmanned ground vehicle (UGV), shown in Fig. 1, using only the internal wheel encoders and a low-cost inertial measurement unit as sensor feedback. During the experimental trials, we subject the vehicle to different terrain types (concrete, grass, mulch) along a fixed desired path, and we compare the reduction in high-frequency vibrations when using our adaptive speed-based control law compared to a traditional fixed velocity control law.

3   Waypoint Tracking Controller

We first present the kinematic vehicle model used as part of our LQR tracking controller. We then make use of the LQR to generate the system inputs to guide our vehicle to the desired waypoints. Our vehicle hardware limitations restricted our usage of torque-based control and a full dynamics model; rather, we commanded wheel velocities from a derived kinematics model.

3.1   Vehicle Kinematics

The classical equation to represent the dynamics of a nonholonomic mobile robot can be obtained from the Lagrangian formulation [6]:

$$M(q)\ddot{q} + V_m(q, \dot{q})\dot{q} + G_m(q) + \tau_d = B(q)\tau - A^T(q)\lambda, \tag{1}$$

where $q \in \mathbb{R}^n$ is the generalized state, $\tau \in \mathbb{R}^r$ is the input vector, $\lambda \in \mathbb{R}^m$ is the vector of constraint forces, $M(q) \in \mathbb{R}^{n \times n}$ is a symmetric and positive-definite inertia matrix, $V(q, \dot{q}) \in \mathbb{R}$ is the matrix of Coriolis and centrifugal forces, $G(q) \in \mathbb{R}^n$ is the vector of gravitational forces, $\tau_d \in \mathbb{R}^n$ is the vector of disturbances, $B(q) \in \mathbb{R}^{n \times r}$ is the input transformation matrix, and $A(q) \in \mathbb{R}^{m \times n}$ is the matrix associated with any constraints (frictional cone, maximum acceleration, etc.).

The above model is widely used in the development of multipurpose controllers, but such controllers generate torque commands, as opposed to velocity commands which are more commonly implemented in commercial robotics. Thus, the use of such a model utilizing torque commands requires additional knowledge of the actuation system of the robot [20], i.e., knowledge regarding the actuator dynamics and model of internal mechanisms relating motor torque to the angular velocity of the wheels. Such information is not readily available for the Jackal UGV. Moreover, the motors on the Jackal have a high gearing reduction which makes torque-based control difficult to implement since the nonlinear friction dynamics, $\eta(t, q, \dot{q})$, become non-negligible at these high gearing ratios:

$$\tau = K_t i + \eta(t, q, \dot{q}).$$

Thus, we will base our controller design on a velocity-based kinematic model of the Jackal where wheel velocity input commands are utilized directly to manipulate system behavior.

We now derive the planar kinematics of an unmanned ground vehicle. In this work, we assume the vehicle maneuvers without slipping, and that the vehicle is based on differential (or skid) steering. Existing skid steering models utilize left and right angular wheel speeds to produce the kinematic and dynamic equations that relate the tire radius and frame size to translational velocities [17]. In contrast, the Jackal UGV comes equipped with its own skid steer controller that takes into account these parameters for direct control of translational velocity in the x and y directions shown in Fig. 2.

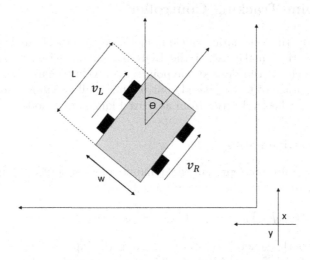

Fig. 2. Kinematic diagram the Jackal unmanned ground vehicle based on differential steering

As a result of the discussion above, most market-available ground robots utilize velocity controllers to track input reference velocities. As such, we developed a velocity controller consistent with the kinematic diagram illustrated in Fig. 2 following the approaches of [2,12]. The resulting kinematics model is as follows:

$$\begin{bmatrix} \dot{x} \\ \dot{y} \\ \dot{\theta} \end{bmatrix} = \begin{bmatrix} \frac{1}{2}(v_L + v_R)\cos\theta \\ \frac{1}{2}(v_L + v_R)\sin\theta \\ (v_R - v_L)/w \end{bmatrix}, \tag{2}$$

such that v_L and v_R are the velocities of the left and right wheels, θ is the heading angle, and $w = 0.4$ m is the vehicle's track width.

The derived model is utilized for direct control of translational velocities and the angular velocity on a two-dimensional plane.

3.2 LQR Tracking Controller Design

We employ a discrete-time linear quadratic regulator (LQR) to navigate towards waypoints by using the current state. We first present vehicle dynamics (2) linearized about heading angle θ_0:

$$\begin{bmatrix} \dot{x} \\ \dot{y} \\ \dot{\theta} \end{bmatrix}_{\theta_0} = \begin{bmatrix} \frac{1}{2}\cos\theta_0 & \frac{1}{2}\cos\theta_0 \\ \frac{1}{2}\sin\theta_0 & \frac{1}{2}\sin\theta_0 \\ -\frac{1}{w} & \frac{1}{w} \end{bmatrix} \begin{bmatrix} v_L \\ v_R \end{bmatrix}. \tag{3}$$

We proceed by discretizing the above linear system dynamics using an exact zero-order hold discretization approach based on exponential matrices [5, p. 99]. The sampling period that we consider in this work is $\Delta t = 0.01$ s. We now introduce the state and control penalty matrices that we have considered in this work:

$$Q = \text{diag}\left([5, 5, 0.5]\right), \quad R = I_{2\times 2}, \tag{4}$$

where Q is the state penalty matrix, and R is the control penalty matrix. These matrices were found in tuning of our simulation and on hardware platform. With this configuration, we prioritize position tracking over heading tracking by more heavily penalizing deviation from the desired positions. For different vehicles, different gain values may be required to obtain control inputs of suitable magnitudes for the actuators. In this work, we have employed a custom finite horizon LQR algorithm of 50 iterations to obtain feedback gain matrix $K \in \mathbb{R}^{2\times 3}$. We relinearize the model and recompute this gain every five seconds.

We implement the controller as follows:

$$\vec{u} = K(\vec{x}_{\text{des}} - \vec{x}),$$

where $\vec{x} \in \mathbb{R}^3$ denotes the full vehicle state, $\vec{x}_{\text{des}} \in \mathbb{R}^3$ corresponds to the current desired waypoint, and $\vec{u} \in \mathbb{R}^2$ corresponds to the desired right and left wheel velocity.

4 High-Frequency Vibration Compensator

As mentioned in the introduction, we wish to regulate the speed of the vehicle to control the severity of high-frequency vibrations. This task naturally poses the question of what constitutes a usable measure of the high-frequency vibration content. We discuss such a measure at the beginning of this section. Given this measure, we then design a compensator to adjust our maximum desired speed.

Given a measure of the high-frequency content, we may then design a model-free controller that outputs a speed limit, which is consequently applied to the control input obtained from the LQR controller in Sect. 3.2.

4.1 High-Frequency Vibration Measure

As mentioned previously, we will draw on linear acceleration readings of a low-cost off-the-shelf inertial measurement unit (IMU) to obtain the strength of high-frequency vibrations. Here, we discuss two candidate measures for the high-frequency content, of which the latter will ultimately be used.

We start with a sampled data signal of the linear acceleration measures. We assume that the IMU is mounted on the vehicle such that the internal z-axis faces normal to the level plane of the vehicle. We first consider the case of a single acceleration signal, expanding to the case of multiple signals afterwards.

Let $a(t)$ be the acceleration signal that we wish to be limited in its high-frequency power. Since we will consider the signal properties in the frequency domain, we will have to obtain sampled time windows of this continuous-time signal, on which we apply a discrete Fourier transform.

Let $a(t)$ be sampled at sampling frequency f_s. We wish to obtain sampled windows of $a(t)$ of duration Δt_{win}, such that each window contains $N_{\text{win}} = f_s \Delta t_{\text{win}} \in \mathbb{N}$ samples. We denote each window by $a_k = \{a(k\Delta t_{\text{win}}), a(k\Delta t_{\text{win}} + 1/f_s), a(k\Delta t_{\text{win}} + 2/f_s), \ldots, a((k+1)\Delta t_{\text{win}})\}$, where $k \in \mathbb{N}$. We may then apply a discrete Fourier transform (DFT) [24, 26] on each of these windows, such that we obtain $\alpha_k[f]$, where $f \in F := \{0, f_s/(2(N_{\text{win}} - 1)), 2f_s/(2(N_{\text{win}} - 1)), \ldots, f_s/2\}$. The latter maximum frequency $f_s/2$ is the Nyquist frequency. The value of $\alpha_k[f]$ denotes the power of the signal for frequencies in the range $[f, f + f_s/(2(N_{\text{win}} - 1))]$.

Multiple Acceleration Signals. Let us now consider the case where we have multiple accelerations that we could like to consider. We denote these n accelerations by $a^{(1)}, \ldots, a^{(n)}$. One approach to combining these signals is by taking a linear combination to obtain

$$a = \sum_{i=1}^{n} b_i a^{(i)},$$

such that $\sum_{i=1}^{n} b_i = 1$ and each $b_i > 0$. This does, however, not preclude the possibility of destructive interference between the various signals, which may make for a signal power of a that is less than that of one of its constituent signals $a^{(i)}$. One could elect to consider $|a^{(i)}|$ instead, but this would not account for the effect of high-frequency sign changes in each signal, which would skew the DFT towards lower frequencies.

A more reasonable approach would be to consider DFT results of the acceleration signal $\alpha_k^{(i)}$, and treat each result as a vector in $\mathbb{R}^{N_{\text{win}}}$. Since each element in each $\alpha_k^{(i)}$ is nonnegative, we can now consider

$$\alpha_k = \sum_{i=1}^{n} \beta_i \alpha_k^{(i)},$$

with the stipulation that $\sum_{i=1}^{n} \beta_i = 1$ and each $\beta_i > 0$. The addition of signal powers is clearly constructive, and allows us to place greater emphasis on some signals by increasing the magnitude of the pertinent β_i coefficients.

Implications of Sampling Parameter Choice. Having obtained the discrete Fourier transform of the acceleration signal for window k, we now make a number of observations on the importance of the sampling rate f_s and window duration Δt_{win}. It is clear from the Nyquist frequency that a higher sampling rate is directly related to a higher observed maximum frequency. In practice, terrain induced vibrations encountered during trials rarely affect frequencies above 20 30 Hz.

With regards to the effect of the window duration, a longer duration allows one to capture more samples, thus allowing for greater resolution in the DFT results. In addition, low frequency values are now distributed much more finely, compared to when Δt_{win} is small. In the case of small Δt_{win}, $\alpha_k[0]$ often accounts for most of the signal's energy, thereby providing less fine-grained control over period high-frequency vibrations.

On the other hand, a longer window duration implies that there is a greater time lag between the DFT results and the time when the terrain features that produced such a frequency response were present. Since control-based on signal frequency content will always be reactive, this feature results in a trade-off between lower frequency resolutions and greater time-delay in producing adequate responses to rapidly changing terrain. In this case, a sliding or moving window approach will not correct for this time lag, since the values obtained in this way will only relate to a process that started time Δt_{win} ago. In addition, an increased window duration will also "blend" new terrain responses with old ones, which would result in a skewed representation of the near-term external vibrations experienced.

4.2   High-Frequency Content Measure

We can now proceed to define a measure of the high-frequency content. We first define a *threshold frequency* $f_{\text{th}} < f_s/2$, which marks the transition from low to high frequencies. Since we wish to encode the effect of all high frequencies, it would be natural to take the integral of the Fourier transform to account for the power of all high frequencies. Since we are working with discrete values, we instead consider a weighted sum of the following form:

$$r_{k,\text{hf}} = \frac{f_s}{2(N_{\text{win}} - 1)} \sum_{f \in F : f \geq f_{\text{ts}}} \alpha_k[f].$$

It should be noted that the magnitude of $r_{k,\text{hf}}$ is directly dependent on the magnitude of the underlying signal $a(t)$. Thus, for a terrain with high amplitude, lower frequency vibrations, a high value of $r_{k,\text{hf}}$ may be obtained, whereas for a terrain with high-frequency vibrations of lower amplitude, a low value of $r_{k,\text{hf}}$ is obtained. This feature makes for a very cumbersome controller design, where one would need to schedule gains based on the signal strength or prior knowledge of the terrain that is to be traversed. Moreover, a controller that is tuned for one vehicle would not be transferable to a different vehicle, since the signal power may be higher or lower. This serves as an alternate solution to the method proposed in this paper.

To avoid these problems, we define the following measure:

$$r_k = \frac{f_s}{2(N_{\text{win}} - 1)} \frac{\sum_{f \in F: f \geq f_{\text{ts}}} \alpha_k[f]}{\sum_{f \in F} \alpha_k[f]}. \tag{5}$$

where k represents the time-step unit. This ratio is agnostic of the signal power, and is much more intuitive to reason about. One would be able to say that 10% of the signal power is permitted to be of a high-frequency, regardless of the vehicle properties. For this reason, we will use this nondimensional measure in our controller design, which we refer to as the *high-frequency content ratio*.

4.3   Velocity Regulator Design

Having defined a measure for the high-frequency external vibrations, we can now proceed to design a model-free velocity regulator that aims to regulate this quantity around a predefined setpoint.

Since the terrain is not known a priori, this controller will be, out of necessity, reactive. We have chosen to design an integral controller with precompensation (PCIC), as described next.

We would like to start out with a nonzero allowable speed defined by the mission, a nominal speed $V_0 > 0$. V_0 serves as the speed the vehicle would like to travel along all terrains given minimal vibrations. In this work, given the limitations of the Jackal UGV, this speed is taken to be 1 m/s in consideration of our UGV's capabilities. We would like to regulate the maximum allowable speed on the different terrains. Applying this controller to more sophisticated robots that can operate at faster speeds is left for future work.

As mentioned previously, a decrease in the vehicle's speed is related to a decrease in the high-frequency vibration content. For this reason, it is most intuitive to decrease the vehicle's maximum permissible speed when the high-frequency vibrations exceed the user-defined threshold, and increase the speed if r_k is too low. We define the high-frequency content ratio set point as $r_0 > 0$. In practice, r_0 will have to be sufficiently large so as to overcome ambient noise in the IMU. Moreover, a larger r_0 allows the vehicle to traverse bumpier terrain at faster velocities. This trade-off must be kept in mind while choosing this constant. In the following, we refer to r as the value r_k that corresponds to the most recent sampling window.

We have elected to regulate r by means of an integral controller that acts on the error $e = r_0 - r$. This choice of controller design, over a proportional controller, stems from a desire to prevent control-induced high-frequency vibrations. In a proportional controller, instantaneous changes are immediately acted upon, forcing jump changes in the maximum speed. In practice, this induces abrupt, undesirable braking action and acceleration.

An integral controller also faces a number of challenges. When the vehicle is stationary, the integral controller will continue accumulating as noise from the IMU is integrated. This in turn leads to the issue of integral windup. We overcome this issue by introducing an integral with projection. The accumulated integration from the current time step is given by I_i and the saturated value, I_o, is given as follows:

$$I_o = \text{proj}_{[-V_0, V_{\lim} - V_0]}(I_i + K_i e \Delta t_{\text{win}}). \tag{6}$$

In (6), $K_i > 0$ scales the output error fed into the integration. A higher gain implies a more aggressive response in the change of the maximum permissible speed. K_i is to be limited in practice when rough terrain is expected during a mission, since controller-induced vibrations are aggravated in such settings.

In (6) we also introduce a projection (or, in this case, 'clamping') operator. The limits of the projection operator are $[-V_0, V_{\lim} - V_0]$, where $V_{\lim}$ is the user-defined limiting speed of the vehicle, such that $V_{\lim} \geq V_0$. The significance of this result will become apparent below.

We obtain the maximum speed from this controller by taking $V_{\max} = I_o + V_0$ at each time step. We proceed by showing how this maximum speed is applied to the control command given by the LQR controller of Sect. 3.2.

Vehicle Speed Regulation. Given a control command $\vec{u}$, we can compute the vehicle's speed by means of (3). We define the commanded vehicle speed as $V_{\text{cmd}} = \sqrt{\dot{x}^2 + \dot{y}^2}$. We identify two cases: the case where $V_{\text{cmd}} \leq V_{\max}$, and the case where $V_{\text{cmd}} > V_{\max}$.

In the former case ($V_{\text{cmd}} \leq V_{\max}$), we apply $\vec{u}$ to the vehicle. This case will most often arise when we are close to our waypoint $\vec{x}_{\text{des}}$ as the error in the desired position fed into LQR controller will be much smaller. In the latter case ($V_{\text{cmd}} > V_{\max}$), we have chosen to scale the velocity command as follows:

$$\vec{u}_{\text{cmd}} = \frac{V_{\max}}{V_{\text{cmd}}} \vec{u}. \tag{7}$$

It is obvious that this scaling results in a commanded velocity of magnitude $V_{\max}$, as is desired.

This form of speed regulation raises the question of waypoint convergence, as the vehicle's velocity may converge to zero before it converges to the waypoint since our error will be small close to the waypoint. In practice, given that r_0 is taken to be greater than the value encountered at ambient noise, and given a sufficiently large limit speed $V_{\lim}$, as well as a sufficiently large gain K_i, the vehicle will converge to the waypoint on all terrains encountered in practice.

The complete controller architecture is given in Fig. 3. We will refer to this architecture as an *LQR-I controller*, short for combined LQR and precompensated integral controller.

Considerations in Tuning. We now discuss a number of practical considerations in tuning the proposed controller. The presented algorithm contains the following tuning parameters:

- State and control penalty matrices Q, R;

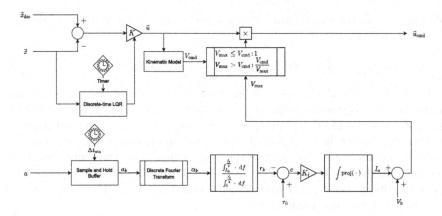

Fig. 3. Block diagram of the combined LQR and precompensated integral controller.

- Nominal and limit velocity V_0, V_{lim};
- High-frequency content ratio setpoint r_0, threshold frequency f_{th}, and integral gain K_i;
- Sampling frequency f_s and window duration Δt_{win}.

The state and control penalty matrices of the LQR controller can be designed beforehand on nominal terrain types. The projection operator of (6) ensures that the vehicle speed will not exceed V_{lim}. In this work, $V_{\text{lim}} = 2\,\text{m/s}$. As mentioned before, V_0 can be assumed to be a mission defined parameter, loosely related to the nominal mission speed, although it could be overruled in the case of sufficiently harsh terrains.

The high-frequency content ratio setpoint r_0 can be tuned heuristically by accounting for ambient sensor noise. In the authors' experience, the most straightforward way is to slightly vary it until the vehicle's perceived motion is sufficiently smooth. The same advice extends to the threshold frequency (f_{th}), although in our experience a value 5 Hz worked for all trials. When taking too high of a value, it is often the case that only sensor noise is accounted for. This phenomenon may be aggravated by a poor choice of Δt_{win}, since a low duration tends to lump most of the signal power in the lowest frequency buckets. Conversely, too high of a window duration causes intolerable time delays in registering the lag of the control signal, as mentioned previously.

The integral gain K_i, as mentioned previously, dictates how aggressively the controller reacts to vibration changes. It is key to properly tune this value, since controller-induced vibrations may quickly arise if this value is too high. In fact, a sufficiently large value of K_i can cause repeated start/stop action at a frequency of $1/(2\Delta t_{\text{win}})$.

Finally, f_s should at least be twice the desired maximum frequency. This could mean that frequencies as low 20 Hz could potentially yield satisfactory performance. Window duration Δt_{win} has been found to yield the best results when taken to be 1 s.

These parameters may require tuning as per user-defined goals and mission objectives. It is important to note that although we developed a controller to operate on the Jackal vehicle, the same control structure could be implemented on other UGVs. Demonstration of this application is reserved for future work.

5 Experiments and Results

We implemented the novel controller on the Jackal UGV performing in various terrains and collected the high-frequency noise and velocity content data. We compared our controller to an open-loop test case to assess performance. We compare high-frequency noise content and velocity command following for paved concrete, grass, and mulch surfaces. Data is collected for transitions between surfaces to illustrate the robustness of the implemented control algorithm and show improved performance over all tested terrain.

For all trials, given the maximum speed of the Jackal is $V_{\lim} = 2$ m/s; we run the Jackal with a nominal speed $V_0 = 1$ m/s. For the following experiments, the vehicle is subjugated to identical terrain. The vehicle first moves through with an open-loop controller with a fixed reference velocity without consideration of the high-frequency content. Second, a closed-loop test is performed with the proposed controller where the maximum permissible velocity is modified in real time to mitigate vibrations impacts. The results form both test runs are compared.

Obviously, performance is altered by the choice of integrator gain. A large gain compensation results in faster changes to calculated $V_{\max}$, which results in more oscillatory behavior. Through experimental trials, a good range of integrator gains was found to be $K_i \in [2, 4]$. The gain parameters were tuned before the experiments for optimal results, however, in practice this tuning is unnecessary since the controller exhibits improved performance over all tested surfaces while $K_i \in [2, 4]$. We set the high-frequency threshold for all experiments to be 5 Hz, while the remaining parameters f_s, Δt_{win}, Q, and R are equal to the values defined in previous sections.

The controller was implemented onboard the vehicle in Python 2.7 on an NVidia Jetson TX2 computer running Ubuntu 16.04 and ROS Kinetic. Low-level controller commands were passed through the appropriate ROS topics provided by Clearpath. The IMU used was a WaveShare 10 DOF IMU sensor board, which was interfaced with the UGV through an Arduino Mega 2560. The controller published input updates 100 Hz, while IMU updates were read 60 Hz.

5.1 Concrete Performance

Like many unmanned ground vehicles, the Jackal is designed with multipurpose maneuverable capabilities. One commonly encountered surface is concrete, and thus to illustrate the capabilities of the LQR-I controller, we first analyze its performance on a tiled concrete terrain. Figures 4 and 5 display the Jackal's performance autonomously navigating over concrete surfaces with open-loop control (fixed movement speed of V_0), and the closed-loop LQR-I controller, respectively.

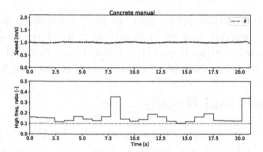

Fig. 4. Open-loop control with nominal velocity $V_0 = 1\,\mathrm{m/s}$.

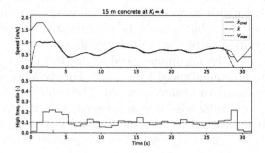

Fig. 5. Closed-loop LQR-I control with nominal velocity $V_0 = 1$ compensating high-frequency vibration content output such that r_k tracks $r_0 = 0.1$. Measured heading speed $\dot{x}$ is closely compared to the heading command $\dot{x}$ and maximum velocity V_{max} set by the controller.

The open-loop manual control results in a steady velocity output that follows the nominal input speed command regardless of high-frequency noise. Given the generally smooth surface of concrete, the high-frequency noise content stays largely steady with the exception of some oscillations created by divots in the path when transitioning from one concrete tile to another. Despite the smooth surface and terrain, the high-frequency noise content ratio in Fig. 4 is consistently above the set threshold $r_0 = 0.1$.

Figure 5 illustrates the enhanced performance of the Jackal as controlled by the proposed LQR-I controller with a tuned integral gain $K_i = 4$. As the vehicle begins moving, a large change in initial speed causes unwanted vibrations. Yet, within approximately 5 s, the controller is able to compensate for the unwanted initial high-frequency noise, decreasing the ratio to below the setpoint. As explained in the previous section, as r_k decreases below the set threshold, the LQR controller increases control action with input $\vec{u}$ as $V_{\mathrm{cmd}} \leq V_{\mathrm{max}}$. This causes the high-frequency ratio to increase until it exceeds its given threshold, at which point the integrator compensates by decreasing the control action to $\vec{u}_{\mathrm{cmd}}$ as outlined in Eq. (7).

Note that the controller exhibits both impressive velocity tracking and high-frequency noise tracking. That is, not only is the high-frequency noise generally kept at the desired threshold, the heading speed also closely follows the heading command and maximum velocity set by the controller. This is largely due to the robust properties of the LQR and integral controllers tasked with producing optimal input commands, i.e., the LQR-I controller exhibits low peak sensitivities at high frequencies [13, 16], and good low frequency command following with high complementary sensitivities at low frequencies. With proper tuning, as the high-frequency noise ratio increases, the integral controller attenuates high-frequency noise with guaranteed convergence with added saturation limits, and as high-frequency noise drops below the predefined threshold, LQR increases the velocity to closely follow the nominal speed V_0. To further test controller performance, we implement the LQR-I controller on grass, where larger vibrations produce more high-frequency noise content.

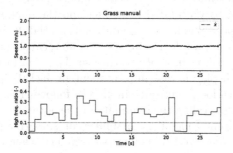

Fig. 6. Open-loop control with nominal velocity $V_0 = 1\,\text{m/s}$.

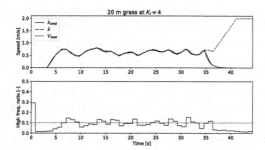

Fig. 7. Closed-loop LQR-I control with nominal velocity $V_0 = 1$ compensating high-frequency vibration content output such that r_k tracks $r_0 = 0.1$. (Video of the controller applied to the Jackal UGV can be found at https://uofi.box.com/s/9lvx16hrrw7kr4ew7og91w4uspid3n02.)

5.2 Grass Performance

To further illustrate the performance and robustness of the proposed control architecture, the Jackal travels on rougher terrains such as grass, where we expect to see significantly larger high-frequency noise [18]. By inspection, one can easily note from Fig. 6 that the open-loop control of the Jackal exhibits a substantially larger amplitude for the high-frequency noise ratio r_k traversing grass in comparison to maneuvering on concrete. Similar to the previous example, the open-loop controller accurately follows the nominal speed of $V_0 = 1\,\text{m/s}$.

This example further highlights the controller performance and ability to reduce unwanted high-frequency noise content while riding the boundary of the maximum permissible vehicle speed. Consistent with previous results, the high-frequency ratio has accurately been reduced to lightly oscillate about $r = 0.1$. At the same time, the heading speed $\dot{x}$ is nearly identical to the controller heading command $\dot{x}_{cmd}$ given by the LQR-I controller and the maximum velocity V_{max} determined by the speed regulator. The only substantial deviation happened as the Jackal slowed to a stop while nearing its waypoint, causing the high-frequency noise content to diminish to 0, which in turn set $V_{max} = 2\,\text{m/s}$.

Figures 5 and 7 display a vast improvement in Jackal's performance when compared to the open-loop controller in closely following a set high-frequency noise content ratio while riding the boundary of the optimal permissible velocity. Given the data shows the controller capabilities on smooth and difficult terrain, we now aim to show improved controller performance when changing from one environment to another.

5.3 Multi-terrain Performance

The final set of experiments contain data consisting of the speed and high-frequency noise ratio as a function of time as the Jackal traverses concrete and mulch over a distance of 15 m. A vertical, red dotted line in Figs. 8 and 9 denotes the distance at which the UGV transitions from concrete to mulch. One can observe from Fig. 8 that the amplitude of the high-frequency ratio is significantly larger when maneuvering on mulch than concrete.

It is important to note that although the open-loop controller follows the nominal velocity closely, there exists no point in time where the high-frequency content is below the set threshold. Conversely, the closed-loop LQR-I controller exhibits the ability to compensate performance to keep r_k near the assigned threshold regardless of changes in terrain while accurately tracking desired heading and maximum velocity commands. Thus, Fig. 9 empirically shows a promising example of the controller being capable of travelling multiple unknown surfaces without a priori knowledge regarding the environment.

In comparison to the previous two examples, the average speed in Fig. 9 is slower, i.e., the controller produces an output velocity that is further from the nominal value $V_0 = 1$. This phenomenon is a result of the following factors: (i) mulch being a particularly uneven surface that causes large high-frequency amplitudes [18] and (ii) the integrator gain $K_i = 2$ being lower than its value in

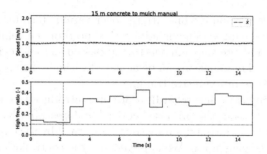

Fig. 8. Open-loop control with nominal velocity $V_0 = 1\,\mathrm{m/s}$.

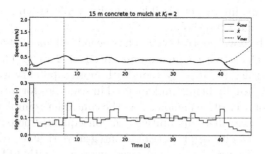

Fig. 9. Closed-loop LQR-I control with nominal velocity $V_0 = 1$ compensating high-frequency vibration content output such that r_k tracks $r_0 = 0.1$

the previous two examples, where $K_i = 4$, which encourages speed maintenance. Since mulch is an unpredictable, uneven surface, the gain K_i is tuned lower to help diminish controller-induced vibrations. In general, a lower K_i gain is advisable when particularly unstructured terrain is expected, although as stated previously, it is not necessary to tune the gain differently depending on the terrain for improved performance.

These three examples illustrate that the proposed controller results in improved UGV performance to mitigate undesirable high-frequency noise regardless of the terrain. Out of all tuning parameters, adopting a lower K_i value appears to be key in handling unstructured terrain, whereas the LQR and frequency threshold settings can remain fixed regardless of the specifics of the current mission.

6   Conclusion and Future Work

This paper presents a controller architecture to attenuate unwanted high-frequency vibrations during transit of a ground vehicle, while preserving way-point tracking capabilities at reasonable speeds. The control system provides a method of maneuvering on unknown terrain while limiting damage to sensitive payloads due to high-frequency vibrations. In a sense, this controller mimics a natural human-like response while driving over various terrains - as the roughness of the terrain changes humans adjust the speed of the vehicle to minimize sudden impulses and prevent damage to the passengers and cargo inside. Furthermore, this speed modulation can be achieved without the need of perception, terrain identification, or extensive training.

Experiments indicate that regardless of noise introduced by unknown obstacles over variable terrain, the proposed control architecture is capable following a desired trajectory at the maximum velocity governed by the experienced vibrations and high-frequency noise. We implemented the controller on a vehicle over concrete, grass, and mulch. Regardless of the high-frequency vibration content, the vehicle was able to perform its primary objective of moving to a desired waypoint at the maximum permissible velocity while staying below the high-frequency content ratio threshold.

In this work, the novel control system has been implemented on the Clearpath Jackal UGV, a vehicle with a maximum speed of 2 m/s. This same control structure can be scaled to larger systems with more capability without requiring much retuning, eventually being implemented in scenarios for medical evacuation and carrying various sensitive payloads. As the controller is implemented on more complex systems, controller capabilities can consequently be enhanced. With a higher maximum velocity, a proportional integral derivative controller can be utilized for faster convergence to the desired waypoint, with added saturation or a pre-filter to account for unwanted overshoot, and a low-pass filter to prevent controller-induced vibrations.

Future work includes the use of a high-fidelity dynamic vehicle model to loosen assumptions on non-slipping movement, which often does not hold on highly unstructured terrain. This generalization would further be supported by a transition to direct torque control, allowing for additional adaptive control capabilities. In addition, model-based filtering techniques may be used to filter out low frequency components of the accelerations, thereby only considering terrain-induced random noise. Finally, terrain classification may be possible using the proposed high-frequency content ratio, as well as new measures based on other sensors, such as piezo-electric haptic feedback and LIDAR sensing.

References

1. Brooks, C., Iagnemma, K., Dubowsky, S.: Vibration-based terrain analysis for mobile robots. In: IEEE International Conference on Robotics and Automation, pp. 3415–3420. IEEE (2005)
2. De La Cruz, C., Carelli, R.: Dynamic model based formation control and obstacle avoidance of multi-robot systems. Robotica **26**(3), 345–356 (2008)
3. Debenedictis, T.A., et al.: The simulation of the whole-body vibration experienced during military land transit. Hum. Factors Mech. Eng. Defense Saf. **2**, Article no. 8 (2018)
4. Filitchkin, P., Byl, K.: Feature-based terrain classification for LittleDog. In: IEEE/RSJ International Conference on Intelligent Robots and Systems, pp. 1387–1392 (2012)

5. Franklin, G.F., Powell, J.D., Workman, D.L.: Digital Control of Dynamic Systems, 3rd edn. Addison-Wesley, Boston (2002)

6. Fukao, T., Nakagawa, H., Adachi, N.: Adaptive tracking control of a nonholonomic mobile robot. IEEE Trans. Robot. Autom. **16**(5), 609–615 (2000)

7. Grzesica, D.: Measurement and analysis of truck vibrations during off-road transportation. In: MATEC Web of Conferences, vol. 211, pp. 13003–13008 (2018)

8. Hoepflinger, M.A., Remy, C.D., Hutter, M., Spinello, L., Siegwart, R.: Haptic terrain classification for legged robots. In: IEEE International Conference on Robotics and Automation, pp. 2828–2833. IEEE (2010)

9. Khurana, A., Nagla, K.S.: Signal averaging for noise reduction in mobile robot 3D measurement system. MAPAN **33**(1), 33–41 (2018)

10. Lu, F., Ishikawa, Y., Shiina, T., Satake, T.: Analysis of shock and vibration in truck transport in Japan. Packag. Technol. Sci. **21**(8), 479–489 (2008)

11. Lv, W., Kang, Y., Zheng, W.X., Wu, Y., Li, Z.: Feature-temporal semi-supervised extreme learning machine for robotic terrain classification. IEEE Trans. Circuits Syst. II Express Briefs **67**(12), 3567–3571 (2020)

12. Lynch, K.M., Park, F.C.: Modern Robotics. Cambridge University Press, Cambridge (2017)

13. Masti, D., Zanon, M., Bemporad, A.: Tuning LQR controllers: a sensitivity-based approach. IEEE Control Syst. Lett. **6**, 932–937 (2021)

14. Ono, M., Fuchs, T.J., Steffy, A., Maimone, M., Yen, J.: Risk-aware planetary rover operation: autonomous terrain classification and path planning. In: IEEE Aerospace Conference, pp. 1–10 (2015)

15. Otsu, K., Ono, M., Fuchs, T.J., Baldwin, I., Kubota, T.: Autonomous terrain classification with co-and self-training approach. IEEE Robot. Autom. Lett. **1**(2), 814–819 (2016)

16. Prakash, A., Parida, S.: LQR based PI controller for load frequency control with distributed generations. In: 2020 21st National Power Systems Conference (NPSC), pp. 1–5. IEEE (2020)

17. Rogers-Marcovitz, F.: On-line mobile robotic dynamic modeling using integrated perturbative dynamics. Technical report, Robotics Institute at Carnegie-Mellon University (2010)

18. Rosenfeld, R.D., et al.: Unsupervised surface classification to enhance the control performance of a UGV. In: Systems and Information Engineering Design Symposium (SIEDS), pp. 225–230 (2018)

19. Rothrock, B., Kennedy, R., Cunningham, C., Papon, J., Heverly, M., Ono, M.: SPOC: deep learning-based terrain classification for mars rover missions. In: AIAA SPACE 2016. American Institute of Aeronautics and Astronautics (2016)

20. Sim, O., Jung, T., Lee, K.K., Oh, J., Oh, J.H.: Position/torque hybrid control of a rigid, high-gear ratio quadruped robot. Adv. Robot. **32**(18), 969–983 (2018)

21. Streijger, F., et al.: Responses of the acutely injured spinal cord to vibration that simulates transport in helicopters or mine-resistant ambush-protected vehicles. J. Neurotrauma **33**(24), 2217–2226 (2016)

22. Vlkovský, M., Veselík, P., Grzesica, D.: Cargo securing and its economic consequences. In: 22nd International Scientific Conference, pp. 129–135. Transport Means, Kaunas University of Technology, Kaunas (2018)

23. Wang, Q., Rajashekara, K., Jia, Y., Sun, J.: A real-time vibration suppression strategy in electric vehicles. IEEE Trans. Veh. Technol. **66**(9), 7722–7729 (2017)

24. Wang, Z.: Fast algorithms for the discrete w transform and for the discrete Fourier transform. IEEE Trans. Acoust. Speech Signal Process. **32**(4), 803–816 (1984)

25. Weiss, C., Frohlich, H., Zell, A.: Vibration-based terrain classification using support vector machines. In: IEEE/RSJ International Conference on Intelligent Robots and Systems, pp. 4429–4434. IEEE (2006)
26. Winograd, S.: On computing the discrete Fourier transform. Math. Comput. **32**(141), 175–199 (1978)
27. Yuvapriya, T., Lakshmi, P., Rajendiran, S.: Vibration suppression in full car active suspension system using fractional order sliding mode controller. J. Braz. Soc. Mech. Sci. Eng. **40**(4), 1–11 (2018). https://doi.org/10.1007/s40430-018-1138-0

Modelling and Simulation of Microrelief Impact on Ground Path Extension

Dana Kristalova[1]([✉]), Tomas Turo[1], Jan Nohel[1], Marian Rybansky[1],
Vlastimil Neumann[1], Jan Zezula[1], and Marek Hütter[2]

[1] University of Defence, Brno, Czech Republic
{dana.kristalova,tomas.turo,jan.nohel,marian.rybansky,
vlastimil.neumann,jan.zezula}@unob.cz
[2] VŠBTU, FBI, Ostrava, Czech Republic
marek.hutter.st@vsb.cz

Abstract. Geographic data are essential input parameters in the decision support systems for the evaluation of whether the planned routes of territory are passable or not. Micro-relief is one of the most notable geographical factors and although its influence is crucial, the micro-relief shapes are as the obstacles often neglected. These obstacles are existed almost in all types of terrain and their quantity and distribution in each specific type of terrain for evaluation of the mathematic model were determined on the bases of carto-metric investigation. Subsequently such system can serve for the optimization of routes of movements, calculating of a coefficient of the deceleration and next parameters and finally for an evaluation of the best solution for manoeuvre of military units according the commander's requirements. The object of this paper is to point out that the calculated route extensions can help in the planning and decision-making process on the "just in time" concept, which should be commonly used in an operational environment.

Keywords: Geographic factors · Micro-relief · Terrain feature modelling · Optimization of routes · Tactical decision support systems; cross-country movement · Off-road vehicle · Off-road navigation · Contemporary operational environment

1 Introduction

The presence of micro-relief objects is the factor, which is very significant for movement of off-road vehicles or UGV to make decision within the purview of decision support system and to a certain extent the considering of this geographic factor appertains to tasks and challenges of the contemporary operational environment [1–3]. Micro-relief shapes influence and limit particular manoeuvre and fall into multi-spectral terrain analyses called Cross-Country Movement [4, 8]. The results of this terrain analyses may be very dissimilar to each other and it depend on a specific determination of military assignment. That means to set a type of terrain, type of unit, a type of vehicle and its technical data, attainable time, length and costs limits and the target for every task [5, 6].

© Springer Nature Switzerland AG 2022
J. Mazal et al. (Eds.): MESAS 2021, LNCS 13207, pp. 93–112, 2022.
https://doi.org/10.1007/978-3-030-98260-7_6

Within the contemporary warfare, it is more than convenient to use various methods of modelling and simulation and to estimate scenarios of possible situations as accurately as possible before a given manoeuvre. Appropriately compiled algorithms can predict the so-called "operational coefficient" according to the parameters assessed. The parameters can be individual terrain components, technical vehicle data or economic or logistical requirements. Basically, it is a point of view, where the right algorithm answers the appropriate question. Without modelling and simulation, these answers would often remain hidden, and sometimes would manifest themselves directly in a given situation. In most of the time, these facts would be negative, and therefore it is greatly important to evaluate the operational coefficient in advance [21–23].

The problem dealing with an absence of mapping of micro-relief shapes is a question of time. Nowadays it is possible to obtain the 3D data of a territory with very high resolution but many systems for off-road navigation do not have these mapping service at its disposal [7]. They have to depend on analogue maps and on some navigator's help or they can apply the 3D map model without high accuracy only. Such situation relates especially on high level operation planning and the results of route planning are unresponsive of this geographic factor.

2 Object of the Research

This paper deals with a modelling of a terrain to determine optimal route for off-road vehicles or UGV. In order to make up commander's mind and settle on a solution the object of a research has to be examined. That means to study the presence of microrelief shapes in different types of terrain and establish the standards of Cross-Country Movement.

The object of a research mentioned below are the targets of a military study and know-how about their mutual impact could serve for determining of the key effect to conducting of military tasks [21–23].

Microrelief

The terrain forms of the "microrelief" can be defined as man-made and natural both elevated and depressed topographic forms that cannot be expressed with regard to its relatively small height differences by use of contour lines or by the means of other principal method of terrain representation [4, 8].

Micro-relief shapes such as: small slopes (terrain steps), rock cliffs, landslides, terraces, erosion forms of watercourses, gullies, craters, holes, embankments, rock groups, boulders and other relief forms created by impact of natural forces and anthropogenic activity can have an important influence on any operations [4, 8].

Within the microrelief, parameters of microrelief forms of watersheds (eventually of profiles of drainage systems) and erosion rills were also identified as impassable obstacles, i.e. those that cannot be overcome in a given place and which need to be bypassed.

Terrain and its Types

The "terrain" is defined as any part of the earth's surface with all the unevenness (created by natural forces or artificially) and with all objects and phenomena found on the earth's surface [8].

The *"types of terrain"* are divided according to terrain objects (forests, settlements, watersheds, etc.) and terrain shapes (given by ruggedness). In terms of tactical characteristics and operational significance, the terrain is evaluated either by morphogenetic properties (types resulting from the action of endogenous and exogenous forces and process of inception) and by morphometric properties (types are determined by relative ruggedness, relative slope ratios, relative elevation, absolute heights and other quantitative indicators).

From a military point of view, relative elevation difference is of the greatest importance. The degrees of height breakdown of the territory (in advance defined by extreme values) are characterised by specified indicators [8].

Dividing by terrain objects:

- *open* - without any terrain objects, or only with unique objects (tree, mast, monument,..);
- *semi-covered* - scattered occurrence of terrain objects;
- *covered* - covered with terrain objects more than half.

Dividing by terrain shapes:

- *Flat* - the area is flat or only slightly undulating;
- *Wavy* - the terrain already has more striking shapes with height differences up to 100 m. The off-road shapes are round, the slopes are gentle, and the angle of the slope is less than 5°;
- *Hilly* - already has distinctive shapes with height differences up to 500 m. As a rule, the ridges follow the direction of the mountain massif, the slopes are steeper, and the valleys are deeper;
- *Mountainous* - includes areas with significant elevation of the earth's surface, usually situated at higher altitudes (i.e. above 1000 m).

Terrain objects and shapes are interrelated and the terrain types mentioned above form different combinations. These combinations have to be embraced in determining the Cross-Country Movement.

Dividing by morphogenetic properties:

- *lowlands;*
- *plains;*
- *basin;*
- *ridges;*
- *mountains, etc.*

Dividing by morphometric properties:

- *plains* - flat areas at any altitude, where relative terrain elevation differences up to a distance of 2 km are less than 30 m, slope slopes up to 1°;

- *uplands* - hills with relative differences from 30 to 150m, slopes are mostly mild, their slopes usually do not exceed 3°;
- *highlands* – area with a relative height difference from 150 to 300 m, slope slopes usually 5° to 10°;
- *mountains* - rocks with height differences of 300 to 600 m, form significant ridges between marked valleys (the ridges usually form a watershed), the slopes are already considerable and range between 10° and 25°;
- *high mountainous* - terrain, often formed by rocky peaks and distinctive ridges between steep valleys, height differences over a distance of 2 km exceed 600 m, slope slopes are greater than 25°.

Analysis of the Cross-Country Movement (CCM)
CCM means the assessment of several geographical factors (GF) and tactical factors together (TF), i.e. that it is global analysis.

These factors and their standards determine 3 levels of Cross-Country Movement:

- *GO;*
- *SLOW GO;*
- *NO GO.*

The level called "GO" means the movement without a loss of speed the level "SLOW GO" means partial deceleration of the speed of the movement and the last level "NO GO" signifies that the movement is not possible. These terms are given. [4, 8].

From the point of view of means of transport (used for a movement) following basic types of terrain are determined:

- *Terrain passable for full track vehicles;*
- *Terrain passable for wheeled vehicles;*
- *Terrain passable for other means of transport;*
- *Terrain passable for infantry troops.*

The impact of all factors is expressed by, **"Coefficient of the deceleration"** (abr. CoD). This multiple coefficient gives degree of deceleration from full speed owing to constituent factors.

The total CoD depends on the partial CoD of the different GF and is given by this formula [4, 8]:

$$C = \prod_{i=1}^{8} c_i \qquad (1)$$

where C *is the total CoD,*
c_i *is partial CoD.*

3   State of the Art

Terrain analysis and route planning and their optimization are still topical theme despite of the fact that a number of mathematical models and terrain tests have been drawn up and conducted [9, 10].

Lot of articles or studies for searching of a right corridor devoted to Air Unmanned Vehicles (UAV) were created [11, 12]. The more complex situation is for off-road vehicles or Unmanned Ground Vehicles (UGV) where the effect of soil condition, vegetation, relief and micro-relief and other factors is more significant and a multiple analysis for a passability of a territory were applicable. The research papers [13, 14] are dedicated to this topic. Traveling salesman problem (TSP) or methods of collision avoidance are elaborated [15–17] but this is an extension of a route searching and optimizing process, nevertheless it could serve as an example of using [21–23].

4   Occurrence of Microrelief and Other Obstacles

The current solutions offer the use of existing data and experimental measurement of the missing variables. Micro-relief, a type of surface, a slope, waters (rivers or lakes), soil conditions and the weather conditions are the main factors, whose reciprocal influence has not been evaluated yet and these factors are of fundamental importance for the movement of vehicles in the field. To create the input dataset, it was necessary to define the type of terrain and perform a carto-metric investigation.

4.1   Assessed Factors

The first assessed factor was microrelief. Embankments, excavations, delves or craters, terrain steps (ascent or descent) and trenches are the most frequent shapes at the territory of the Czech Republic. These shapes have very important parameters for CCM, that are subject of interest, for example slope gradients, height or width.

The target was to obtain the average value of lengths, numbers and heights of shapes separately for each type of terrain. The shapes were divided into 4 categories – according to length interval in the following way:

- 0 m < Category A < 100 m
- 100 m < Category B < 500 m
- 500 m < Category C < 1000 m
- 1000 m < Category D

The hollowed microrelief erosion rills, watesheds (rivers, lakes), railways and slopes have been measured too, but for methods of searching and optimizing of routes of off-road vehicles these factors do not express the primary aim of algorithms. For the added factors - as profiles of drainage systems, erosion rills and railways - the other advanced models and algorithms were created. The data mentioned above were obtained by digitalization and using of ArcGIS from another data sources – from TM in scale 1:25 000. The data about micro-relief were the major subject of an interest in [8].

4.2   Types of Terrain

Types of terrain" are very significant input parameter what was established by carto-metric investigation. The distribution and frequency of microrelief shapes as obstacles are determined by the type of terrain and the set of input values could be determined from a large number of statistical pronunciations [8].

The most important types of terrain (in the Czech Republic) are: plains (4,5%), uplands (50,1%), highlands (33,8%), mountains (10,8%) and high mountains (0,8%). Due to the low incidence of high mountains, this factor was omitted in further modelling. The frequency and parameters of obstacles within each terrain type are given in the following chapter. See Fig. 1.

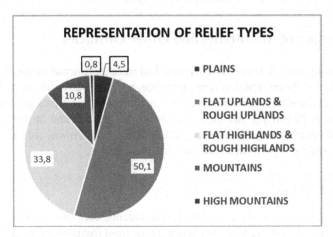

Fig. 1. Percentage representation of terrain types

5 Carto-Metric Investigation and Formation of Data Set

The input data was evaluated on the basis of mathematics and statistical methods which are used in cartography. The basic file was formed as a selection from all the maps of the Czech Republic. The presence of microrelief forms on area of movement was determined by use of the basic topographic maps (TM) in scale 1:10 000 (depicted forms as a rule of height difference over 0.5 m). These TMs 1:10 000 are not in using today (the editorial deadline was in 1958–1964), but they have been evaluated as the best source of this factor. See Fig. 2.

Substitute source, which is able to obtain data from, is the Digital Terrain Model (DTM) in scale 1:25 000 (depicted forms of height difference over 2 m). This base was used to create the input data file of microrelief, watersheds and communication (railways). The accuracy of microrelief shapes from this data source was not detailed - microrelief data on maps are very generalized due to their small sizes, so the occurrence of this factor is very inaccurate, but it served for a comparison of both models.

The task of the carto-metric investigation was to obtain the length of the microrelief shapes, their numbers in the squares of 10 by 10 km and the average height of these shapes.

Fig. 2. The special tool "opisometer" was used to measure the lengths of microrelief shapes on maps in scale 1: 10 000.

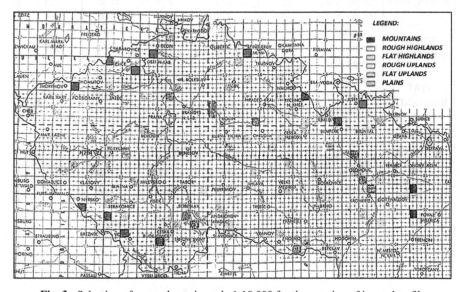

Fig. 3. Selection of maps sheets in scale 1:10 000 for the creation of input data file

The measurement of all the statistical units and statistical characters of basic file, however, is not tolerable and so there is chosen a sufficiently representative file (= set of maps). The whole area can be assessed according to the quantitative and spatial characteristics. See Fig. 3.

The first selection input file was composed of 320 pieces of tm 1:10 000. A square of 2 * 2 km was measured on each map, i.e. 1280 km^2.

The second set consisted of 102 sheets of these maps and each territory was examined for at least five representative maps of 1:25 000 for each type of relief. (Four map sheets of 1:10 000 record 4 * 5 km^2 and this is equal to the space of one map of 1:25 000. For 5 sample maps 1: 25 000, it is a total of 100 km^2 for each individual type of relief. In total, more than 600 km^2 were measured.)

As a result of carto-metric investigation the input parameters of data set were quantified. See Table 1.

Table 1. Input data file as a result of carto-metric investigation

TYPE OF RELIEF	PARAMETERS	CATEGORY OF MICRORELIEF SHAPES				LENGTH OF WATERSHEDS km/1km²
		within 100 m	within 500 m	within 1 km	over 1 km	
MOUNTAINS	Number	312	247	22	15	1,1
	Length /m/	55	213	714	1648	
	Height /m/	1,5	1,5	2,2	2,0	
ROUGH HIGHLANDS	Number	707	482	40	10	1,34
	Length /m/	62	213	779	1074	
	Height /m/	1,6	1,8	2,0	2,2	
FLAT HIGHLANDS	Number	387	329	25	6	0,67
	Length /m/	63	170	620	1148	
	Height /m/	1,4	1,3	1,8	1,6	
ROUGH UPLANDS	Number	382	316	36	12	0,89
	Length /m/	60	214	629	1366	
	Height /m/	1,5	1,5	1,8	1,6	
FLAT UPLANDS	Number	272	205	17	9	0,7
	Length /m/	56	209	664	1622	
	Height /m/	1,5	1,7	2,1	2,4	
PLAINS	Number	65	85	27	19	0,95
	Length /m/	85	175	580	1343	
	Height /m/	1,7	1,5	1,7	1,7	

6   Approach to the Solution

6.1   The Key Idea

The main idea of a movement is formed as follows: If a vehicle cannot overcome a microrelief form, it has to by-pass it and the vehicle route then lengthens. Even if such an event should not mean a vehicle deceleration on every occasion, the resulting effect on overall time of movement will be similar to the situation that a vehicle overcomes obstacles on direct route but with decreased speed by impact of relief gradient or of the other geographic factor. On the vehicle particular movement (lengthened), caused by obstruction by-passing, the following parameters of impassable microrelief forms have the impact:

- number of microrelief forms;
- length of microrelief forms;
- orientation of microrelief forms with regard to vehicle path axis;
- overall structure (space distribution of microrelief forms).

Several models (mentioned bellow) were created on the basis of values gathered by a carto-metric investigation. Different interconnections were sought and different criteria were applied.

6.2 Establishment of a Basic Simulation Model

It is necessary to convert the obtained values into a vector format (for example if the optimization is solved in program language C++) or into a bitmap format, which is the search of optimized routes performed using instruments of ArcGIS.

Table 1 as input values for a model of the microrelief has been accepted. However, this database can be read from any other source, if needed.

The computational algorithm of the route works on the principle of iterative search of minimal anti-collision junctions on the destination line (from the current point) approximated by 2D vectors. and algorithm to prevent possible intersect of microrelief units was resolved;

Gaussian normal distribution was used for the representation of shapes and basic train area is the square of the size 10 by 10 km;

The program saves each microrelief shapes and their coordinates (each shape is characterised by four values - these are the coordinates of two points in the plane – $[x_i, y_i]$ and $[x_j, y_j]$, where the set of all points that lie on their shortest connector defines an approximation to the theoretical microrelief) in the field (which is a sequence of consecutive variables of that type) with an accuracy of one meter.

7 Models and Their Resulting Effect on a Movement

For the calculation of the elongation of the route it is appropriate to consider, for which vehicle the simulation is created. It is assumed that ambulances or other operational vehicles are equipped with GPS devices, which can maintain the direction to the target point in the current time.

7.1 The Model Nr. 1 – Non-optimized

The Model Nr. 1 is for a vehicle equipped with GPS devices. This vehicle searches the direct route, if it is intersected by a certain shape, it starts bypassing along the edge of the shape (the shorter end), and after finding that it is at the end of the obstacle, it again will orient to the target point. It continues in the same way.

The Description of Model Calculation

1. From the starting point (Z) $(Z = S$ if the algorithm is at the beginning) the shortest connection is led to the destination point (C), (in our case always 10 km away from S).
2. Determining the nearest m_i that intersects this shortest line (the nearest mi is found by searching the whole field, while detecting the intersections P_i [x_p,y_p] of the vector $d_i = SC$ with m_i.

Collision (intersection) detection of 2 random **vectors with indices i, j** is given by this formula [8, 21]:

$$
\begin{aligned}
c_1 &= (y_{i1} - y_{i2})x_{j1} + (x_{i2} - x_{i1})y_{j1} - (y_{i1} - y_{i2})x_{i1} - (x_{i2} - x_{i1})y_{i1} \\
c_2 &= (y_{i1} - y_{i2})x_{j2} + (x_{i2} - x_{i1})y_{j2} - (y_{i1} - y_{i2})x_{j1} - (x_{i2} - x_{i1})y_{j1} \\
c_3 &= (y_{j1} - y_{j2})x_{i1} + (x_{j2} - x_{j1})y_{i2} - (y_{j1} - y_{j2})x_{j1} - (x_{j2} - x_{j1})y_{j1} \\
c_4 &= (y_{j1} - y_{j2})x_{i2} + (x_{j2} - x_{j1})y_{i2} - (y_{i1} - y_{i2})x_{j1} - (x_{j2} - x_{j1})y_{j1}
\end{aligned}
\tag{2}
$$

where c_i – *parameters of collision.*
$x_i,\ y_i$ – *components of I vector*
$x_j,\ y_j$ – *components of J vector.*
If c1 and c2 have different signs and at the same time c3 and c4 have different signs as well, then vectors with i,j indices (defined by their entry points) intersect. In such cases, the parameter of this **vector (t)** is calculated.

The calculation of the **parameter t_i** (the parametr t_i appertains to vector d_i) and coordinates of the intersection P of both vectors i,j is this:

$$
\begin{aligned}
t_i &= \frac{(y_{j2}-y_{j1})(x_{i1}-x_{j1})+(x_{j2}-x_{j1})(y_{j1}-y_{i1})}{(x_{j2}-x_{j1})(y_{i2}-y_{i1})-(y_{j2}-y_{j1})(x_{i2}-x_{i1})} \\
x_p &= (x_{i2} - x_{i1})t_i + x_{i1} \\
y_p &= (y_{i2} - xy_{i1})t_i + y_{i1}
\end{aligned}
\tag{3}
$$

The obstacle m_i that is in collision with the $d_i = SC$ vector with the lowest parameter value (t) is sought, if any. If no m_i intersects this connector (ZC), the algorithm ends, which means the final path finding from point S to point C. If not, the algorithm continues with step 3.

3. If the path is crossed by a microrelief shape, the algorithm determines the "diversion" of the obstacle by this way:

- by calculating the distance from the starting point **Z** (Z - is meant here as the starting point from which the path is led towards point C_i);
- it is determined the route through the individual vertices V_i of the microrelief shape m_i (of this one, that intersects the vector **ZC** with the lowest parameter value **t**) to **C**. The shorter path that leads over the vertex V_i continues. This means that the **Z** point in the next step is assigned the coordinate of the definite vertex (V_{i1} or V_{i2}) through which this distance is shorter. See Fig. 4.

4. In this step, the algorithm goes back to step 1, where the shortest connector from point Z (in this case, it will be one of the vertices of some mi shape) to C (destination point) is determined again, and the path search procedure is repeated until we are at the destination.

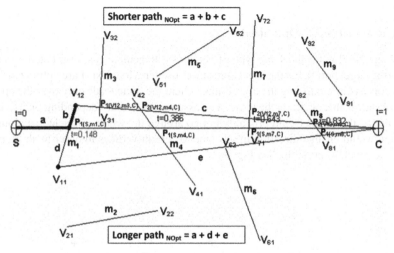

Fig. 4. Criterion for selecting the vertex of a microrelief shape in a non-optimized process

Six thousand attempts for each of the six types of terrain were carried out. The obstacles were newly (randomly) placed on the basis of a table of input values for each passage. The extension of the vehicle's movement route due to microrelief was implemented.

The total elongation is **about 5–8 c/o, it is** depending on the morphometric type. See Table 2.

This algorithm for simplicity calculates that once it encounters an obstacle, in this case m_i, it proceeds along this obstacle to one of its peaks (edges) and from there directly to the destination point, this of course prolongs the path. However, depending on his visual effects, the driver of the vehicle would probably change direction to a certain peak in advance in order to shorten the journey as much as possible. This model can therefore be optimized.

Table 2. Percentage route extension for non-optimalized model (average, maximum and minimum)

Type of relief	Path elongation		
	Mean value	Maximal value	Minimal value
Plains	7,9	10,9	3,6
Flat uplands	7,0	12,7	3,7
Rough uplands	7,4	11,2	4,6
Flat highlands	6,9	9,1	3,7
Rough highlands	5,4	6,6	4,6
Mountains	7,6	9,3	5,6

7.2 The Model Nr. 2 - Optimized

The Model Nr. 2 is for the same type of vehicles (the vehicle is still oriented by using GPS to the target point), but the input algorithms used in Model Nr. 1 are optimized. This model consists in modifying the algorithms, where the route leads only over the apexes of the obstacles that occur on the route. The assumption is that - depending on its visual capabilities - the driver would probably not go around the whole obstacle just along its edge, but would change direction in advance to certain vertices in order to shorten the journey as much as possible. See Fig. 5.

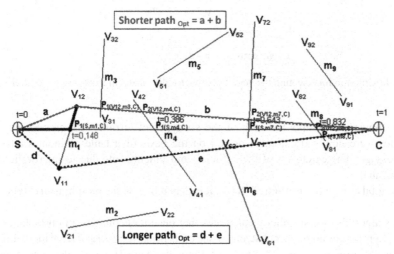

Fig. 5. Criterion for selecting the vertex of a microrelief shape in an optimized process

To Find an Optimized Path
Steps 1–4 remain the same as for process in Model Nr.1.

5. The algorithm examines whether the path intersects at least one shape m_i (otherwise it ends – the path is optimal and at the same time the shortest possible).

6. If so, it will check whether it is possible to connect any two found vertices directly so that they do not intersect with obstacle m_i (the number of these combinations is equal to a two-element combination of N)

$$\binom{N}{2} = \frac{N!}{(N-2)!2!}$$ (4)

where N – *the number of vertices.*

If it finds two vertices that meet this condition, it discards all points (mi) that lie between this connector (vertexes linked by the path). See Fig. 6.

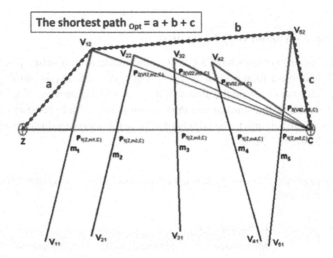

Fig. 6. The optimization principle is demonstrated on the following figure (dashed line)

As a final step, the length of the entire route, is calculated along edge points. The calculation of the length of the path along the edge points (s - is the number of edge points) is given by the relationship:

$$D_c = \sum_{n=1}^{S-1} \sqrt{(x_{n+1} - x_n)^2 + (y_{n+1} - y_n)^2}$$ (5)

where D_c – *path length,*
S – *the number of vertices in the path,*
x,y – *x,y axes of the particular vertex,*
n – *index within the set of vertices.*

This **optimized** process is characterized by shortening the default path on average up to an extension of **1–1.4** percent due to the shortest possible path (this depends on the number and type of micro-relief shapes), compared to the **5 to 8** percent extension that is characteristic of the default **non-optimized path** (the extension again depends on the type of relief). See Table 3.

Table 3. Route extension for the optimized model (average, maximum and minimum)

Type of relief	Path elongation [%]		
	Mean value	Maximal value	Minimal value
Plains	1,4	1,8	1,0
Flat uplands	1,3	1,8	1,1
Rough uplands	1,4	1,7	1,2
Flat Highlands	1,2	1,5	0,9
Rough highlands	1,1	1,3	0,9
Mountains	1,3	1,6	1,0

7.3 Model Nr. 3 -Without GPS Navigation

Unfortunately, not all vehicles have a GPS or other navigation instruments, if they are equipped only with instruments such as a gyroscope, or gyrocompass the length of the distance-size detour depends on the ability of the crew to keep moving. The crew knows only the coordinates of the starting and destination point and a bypassing of obstacles leads to change of the azimuth to the target point. Here the average value of the elongation of the route is **12–15%** (Table 4).

Table 4. Percentage route extension for non-optimized model without navigation device (average, maximum and minimum)

Type of relief	Path elongation		
	Mean value	Maximal value	Minimal value
Plains	14,3	22,5	10,3
Flat uplands	14,1	21,2	10,9
Rough uplands	15,1	19,8	10,3
Flat highlands	12,1	18,1	9,9
Rough highlands	14,1	19,0	9,2
Mountains	13,7	17,4	10,0

7.4 Model Nr. 4 – Joint Factors

The advanced models of optimization were created similarly. Input data were gathered using digitalization of layers of microrelief, waters and railways from the TM 1:25 000. The obstacles were selected carefully, the limits of CCM and NO GO obstacles for the Tank T-72 were considered.

The area of an interest was surrounding of Tišnov town (TM M-3393-D-d) and it was a part of the defence research project. The first simulation model was for influence of microrelief only but the second was for an influence of microrelief, waters and railways together. It is obvious that path optimization has very satisfactory results.

The values of elongation of routes are stated in Table 5.

Table 5. Values of elongation of routes

Type of model	Elongation of route (non-optimized) [%]	Elongation of route (optimized) [%]
Model of "Microrelief"	7,40	1,19
Model of "Microrelief + Watershed + Railways"	14,91	3,79

8   Analysis and Discussion

8.1   Interdependence of the Parameters Evaluated

The carto-metric survey shows the density of microrelief shapes occurrence – this is evaluated in km/km^2. The definition of types of relief is given by morphometric parameters like the size of elevation or value of inclination of slopes and it was supposed that the density of microrelief shapes would be the crucial parameter.

See Table 6.

Table 6. The table shows the density of obstacles

Type of relief	Density of microrelief shapes
Plains	0,71
Flat uplands	0,70
Rough uplands	1,50
Flat highlands	1,17
Rough highlands	1,75
Mountains	1,14

However, the density of microrelief obstacles is not a very definite indicator for calculating of route extensions, as was found by simulation models. It has been evaluated that the length of individual shapes has the most significant influence.

Longer shapes have the greatest effect on CCM. Impact of obstacle size on route extensions indicate the Table 7.

Table 7. Impact of obstacle size on route extensions for an optimized and non-optimized model

Length of obstacle [m]	Number of shapes			
	1000	2000	1000	2000
	Elongation of route (Optimized) [%]		Elongation of route (Non-Optimized) [%]	
100	3,7	19,8	15,6	45,8
200	7,7	36,0	20,5	58,2
300	13,5	47,0	28,0	73,3
400	19,5	56,2	37,1	80,3

The 2D and 3D charts (located below, see Figs. 7 and 8) are a final example of a statistical calculations and they show how the length of microrelief shapes and their number affects the overall elongation of the route. The 2D graph also indicates a comparison of the effectiveness of the optimized and non-optimized calculation process as a percentage.

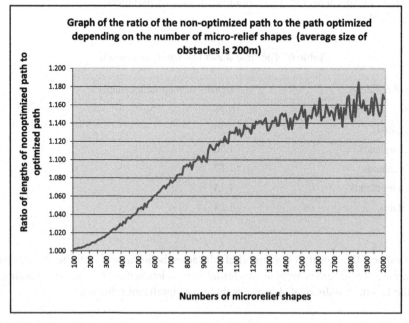

Fig. 7. The 2D graph shows the comparison of the effectiveness of the non-optimized and optimized calculation procedure in finding a path complicated by obstacles. (An average length of obstacles is 200 m. The area of the test area is 10 × 10 km.)

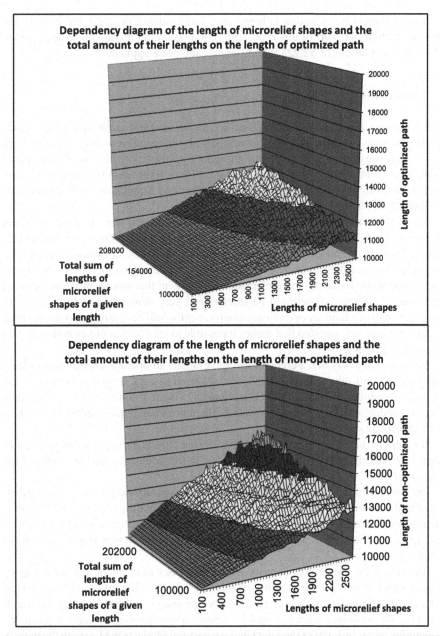

Fig. 8. The 3D graphs of the mutual effect of the number of obstacles, their sizes and the length of the distance travelled. These charts illustrate the dependence of path length (optimized – above and non-optimized - below) on the length of microrelief shapes whose total (sum) does not exceed a given value.

This statistical investigation is based on a large number of generated models and their calculations. As part of this task, 1,536,000 microrelief models were generated to which the calculations of 52,224,000 different paths were applied.

9 Conclusion

As it is obvious, the significance of effective automation of operational planning aspects in different areas (as a decision support component) increases constantly. It is necessary to say, that automation and optimization include a great potential in of operational tasks. These procedures are closer to the human high-level reasoning instead of low-level engineering problems and it seems Modelling and Simulation methods could be successfully applied [18, 21].

Much projects, studies and other work has already been done to build a system that would have different algorithms available to evaluate the influence of particular - specific input parameters, whether tactical, technical, economic, logistical or otherwise, but the research is definitely not finished yet. The variability of interactions of the factors (mentioned above) is large, and the creation of a system that would, moreover, predict the common influence of selected parameters is a very demanding task. A decision-making system in a military operational environment would certainly gladly accept such a solution. However, any step that makes it possible to get close to this complex system is certainly worthy.

It is possible to say that a crucial aspect to evaluate an operational planning dealing with OPFOR and other players (e.g. civilians, neutral units, suspect ones) is tightly related to the ability to evaluate their behaviours. Owing to this assertion to create some effective models and behaviour that reproduce the actions/reactions based on the different boundary conditions is necessary.

It is certain that use of Intelligent Agents driving objects during simulation magnify largely the effectiveness of simulation approach in this context as well as in other joint scenarios [19].

Presented solution shows the possible approach in operational problem solution dedicated to the air manoeuvre optimization in operational conditions [20] with undefined starting and destination point, what means new dimension of options and calculations leading to a higher decision area then problems with selected constraints and known initial inputs.

The aim of the article is to describe the methodology of evaluation of the influence of microrelief shapes on the mobility of military vehicles. The article is based on the statistical determination of the extent of microrelief shapes in the territory of the Czech Republic. These shapes were detected in various geomorphological relief types using topographic maps of 1:10 000 and the Digital Elevation Model (DEM 5) with the density of 1×1 m.

The article describes both the methodology of calculations of possible collisions of the vehicle chassis with the terrain, as well as the calculation of the optimal route of the vehicle avoiding micro-relief obstacles. These models are closely correlated with the factors under assessed and their parameters, and in the field of the use of non-manned equipment, they are a necessary step leading to a successful outcome of the military task, but also in crisis management operations. The algorithms of the route optimization (for communication and outside of them) may not be intended only for the war purposes because finding optimal routes outside the communication may be relevant also in civilian crisis situations, or the provision of assistance when natural disasters such as floods, fires, storms etc.

Micro-relief shapes, types of surface, slopes, soil conditions and the weather conditions are the main factors, whose reciprocal influence has not been evaluated yet and these factors are of fundamental importance for the movement of vehicles in the field. Therefore, the current state of the supporting documents, which contain these data was assessed. The data about micro-relief were the subject of an interest in [8]. Several models were created on the basis of values gathered by a carto-metric investigation.

Further research will look for the correlation between synthetic model and real terrain in terms of validation of some previous assumptions about terrain categories and also it will look for the high-level generic model for operational purposes to upgrade the estimated total sum of the particular paths for logistic planning and preparation.

References

1. Hodicky, J., Frantis, P.: Decision support system for a commander at the operational level. In: Dietz, J.L.G. (ed.) Proceedings of the International Conference on Knowledge Engineering and Ontology Development, KEOD 2009, Funchal, Madeira, October 2009, pp. 359–362. INSTICC Press (2009). ISBN: 978-989-674-012-2
2. Nohel, J., Stodola, P., Flasar, Z.: Combat UGV support of company task force operations. In: Mazal, J., Fagiolini, A., Vasik, P., Turi, M. (eds.) MESAS 2020. LNCS, vol. 12619, pp. 29–42. Springer, Cham (2021). https://doi.org/10.1007/978-3-030-70740-8_3
3. Mokrá, I.: A model approach to the decision-making process. In: Conference Proceedings 3, Applied Technical Sciences and Advanced Military Technologies, vol. 3, no. 1, pp. 278–281 (2012). ISSN: 1843-6722
4. Rybanský, M.: Cross-country movement - the impact and evaluation of geographical factors. The Czech Republic, Brno, p. 114 (2009). ISBN: 978-80–7204-661-4
5. Kristalova, D.: Vliv povrchu terénu na pohyb vojenských vozidel (The Effect of the Terrain Cover on the Movement of Military Vehicles), The Ph.D. thesis (in Czech). The Univerzity of Defence, Brno, The Czech Republic, p. 318 (2013)
6. Mazal, J., Stodola, P., Hrabec, D., Kutěj, L., Podhorec, M., Křišťálová, D.: Mathematical modeling and optimization of the tactical entity defensive engagement. Int. J. Math. Models Methods Appl. Sci. 9(summer 2015), pp. 600–606 (2015). ISSN 1998-0140
7. Dohnal, F., Hubacek, M., Simkova, K.: Detection of microrelief objects to impede the movement of vehicles in terrain. ISPRS Int. J. Geo Inf. 8, 101 (2019)
8. Zelinkova, D.: The analysis of the obtaining and using of the information for evaluation of CCM, (Analýza získávání a využitelnosti informací pro vyhodnocení průchodnosti území), Diploma Thesis (in Czech), VA Brno (2002)
9. Mazal, J.: Real time maneuever optimization in general environment. In: Brezina, T., Jablonski, R. (eds.) Recent Advances in Mechatronics, pp. 191–196. Springer, Heidelberg (2010). https://doi.org/10.1007/978-3-642-05022-0_33, ISBN 978-3-642-05021-3
10. Křišťálová D., et al.: Geographical data and algorithms usable for decision-making process. In: Modelling and Simulation for Autonomous Systems, pp. 226–241. Springer, Roma (2016). ISSN 0302-9743, ISBN 978-3-319-47604-9

11. Drozd, J., Stodola, P., Křišťálová, D., Kozůbek, J.: Experiments with the UAS reconnaissance model in the real environment. In: Mazal, J. (ed.) MESAS 2017. LNCS, vol. 10756, pp. 340–349. Springer, Cham (2018). https://doi.org/10.1007/978-3-319-76072-8_24

12. Stodola, P., Drozd, J., Nohel, J.: Model of Surveillance in Complex Environment using a Swarm of Unmanned Aerial Vehicles. In: Jan Mazal. Modelling and Simulation for Autonomous Systems. MESAS 2020. Notes in Computer Science. Cham: Springer, 2021, roč. 12619, p. 231–249. https://doi.org/10.1007/978-3-030-70740-8_15, ISSN 0302–9743. ISBN 978–3–030–70739–2

13. Kristalova, D.: An effect of sandy soils on the movement in the terrain. In: Hodicky, J. (ed.) MESAS 2014. LNCS, vol. 8906, pp. 262–273. Springer, Heidelberg (2014). https://doi.org/10.1007/978-3-319-13823-7_23, ISBN 978-3-319-13823-7

14. Křišťálová D.: Evaluation of the data applicable for determining the routes of movements of military vehicles in tactical operation. In: The Complex Physiognomy of the International Secuirity Environment, , p. 197–203. "Nicolae Balcescu" Land Force Academy Publishing House, Sibiu (2015). ISBN 978-973-153-215-8

15. Tsourdos, A., White, B., Shanmugavel, M.: Cooperative path planning of unmanned aerial vehicles, p. 214. Wiley (2010). ISBN: 978-0-470-74129-0

16. Duan, H.B., Ma, G.J., Wang, D.B., Yu, X.F.: An improved ant colony algorithm for solving continuous space optimization problems. J. Syst. Simul. **19**(5), 974–977 (2007)

17. Kress, M.: Operational Logistics: The Art and Science of Sustaining Military Operations. Springer, Heidelberg (2002). https://doi.org/10.1007/978-3-319-22674-3

18. Rybar, M.: Modelovanie a simulacia vo vojenstve. Ministerstvo obrany Slovenskej republiky, Bratislava (2000)

19. Bruzzone, A., Massei, M.: Simulation-Based Military Training. In: Mittal, Saurabh, Durak, Umut, Ören, Tuncer (eds.) Guide to Simulation-Based Disciplines. SFMA, pp. 315–361. Springer, Cham (2017). https://doi.org/10.1007/978-3-319-61264-5_14

20. Mazal, J., Stodola, P., Procházka, D., Kutěj, L., Ščurek, R., Procházka, J.: Modelling of the UAV safety manoeuvre for the air insertion operations. In: Modelling and Simulation for Autonomous Systems, MESAS 2016, pp. 337–346. Springer International Publishing, Rome (2016). https://doi.org/10.1007/978-3-319-47605-6_27, ISSN 0302-9743. ISBN 978-3-319-47604-9

21. Mazal, J., Rybanský, M., Bruzzone, A., Kutěj, L., Scurek,R., Foltin, P., Zlatník, D.: Modelling of the microrelief impact to the cross country movement. In: Bottani, E., Bruzzone, A.G., Longo, F., Merkuryev, Y., Piera, M.A. (eds.) Proceedings of the 22nd International Conference on Harbor, Maritime and Multimodal Logistic Modeling & Simulation, HMS, vol. 22, pp. 66–70 (2020). ISSN 2724-0339. ISBN 978-8-885-74146-1

22. Mazal, J, Bruzzone, A, Kutěj, L, Scurek, R, Foltin, P., Zlatník, D.: Optimization of the ground observation (2020). ISSN 2724-0339, ISBN 978-88-85741-46-1

23. Mazal, J, Bruzzone, A., Turi, M, Biagini, M., Corona, F., Jones, J.: NATO use of modelling and simulation to evolve autonomous systems. In: Complexity Challenges in Cyber Physical Systems: Using Modeling and Simulation (M&S) to Support Intelligence, Adaptation and Autonomy, pp. 53–80. John Wiley & sons, Hoboken (2019). ISBN 978-1-119-55239-0

Integrating Real-Time Vehicle and Watercraft Modeling and Simulation Tools for Analysis of Amphibious Operations

John G. Monroe[1]([✉]), Keith Martin[2], Mark Ewing[3], Morgan Johnston[2], Mary Claire Allison[2], Zachary Aspin[1], Collin Davenport[1], Gary Lynch[2], David P. McInnis[1], and Tom McKenna[2]

[1] Geotechnical and Structures Laboratory, U.S. Army Engineer Research and Development Center (ERDC), 3909 Halls Ferry Road, Vicksburg, MS 39180-6199, USA
John.G.Monroe@erdc.dren.mil
[2] Coastal and Hydraulics Laboratory, U.S. Army ERDC, Vicksburg, USA
[3] Department of Electrical and Computer Engineering, Mississippi State University, 406 Hardy Road, 216 Simrall Hall, Starkville, MS 39762, USA

Abstract. Amphibious operations are complex, multi-domain problems that occur in an unpredictable environment. As such, they require knowledge and understanding of the battlespace to be successful. This report describes the development and demonstration of a proof-of-concept tool that combines existing simulation capabilities for watercraft and ground vehicles. The new multi-domain co-simulation environment allows for the modeling and simulation of amphibious operations. During operation, the ship simulator half of the Ship-to-Shore proof-of-concept tool controls an amphibious vessel during its approach to the beach. Once at the shore, primary simulation control passes to the vehicle simulator. Then the ground vehicle is either manually or autonomously maneuvered from the vessel bay, across the beach, and further inland. This combined capability provides a novel environment for ship-to-shore mission assessment. This enables enhanced planning, rehearsal, and decision support prior to mission execution.

Keywords: Amphibious operations · Modeling and simulation · Ground vehicles · Watercraft

1 Introduction

Military history shows that understanding the battlefield and having the ability to rehearse in a similar environment often means the difference between operational success and failure. This paradigm is especially true of maritime and amphibious operations where the environment and surface conditions are complex and unpredictable. Modern modeling and simulation (M&S) tools provide the capability to model the interaction between the physical environment and military systems. M&S thus enables the visualization of possible issues before mission execution. Incorporating that understanding into virtual rehearsals supports identifying hazards and exploring possible mitigation plans during the assessment of proposed operational maneuvers.

This is a U.S. government work and not under copyright protection in the U.S.; foreign copyright protection may apply 2022
J. Mazal et al. (Eds.): MESAS 2021, LNCS 13207, pp. 113–126, 2022.
https://doi.org/10.1007/978-3-030-98260-7_7

1.1 Multi-domain Operations

Multi-Domain Operations (MDOs) began as a concept developed jointly by the U.S. Army and the U.S. Marine Corps (USMC). The goal of this joint warfighting concept is to maintain superiority with regard to peer and near-peer competitors. This is partially accomplished by synchronizing theater arrival to overcome a near-peer's layered, anti-access defense [1]. Using MDO methods, U.S. forces present multiple difficulties to a competitor in air, space, land, sea, and cyberspace. Future maritime and amphibious operations will demand rapid planning and execution to provide an asymmetric advantage to U.S forces and allies.

1.2 M&S for MDO

Leveraging M&S to inform tactics will provide decision support to forward commanders and accelerate the operational planning timeline. The work discussed in the present study is relevant to two domains – sea and land. The proof-of concept simulation technology in this work could support planning and rehearsal of amphibious landings in a single area of operation (AO) or multiple locations along a coast.

An entirely new simulation tool can take years of effort at great expense to reach maturity. Instead, this work brought together two existing tools, the Ship/Tow Simulator (STS) and the Autonomous Navigation Virtual Environment Laboratory (ANVEL) over a period of a few months to provide a usable co-simulation solution to a complex, multi-domain problem. The Ship-to-Shore (S2S) tool fits within a broad set of ERDC MDO-related capabilities but provides valuable benefits in and of itself.

During operation, the ship simulator half of the S2S proof-of-concept tool controls an amphibious vessel during its approach to the shore. Once at the shore, the vessel ramp drops, and ANVEL then controls the vehicle – maneuvering from the vessel bay, across the shore, and then further inland. This combined capability provides a novel ship-to-shore mission assessment environment.

2 Existing Simulation Tools

2.1 Ship/Tow Simulator

Since the 1980s, the ERDC STS has served as a vital engineering tool to evaluate channel design for the U.S. Army Corps of Engineers (USACE). There are many ship simulators across the U.S., but all concentrate on training mariners. Conversely, the ERDC STS, located in the Coastal and Hydraulics Laboratory (CHL), is primarily a predictive engineering tool. A multi-disciplinary team of CHL engineers and scientists manage the STS.

The ERDC STS consists of three full mission ship bridges that feature hardware to replicate an actual ship bridge. The STS bridges include 11 floor-to-ceiling screens, an Electronic Chart Display and Information System (ECDIS), a binocular channel, and a radar. The bridge controls allow pilots to operate rudders, thrusters, and throttles and give tug commands via radio. Simulations occur in real-time, and the bridges can operate independently or be connected to capture ship-to-ship interaction. The ERDC STS can simulate the navigational conditions of ports, harbors, inland waterways, and any other maritime environment. Figure 1 shows a craftmaster piloting a Landing Craft Utility (LCU) in the STS.

Fig. 1. Craftmaster piloting LCU toward the shore in one of the STS bridges.

For a typical civil works project, channel design or modifications are evaluated prior to construction with no risk by replicating the area of interest (AOI) in a virtual environment. The virtual harbor or waterway includes three main simulation components: environmental visuals (i.e., scene), vessel performance, and hydrodynamics. A database is created to represent the existing conditions in the AOI. Pilots familiar with the AOI will then use the STS to assess the existing-conditions database to determine whether any modifications are required to create a more realistic representation of the area. Particular attention is given to vessel response and water currents. Any identified areas of concern are addressed and then retested. Once the pilots have validated the existing conditions database, a copy is modified to reflect future conditions. Then, pilots begin testing the proposed modifications thoroughly to ensure there are no navigational concerns.

The concept of using ship simulation technology in support of amphibious operational planning was initially explored in 2015 [2]. Using Cook Inlet, AK, as an AO, a simulated amphibious landing was conducted at Anchorage, AK, using the LCU 1646. Anchorage was selected due to the similarity in conditions it shares with Inchon, Korea, which witnessed the last contested landing operation by the U.S. military. During this study, the operational parameters of axes of assault, timing, and lighting were examined. The data resulting from this study included the position, heading, and speed of the LCU and other parameters affecting the transit of the LCU. These data were used by the USMC subject matter experts (SMEs) to successfully develop an operational landing plan.

2.2 Autonomous Navigation Environment Laboratory

A major task in developing autonomous unmanned ground vehicle (UGV) systems is creating robust autonomy algorithms that exhibit reliable performance in austere communication or environmental conditions. Inclement weather, harsh or poor lighting, sensor quality, and surface conditions can all adversely affect the decision-making of an autonomous UGV. Thus, the behavior of an autonomous system must be understood or anticipated for edge case conditions. M&S tools that augment physical testing are critical to the autonomous UGV testing and evaluation (T&E) process.

For over a decade, the Mobility Systems Branch (MSB) in ERDC's Geotechnical and Structures Laboratory (GSL) has been developing and using a suite of government-owned M&S tools to assist in autonomy development and risk reduction. These tools comprise the Virtual Autonomous Navigation Environment (VANE) [3] and can be configured to operate independently or in a co-simulation. VANE provides UGV vehicle-terrain and sensor-environment simulation capabilities – be it in a high-fidelity high-performance computing (HPC) environment or in real-time desktop-based simulations.

For real-time vehicle-terrain simulation, GSL-MSB has historically used ANVEL but has begun to incorporate Unreal Engine 4 (UE4) [4]. The core of VANE's sensor-environment simulations is the Environment Sensor Engine (ESE) [5], which was referred to as VANE until the definition was broadened to encompass all aspects of ground vehicle simulations. VANE also leverages parts of open source simulation projects such as Project Chrono [6] or Simulation of Urban Mobility (SUMO) [7]. Figure 2 shows VANE configured for software-in-the-loop simulation for autonomy development and evaluation.

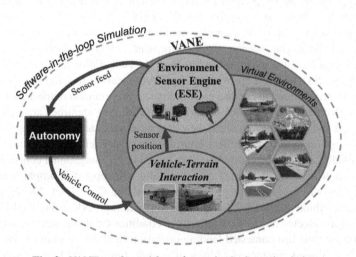

Fig. 2. VANE configured for software-in-the-loop simulations.

The VANE::ANVEL tool was developed specifically for research, development, testing, and evaluation of UGVs. Users of ANVEL build complete end-to-end, systemic models of their intelligent vehicle system, place the models in a virtual environment, and perform interactive testing while tracking key variables and collecting data from virtual sensors. Several Army robotics programs have made extensive use of ANVEL to enable early autonomous algorithm development before physical systems have been available for testing.

3 Motivation for Amphibious M&S

In 21st-century warfare, it is imperative that the U.S. military maintain rapid and asymmetric movement throughout the Pacific. The challenges presented by modern Anti-access/Area-denial (A2/AD) capabilities go far beyond those encountered during Operation Chromite at Inchon, Korea or even Normandy, France. In a testimony to the House Subcommittee on Seapower and Projection Forces, Bradley Martin of the RAND Corporation noted:

> Amphibious operations have always assumed the need to overcome an opposing force and to establish a degree of battlespace dominance before attempting operations, but the reach and lethality of modern weapons systems make aspects of amphibious operations particularly challenging today [8].

Regardless of the preliminary effort required to make amphibious operations even feasible by gaining air and sea superiority in an AO, landing forces will still face significant hazards posed by the natural environment. As the examples of Normandy and Inchon show, knowledge of the battlespace is critical to success.

In the approach to the shore, water can obscure hazards to the landing craft while also being a hazard itself. As was seen at Inchon, knowledge of sea conditions is critical to know when and where a landing can be made. Once ground vehicles disembark, the slopes and soft soil that often face amphibious troops are among the most challenging terrains in terms of vehicle mobility – even without any enhancements from opposing forces. These factors make the ability to simulate multi-domain environments critical in the planning and preparation stages of amphibious operations. Integration of M&S at both the strategic and tactical levels can enable the rapid assessment of operation feasibility and allow the visualization of an operation before mission execution. Incorporating the understanding gained by virtual trials allows for the assessment of proposed operational maneuvers to identify hazards and explore possible mitigation plans.

The STS enables a craftmaster to test proposed operational maneuvers for getting a force to the shore, and co-simulation with ANVEL for vehicle-terrain interactions can enable rehearsal or analysis of short- to medium-range land maneuvers with regard to timing, space constraints, route planning, etc. ANVEL does not natively address more computationally intensive mobility issues such as trafficability, soil deformation, or long-range maneuver; but such tools exist at ERDC if a comprehensive study is desired.

Going beyond the expected utility of amphibious simulations, the context of Inchon exposes another crucial benefit. LTG Jack C. Fuson saw combat in three wars. He served with an engineer amphibious brigade in WWII conducting numerous amphibious beach assaults in the Pacific. He was a port commander during the Korean and Vietnam Wars, and at the end of his career, he served as the Army Deputy Chief of Staff for Logistics. With that breadth and depth of experience, LTG Fuson said in 1994,

> If war were to occur, in all probability the [Army] Transportation Corps would be called upon to duplicate the actions of the engineer amphibious units in World War II. ... Looking back on my firsthand experience with the difficulties in learning how to accomplish this mission in World War II, I know that it would be a slow, costly, and difficult job to reinvent such capabilities in the future (qtd. in [9]).

The success of Operation Chromite in September 1950 largely depended on the fact that most of the planners and operators had amphibious landing experience from the island-hopping campaigns in the Pacific during WWII. That type of institutional knowledge has since passed out of operational memory and into textbooks. The U.S. military conducts training that includes amphibious operations, but these are expensive and dangerous, as shown by the tragic accident in August 2020 when seven Marines and a sailor drowned when their Assault Amphibious Vehicle (AAV) sank off the California coast during a training exercise [10]. M&S can augment physical training and rehearsals, which supports safe and smooth ship and craft operations.

4 Co-simulation Implementation

In the S2S simulation environment, each constituent M&S tool retains supremacy over its own traditional domain. Thus, the primary new development required to enable the proof-of-concept co-simulation tool was the creation of a communication bridge between STS and ANVEL to synchronize craft position in the two simulators. This chapter will discuss this development and the creation of a correlated virtual environment in which the co-simulation occurred during a demonstration of the S2S proof-of-concept.

Figure 3 shows a high-level overview of the co-simulation architecture. The STS software K-Sim [11], running in CHL Bridge-A, passed the position and orientation of the LCU to an ANVEL application programming interface (API) script that was controlling four instances of ANVEL. The four instances consisted of one master instance that was responsible for simulating the vehicle performance and vehicle-terrain-interaction and three viewer instances, each connected to a screen in Bridge-B. Each instance ran on a separate laptop for the sake of efficiency. No other hardware or software in Bridge-B was required. For simplicity, these four instances of ANVEL will be referred to as $ANVEL_M$, $ANVEL_{V1}$, $ANVEL_{V2}$, and $ANVEL_{V3}$. When referred to collectively, the viewer instances will be termed $ANVEL_{V1-3}$ (or $ANVEL_{V1-N}$, since exactly three viewers is not necessary).

Fig. 3. Amphibious co-simulation architecture.

The co-simulation took place in two stages. In the first stage, the ANVEL simulation was a passive mimic of K-Sim until the LCU reached the shore. At that point, the second stage started, and $ANVEL_M$ took control of the vehicle model as it was manually driven onto the beach and then inland. During the initial demonstration of the S2S tool, the vehicle was controlled manually. However, ANVEL retains the capability to integrate autonomy for software-in-the-loop simulations. Thus, an S2S co-simulation of UGVs disembarking from a landing craft would be possible should it be required.

4.1 Ship/Tow Simulator Modifications

K-Sim does not have a native API like ANVEL. Thus, K-Sim developer Kongsberg added new communication hooks to the software in the form of a Distributed Interactive Simulation (DIS) interface [12]. This enabled synchronization of the two sides of the amphibious simulation by allowing ANVEL to receive and replicate the position of the LCU being piloted in K-Sim. To achieve a faster turn-around of the integration of K-Sim with ANVEL, Kongsberg leveraged the previous development efforts of the integration of its Polaris Simulator with the Mariners Skills Suite.

Kongsberg also provided a DIS proxy application developed in this previous effort to enable the GSL teams to test the communication protocols in the absence of an active K-Sim connection. The DIS proxy replicated the effect of passing Entity State Protocol Data Units (PDUs) to and from K-Sim. This tool allowed the ANVEL API script to be developed and tested using a recording of the LCU motion instead of requiring the full simulation configuration.

4.2 ANVEL API Code

Unlike K-Sim, no direct modification of or addition to the ANVEL software was required. Instead, all new development was external to ANVEL and utilized the native ANVEL API to communicate with and control the ANVEL simulation. Three Python files contained the code that controlled all aspects of the ANVEL simulation. These were *ANVEL_Startup.py*, *classAnvelAPI.py*, and *Ship2Shore.py*. ANVEL_*Startup* was used only as a quick reboot option following the crash of one of the ANVEL instances, which happened occasionally during development and testing. *Ship2Shore* contained all the code for communicating with K-Sim and coordinating vehicle and view motion in $ANVEL_M$ and $ANVEL_{V1-3}$. It also called *classAnvelAPI*, which contained the communication protocols to establish a connection between the core logic of Ship2Shore and the APIs of the ANVEL instances.

Below is a categorized list of all *Ship2Shore* classes with brief descriptions. As seen from the list, *Ship2Shore* handles setting up all ANVEL instances, getting and converting position data from K-Sim, passing that information to $ANVEL_M$, and calculating and sending the appropriate view locations and orientations for $ANVEL_{V1-N}$.

- *Supporting Code*

    - CoordinateConversions

        Used to convert between the World Geodetic System (WSG) 84 Cartesian (K-Sim) the local (x, y) ANVEL coordinate system.

- Utilities

A collection of static methods used for various tasks including finding, starting, or killing operating system (OS) processes; joining threads; and displaying log or debug messages

- *Communication with K-Sim*

  - KSimOffset

  Data structure that defines calibration for coordinate transformation between K-Sim and ANVEL

  - KSimComms: inherits KSimOffset

  Used by ANVEL API code to send and receive live data with K-Sim

  - KSimReplay: inherits KSimOffset

  Used by ANVEL API code to send and receive data saved from K-Sim

- *ANVEL Multi-instancing*

  - ViewVeh: inherits AnvelObject (see below)

  Adds an ANVEL vehicle in an ANVEL viewer instance

  - ViewANVELs: inherits Utilities

  Instances of ANVEL that receive vehicle and view commands from ANVEL$_M$ for showing ANVEL simulation on multiple bridge screens

  - AnvelInstance: inherits Utilities

  Contains methods for controlling vehicles and world view for ANVEL$_M$ and ANVEL$_{V1-3}$

- *Conducting Amphibious Simulation*

  - AnvelObject: inherits Utilities

  Methods for adding, removing, and positioning ANVEL objects

  - MasterVeh: inherits AnvelObject

  Adds an ANVEL vehicle in ANVEL$_M$

– MasterANVEL: inherits Utilities

Instance of ANVEL that runs vehicle simulation and communicates with K-Sim

– AmphibiousLanding: inherits Utilities

Controls ANVEL simulation during Stage 1 (LCU moving), Stage 2 (vehicle moving), and the transition between them
Supports both multi-ANVEL simulation (i.e., $ANVEL_M$ and $ANVEL_{V1-N}$) and single ANVEL simulation (i.e., $ANVEL_M$ only)

External scripts communicate with the ANVEL API over a TCP/IP (Transmission Control Protocol/Internet Protocol) connection. *Ship2Shore* runs on the primary ANVEL computer (i.e., the one with $ANVEL_M$) and communicates with $ANVEL_M$ and $ANVEL_{V1-N}$ over multiple socket connections—one connection per instance of ANVEL. The Ship2Shore code could run on a computer not running ANVEL and still communicate with $ANVEL_M$ and $ANVEL_{V1-N}$ over IP. However, having it on the same machine as $ANVEL_M$ reduces the overall communication time, which helps the simulation to stay real-time.

Many of the *Ship2Shore* classes and almost all of *classAnvelAPI* stemmed from existing ANVEL API scripts originally developed for UGV simulations. The two entirely new functionalities created to enable the S2S co-simulation were the network communication with K-Sim and a multi-instancing approach to ANVEL simulations to utilize multiple screens in one of the CHL bridges. The latter of these was not strictly necessary for the co-simulation, as only $ANVEL_M$ actually performs vehicle simulations, and $ANVEL_{V1-3}$ only duplicate the motion for viewing purposes. However, expanding the vehicle field of view by using multiple screens increases driver immersion.

4.3 Python DIS

The ANVEL API already provides a clean interface into the data values of simulated entities, but K-SIM does not provide such an interface. In order to integrate the two simulation platforms, the DIS standard was selected to facilitate data transport. DIS utilizes a User Datagram Protocol (UDP) in the transport layer and several PDUs to encode the data passed between simulations. Since there is no acknowledge/not-acknowledge phase within the UDP, the protocol is ideal for real-time simulations, since it prevents situations in which a process is iteratively trying to receive data that could be several simulation cycles old.

For simplicity, the only PDU used for MDO was the Entity State PDU (Fig. 4). The Entity State PDU contains the information that describes the vessels of interest. The bytes in the PDU are identified starting with the first byte, referred to as "Byte 0." The bytes of interest are highlighted blue in Fig. 4 but are also given in the list below. The Entity State PDUs were exchanged between K-SIM and ANVEL at 60 Hz, which led to a smooth presentation of the LCU moving in ANVEL.

- Bytes 20–27 contain the description of the vessel.
- Bytes 48–71 contain the location of the vessel as double-precision Cartesian representations of the x, y, and z coordinates in meters with reference to WGS 84.
- Bytes 72–83 contain the floating point representation of the yaw, pitch, and roll of the vessel in radians.
- Bytes 84–87 are used to identify whether the LCU ramp was in the up or down position.

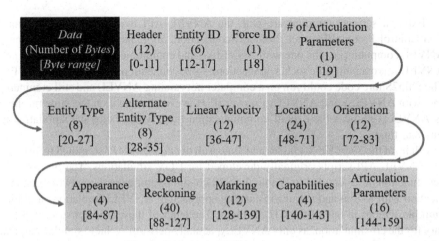

Data (Number of Bytes) [Byte range]	Header (12) [0-11]	Entity ID (6) [12-17]	Force ID (1) [18]	# of Articulation Parameters (1) [19]

Entity Type (8) [20-27]	Alternate Entity Type (8) [28-35]	Linear Velocity (12) [36-47]	Location (24) [48-71]	Orientation (12) [72-83]

Appearance (4) [84-87]	Dead Reckoning (40) [88-127]	Marking (12) [128-139]	Capabilities (4) [140-143]	Articulation Parameters (16) [144-159]

Fig. 4. Entity state PDU byte structure.

4.4 ANVEL Multi-instancing

In the STS bridges, each of the eleven floor-to-ceiling screens connects to a separate computer. During normal STS operation, an instance of K-Sim runs on each computer while a twelfth instance coordinates the overall simulation and communicates with other bridges if a multi-craft simulation is being conducted. The result is that the eleven screens display a unified simulation in real-time that is visually seamless down to the water ripples from one screen to the next. The ANVEL portion of the S2S proof-of-concept was constructed in a similar fashion.

The *Ship2Shore* Python script includes classes that allow ANVEL$_M$ to communicate with K-Sim to get craft position, perform simulations of a single vehicle, and communicate the craft and vehicle pose to ANVEL$_{V1-3}$. ANVEL$_M$ also communicates individual view orientations to ANVEL$_{V1-3}$ to display the ANVEL simulation on multiple (i.e., three) bridge screens, with each screen connected to a workstation laptop running one of the dependent ANVEL instances. The Ship2Shore code supports an arbitrary number of dedicated ANVEL viewing instances—including none, in which case the ANVEL$_M$ is responsible for both simulation and display of the ground vehicle.

To send view information from $ANVEL_M$ to $ANVEL_{V1-N}$ with as little delay as possible, the ViewANVELs class assigns the various $ANVEL_V$ connections to different threads on the master computer. This allows ViewANVELs to send the view update commands in parallel. If there are more view instances than available threads, some or all of the threads are assigned more than one $ANVEL_V$ connection. Although all threads would still execute in parallel, a thread with more than one $ANVEL_V$ instance will send commands to its members in series. However, since there were only three instances of $ANVEL_V$ during the S2S demo, each connection held a dedicated communication thread.

Originally, ANVEL was installed on the eleven bridge computers, creating $ANVEL_{V1-11}$, and $ANVEL_M$ communicated with all of them from a separate laptop. However, this approach resulted in three problems. First, the bridge computers were less powerful than the workstation laptops and failed to run ANVEL in real-time. Second, there was noticeable frame desynchronization between some of the eleven screens due to the way $ANVEL_M$ passes view orientations to $ANVEL_{V1-N}$. Finally, because the ANVEL view(s) is locked to the ground vehicle, which is shown in the 3rd person, small bumps and tilts of the vehicle caused large and violent motion on the screens farthest from the center. Additional development could alleviate this last issue by setting the view to mimic only the vehicle's rotation about the z-axis (in addition to position), instead of mimicking roll, pitch, and yaw. However, since the physical configuration of the bridge screens limits the vehicle view to the 3rd person, as opposed to a 1st person view from inside the vehicle cab, allowing the use of all eleven screens would not have added any useful information to the simulation visuals.

4.5 Virtual Environment Development

For a demonstration of the S2S proof-of-concept, a virtual representation of Mile Hammock Bay (MHB) in North Carolina, USA was developed. The geometry of both the K-Sim and ANVEL scenes was based on off-the-shelf light detection and ranging (LIDAR) from the National Oceanic and Atmospheric Administration (NOAA), but the K-Sim scene required additional data to define typical wind and current conditions that would affect the LCU. The assets in the two scenes were precisely correlated, and the coordinate transformation class in the ANVEL API script ensured that where the LCU landed in K-Sim and where the vehicle disembarked in ANVEL matched. Figure 5 shows a satellite view of the MHB site and the LCU and vehicle paths for the S2S demonstration.

The NOAA aircraft-based LIDAR data allowed the MHB site to be modeled without an in situ data collection. However, because of the relatively low resolution of the point cloud, all routes intended for ground vehicle travel had to be smoothed using ANVEL's terrain editing tools. Additionally, to simplify the scene for better performance, impassible forests at a distance from the landing area and the planned vehicle route were represented by two-dimensional walls with tree textures. Supplementary three-dimensional tree models placed in front of the walls and near the vehicle route produced the illusion of an actual forest. An AO satellite image provided the approximate locations for tree and vegetation placement as well as terrain surface types and locations.

The landing/dock area was imported into ANVEL directly from K-Sim to ensure that the geometry matched exactly during the landing procedure. However, further inland, the

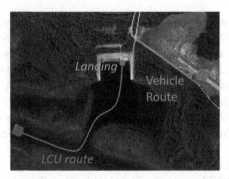

Fig. 5. Satellite image of Mile Hammock Bay with approximate demonstration path.

ANVEL scene includes buildings, static vehicles, and traffic barriers not present in the K-Sim version, as these objects were not visible from the LCU. Although the physical MHB site is devoid of such objects, during the S2S demonstration they acted to define the travel path for the ground vehicle and show the versatility of the M&S environment. M&S allows mission rehearsal to be conducted in a virtual environment both as it exists in the real world and as it may be configured in the future.

4.6 Demonstration

This proof-of-concept S2S M&S tool was exhibited alongside many other ERDC sensing and analysis tools at the Multi-domain Operations Demonstration held at the ERDC Waterways Experiment Station (WES) in August 2019. The event was attended by leaders and stakeholders from across the Army as well as from a NATO ally. When a group of visitors would arrive at the S2S part of the event, the demonstration would begin in one of the STS bridges. There a replay of an LCU landing was shown while an ERDC SME described STS and its role in the overall simulation. Although a replay was utilized for repeatability, the LCU position data was still being streamed to the ANVEL simulation in an adjacent bridge. The group would then walk into the ANVEL bridge as the LCU approached the shore (still being controlled on the ANVEL side by the data coming from the STS replay) and listen to another ERDC SME describe ANVEL and its part in the demonstration (as shown in Fig. 6). Meanwhile, the LCU would hit the beach, and another ERDC SME would take control of the vehicle and drive it inland, as shown in Fig. 5. Each demonstration went smoothly over the course of the event, and there were no communication issues between STS and ANVEL$_M$ or between ANVEL$_M$ and ANVEL$_{V1-3}$.

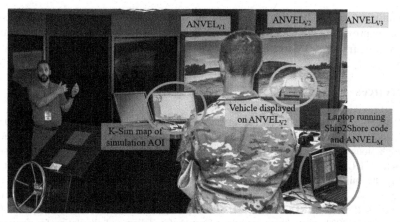

Fig. 6. ANVEL portion of S2S demonstration with annotations.

5 Conclusion

The S2S technology developed in this study could potentially play a significant role in the preparation and rehearsal stages of MDO. S2S M&S incorporates hydrodynamics (including waves), ship motion, vehicle performance, and terrain parameters. This enables the analysis of a landing operation from its origination in the well-deck(s) of a ship(s) stationed below the horizon to the beach to the final upland objective. By combining both sea and land domains into a single co-simulation tool, planners can assess risk and determine the timing and logistics of littoral and upland operations in real-time. This analysis allows planners to develop operational plans based on potential weather, sea, and land conditions and gives operational commanders significantly more data than previously available to make go/no-go decisions. Although simulations of various plans would need to be run in real-time, at least for the STS given the human-in-the-loop requirement, playback of recorded simulations could be played back at faster-than-real-time or condensed down to GIS-based reports for planners' analysis.

ANVEL and the STS are both valuable tools in their respective domains. However, ERDC SMEs created a new Ship-to-Shore M&S capability by combining these existing tools into a co-simulation architecture. The S2S proof-of-concept represents one of the ways the ERDC M&S tools are evolving to engineer and win the future fight. These technologies are outside the scope of this report, but [1] introduces how various ERDC capabilities can work together to support MDO. Many possible improvements are being explored to transform the proof-of-concept into a robust simulation tool for MDO. These improvements include expanding the types of operations that can be simulated, increasing the size of operations simulated, and improving the technology itself through research and development (R&D) initiatives.

In addition to the air and sea superiority that modern A2/AD capabilities require an invasion force to obtain, future maritime and amphibious operations will demand rapid planning and execution to maintain an asymmetric advantage. ERDC researchers have a broad range of expertise that has resulted in many cutting-edge tools that can support the planning and rehearsal of amphibious operations. These capabilities would allow commanders to assess hazards throughout the ship-to-shore cycle and promote mission success by providing knowledge and virtual experience of the battlespace in a tactically relevant time frame.

References

1. Boone, N.: Engineering the Theater. Army AL&T Magazine 2020 (Winter), 47–53 (2020)
2. Cialone, M.A., et al.: Analysis of the effect of environmental conditions in conducting amphibious assaults using a ship simulator/vessel-response model proof-of-concept study. ERDC/CHL TR-17-4, Vicksburg, MS: Engineer Research and Development Center. https://apps.dtic.mil/dtic/tr/fulltext/u2/1037449.pdf
3. Kieffer, C., Holland, M., Monroe, J.G.: Modeling and simulation for unmanned ground vehicles. Power of ERDC Podcast, U.S. Army Engineer Research and Development Center, August 2021. https://poweroferdcpodcast.org/modeling-and-simulation-for-unmanned-ground-vehicles/
4. Epic Games: Unreal Engine (2019). www.unrealengine.com
5. Goodin, C., et al.: Unmanned ground vehicle simulation with the virtual autonomous navigation environment. In: 2017 International Conference on Military Technologies (ICMT), Brno, Czech Republic. IEEE (2017). https://doi.org/10.1109/MILTECHS.2017.7988748
6. Tasora, A., et al.: Chrono: an open source multi-physics dynamics engine. In: Kozubek, T., Blaheta, R., Šístek, J., Rozložník, M., Čermák, M. (eds.) HPCSE 2015. LNCS, vol. 9611, pp. 19–49. Springer, Cham (2016). https://doi.org/10.1007/978-3-319-40361-8_2
7. Lopez, P., Behrisch, M., Bieker-Walz, L., Erdmann, J., Flötteröd, Y.-P.: Microscopic traffic simulation using SUMO. In: 21st IEEE Intelligent Transportation Systems Conference (ITSC) (2018)
8. Martin, B.: Amphibious operations in contested environments: insights from analytic work. RAND Corporation, Santa Monica (2017). https://www.rand.org/pubs/testimonies/CT476.html. Accessed 13 Nov 2020
9. Boose, D.W., Jr.: Over the Beach: US Army Amphibious Operations in the Korean War. Combat Studies Institute Press, Fort Levenworth (2008)
10. Harkins, G.: Marine AAV hit rough seas, rapidly took on water before sinking. Miltary.com 2020. https://www.military.com/daily-news/2020/08/03/marine-aav-hit-rough-seas-rapidly-took-water-sinking.html. Accessed 13 Nov 2020
11. Kongsberg: K-Sim navigation: simulator system maximizing performance. Kongsberg (2020). https://www.kongsberg.com/globalassets/digital/maritime-simulation/k-sim-navigation/docs/k-sim-navigation-brochure.pdf. Accessed 21 Jan 2020
12. IEEE. 2012: "IEEE 1278.1-2012 - IEEE Standard for Distributed Interactive Simulation--Application Protocols". IEEE Std 1278.1-2012 (Revision of IEEE Std 1278.1-1995)

A Design of a Global Path Planner for Nonholonomic Vehicle Based on Dynamic Simulations

Roman Adámek[1]([✉]) [iD], Matej Rajchl[1] [iD], Václav Křivánek[2] [iD], and Robert Grepl[1] [iD]

[1] Faculty of Mechanical Engineering, Brno University of Technology, Technická 2896/2, 616 69 Brno, Czechia
roman.adamek@vut.cz
[2] Faculty of Military Technology, University of Defence, Kounicova 65, 662 10 Brno, Czechia

Abstract. In this paper, an algorithm for global path planning of nonholonomic unmanned ground vehicle (UGV), which moves through diverse terrain, is presented. The proposed algorithm utilizes a simplified dynamic model of the vehicle to verify the passability of a planned movement from node to node and to calculate the duration of this movement and the actual velocity of the vehicle in these nodes. The algorithm operates on a grid map that represents terrain elevation obtained from a triangular irregular network (TIN) map. This map was received from publicly available aerial laser scan data of a land surface in Central Europe. The final path is optimised from the point of view of travel time and respects the nonholonomic constraints of the UGV. This approach ensures that the obtained path is feasible, not only considering the geometric constraints of the vehicle but also its physical limits, i.e., maximum applied torque, maximum velocity, steering limit, etc.

Keywords: Global path panning · Nonholonomic vehicle · UGV · Elevation map · Hybrid A*

1 Introduction

The application of mobile robots in industrial and commercial settings has become increasingly common in recent years. Mobile robots have the ability to move through the environment autonomously and therefore save time and effort when completing mundane or dangerous tasks. We have seen recent applications of UGVs in cooperation with unmanned aerial vehicles (UAV) conducting

This research was funded by the Faculty of Mechanical Engineering, Brno University of Technology under the project FSI-S-20-6407 "Research and development of methods for simulation, modelling a machine learning in mechatronics" and University of Defence, Development program "DZRO Military Autonomous and Robotic Systems".

© Springer Nature Switzerland AG 2022
J. Mazal et al. (Eds.): MESAS 2021, LNCS 13207, pp. 127–144, 2022.
https://doi.org/10.1007/978-3-030-98260-7_8

search and localization tasks [12], another application might be the use of UGVs in military services [14] or in rescue operations and disaster robotics as described by Jorge et al. [7]. Accomplishing these challenging tasks usually require more advanced algorithms which obtain data from various sources and the task of navigation is a more complicated one.

In general, autonomous navigation in an environment with some type of uncertainty (e.g., unknown obstacles, different weather conditions, terrain conditions, etc.) requires two different consecutive motion planning algorithms – global and local path planners. The point of a global path planner (GPP) is to find an approximate path through the environment which seems to be feasible on a large scale using more general data which can be offline or online map data. On the other hand, the local path planner uses sensors that provide data from immediate proximity. These sensors can be, for example, LIDAR, cameras, ultrasonic sensors, radars, IMUs or any other sensor used for typical SLAM algorithms. Many of the experimental vehicles already contain all of the above sensors (e.g. [4,13]), which makes the task easier. The local planner then finds a path through the environment which not only avoids obstacles that do not have to be present on a global map but also can react to changes in the environment (e.g., moving obstacles).

In further sections, we propose a global path planning algorithm for an UGV which uses offline map data containing digital terrain and digital surface models of the environment. This data is usually well known and is fairly accurate to allow for good global path planning on a large scale. In this paper, we will use map data with a resolution of one meter and size ranging from one to two kilometers square. Size of the map data depends on the path planning goal as well as on the computational hardware and the time window in which the algorithm needs to find the path.

1.1   Review of the Literature

Designing a GPP for a UGV moving through a difficult terrain is a complex task involving selection and preparation of map data which will be used as an input to the path planner. This preparation might include initial traversability assessment and obstacle detection to already exclude some areas of the map which might not be searched by the path planning algorithm. The algorithm itself is mostly an enhanced version of standard algorithms which complies with nonholonomic constrains of the vehicle and further evaluates its stability in terrain and respects dynamics of the vehicle as well.

Map data can be obtained, for example, in the form of elevation maps combined with maps of roads, settlements, waterways, vegetation, and so on, or in the form of aerial scans of the ground for example from a UAV. Rybanský et al. describe the process of combining multiple maps to characterize the traversability in each location of the map [22]. The optimal path is then planned using traction coefficients of the vehicle and geometric coefficients. Another approach for classifying traversability using deceleration coefficients is presented in [10]. A very similar topic is covered in [1] where the traversability over microreliefs is determined by the geometric features of the vehicle. Situation where the map is obtained from a UAV working in collaboration with UGV and exploring the terrain ahead to plan the optimal path is described in [21]. Another application of map data in the form of the pointcloud obtained from UAV is in [3]. Where the traversibility of the map is classified only by geometrical features, for example, slope, elevation difference and roughness.

Many of the typical GPP algorithms use A* [5] or RRT algorithm [17] to find a path between two points resulting in a holonomic path which does not have to be suitable for a UGV especially if the vehicle in question is not a small one. Some of the typical nonholonomic planning algorithms are Hybrid A* [2,19,29], RRT* [11], algorithms using Reeds Shepp curves [8] or Dubins path [30]. A different approach is energy-optimal trajectory planning described in [28] or in [15] a length-optimal trajectory algorithm for an UAV is presented. These path planners ensure that the resulting path will be suitable for a vehicle with given kinematics and nonholonomic constrains. However, even this may not be sufficient for a vehicle moving through a challenging terrain where other factors come into play, like limitations of the vehicle powertrain, wheel slipping, losing contact with the ground, tipping over, sliding of the vehicle on a slippery terrain or getting stuck in mud or sand and so on.

Another step to make these nonholonomic path planners more suitable for planning a path through a non-planar terrain is to add some information about the terrain traversability. This information might be a simple one, for example, the limitation of the maximal slope angle as in [16] where Muñoz et al. present their approach based on A* algorithm to plan a path in digital terrain map (DTM). More sophisticated traversibility classification methods are used in [11] and [3]. In the first mentioned one, Krüsi et al. use RRT and RRT* planners in combination with traversibility classification based on the surface roughness and maximal roll and pitch angles. In the latter, Fedorenko et al. use only RRT and the same traversibility classification methods as in the previous work to plan a path in the pointcloud data obtained from a UAV. The Hybrid A* algorithm is used by Thoresen et al. [29] in combination with the fast marching grid heuristic to plan a path in already known traversibility map.

Path planners that are able to combine kinematic and dynamic constrains belong to the category of kinodynamic planners. These planners may either improve the previously mentioned nonholonomic path planners (PRM [9], RRT [18]) or come up with their own unique approach of path generation as in [27]. They also greatly differ in a complexity of the dynamic model and influences that it takes into account depending on the targeted application and computational power. Some researchers simplify the vehicle to the point mass [9,25] therefore assume that all wheels are in permanent contact with the ground and that there is no slip between the wheel and the ground. Others take into consideration the whole 3D dynamics and six degrees of freedom motion of the vehicle [27]. Another difference is in the way how the surface is represented Shiller et al. in [25] represents the surface as a B patch where as Kobilarov et al. in [9] assumes that the surface is composed of angled planar sections. Some works take into account other influences such as different soil types and their interaction with the tires [26].

Based on these related works and our goal, which is to create a global path planner able to find a path in considerable large areas based on aerial laser scans of the terrain, we will extend the Hybrid A* algorithm by using a simplified dynamic model of the vehicle to asses the traversability of the terrain and to obtain a near time optimal path. This approach will secure that the resulting path meets kinematic and dynamic constrains and it is easily modifiable to account for various influences. This method might be also suitable for the enhancement of the current applications which already use Hybrid A* algorithm.

2   Implementation

This section describes the input conditions for our research, limits, theoretical basement and the final algorithm of the solution.

2.1   Terrain Map

To be able to perform the global path planning, map data is required. In this case, two types of map data were obtained – digital terrain model (DTM) and digital surface model (DSM), both in the triangular irregular network (TIN) format. DTM in the version called DMR 5G contains data specifying only the terrain profile without any structures, trees and obstacles. DSM in the version called DMP 1G contains a height profile including the latter. Difference between these two types of maps is depicted in the Fig. 1. DMR 5G and DMP 1G maps are published by Czech Office for Surveying, Mapping and Cadastre (ČÚZK).

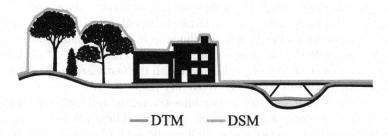

Fig. 1. DTM vs. DSM map comparison

These types of maps can be obtained from UAV drones, enforcing the idea of cooperation between UAV and UGV. The data itself is obtained using a LIDAR scan of the ground, obtaining the DSM and interpolating between the lowest points to produce the DTM. More advanced algorithms can also use radar or sonar to improve the quality of mapping or in the latter case to also get the DTM data of underwater sections.

An essential step that affects the final result is the postprocessing of the data. The main goals of the postprocessing are:

- grid interpolation,
- obstacle detection.

Since the raw data is in the TIN format, it needs to be transformed into an uniform grid which is more convenient for path planning. This transformation is done via linear interpolation. This interpolation introduces an error that causes an inaccurate representation of the true terrain. However, this effect is negligible considering the size of the search area and the fact that we are designing a global planner, in which case some inaccuracies are permissible. The error also decreases with the grid resolution, which was set to 1 m.

To detect obstacles, we use the fact that the DSM map contains the obstacles and the DTM map does not. Therefore, we subtract the DTM map from the DSM and locations with nonzero elevation, with some tolerance margin, correspond to the location of obstacles. The resulting map with separated obstacles can be seen in the Fig. 2.

Since the input map data contains information only about the elevation of the terrain, we assume that the surface of the terrain is smooth and uniform throughout the map. Although this assumption might be violated in many cases additional map data can be included with minor modifications of the path finding algorithm.

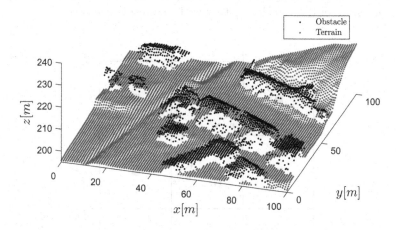

Fig. 2. Obstacle classification based on DTM and DSM subtraction

2.2 Hybrid A* Algorithm

The proposed path planner uses a modified Hybrid A* algorithm which is based on the standard A* algorithm [6]. The A* algorithm is well known and its grid-based version will be only briefly described to clarify used notation.

A* algorithm works with two sets of nodes: open O and closed C. Each iteration of the algorithm starts with a selection of the most promising node out of the open set. At the beginning, only the start node is in the open set. This node is selected based on its total cost $f = g + h$ consisting of two parts, where g cost represents the total cost required to get from the start node to the current node. The second part is the heuristic cost h which represents the cost estimate of getting from the current node to the goal. The better this heuristic cost is, the fewer nodes are required to be expanded and the faster the whole algorithm is. This node is then expanded to find further possible movements. In a grid-based A*, the expansion is either to four neighbouring cells or to all eight adjacent cells. These child nodes are then checked for collision with obstacles and checked if they are already in the close set or in the open set with a higher cost. Then they are added to the open set and the whole process repeats unless the goal is reached or there are no more nodes to expand.

The original Hybrid A* algorithm described in [2] was designed for path planning of autonomous nonholonomic vehicles in DARPA Urban Challenge in 2007. It improves the A* algorithm in a way that the expansion is done not only for four or eight neighbouring cells but by motion primitives generated using a kinematic model of the respective vehicle as can be seen in the Fig. 3 assuming movement with constant velocity. It, therefore, transforms the input in the joint space of the nonholonomic vehicle to the search space (x, y, ψ). Where x and y is a Cartesian position of the vehicle and ψ is the yaw angle of the vehicle. Each node of the graph, therefore, stores not only its discrete state but also the continuous state of the vehicle together with the respective cost.

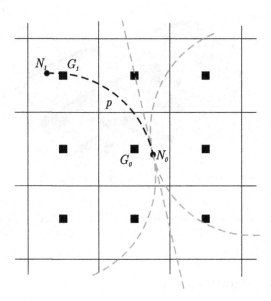

Fig. 3. Node expansion for Hybrid A* algorithm

Choosing a node for expansion is done in the same way as in the case of original A*. Only cost functions are modified.

g function represents the distance between the current node and the child node, which also includes penalties for reversing and changing the direction of movement.

h function original algorithm uses two heuristic functions. The first function calculates the distance from a current state to the goal state, ignoring obstacles but respecting the nonholonomic nature of the vehicle.

The second heuristic function provides the shortest distance from a current state to the goal state, ignoring nonholonomic constraints but considering obstacles. The resulting h cost is taken as the maximum of both outputs of the upper mentioned heuristic functions.

Another addition to the standard A* algorithm is the analytical expansion, where the Reed-Shepp curve is calculated from the current state to the goal and the node is added only if the calculated path is obstacle-free. This addition improves the precision and performance of the algorithm. It should be noted that this analytical expansion is calculated only for some nodes and the frequency of this calculation is increasing towards the goal.

2.3   Nonholonomic Path Planner with Dynamic Model

The proposed algorithm for global path planning in an elevation grid map is based on the Hybrid A* algorithm described in the previous section. Unlike the original algorithm, this method does not expect the vehicle to move with a constant velocity but the velocity is based on a simplified dynamic model. Therefore, the search space of the algorithm is (x, y, ψ, v), where v is the longitudinal velocity of the vehicle.

Each expanded node N of the graph holds the following information:

$$N = (\tilde{x}, \tilde{y}, x, y, \psi, v, d, \delta, g, f) \tag{1}$$

where $\tilde{x}$ and $\tilde{y}$ are discrete vehicle coordinates corresponding with centers of individual grid cells G of the map, d is a vehicle direction, where 1 represents forward movement and -1 reverse movement and δ is a steering angle. We also denote $\boldsymbol{n} = (x, y)$ as a vector of continuous positions of the node N.

Expansion

The purpose of the expansion is to find new possible movements of the vehicle from the current node N_0. A node with the lowest cost f is chosen from the open set O for expansion. Children of this node lie on the end of the motion primitives. These primitives are generated based on a bicycle kinematic model [20]. This model was chosen as a simplification of the Ackermann model, where the front and the rear wheels of the bicycle lie in the middle of the front and rear axis of the four-wheel vehicle. Each motion primitive has the same length p and is calculated for a fixed steering angle δ of the front wheel. The range of these steering angles is a user input into the planner.

Calculation of the motion primitive from node N_0 to the child node N_1 is covered by Eqs. (2)–(7) and depicted in the Fig. 4. The continuous vehicle position $\boldsymbol{n_1}$ in node N_1 is obtained from the Eq. (6) and its yaw angle is based

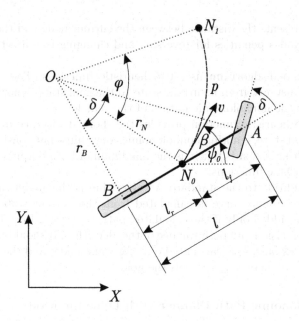

Fig. 4. Kinematic bicycle model

on the Eq. (7). These equations differ if the vehicle is going straight or if it is turning. Position in node N_1 is derived as position n_0 in node N_0 to which the difference between positions n_0 and n_1 in local vehicle coordinate system is added. This difference is rotated by angle ψ_0 to the global coordinate system using rotation matrix $\boldsymbol{R}$. Calculation of the yaw angle is straightforward; if the vehicle is not turning, the yaw angle is preserved. Otherwise, the turning angle φ is added. This turning angle depends on the radius of curvature r_N which is based on the simple geometrical formulas (2) and (3).

$$r_B = \frac{l}{\tan \delta} \tag{2}$$

$$r_N = \sqrt{l_r^2 + r_B^2} \tag{3}$$

$$\varphi = p/r_N \tag{4}$$

$$\boldsymbol{R} = \begin{bmatrix} \cos \psi_0 & -\sin \psi_0 \\ \sin \psi_0 & \cos \psi_0 \end{bmatrix} \tag{5}$$

$$n_1 = \begin{cases} n_1 + \boldsymbol{R} \begin{bmatrix} r_N \sin \varphi \\ r_N(1 - \cos \varphi) \end{bmatrix} & \text{if } \delta \neq 0 \\ n_1 + \boldsymbol{R} \begin{bmatrix} p \\ 0 \end{bmatrix} & \text{otherwise} \end{cases} \tag{6}$$

$$\psi_1 = \begin{cases} \psi_0 + \varphi & \text{if } \delta \neq 0 \\ \psi_0 & \text{otherwise} \end{cases} \tag{7}$$

G Function

Unlike most of the derivations of the A* algorithm, the output of our g function is not the distance between two nodes, which might be penalized in numerous ways, but time. This time is calculated based on a mass point equation of motion (10).

We will use the simplified dynamic model of the vehicle where the vehicle is considered as a point mass. Due to the coarse resolution of the map data and lack of any information about the surface of the terrain, we will neglect the assessment of the vehicle stability and we will assume no slip and permanent contact between the wheels and the ground.

Based on our kinematic model of the bicycle, the path between two consecutive nodes can be either a straight line or a circular arc. Since each node N may have a different elevation, based on its corresponding node G, the path between nodes might be inclined. This transforms the circular arc path into the part of a helix as can be seen in the Fig. 5. In the same way, the straight-line path might be angled towards the horizontal plane.

The helical path can be transformed into the orthogonal triangle provided we neglect the wheel slip and centripetal forces. In this orthogonal triangle, the legs represent the length of the motion primitive p and the elevation difference ΔH, the hypotenuse is the actual length of the slope s. Using this simplification, both turning and going straight can be assessed in the same way.

To calculate the length of the slope s the Eq. 8 is used, where ΔH is the elevation difference between grid cells G_1 a G_0. The slope angle is given by Eq. 9.

$$s = \sqrt{\Delta H^2 + p^2} \tag{8}$$

$$\theta = \text{atan}\frac{\Delta H}{p} \tag{9}$$

The following equations of motion are based on the Fig. 6. Since we neglect the rotation of the wheel and consider the whole vehicle as a point mass, the only relevant equation of motion for our case is Eq. (10) in direction of X_l of the local coordinate system. Members of Eq. (10) are calculated based on Eqs. (11)–(15). Symbols used in the following equations are described in the Table 1. There are two types of dissipative forces in our model. The first one is the torque M_f representing rolling resistance and the second one is F_b which depends on the velocity and represents all other losses.

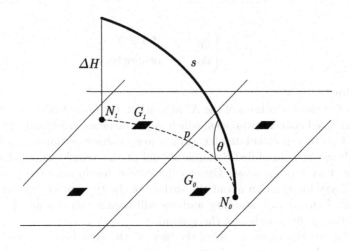

Fig. 5. Vehicle path

Table 1. Table of symbols

Symbol	Description
m	Vehicle mass
a	Vehicle acceleration
M_m	Driving torque
F_m	Driving force
M_f	Rolling resistance torque
r	Wheel radius
F_g	Vehicle weight
F_N	Normal force
v_0	Initial velocity
F_b	Resistive force
ξ	Coefficient of rolling resistance
t	Time

$$X_l : ma = 4F_m - 4\frac{M_f}{r} - F_g \sin\theta - F_b \tag{10}$$

$$F_m = \frac{M_m}{r} \tag{11}$$

$$M_f = F_N \xi \tag{12}$$

$$F_N = \frac{F_g \cos\theta}{4} \tag{13}$$

$$F_g = mg \tag{14}$$

$$F_b = bv^2 \tag{15}$$

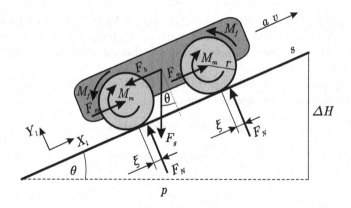

Fig. 6. Free-body diagram of a vehicle on a slope

We can numerically solve Eq. (10) to obtain the position and velocity throughout time. Based on these values, we can determine if the vehicle reached the node N_1, how long did it take and what was the velocity of the vehicle at that time. This velocity will be used for the calculation of g cost from the node N_1 to its child nodes. The resulting g cost corresponds to time required to move from N_0 to N_1. If the node N_1 is unreachable, g cost is set to a high value.

H Function. Since the g cost represents time, the output of the h function has to be time as well. The original H* algorithm uses two heuristic functions as described in Sect. 2.2. These heuristics are well suited for the conditions of the autonomous car moving through mostly flat areas with known obstacles. In our case, the classified obstacles are only obstacles resulting from the difference of DSM and DTM maps, these are vegetation, buildings and other man-made structures, as described in Sect. 2.1. Therefore, all terrain obstacles such as cliffs, ridges, hills, etc. are not considered as obstacles. Since we designed our path planner for a vehicle moving through challenging terrain where terrain obstacles are dominant, the original heuristics were not very beneficial.

Our heuristic function (16) calculates time required to move from the child node to the goal as an Euclidean distance divided by the maximum allowable velocity $v_{\max}$ of the vehicle. The Euclidean distance is a spatial distance considering also the elevation of the child node and the goal.

$$h = \frac{\sqrt{(n_{\mathrm{g}} - n_1)^2 + (H_{\mathrm{g}} - H_1)^2}}{v_{\max}} \tag{16}$$

3   Experiments

We decided to test our algorithm in locations with a challenging terrain where our dynamic model can be fully evaluated and where it can be also compared to a known optimal path. Therefore, locations of quarries were chosen since they provide a difficult terrain and roads for mining vehicles which can be used for comparison. We used DSM and DTM maps of three quarry sites located in the Czech Republic for testing, namely quarries in Jakubčovice nad Odrou, Mokrá–Horákov and Dolní Kounice.

For the purpose of the following tests, a 4WD (wheel drive) vehicle with the following parameters was chosen (Table 2).

Table 2. Vehicle parameters

Parameter	Value
m [kg]	500
l [m]	1.5
l_r [m]	0.75
M_m [Nm]	65
r [m]	0.2
ξ [m]	0.02
b [kgm^{-1}]	50

During each step of expansion, five motion primitives with length p of 2 m were generated for both forward and reverse direction with maximal steering angle δ of 40°.

The algorithm was programmed in MATLAB® and the experiments were conducted on the PC running Ubuntu 20.04 with Intel® Xeon™ CPU E3-1245 v3 @ 3.40 GHz and 16 GB of RAM.

Map data from the query in Dolní Kounice were used for the first experiment. Locations of the start and goal nodes together with a found path and expanded nodes are shown in Fig. 7a. For comparison, there is an aerial view of the quarry in the Fig. 7b. The aerial view does not match with the map data due to different dates of capturing data. It can be seen that there are classified obstacles at the bottom of the quarry near the start node and that the start node is surrounded by steep slopes from three sides. Therefore, the only way to the goal leads through the obstacles and via a side road on the left side of the quarry. Calculation of the path took 17 s and 78 956 nodes were expanded.

The second experiment presented in this paper was conducted with data from the quarry in Mokrá–Horákov (see Fig. 8). This quarry has an interesting layout with two quarries separated by a narrow passageway. Calculation of the path took 114 s and 474 139 nodes were expanded.

The last experiment presents the influence of the driving torque M_m on the planned path. This experiment was conducted on map data from a quarry in Jakubčovice nad Odrou. The experiment was conducted for three different settings of the driving torque with the same position of the start and goal nodes in all three conditions. The lowest driving torque was set to a value which still enables the vehicle to successfully find a path. The middle value was the same as in the previous experiments and the highest setting was set to more than double of the standard driving torque.

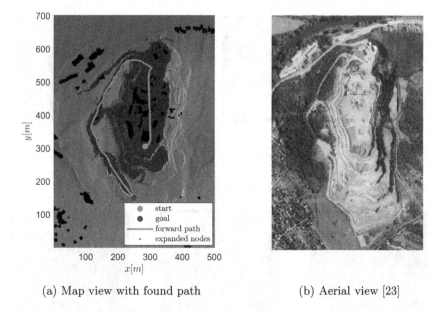

(a) Map view with found path   (b) Aerial view [23]

Fig. 7. Quarry in Dolní Kounice

4   Results

The result of the first experiment can be seen in the Fig. 7a. The proposed planner was able to find a way through the bottom of the query and successfully avoided all classified obstacles. It also found a side road leading towards the top of the query while avoiding cliffs and steep slopes. This evaluation was done solely based on the simplified dynamic model of the vehicle.

In the second experiment, the planner found a way through the narrow section separating both quarries and precisely followed the same road towards the goal which is regularly used by mining trucks. The result of this experiment is in the Fig. 8a. Finding a way through narrow sections is a difficult task for all planners, especially in our case where the movement of the vehicle is constrained by the kinematic model and by the number of generated motion primitives. This causes that the vehicle cannot reach an arbitrary yaw angle and therefore it has to alternate between left and right turns to reach a goal in certain directions.

Results of the last experiment are shown in the Fig. 9. It can be seen that if the driving torque is set low, it will result in the situation shown in the Fig. 9a. In this case, the vehicle was able to reach the goal but it had to take a longer "zigzag" path while going up the hill to minimize the slope angle and therefore the torque required to move. The middle Fig. 9b depicts a situation where the vehicle has a reasonably set driving torque. The planned path follows the existing road and tends to the goal. In the last presented case shown in the Fig. 9c, the driving torque is set too high which results in the planned path going almost directly towards the goal. This result is correct but may not be feasible for a

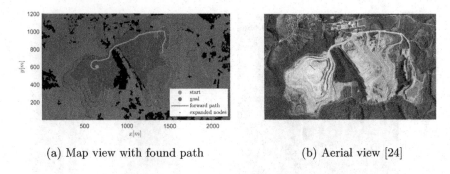

(a) Map view with found path   (b) Aerial view [24]

Fig. 8. Quarry in Mokrá–Horákov

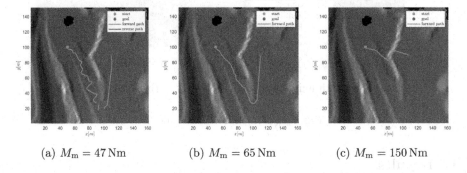

(a) $M_\mathrm{m} = 47\,\mathrm{Nm}$ (b) $M_\mathrm{m} = 65\,\mathrm{Nm}$ (c) $M_\mathrm{m} = 150\,\mathrm{Nm}$

Fig. 9. Comparison of different driving torque settings

real vehicle since our model does not account for the tipping over of the vehicle. If the driving torque was set even higher, the resulting path would be a straight line connecting the start and goal nodes.

5   Discussion

Based on the results presented in the previous subsection, the algorithm proved to be capable of finding a global path while respecting all constraints. The aim of this paper is to present an idea of extending the Hybrid A* algorithm by a simplified dynamic model of the vehicle which is used for traversability assessment and obtaining a near time optimal path.

To improve the performance of the planner and preserve the simplicity of the whole algorithm, some simplifications of the vehicle model were made. These simplifications are acceptable since the proposed planner is meant to be a global planner and all errors resulting from the simplifications should be handled by the local planner of the vehicle. Nevertheless, there is still space for further improvements described below.

The whole algorithm can be improved by incorporating the findings of other researchers. For example, the used grid map can be expanded by additional layers in [22] which will provide information about settlements, roads, waterways, vegetation, soils, etc. This information can be incorporated either directly into the dynamic model or as additional heuristics.

Another improvement might be to smooth out and post-process the found path as in [2]. Since the expansion of each node is only to several discrete locations, the resulting path is usually not smooth and might frequently change the steering angle.

Further improvements can be directed towards the dynamic model of the vehicle which could take into account other influences mentioned for example in [27] and [9]. It could also consider the real center of gravity of the vehicle and its dimensions, so any tip-over of the vehicle could be penalised, this would prevent the situation shown in the Fig. 9c. Addition of side forces could help to better describe the turning motion of the vehicle.

An addition to the dynamic model could be a collision model of the vehicle as presented in [1] which could better assess collisions between a vehicle and microrelief objects.

For the final application the algorithm has to be greatly optimised and rewritten to C++ to improved its performance. Another possibility is to parallelize certain parts of the algorithm, especially the expansion section can benefit from it.

6   Conclusion

A kinodynamic global path planner has been presented in this paper. This path planner operates on a grid map with obstacles and elevation data in each cell of the grid. This grid map is based on TIN data obtained either from a UAV or publicly available DTM and DSM maps.

A modified Hybrid A* algorithm was used as a graph search method, where the total cost of movement from the start node to the goal node is expressed a time. Therefore, the resulting path is time optimal. The evaluation of the cost required to move from one node to the other was based on the dynamic simulation of the simplified dynamic vehicle model. This ensures that the resulting path not only complies with the nonholonomic constraints of the vehicle but also with the limitations of the vehicle powertrain. Inclusion of the dynamic model of the vehicle which is numerically solved and which can be further expanded to include other influences into the simulation is the main innovation of this work.

The proposed planner was tested on map data of several quarries which provide challenging terrain for testing the algorithm with numerous terrain obstacles. These terrain obstacles are not classified as physical obstacles and have to be identified by the dynamic simulation. The planner was able to find paths that correspond to the regular roads used in these quarries. These roads can be considered time-optimal and is proven that these roads are traversable for nonholonimic mining vehicles.

References

1. Dohnal, F., Hubacek, M., Simkova, K.: Detection of microrelief objects to impede the movement of vehicles in terrain. ISPRS Int. J. Geo-Inf. **8**(3), 101 (2019). https://doi.org/10.3390/IJGI8030101
2. Dolgov, D., Thrun, S., Montemerlo, M., Diebel, J.: Practical search techniques in path planning for autonomous driving introduction and related work. Technical report (2008). www.aaai.org
3. Fedorenko, R., Gabdullin, A., Fedorenko, A.: Global UGV path planning on point cloud maps created by UAV. In: 2018 3rd IEEE International Conference on Intelligent Transportation Engineering, ICITE 2018, pp. 253–258. Institute of Electrical and Electronics Engineers Inc., October 2018. https://doi.org/10.1109/ICITE.2018.8492584
4. Grepl, R., Vejlupek, J., Lambersky, V., Jasansky, M., Vadlejch, F., Coupek, P.: Development of 4WS/4WD experimental vehicle: platform for research and education in mechatronics. In: 2011 IEEE International Conference on Mechatronics, ICM 2011 - Proceedings, pp. 893–898 (2011). https://doi.org/10.1109/ICMECH.2011.5971241
5. Gunawan, S.A., Pratama, G.N.P., Cahyadi, A.I., Winduratna, B., Yuwono, Y.C.H., Wahyunggoro, O.: Smoothed a-star algorithm for nonholonomic mobile robot path planning. In: 2019 International Conference on Information and Communications Technology (ICOIACT), pp. 654–658 (2019). https://doi.org/10.1109/ICOIACT46704.2019.8938467
6. Hart, P.E., Nilsson, N.J., Raphael, B.: A formal basis for the heuristic determination of minimum cost paths. IEEE Trans. Syst. Sci. Cybern. **4**(2), 100–107 (1968). https://doi.org/10.1109/TSSC.1968.300136
7. Jorge, V.A.M., et al.: A survey on unmanned surface vehicles for disaster robotics: main challenges and directions. Sensors **19**(3) (2019). https://doi.org/10.3390/s19030702. https://www.mdpi.com/1424-8220/19/3/702
8. Kim, J.M., Lim, K.I., Kim, J.H.: Auto parking path planning system using modified Reeds-Shepp curve algorithm. In: 2014 11th International Conference on Ubiquitous Robots and Ambient Intelligence, URAI 2014, pp. 311–315 (2014). https://doi.org/10.1109/URAI.2014.7057441
9. Kobilarov, M., Sukhatme, G.: Near time-optimal constrained trajectory planning on outdoor terrain. In: Proceedings of the 2005 IEEE International Conference on Robotics and Automation, pp. 1821–1828 (2005). https://doi.org/10.1109/ROBOT.2005.1570378
10. Kristalova, D., et al.: Geographical data and algorithms usable for decision-making process. In: Hodicky, J. (ed.) MESAS 2016. LNCS, vol. 9991, pp. 226–241. Springer, Cham (2016). https://doi.org/10.1007/978-3-319-47605-6_19
11. Krüsi, P., Furgale, P., Bosse, M., Siegwart, R.: Driving on point clouds: motion planning, trajectory optimization, and terrain assessment in generic nonplanar environments: driving on point clouds. J. Field Robot. **34** (2016). https://doi.org/10.1002/rob.21700
12. Lazna, T., Gabrlik, P., Jilek, T., Zalud, L.: Cooperation between an unmanned aerial vehicle and an unmanned ground vehicle in highly accurate localization of gamma radiation hotspots. Int. J. Adv. Robot. Syst. **15**(1), 1729881417750787 (2018). https://doi.org/10.1177/1729881417750787
13. Ligocki, A., Jelinek, A., Zalud, L.: Brno urban dataset - the new data for self-driving agents and mapping tasks, pp. 3284–3290 (2020). https://doi.org/10.1109/ICRA40945.2020.9197277

14. Matejka, J.: Robot as a member of combat unit a utopia or reality for ground forces? Adv. Mil. Technol. **15**(1), 7–24 (2020). https://doi.org/10.3849/aimt.01332

15. Mazal, J., Stodola, P., Procházka, D., Kutěj, L., Ščurek, R., Procházka, J.: Modelling of the UAV safety manoeuvre for the air insertion operations. In: Hodicky, J. (ed.) MESAS 2016. LNCS, vol. 9991, pp. 337–346. Springer, Cham (2016). https://doi.org/10.1007/978-3-319-47605-6_27

16. Muñoz, P., R-Moreno, M.D., Castaño, B.: 3DANA: a path planning algorithm for surface robotics. Eng. Appl. Artif. Intell. **60**, 175–192 (2017). https://doi.org/10.1016/j.engappai.2017.02.010. https://www.sciencedirect.com/science/article/pii/S0952197617300337

17. Noreen, I., Khan, A., Habib, Z.: A comparison of RRT, RRT* and RRT*-smart path planning algorithms. IJCSNS Int. J. Comput. Sci. Netw. Secur. **16**(10), 20 (2016)

18. Pepy, R., Lambert, A., Mounier, H.: Path planning using a dynamic vehicle model. In: 2006 2nd International Conference on Information Communication Technologies, vol. 1, pp. 781–786 (2006). https://doi.org/10.1109/ICTTA.2006.1684472

19. Petereit, J., Emter, T., Frey, C.W., Kopfstedt, T., Beutel, A.: Application of hybrid A* to an autonomous mobile robot for path planning in unstructured outdoor environments. Technical report (2012)

20. Rajamani, R.: Vehicle Dynamics and Control (2012). https://doi.org/10.1007/978-1-4614-1433-9

21. Ropero, F., Muñoz, P., R-Moreno, M.D.: TERRA: a path planning algorithm for cooperative UGV-UAV exploration. Eng. Appl. Artif. Intell. **78**, 260–272 (2019). https://doi.org/10.1016/J.ENGAPPAI.2018.11.008

22. Rybansky, M., Hofmann, A., Hubacek, M., Kovarik, V., Talhofer, V.: Modelling of cross-country transport in raster format. Environ. Earth Sci. **74**(10), 7049–7058 (2015). https://doi.org/10.1007/s12665-015-4759-y

23. Seznam.cz a.s., Microsoft Corporation, www.basemap.at, EUROSENSE s.r.o., GEODIS Slovakia s.r.o., OpenStreetMap: Mapy.cz - Dolní Kounice (2018). https://mapy.cz/letecka?x=16.4515236&y=49.0741100&z=17

24. Seznam.cz a.s., Microsoft Corporation, www.basemap.at, EUROSENSE s.r.o., GEODIS Slovakia s.r.o., OpenStreetMap: Mapy.cz - Mokrá-Horákov (2018). https://mapy.cz/letecka?x=16.7587009&y=49.2308069&z=16

25. Shiller, Z.: Obstacle traversal for space exploration. In: Proceedings 2000 ICRA. Millennium Conference. IEEE International Conference on Robotics and Automation. Symposia Proceedings (Cat. No.00CH37065), vol. 2, pp. 989–994 (2000). https://doi.org/10.1109/ROBOT.2000.844729

26. Shiller, Z., Mann, M.P., Rubinstein, D.: Dynamic stability of off-road vehicles considering a longitudinal terramechanics model. In: Proceedings 2007 IEEE International Conference on Robotics and Automation, pp. 1170–1175 (2007). https://doi.org/10.1109/ROBOT.2007.363143

27. Singh, A.K., Krishna, K.M., Saripalli, S.: Planning non-holonomic stable trajectories on uneven terrain through non-linear time scaling. Auton. Robots **40**(8), 1419–1440 (2016). https://doi.org/10.1007/s10514-015-9505-5

28. Stodola, M., Rajchl, M., Brablc, M., Frolík, S., Křivánek, V.: Maxwell points of dynamical control systems based on vertical rolling disc-numerical solutions. Robotics **10**(3), 88 (2021). https://doi.org/10.3390/ROBOTICS10030088. https://www.mdpi.com/2218-6581/10/3/88/htm

29. Thoresen, M., Nielsen, N.H., Mathiassen, K., Pettersen, K.Y.: Path planning for UGVs based on traversability hybrid A*. IEEE Robot. Autom. Lett. **6**(2), 1216–1223 (2021). https://doi.org/10.1109/LRA.2021.3056028

30. Štefek, A., van Pham, T., Křivánek, V., Pham, K.: Energy comparison of controllers used for a differential drive wheeled mobile robot. IEEE Access **8**, 170915–170927 (2020). https://doi.org/10.1109/ACCESS.2020.3023345

Robotic Snap-fit Assembly with Success Identification Based on Force Feedback

Filip Radil$^{(\boxtimes)}$, Roman Adámek , Barnabás Dobossy , and Petr Krejčí

Institute of Solid Mechanics, Mechatronics and Biomechanics,
Faculty of Mechanical Engineering, Brno University of Technology,
Technická 2896, Brno, Czech Republic
182636@vutbr.cz

Abstract. Snap-fit assembly is a standard method in manufacturing to join mainly plastic parts together without any additional processing. However, most of these assemblies are carried out by human workers as they are able to recognize whether the operation is executed correctly. This work aims to improve a robotic snap-fit assembly and replace humans in the process. It is achieved by force measuring using the sensor at the end of the robotic arm. For the purpose of the work, the custom made snap joint was designed in various variants. A set of features was established to enable the classification of obtained signals. The features were tested on a created dataset consisting of measured signals of the four primary cases that may occur during the assembly. This solution provides a possible expansion to create a framework with a selected classification algorithm for the autonomous classification of measured signals.

Keywords: Snap-fit assembly · Industrial robot · Force feedback · Classification

1 Introduction

Currently, there is a significant trend to automate as many processes as possible in any field, particularly in the manufacturing industry. This happens mainly due to the customer's requirements, which are generally speed, quality, and lately customizability. The critical feature in all manufacturing processes is hence flexibility. A solution to this challenge is partially provided by industrial robots that proves the steep growth of their implementation [1].

However, most industrial robots are programmed with a fixed path and controlled positionally. This method is relatively simple and reliable but can be time-consuming, and it must be adjusted manually for every specific case. In addition, some processes can not even be fully automatized by this technique because of a lack of specific attributes that human workers naturally have, for example, tactile perception. Despite the development of intelligent robotic methods in recent years, which are based on learning from collected data or human collaboration, many challenges remain, not including those out of the

© Springer Nature Switzerland AG 2022
J. Mazal et al. (Eds.): MESAS 2021, LNCS 13207, pp. 145–157, 2022.
https://doi.org/10.1007/978-3-030-98260-7_9

industry field. This work aims to improve a complex robotic operation called snap-fit and take this process one step further to full automation.

In Sect. 2, the snap-fit assembly operation is described, and related works are mentioned. Section 3 presents the conducted experiment, including setup description, dataset creation and proposed features enabling signals distinction. Following Sect. 4 presents the experiment results, Sect. 5 provides the discussion and finally, Sect. 6 contains the conclusion.

2   Snap-fit Assembly

Snap-fit assembly is a standard method in the manufacturing industry for joining parts together simply, rapidly and without any additional parts processing. Another advantage is that the joints can be designed as detachable. The method is used mainly for plastics because a locking mechanism is based on the material's elasticity. An assembly can be described as a phases sequence. Firstly, parts are engaged and pushed against themselves (Fig. 1a). Secondly, the snap part is deflected (Fig. 1b) and inserted into the opposite part. Finally, the snap part is elastically recovered to the original position (Fig. 1c), which causes the parts to lock.

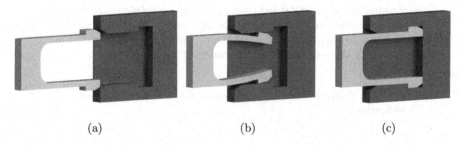

(a)   (b)   (c)

Fig. 1. The Snap joint assembly phases (a) Engaging, (b) Deflection, (c) Completion

2.1   Operation Description

Although the principle is always the same, there are many design possibilities. Whether it is in the locking mechanism, determining the difficulty of disassembly (hook, stud or bead) or in a joint shape itself. The most basic and used shape is the cantilever snap joint (Fig. 1). Dominantly flexural load enables high-speed and straightforward design. Thus, it can be found in a broad spectrum of products, ranging from cars, appliances to furniture. Other cases are, for example, torsion, annular (Fig. 2b), or u-shaped snap joints (Fig. 2a).

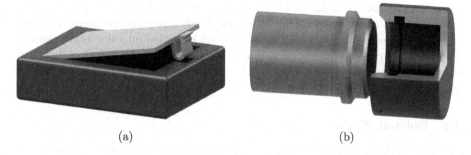

(a)   (b)

Fig. 2. (a) U-shape snap joint example, (b) Annular snap joint example

A human performing this operation can intuitively determine whether it was successful because of his tactile sense and the possibility of hearing the clicking sound. A problem arises when the task is automated and carried out by robots. The locking mechanism is often in the internal structure of the part, and for this reason, a visual confirmation can not be achieved. Therefore there is an effort to imitate a human tactile sense by measuring force signal during the assembly and by that replace humans in the process.

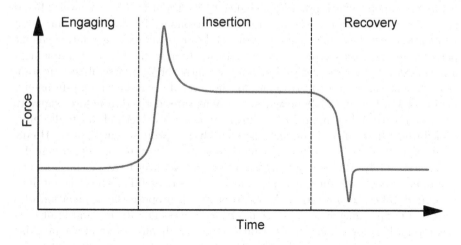

Fig. 3. Theoretical force signal from cantilever snap joint

The measured force signal will be affected by the joint type, locking mechanism type, used material, manufacturing quality and tolerances, robot positioning error and other external factors. Nevertheless, with operation knowledge, the approximate signal shape can be theoretically deduced (Fig. 3). In the beginning, the force is supposed to stay at zero levels as the robot approaches the part without resistance. After the engagement of the parts, the force should steeply rise and immediately drop slightly, which is caused by the parts settling and static friction overcoming. During the insertion, the snap part slides with a constant resistance until elastic recovery, when the significant drop in signal is anticipated, followed by a gap where the force values are negative. The gap is caused by momentary retraction as the part is settled into the final position. The assembly phases correspond to the sequences in a signal, and thus it should be possible to trace those phases in the signal back and identify the required information.

2.2 Related Works

As the snap-fit assembly is such a standard, several works focus on this issue. Rusli et al. [18] investigate the tactile feedback from a cantilever snap-fit joint, but obtained by a manual assembly. A study made by Genc et al. [5] is not focusing on fault classification but a type of snap-fit joint classification. Rodriguez et al. [13] is using the force signature captured during the contact phase of the assembly process and detect failure by that.

Huang et al. [9] model both the fault-free and faulty cases of assembly by different switched linear models with known switching sequences, bounded parameters, and external disturbances and use it for the FDI (Fault detection and isolation) algorithm for electrical connectors. Hayami et al. [6] attempts to predict failure by classifying data from a 6D force/torque sensor. It predicts the fault before it happens, which provides a chance to the robot for error recovery. Rojas et al. [16] design a five-layered hierarchical taxonomy for cantilever-snap joint verification based on increasingly abstract layers that encode relative-change patterns in the task's force signatures. In another work [17], they implement a novel failure characterization scheme for cantilever snap assemblies. In addition, they present a control strategies for more reliable cantilever snap-fit assembly in [14,15]. Luo at al. [12] provides a set of features derived from force signal to SVM and by that determine failure during an assembly. Methods for fault detection during the electrical connectors assembly are proposed in [7,8] by Huang et al. Koveos et al. [11] propose a methodology for fast robot deployment for snap-fit assembly. Measured forces are processed by wavelet transformation, and a mother wavelet containing the information of a successful assembly is obtained by human-robot collaboration. Di Lello et al. [3] propose the application of a Bayesian nonparametric time-series model to process monitoring and fault classification while force measuring. Doltsinis et al. [4] aim to create a machine learning framework that enables the classification of many different joint types. Finally, Chan et al. [10] classify snap-fit assembly without force sensor. Data are obtained from joints torque. Similarly, Karlsson et al. [2] implement external sensors and compare their results with the inbuilt robot wrist sensor.

A slightly different approach for investigation of battery insertion into the holder is presented by Rónai et al. [19]. Model and simulation determine the so-called trip point, and then the force-controlled robot performs the assembly. However, similar approaches were used even in other fields. Vejlupek et al. [20] presents a method for the testing of the automotive fuel pump with DC motor based on current and magnetic field measurements and subsequent signal processing.

The field has already become relatively well researched, but some improvements are still possible. Deployment of the current methods could be complex, and they are usually one-joint-type purposed. The signals are classified into successful or unsuccessful cases, although it could be useful to reveal why the assembly failed. A reliable but straightforward method requiring short commissioning time is needed for possible industry usage.

3   Experiment

The main experiment objective is to determine if it is possible to obtain a force signal corresponding to the signature shape (Fig. 3) by a measurement provided by a regular built-in sensor at the robotic arm. If so, propose a method for recognising correct and incorrect attempts, but preserve the method feasible and straightforward.

3.1   Setup

The custom snap-fit joint was designed and manufactured by 3D printing. Even if it is shaped as an annular snap joint, due to the slits, the strain is mainly flexural, and thus it is considered a cantilever joint. Unlike the planar cantilever, the part's orientation during the assembly can be arbitrary, the part is settled smoother, and this model offers more possibilities for modification.

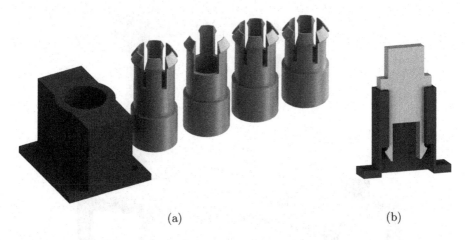

(a)   (b)

Fig. 4. (a) Annular cantilever joints, (b) Planar cantilever joint

The snap part was manufactured in the four variations (Fig. 4a). One standard piece, one piece of the same length but with a few missing teeth and two shortened pieces, each of different lengths. These parts simulate the faults possibly occurring during the assembly. This set is used for the data collection. In addition, one sample of the planar cantilever joint was manufactured (Fig. 4b) to prove that it is possible to measure the same signature signal even with another joint type.

The robot for the task is a UR5 e-series from Universal Robots company. It was equipped with the gripper OnRobot RG2 which has modified fingers that correspond to the circular shape of the parts (Fig. 5). The data from a built-in force sensor FT 300 can be read together with the TCP (tool centre point) position over the TCP/IP protocol. All data are directly stored on a client PC, where they are immediately processed in Matlab.

3.2   Dataset Creation

The robot's task consists of the movement from a home position towards the chosen snap part. The robot continues with the grasped part to the statically placed hole. After the approach, the force measuring starts, and the robot is performing the assembly, i.e. inserts the part into the hole, pushes it through and continues in the motion 2 mm. Measurement continues until the movement stops and the gripper is opened. The force signal length is 14 s at an insertion speed of 0.005 m/s. After that, the robot returns to the home position. It is essential to mention that the robot is controlled positionally, as the most industrial manipulators. It does not have any other feedback whether the operation was executed successfully.

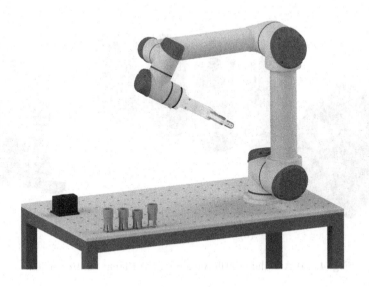

Fig. 5. Robot task setup

The four primary cases, which could occur during the assembly, were considered to create a complex dataset. The first case is naturally when the joint is correctly assembled. It was carried out thirty times and labelled as *"Correct"*. For the following operation, the missing teeth snap part was used. This represents the case that can happen in a real environment by faulty manufacturing or, for instance, parts careless handling. It was executed twenty times and labelled as *"Missing teeth"*. Another twenty measurements (ten times each part) were executed with the shortened snap parts. This is supposed to represent a case when the robot reaches the final position, but the joint is not fully assembled. This may occur when the part is faulty manufactured or incorrectly grasped by a robot. Identification of these cases brings up the major challenge of this work. Those samples were labelled as *"Incorrect"*. Finally, five measurements were not finished, and these represent a case when the assembly is not finished due to the jam of the part in some phase. These were labelled as *"Blocked"*.

3.3   Features Selection

The signal peak value was selected as the first and the most prominent feature, primarily because of the *"Blocked"* signals. The robot has a specific maximum force that it can apply, and when a part is jammed, the robot reaches the limit and then stops. The *"Blocked"* signals have thus significantly higher maximum value than those correct samples.

On the contrary, the next feature is the minimum value of force. In this case, not the entire signal is processed, but only a section. It is determined as the z-axis position interval measured in the robot base coordinate system. The interval depends on the final insertion position and the snaps part teeth length. The force signature drop and gap are supposed to be located in this section, and the incorrectly assembled joints miss this phenomenon.

The third and fourth features are a minimal and maximal value of the computed numerical derivative of the same position interval as the feature above. The *Correct* samples contain the drop followed by the increase. Hence, both derivative values are supposed to be high.

The last one contains computed mean values of the signal. The *"Missing teeth"* samples put less resistance against the movement than the *"Correct"*, and thus the measured force is supposed to be lower.

4   Results

One of the objectives was to determine if it would be possible to distinguish each assembly phase in the signal measured by the robot sensor. In Fig. 6, there is an example of a measured signal of the annular cantilever joint and the planar joint. The sample of the planar joint reaches lower force values, but it corresponds to the shape of the annular joint, and both correspond to the theoretical prediction (Fig. 3).

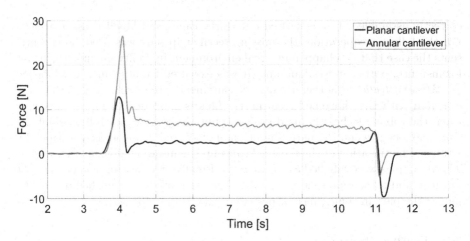

Fig. 6. The signal of the correctly assembled planar and annular cantilever joint

The results of the measured *"Correct"* signals can be seen in Fig. 7. The signal preserves its shape in all samples, proving a small standard deviation value. Thus, classification by shape features should be possible. The *"Incorrect"* data do not show any particular trend and yield a higher standard deviation.

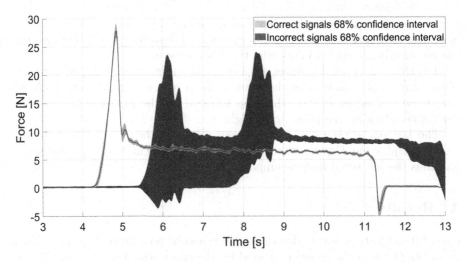

Fig. 7. 68% confidence intervals for dataset signals

The features results and their effectiveness can be seen in the graphs in Fig. 8 and Fig. 9. The first graph (Fig. 8a) shows the peak force value of the whole signal. The *"Blocked"* samples are evidently well separated from the others in the dataset, thus potentially easily classifiable.

The data shown on graph Fig. 8b (minimum force) indicate slight separation. The *"Correct"* and the *"Missing teeth"* samples are low. However, the values in the *"Incorrect"* signal drop down at the end, causing the minimum value to be also zero or less. Despite that, the minimum value can be helpful for classification in combination with the other features. The *"Blocked"* signals are not displayed because the robot did not reach the position in the interval where the minimum value is computed.

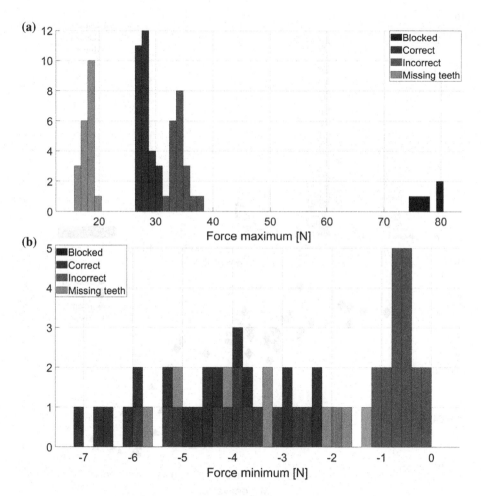

Fig. 8. (a) Force maximum feature (b) Force minimum feature

The Fig. 9a shows a computed mean value. The primary purpose is the segregation of the *"Correct"* and the *"Missing teeth"* signals, and that is happening according to the figure. The *"Incorrect"* samples can reach arbitrary values as it depends on how positionally shifted they are. The peak value of the *"Blocked"* samples is much higher than others, and therefore the mean value is also higher.

The Fig. 9 b shows the minimum and the maximum derivative features in one graph. Their combination segregates the *"Incorrect"* and the *"Correct"* data well. It is mainly due to the gap at the end of the signal, which the "incorrect" samples lack. As the *"Missing teeth"* joints produce a low force, they are not strictly separated from others. The "blocked" data are not displayed again as they are not included in the required position interval.

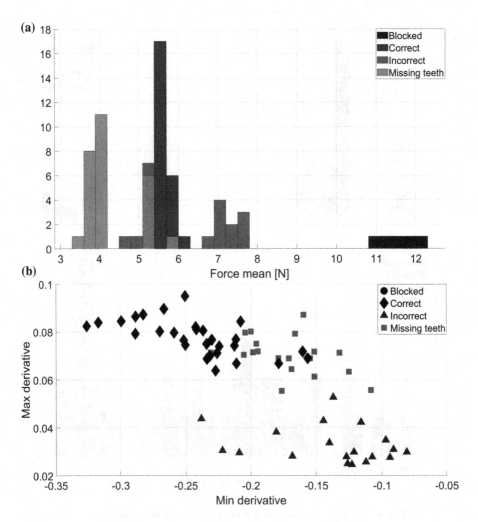

Fig. 9. (a) Mean feature (b) Minimum and maximum derivative feature

5   Discussion

The results show that it is possible to measure a force during the snap-fit assembly only by a robot build-in sensor and obtain a signal in which the assembly phases can be recognized. The measuring was verified by acquiring the force signal from the planar cantilever joint assembly that produces the same characteristic shape differing only by the force magnitude. The measured signals contain slight noise that can be easily filtered.

The force signal measured during the snap-fit assembly is characterized primarily by the steep drop followed by a gap where the values are negative. The gap is created when a material recovery pulls the snap part into the final position. To make this section more significant, a specific design element was used. The large radius was added at the snap part's locking edge, causing gradual retraction of the part into the hole. This radius is usually smaller at generally used joints, and the measured gap would be thus narrower but presumably still distinguishable and applicable for classification.

The dataset was created to cover the four main events that can occur during the robotic assembly of this specific joint. Other types may require collecting more data and establishing more or different cases. For example, the planar cantilever joint can be narrow, and the snap part can be placed outside the hole. The number of samples in the dataset is enough to test whether the chosen features serve their purposes.

The features results are promising. They all enable the distinction of specific data and could potentially serve for classification by their combination. Slightly misleading could be the measured *"Incorrect"* data. The samples are the results of only two lengths of the snap part. Theoretically, the part can be shifted by any distance, but the only feature that could be affected is the mean feature, and it should not produce complications. The *"Incorrect"* signals are finished by a steep drop caused by the robot relief. This affects the features computed from the final position interval (minimum value and derivative). Isolation of the interval where the signal gap is located would presumably produce more evident results. However, a manual specification of this interval would make the method less versatile, and an automatic detection would unnecessarily complicate the method.

6   Conclusion

This paper presents a proposed method for force acquisition and its processing during a robotic snap-fit assembly. The experiment revealed that the force signal measured by the robot's sensor is indeed shaped as the theoretically predicted signature shape. After acquiring more signal samples of four cases that may occur during the assembly, the dataset was created, and the chosen features were extracted. The results show that feature values enable distinction between signals.

This work is supposed to be followed by the subsequent work exploring a classification algorithm choice. This was also the purpose of the features extraction, which an algorithm will exploit.

Acknowledgement. This research was funded by the Faculty of Mechanical Engineering, Brno University of Technology under the projects FSI-S-20-6407: "Research and development of methods for simulation, modelling a machine learning in mechatronics", and FV-21-26: "Robotic assembly using force interaction".

References

1. Ifr presents world robotics report 2020 (2020). https://ifr.org/ifr-press-releases/news/record-2.7-million-robots-work-in-factories-around-the-globe
2. Chan, Y., Yu, H., Khurshid, R.P.: Effects of force-torque and tactile haptic modalities on classifying the success of robot manipulation tasks. In: 2019 IEEE World Haptics Conference (WHC), pp. 586–591 (2019). https://doi.org/10.1109/WHC.2019.8816131
3. Di Lello, E., Klotzbücher, M., De Laet, T., Bruyninckx, H.: Bayesian time-series models for continuous fault detection and recognition in industrial robotic tasks. In: 2013 IEEE/RSJ International Conference on Intelligent Robots and Systems, pp. 5827–5833 (2013). https://doi.org/10.1109/IROS.2013.6697200
4. Doltsinis, S., Krestenitis, M., Doulgeri, Z.: A machine learning framework for real-time identification of successful snap-fit assemblies. IEEE Trans. Autom. Sci. Eng. 1–11 (2019). https://doi.org/10.1109/TASE.2019.2932834
5. Genc, S., Messler, Robert W., J., Gabriele, G.A.: A hierarchical classification scheme to define and order the design space for integral snap-fit assembly. Res. Eng. Design **10**(2), 94–106 (1998). https://www.proquest.com/scholarly-journals/hierarchical-classification-scheme-define-order/docview/2262571747/se-2?accountid=17115. copyright - Research in Engineering Design is a copyright of Springer, (1998). All Rights Reserved. Accessed 24 July 2019
6. Hayami, Y., Wan, W., Koyama, K., Shi, P., Rojas, J., Harada, K.: Error identification and recovery in robotic snap assembly (2021)
7. Huang, J., Di, P., Fukuda, T., Matsuno, T.: Model-based robust online fault detection for mating process of electric connectors in robotic wiring harness assembly systems. In: 2007 International Symposium on Micro-NanoMechatronics and Human Science, pp. 556–563 (2007). https://doi.org/10.1109/MHS.2007.4420916
8. Huang, J., Fukuda, T., Matsuno, T.: Model-based intelligent fault detection and diagnosis for mating electric connectors in robotic wiring harness assembly systems. IEEE/ASME Trans. Mechatron. **13**(1), 86–94 (2008). https://doi.org/10.1109/TMECH.2007.915063
9. Huang, J., Wang, Y., Fukuda, T.: Set-membership-based fault detection and isolation for robotic assembly of electrical connectors. IEEE Trans. Autom. Sci. Eng. 1–12 (2016). https://doi.org/10.1109/TASE.2016.2602319
10. Karlsson, M., Robertsson, A., Johansson, R.: Detection and control of contact force transients in robotic manipulation without a force sensor, pp. 1–9 (2018). https://doi.org/10.1109/ICRA.2018.8461104
11. Koveos, Y., Papageorgiou, D., Doltsinis, S., Doulgeri, Z.: A fast robot deployment strategy for successful snap assembly (2016). https://doi.org/10.1109/IRIS.2016.8066070

12. Luo, W., Rojas, J., Guan, T., Harada, K., Nagata, K.: Cantilever snap assemblies failure detection using SVMs and the RCBHT. In: 2014 IEEE International Conference on Mechatronics and Automation, pp. 384–389 (2014). https://doi.org/10.1109/ICMA.2014.6885728
13. Rodriguez, A., Bourne, D., Mason, M., Rossano, G., Wang, J.: Failure detection in assembly: force signature analysis, pp. 210–215 (2010). https://doi.org/10.1109/COASE.2010.5584452
14. Rojas, J., et al.: A constraint-based motion control strategy for cantilever snap assemblies. In: 2012 IEEE International Conference on Mechatronics and Automation, pp. 1815–1821 (2012). https://doi.org/10.1109/ICMA.2012.6285097
15. Rojas, J., et al.: Probabilistic state verification for snap assemblies using the relative-change-based hierarchical taxonomy. In: 2012 12th IEEE-RAS International Conference on Humanoid Robots (Humanoids 2012), pp. 96–103 (2012). https://doi.org/10.1109/HUMANOIDS.2012.6651505
16. Rojas, J., et al.: A relative-change-based hierarchical taxonomy for cantilever-snap assembly verification. In: 2012 IEEE/RSJ International Conference on Intelligent Robots and Systems, pp. 356–363 (2012). https://doi.org/10.1109/IROS.2012.6385604
17. Rojas, J., Harada, K., Onda, H., Yamanobe, N., Yoshida, E., Nagata, K.: Early failure characterization of cantilever snap assemblies using the PA-RCBHT. In: 2014 IEEE International Conference on Robotics and Automation (ICRA), pp. 3370–3377 (2014). https://doi.org/10.1109/ICRA.2014.6907344
18. Rusli, L., Luscher, A., Sommerich, C.: Force and tactile feedback in preloaded cantilever snap-fits under manual assembly. Int. J. Ind. Ergon. 40(6), 618–628 (2010). https://doi.org/10.1016/j.ergon.2010.05.005. https://www.sciencedirect.com/science/article/pii/S0169814110000533
19. Rónai, L., Szabó, T.: Snap-fit assembly process with industrial robot including force feedback. Robotica 38(2), 317–336 (2020). https://www.proquest.com/scholarly-journals/snap-fit-assembly-process-with-industrial-robot/docview/2336208309/se-2?accountid=17115. copyright - © Cambridge University Press 2019; Last updated - 2020-01-13
20. Vejlupek, J., Grepl, R., Matějásko, M., Zouhar, F.: Automotive fuel pump fault detection based on current ripple FFT and changes in magnetic field, 2013(3), 130–138 (2013)

Measuring Multi-UAV Mission Efficiency: Concept Validation and Enhanced Metrics

Julian Seethaler[✉] [ID], Michael Strohal, and Peter Stütz

Institute of Flight Systems, University of the Bundeswehr Munich, 85577 Neubiberg, Germany
{julian.seethaler,michael.strohal,peter.stuetz}@unibw.de

Abstract. Compositions of future airborne forces with highly automated and/or autonomous components shall be assessed in terms of mission effectiveness and efficiency through a modelling and simulation process. To this end, we have proposed a systematic metric derivation and measurement scheme in previous work. Now, its evaluation and validation were conducted through subject-matter experts (SMEs), who agree with the assessment outcomes and attest to their meaningfulness when comparing their judgements of the simulated example mission runs with the generated results.

In this paper, a representative vignette for reconnaissance by unmanned aerial vehicles (UAVs) is established together with different mission execution plans to enable and test a more general use of the process in a multi-agent simulation testbed. Additionally, a complementary concept for the derivation of suitable force compositions from archetypes, which represent specific categories of aircraft, is put forward based on a morphological approach. These force compositions' capabilities were matched to the requirements of the respective mission vignette and subsequently simulation-based comparative assessments were conducted. These assessments relied on criteria hierarchically derived from the specific vignette's mission tasks and objectives. Therefore, we propose metrics for mission success and effort sensitive to the number of entities and the composition of force packages.

We conducted simulation of several variants and replications to cover different random variables and mission execution plans to generate assessments for different force compositions and mission plans. Eventually, the derivation of a simplified metamodel for the estimation of mission efficiency using the most influential input variables was attempted.

Keywords: Agent-based simulation · Measures of effectiveness · Measures of performance · Metrics · Modeling and simulation · Systems effectiveness · Systems of systems · UAV

1 Introduction

Performance indicators, also known as metrics, are required for the objective and quantified assessment of the performance, utility, and effectiveness of civil and military aerial systems. In the past, comprehensive methods for the assessment of single military aircraft [1, 2], and metrics and predictors for performance in air combat [3–6] have been presented by various authors.

© Springer Nature Switzerland AG 2022
J. Mazal et al. (Eds.): MESAS 2021, LNCS 13207, pp. 158–179, 2022.
https://doi.org/10.1007/978-3-030-98260-7_10

However, in the context of cooperation of multiple manned and/or unmanned aerial vehicles (UAVs) and force compositions with automated and/or autonomous components, it is insufficient to examine only the parameters of a single aircraft by *measures of performance* (MoP) for single aircraft, e.g., range or sensor quality. This is due to the numerous interactions with the mission environment, including adversaries, and in between the entities that make up a cooperative system (also called system-of-systems or force package). In our previous work [7] this was confirmed by subject-matter experts (SMEs) and is also recognized in naval context [8]. Thus, we have proposed to use *measures of effectiveness* (MoE) for evaluating overall mission results in a way that is agnostic to the number and type of aircraft. Aggregating these MoEs over several mission executions enables to obtain a probabilistic measure of capability or *system performance potential* (SPP) for the force package.

Generally, combining measures of *capability*, *reliability* and *availability* results in an overall *systems effectiveness* probability [9], which in turn yields an average measure of efficiency when related to cost and effort. In a mission-specific perspective, we call this *mission efficiency*, i.e., the considered system-of-systems' effectiveness versus generalized cost.

As a test case, a simulation of a reconnaissance mission executed by a group of UAVs (also see Fig. 2 for the vignette illustration) is used here to evaluate a set of MoEs (see Table 4) such as number of detected targets and duration of detectability. The aggregated measures are used to judge which configuration of the UAV force package is the best for this specific mission.

The remainder of this paper is structured as follows: In Sect. 2 we first review the basic concept for gaining a comprehensive assessment of a single mission's execution, and then from several executions a more general measure for the effectiveness and efficiency of an aerial force composition. Then, in Sect. 3 the conducted simulation experiment is presented along with the proposed mission-related metrics. Section 4 gives the experimental results. They are discussed in Sect. 5. Finally, Sect. 6 gives concluding remarks and an outlook for future research.

2 Concept and Methods

2.1 Mission Assessment Process

Figure 1 gives an overview over the process flow we propose for comprehensive assessment of a mission. It consists of the three main blocks of metric derivation, simulation and measurement, and finally validation.

Metric Derivation
Metric derivation aims at finding a total objective function, often called J or SPP, for a specific mission. We previously [7] have proposed a systematic process to derive mission-based criteria and weights relying on SMEs. Its first step is the decomposition of the mission into relevant and measurable elementary criteria in a hierarchical requirements and criteria breakdown structure by brainwriting and moderated group discussion. The second step is to determine the relative importance of these criteria represented by

weights, which are obtained through pairwise comparison using linguistic variables in the *Fuzzy Analytic Hierarchy Process* (FAHP) [10].

Simulation
The assessment of one or more aerial systems is mainly conducted in constructive simulation during early phases of systems development. The simulation must adequately represent significant interactions between opponent and cooperative forces in the scenario and model all relevant capabilities of the involved force package(s).

For a technical systems evaluation, a potentially wide state space of *variants* of mission plans, behavior, doctrine, and generally everything that a human decision maker could influence in an actual mission should be covered to highlight the impact of the system's design. Furthermore, multiple simulation *replications* (repeat simulation runs of the same variant but different random seeds) are usually required to account for effects of randomness and probabilistic modelling, e.g., concerning missile hit rate.

Measurements specific to the previously defined criteria must be made during the simulation. The results are afterwards accumulated through multi-criteria decision analysis (MCDA) methods to gain the resulting mission-based effectiveness and efficiency rating(s).

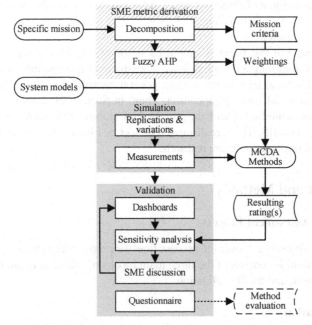

Fig. 1. Concept overview: SME metric derivation process, measurements in simulation, and validation.

We have presented first results of this approach, which considered MoEs for the effect chain, cost, side effects, time, and gained experience for an air-to-ground mission by UAVs, in a previous version of a metrics testbed simulation [11].

Technical details about the current experimental setup and metrics testbed in an agent-based simulation are given in Sect. 3. In addition to the mission-based capability simulation, probabilistic models for reliability and availability are used.

Aggregation

To get the overall assessment result, i.e., fuse all elements in the expert-derived criteria breakdown tree into one mission MoE, the individual criteria results must be normalized and hierarchically aggregated according to their respective weights. This is achieved by application of an MCDA method.

Generally, MCDA methods can be divided into compensatory methods, where deficits in some criteria can be fully compensated by good results in other criteria, and non-compensatory or outranking methods, that do not allow this [12, 13]. The first category includes, among others, the *Hierarchical Weighted Sum* (HWS) [14] – which can easily be transformed into a *Simple Additive Weighting* (SAW) formulation, also known as *Weighted Sum* (WS) –, *Multi-Attribute Utility Theory* (MAUT), and *Technique for Order of Preference by Similarity to Ideal Solution* (TOPSIS) [15]. MAUT requires more input from stakeholders in the form of utility functions, TOPSIS ranks by Euclidian distances to a (notional) ideal alternative. A prominent example for the non-compensatory type is *Preference Ranking Organization Method for Enrichment Evaluations* (PROMETHEE) [16], which uses preference functions to calculate dominance flows of the given alternatives.

Validation

Validation is the last step in the concept. For this, to ensure consistency, the same SMEs are consulted who previously have derived the criteria breakdown. Here, all relevant mission data are presented to the SMEs in "dashboard"-type collections of diagrams along with the measurement results on all criteria decomposition levels. Additionally, full mission runs are replayed graphically. Local sensitivity analysis outcomes are displayed, giving insight how altering a criterion's weight influences the total effectiveness rating.

In a moderated group discussion, the SMEs judge whether they agree with the overall assessment and whether they see their previously derived criteria adequately represented. This is supplemented by a questionnaire, partially given in Table 1, for the involved SMEs for evaluating the concept itself, its benefits (e.g., transparency, structure, completeness), drawbacks, and required effort.

2.2 System Assessment

As mentioned, while the derived and weighted metrics allow for the assessment of a single mission according to Fig. 1, the fair and reliable judgment of an aerial system via modeling and simulation must consider various dimensions in the parameter space.

Missions, implemented in the form of representative vignettes, generally can be approached in various ways and the respective plan of action depends on the employed entities and vice versa. Therefore, action plans and concepts of operations (CONOPS) accounting for doctrine and rules of engagement (RoE), which must be reflected in the agents' rule base, entail simulating several variants of the same mission.

In the next step, for all these mission plans replications are necessary to account for all kinds of randomness, including but not limited to partially random decisions as well as effects simulated probabilistically such as communications reliability, weapon hit probabilities, and environmental conditions.

Adding another layer of complexity, one finds that for a constant force package usually a set of mission types can be relevant. Thus, multiple SME-selected vignettes are utilized, all of which include random replications and mission plans respectively.

Finally, when wishing to compare force compositions with various capabilities, this gives another dimension in the parameter space, meaning in principle all mentioned input variations must be simulated for all those differently composed systems.

Force Compositions

As a prerequisite, the capabilities of the force packages must match the vignettes' require-

ments to enable successful mission execution. Therefore, only a limited number out of all possible alternative multi-aircraft compositions comes into consideration. Hence, we employed the following approach for the a priori determination of relevant force compositions: It is comprised of first identifying *archetypes*, then combining these into promisingly composed force packages. This preselection also serves as part of the *design of experiment* (DoE) process, ensuring focus on relevant alternatives.

Identification of Archetypes

Every possible aircraft in a cooperative system is part of a class, such as medium-altitude long-endurance (MALE) UAV, 4th or 5th generation fighter, or unmanned combat aerial vehicle (UCAV). Each of these categories is epitomized by an archetype, representing the typical properties of all elements in its class. For each, short profiles including their specific capabilities and performance envelope properties were prepared. E.g., the MALE archetype is characterized by flight altitudes up to 9000 m, typical cruise at Mach 0.15 to 0.16 (maximum at about Mach 0.2), long endurance of 24 to 40 h, and a maximum range of over 4000 km. It is equipped with a main high-resolution electrooptical sensor, and line-of-sight (LOS) and satellite communications to receive commands from and transmit data to a ground control station.

Archetypes are also the basis for the agents modelled in the simulation.

Force Compositions

We employ *morphological analysis* [17, 18] using a morphological box (also known as Zwicky box) or a cross-impact matrix to derive viable force compositions from the archetypes. In these tools, archetypes can be matched based on their capabilities as noted in their respective profiles, as they should complement each other regarding the requirements of the intended mission(s).

Ruling out irrelevant force compositions and focusing on promising ones is a simple but required part of sampling stratification and DoE that can be accomplished when generating homogenous and heterogenous force compositions in this way. Further heterogeneity can also be introduced by differences in mission equipment (payload).

3 Experiment

In the experiment, we applied the described process for a comprehensive assessment in a specific mission vignette. Several variants and replications were executed in a multi-agent simulation environment to test the proposed MoEs by applying them to different force compositions and mission plans.

MoEs for reconnaissance, effort, time, gained experience, and side effects were quantitatively measured and used to rank alternative force compositions and mission plans. The existing [7, 11] metrics were enhanced mainly by lower-level criteria for target detection and time-integrating criteria for detectability. Table 4 gives more details regarding MoEs, (sub-)criteria and weights.

The development of the experimental setup and the used mission vignette was accompanied by SMEs as described in Sects. 3.1 and 3.2.

3.1 Expert Validation of the Experimental Setup

Previously, we have implemented the concept described above and applied it to a simplified air-to-ground operation with three different force packages [11]. The resulting dashboards, generated assessments, and video replays of all mission executions were presented to the same SMEs who also [7, 11] had derived the criteria tree for the mission.

Based on this information the experts were now asked to answer a questionnaire mostly consisting of statements with a five-point Likert scale [19] as shown in Table 1, supplemented by open free-text questions. The main questions asked whether they agree with the overall assessment results, about the realism and fidelity of the simulation, usefulness of the method and its supposed advantages (transparency, structure, completeness). Furthermore, the SMEs were asked to predict the assessment ranking order of four alternative cooperating forces in the given air-to-ground operation based on their respective properties.

3.2 Vignette Definition and Mission Plans

For the simulation experiment in this paper, vignettes and mission plans were created, and the metrics testbed simulation and data logging in the simulation were upgraded according to the SME feedback given to the questionnaire described in Sects. 3.1 and 4.1.

Vignettes

Vignettes are concrete descriptions and specifications of scenarios, used as "recipes" to implement typical missions in the simulation environment. To define proper vignettes, example scenarios of the relevant mission types [20] such as close air support (CAS), imagery intelligence (IMINT), or suppression/destruction of enemy air defense (SEAD/DEAD) were proposed to SMEs, who ordered them by perceived relevance. Accordingly, vignettes were established as relevant specifications to be used in simulation.

Table 1. Statements on the questionnaire rated by the SMEs. The last column indicates the overall favorability of the responses towards the presented approach as positive (+), neutral (o), or negative (−).

No.	Question or statement	Answers +/o/−
1.1	In the presented scenario the use of a composite force promises advantages compared to a single aircraft	+
1.2	The executions and results of the missions are realistic	−
1.3	I agree with the generated assessments of the considered mission results	o
1.4	The generated assessments are in agreement with my weightings of the criteria	+
1.5	The generated ranking of the compared systems appears to be correct	+
1.6	My previous approach would have generated the same ranking of the compared systems/force packages	o
1.7	I think the demonstrated method makes sense	+
1.8	The presented approach is suitable to determine the best of several mission executions	+
1.9	The presented approach is suitable to determine the best of several force packages	+
1.10	The presented tool and its results enable relevant conclusions, which would have been impossible with the previous approach	+
1.11	The presented tool and its results are useful for decisions in mission planning	o
1.12	The presented approach supports weighing different interests/aims of one or more decision makers against each other	+
1.13	The demonstrated method covers more information depth than the previous approach	+
1.14	I would like to use the demonstrated method as an element of future analyses/assessments	o+
1.15	I prefer my previous approach	+

IMINT, i.e. cooperative aerial reconnaissance [21, 22], was selected as a relevant mission type and therefore was used for the more detailed metrics and simulation implementation in this experiment. The IMINT vignette (see Fig. 2 for a sketch) specifies a square area of operation of eight-by-eight kilometers, in which six ground targets, specifically light trucks, are moving from the north towards the center of the area at about 40 km/h on various paths. Some surface-to-air threat is present in the form of two short-range Man Portable Air Defense System (MANPADS) locations, positioned to the south and south-west of the area's center. These also serve as acoustic observation points.

In our experiment, homogenous force compositions of the MALE archetype were considered. The UAVs, which are equipped with one main electrooptical sensor, successively enter the zone from the south-west. Their goal is to reconnoiter the whole area and detect all of the present trucks. All types of air-to-ground effect are prohibited by RoE. The electromagnetic spectrum is given as uncontested.

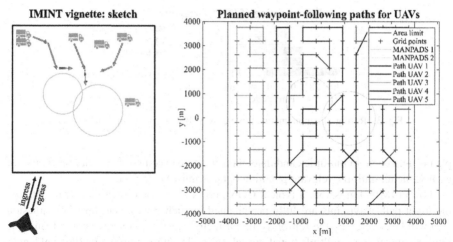

Fig. 2. Left: symbolic sketch of the IMINT vignette (not to scale) with movement paths of the ground targets indicated by red arrows and MANPADS engagement zones represented by red circles. Right: example for coverage path planning with waypoint grid over the whole area of operations for five UAVs. (Color figure online)

The statements in Table 2 were given to the SMEs for evaluation of this IMINT vignette. The experts' ratings on a five-point Likert scale can be seen in Fig. 3. Overall, the positive evaluation of the vignette regarding its realism, relevance and representativeness legitimizes its use as a typical case in the experiment.

Table 2. Statements on the IMINT vignette rated by the SMEs. The last column indicates the overall favorability of the responses towards the specified vignette as positive (+), neutral (o), or negative (−).

No.	Question or statement	Answers +/o/−
2.1	The specific IMINT vignette is realistic	+
2.2	In the described scenario the use of a composite force promises significant advantages compared to a single aircraft	+
2.3	The scenario represented by the specific vignette is relevant in real deployment	+
2.4	The specific vignette is representative for the general mission type IMINT	+
2.5	The mission described in the vignette is usually executed by a cooperative force package	o

Mission Plans

For any given mission vignette and the selected force compositions (in the experiment consisting of the MALE archetype) relevant and tailored action plans for mission execution must be developed. Mission plans must strive to optimally capitalize on the respective force's abilities in the scenario, otherwise the mission outcome will not accurately reflect the technical possibilities of the system-of-systems. However, often the optimal

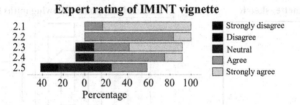

Fig. 3. Rating of statements in Table 2 by SMEs on five-point Likert scale.

plan is unknown, so multiple candidate plans must be executed in simulation. Thus, for each vignette, multiple relevant planning options must be devised.

For an IMINT mission, a typical CONOPS is a coverage path plan [21, 22], i.e., the flight path waypoints are fully pre-planned in such a way that the whole mission area will be covered by the sensor(s). In our metrics testbed, coverage path planning is implemented using a multiple travelling salesmen algorithm minimizing the total length of the flight path(s). For an example planning result see Fig. 2.

Variants

In the realized coverage path plans the waypoints of the paths depend on the number of UAVs in the force package, but the plans in the experiment also differ in cruise altitude. The chosen altitude in an IMINT mission is the result of a trade-off between reducing the aircraft's own detectability and increasing the sensor footprint with increased altitude, or better ground sample distance (GSD) with lower altitudes (for constant focal length of the optics). Higher detection probabilities can also be provided by different equipment, in the considered case by higher resolution electrooptical sensors.

Thus, we vary the following parameters: number of UAVs, planned cruise altitude, and optical sensor resolution. The used values of these are given in Table 3.

For each variant, i.e., one of the eight resulting combinations of parameters, a constant number of replications with different random seeds were simulated to allow for differences in random number generation and minor deviations in the automated coverage path planning.

Table 3. IMINT vignette parameters of the different variants

Parameter	Lower bound	Upper bound	Units
Number of UAVs	3	5	–
Cruise altitude	4000	5000	m
Sensor resolution	4000 × 3000	6000 × 4500	Pixel

3.3 Simulation Environment

The metrics testbed is a multi-agent based simulation environment, which was extended from our previous proof-of-concept [11]. Aircraft, ground targets, and surface-to-air missile (SAM) sites or MANPADS are each separate agents. Figure 4 gives an architectural overview of the current environment setup.

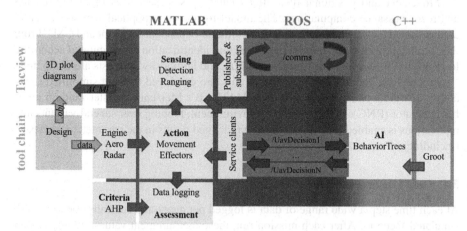

Fig. 4. Architectural overview of the metrics testbed simulation environment

One of the major enhancements is the use of *Behavior Trees* (BT) [23], one for each UAV agent reflecting the specific capabilities and the doctrine, following the overall plan of action. The BT are implemented using a C++ library and can be visually edited via *Groot* [24]. They are connected to the main MATLAB [25] environment by a Robot Operating System (ROS) [26] service for each aircraft. ROS subscribers and publishers are also used to model directed and broadcast communications between the UAV agents.

Furthermore, visualization by MATLAB plots has been improved and additionally extended by using the software *Tacview* [27] for 3D graphics. Tacview can replay a recorded mission from an Air Combat Maneuvering Instrumentation (ACMI) text file or display live telemetry via TCP/IP. Aerial, naval, and ground vessels are rendered from included or custom *obj*-files.

The aircraft models are generated in a separate tool chain using design synthesis code yielding aircraft performance data, i.e., aerodynamics (lift-drag polars) and propulsion data (characterized by thrust and specific fuel consumption over altitude, Mach number and degree of throttling). These models are used in the 2.5D dynamics simulation, which includes an international standard atmosphere (ISA) model for altitude dependencies. The aircraft models also include observability data, namely radar cross-section (RCS) over spherical coordinates and an acoustic profile over sonic frequencies and engine degrees of throttling.

The MANPADS locations are used as observation points to get the attenuated sound pressure levels (distinct per aircraft and cumulative for the force package), and the according loudness and probability of perception via the hearing curve [28]. It is assumed that a single hit from any surface-to-air missile takes out the concerned aircraft. The reflected radar power is determined from attitude-dependent RCS and distance via radar equation and is compared to the background noise to determine detection.

Movement and positional data after each time step serves as input for sensing and reconnaissance computations. The model for the electrooptical sensors currently assumes fixed view, resulting in a platform altitude-dependent footprint and GSD. From this, probabilities for detection, recognition, and identification are calculated according to Johnson [29]. Additionally, brightness, hue, and luminosity contrasts are computed.

Probabilistic models, e.g., for missile hits, are employed through simplified probability distributions in conjunction with MATLAB's Mersenne Twister random number generator (RNG). In the simulation environment, looping over several variants and replications is enabled for exploring user-predefined points in the parameter state space (including different random seeds).

3.4 Data Logging and Measurements

At each time step a wide range of data is logged per agent and centrally for the whole simulated scenario. After each mission run, these measurement values are aggregated according to the previously defined MoE criteria for benefit and effort (including cost). Then, the mission assessment is calculated by one of the provided aggregation methods using the weights which have previously been derived by FAHP in a separate graphic user interface (GUI).

Metrics

The criteria, which must relate to the mission goals and ideally should be connected to the expected strengths and weaknesses of the force composition are quantified by the application of metrics to the logged data. For this experiment, we enhanced our previous hierarchical criteria tree [7, 11] by some lower-level criteria, mainly by more detailed metrics for detectability and target detection. The applied metrics are presented in Table 4.

For the IMINT vignette, the main MoE concerns reconnaissance. This is quantified by the sub-criteria of the ratio of area covered by the sensor footprints versus the total operation area, but also by the number of unique ground targets detected, recognized, and identified according to the aforementioned Johnson criteria (while contrast is assumed to be constant over the mission area).

Effort is quantified by monetary cost, e.g., from fuel and ammunition used. But also the data rate required to transmit the generated sensor images is now considered as an effort item. It also represents the amount of data storage needed.

Time plays a significant role for the assessment of a mission's execution. Considered aspects are the total mission duration and the cumulated operating time of the individual force components. Furthermore, time under threat means the length of time, during which some UAV was within effector range of any opponent. Detectability time was chosen as measure of survivability (which can be decomposed into detectability i.e. signature, susceptibility, and vulnerability [30]) over the whole mission. It sums up all time steps, when any UAV was detectable (over a probability threshold) by either radar or acoustically at the predefined observation points.

Lastly, experience gained by involved personnel in planning, operating, and maintenance as a result of the mission is seen as a benefit. For a simple approach to quantify the typical increase in productivity due to experience gained based on mission time see [11].

Table 4. MoEs for the IMINT vignette, with their criteria and sub-criteria: Global weight indicates the priority of the respective criterion for the total mission effectiveness assessment. Values are rounded to four significant digits.

MoE	Criteria and sub-criteria		Global weight	Normalization
Reconnaissance	Total area surveyed (completion ratio)		0.2294	Mission area
	Number of detected targets		0.1147	Total targets
	Number of recognized targets		0.1147	Total targets
	Number of identified targets		0.2294	Total targets
Effort	Communications	Required data rate	0.0136	Maximum variant
	Monetary cost	Cost of lost UAVs	0.1220	Maximum possible
		Maintenance cost		
		Operational cost		
		Fuel used		
		Ammunition used		
		Flares/chaff used		
Time	Mission time		0.0095	Maximum variant
	Operating time		0.0068	Mission time
	Time under threat		0.0794	Mission time
	Detectability time	Acoustic (time over threshold)	0.0143	Mission time
		Radar (time over threshold)	0.0143	
Experience gained	Mission time-based		0.0520	Maximum variant
Side effects	Collateral damage		0	Maximum variant
	Accidental benefits			

More metrics have been taken into account as candidates, but not deemed useful here. For instance, the average distance between the UAVs is not relevant, because the path planning generates almost independent paths without formations. Effect, which would be quantified by missiles launched, missile hits, and number of destroyed targets, is not permitted in the IMINT vignette. Also, no collateral damage nor positive side effects are possible with unarmed UAV in this scenario, so the weight of the side effects criteria has been set to zero.

Some MoP or intermediate measures were tried out for the extended criteria tree, but not used in the overall operational MoE, as the information is either irrelevant for mission success or already included in other measures. Among these are: communications reliability in terms of number of received and sent broadcasts, number of evasions versus another friendly UAV, number of evasions from a surface-to-air threat, number of online re-plans, peak cumulative loudness, and cumulative reflected radar power for all aircraft over the mission.

Display of Logged Data and Aggregation
Dashboards with all relevant diagrams, e.g., observability over time, communications, and cost, can be plotted for every mission run for each agent and as a synopsis for the whole force package. These are important for presenting an overview to SMEs, enabling them to validate the simulation and the assessment results. Additionally, the MCDA results at each criteria breakdown node can be shown graphically as spider plots (as seen in [11]).

The MCDA methods HWS, MAUT, TOPSIS, and PROMETHEE were implemented for aggregation of the measurement results. The importance of the respective criteria normalized over the total mission are given as global weights in Table 4, as derived by FAHP.

HWS was selected for use in this experiment because it requires the least stakeholder input and SMEs can most easily retrace the calculations. Also is has the lowest computation time when all mission runs (variants and replications) are compared separately. Alternatively, measurements could also be averaged over the replications per variant.

3.5 Reliability and Availability Models

In a separate module from the mission-based capability assessment, reliability and availability probabilities can be calculated. Here, for each UAV (i.e. component of the force composition) failure rate is modeled as a piecewise function, i.e. a segmented distribution, because it is seen as inappropriate to use the same analytical function over the whole life cycle [31].

3.6 Regression Analysis

In order to quickly estimate mission effectiveness without simulation, a simplified meta-model was to be derived by regression analysis. The properties of the force compositions and the plans of action, namely number of UAVs, cruise altitude, and number of pixels, were used as predictors. The overall assessment result, the SPP, from all replications is the response variable.

Inputs were used from all replications. Additionally, SPP was set to zero for zero UAVs (with all other parameter variants), as the mission cannot be completed without using any UAV.

4 Results

4.1 Expert Validation of the Approach

After watching the replays and examining the dashboard data presentation of the previous air-to-ground operation simulation [11], consensus among the SMEs yielded that the generated assessments of the mission runs accurately reflect the mission effectiveness of the simulated force compositions.

Figure 5 gives the SME Likert scale ratings of the evaluation statements in Table 1, which also provides an indication whether the SMEs' answers can be regarded as favorable, neutral, or disapproving towards our approach. Overall, the approach was rated quite favorably. However, the realism of the simulated mission executions was seen as insufficient. This also resulted in partial disagreement with the assessment results, even though the experts stated that the generated ranking appears correct. The SMEs also agreed that the overall method makes sense and covers more information depth than their previous approaches. These criticisms against the simulation, which deemed increased model fidelity necessary and highlight the importance of validating the models and simulation themselves, were addressed in the updated metrics testbed (see Sect. 3.3) before the simulation experiments described in this paper were conducted.

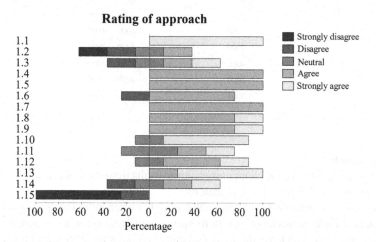

Fig. 5. Rating of statements in Table 1 by SMEs on five-point Likert scale.

The SMEs overwhelmingly stated that they preferred the new method over their existing approaches, which are usually less structured and more based on intuition as we have previously shown [7], and most experts would want to use it as element of future analyses. Some of these SMEs suggested developing a simplified metamodel, that would be useful for generating effectiveness estimates faster (see statement 1.14).

The median predicted ranking order by the SMEs slightly differed from the assessment outcomes. The resulting best two alternatives were in agreement with the prediction, but the last-ranked force packages were interchanged. The prediction also had a significant spread among the individual SMEs, whereas after viewing recordings of the simulations, the dashboards and assessment result diagrams, all SMEs unanimously agreed with the ranking generated by the presented method.

4.2 Simulation Outcomes

Table 5 gives the eight simulation variants (conducted after the upgrades mentioned in Sects. 3.3 and 4.1) with their differing input parameters and their effectiveness ranking by SPP computed via HWS after the replication data was averaged by arithmetic mean. Variant 8 with the highest number of UAVs, the higher planned cruise altitude and the higher sensor resolution was ranked best, followed by variant 6, i.e., the same force composition but with lower flight altitude.

Table 5. Presented variants and resulting effectiveness ranks from aggregation by HWS (data averaged over replications by arithmetic mean)

Variant	Number UAVs	Cruise altitude [m]	Sensor [pixel]	HWS rank
1	3	4000	4000 × 3000	8
2	3	4000	6000 × 4500	5
3	3	5000	4000 × 3000	7
4	3	5000	6000 × 4500	3
5	5	4000	4000 × 3000	6
6	5	4000	6000 × 4500	2
7	5	5000	4000 × 3000	4
8	5	5000	6000 × 4500	1

Criteria Results

Figure 6 gives boxplots of the elementary criteria results from the eight variants including the spread and median from the replications.

Obviously, the cumulative operating time is higher when more UAVs are used. However, higher altitude means larger sensor footprint, so the mission area is scanned in lower time and some UAVs can leave the area of operation earlier after reaching their respective last waypoint.

Time under threat reflects on the waypoint planning (which induces slight differences between the replications) and evasion of detected surface-to-air threats. Overall, a lower number of UAVs tends to lead to less cumulative time during which a UAV is too close to such surface-to-air threat. No UAV was actually shot at, though.

The number of targets detected is not independent from the surveyed portion of the mission area, so more numerous force compositions have an advantage here, while there is also a slight benefit from higher cruise level.

GSD difference between cruise altitudes plays no actual role here, but clearly the lower resolution optical sensor prevents alternatives 1, 3, 4 and 7 from crossing the Johnson criteria threshold for target recognition. Thus, the reconnaissance MoE of the force compositions with higher resolution sensors is higher than for their counterparts with the same number of UAVs and same cruise altitude.

Unsurprisingly for the low distance to the radar ground station, all variants are detectable by radar over the whole mission time. The acoustic detectability at the MAN-PADS locations, however, shows that the package of five UAVs will always be audible during the full mission, but for three UAVs, especially at higher altitudes (variants 3 and 4), there are some time steps when the total loudness at the observation points does not lead to detectability.

The differences in monetary cost can be attributed mainly to the number of UAVs, and to a much smaller extent to the difference in planned cruise altitude leading to variation in fuel consumption. The required network data rate for a ground control station to receive live imagery reflects the sensor resolution and the number of sensors, i.e., the number of UAVs.

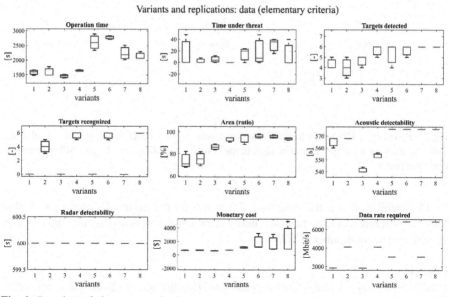

Fig. 6. Boxplots of elementary criteria measurements over the eight variants (also see Table 5) showing their spread and median in the respective replications.

MoE Results

Figure 7 then shows boxplots of the higher-level MoE results (where higher SPP percentage is always better) of the eight variants listed in Table 5, again with some spread over the replications per variant.

The respective mean overall SPPs in the vignette give the ranking of the alternative variants. Main factor is the reconnaissance MoE, which ranks variants 6 and 8 the highest due to the ratio of area scanned and number of correctly recognized ground targets. Variant 6 with the lower cruise altitude has a wider spread in this metric, however.

Effort shows almost no spread over the replications. The alternatives with highest number of UAVs and highest sensor resolution rank last at this MoE, as already indicated in Fig. 6 regarding monetary cost and required data rate.

The time MoE shows rather wide spreads in most variants. The indicated median values tend to show better SPP for higher number of UAVs.

Experience gained directly relates to cumulative operating time as suggested in [11], and thus is obviously higher for greater number of involved UAVs.

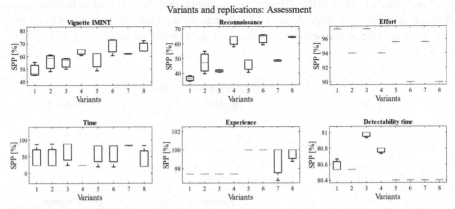

Fig. 7. Boxplots of HWS-based effectiveness assessment results over the eight variants (also see Table 5) showing their spread and median in the respective replications. Higher SPP percentage is always better.

Detectability time is the same over all variants for the radar sub-criterion, so in this MoE the difference in acoustic detectability shows, preferring the combination of higher altitude and lower number of UAVs.

Efficiency

Finally, efficiency was calculated according to its definition as effectiveness related to cost:

$$E = \frac{SPP}{cost} \tag{1}$$

The results are again given as boxplots in Fig. 8. Clearly, the variants with only three UAVs are more cost efficient. If efficiency is calculated by relating SPP to generalized effort (in terms of $100\% - SPP_{effort}$) this general finding does not change.

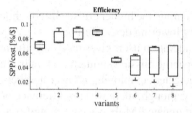

Fig. 8. Efficiency results of the eight variants with boxplots showing the spread and median

As a caveat: efficiency results depend on whether the required data rate criterion is included, and if so, also on its normalization. Considering only monetary cost yields a significantly different ranking than when relating SPP to generalized effort but can be accepted if the network is assumed to provide the bandwidth anyways.

4.3 Metamodel

Stepwise least squares regression analysis with a linear-quadratic model accepting terms based on a significance threshold at $p \leq 0.05$ gives the analytic estimation formula for mission effectiveness:

$$SPP = 32.705 \cdot n_{UAVs} - 4.1697 \cdot n^2_{UAVs} + 3.6843e^{-15} \qquad (2)$$

This model does not use any terms considering cruise altitude, nor sensor resolution. However, it yields the expected diminishing returns from increasing number of UAVs. Also, it predicts a tipping point, where more UAVs actually lower mission effectiveness.

5   Discussion

5.1   Expert Validation of the Approach

Expert judgment shows the validity of our approach shown in Fig. 1. It points out the generated assessments are overall congruent with the SMEs' opinions on the mission executions. The feedback also highlights the advantage of a structured and transparent mission-based method over any previous approaches at assessing aerial systems of systems. Criticism against the testbed simulation have already been addressed by enhancing the model fidelities.

5.2   Enhanced Simulation

Force Compositions in the Specific Vignette
Main measure in the specific IMINT vignette is, of course, the reconnaissance MoE. The given results indicate that equipping the UAVs with the lower resolution electrooptical sensor should be avoided. An even lower cruise altitude could compensate for that by increasing GSD and in turn target recognition probabilities, but at the same time would decrease the footprint area, leading to increased mission time.

In terms of mission planning, the higher cruise altitude should be preferred, reducing operating time and slightly lowering detectability.

The detectability consideration itself gives more specific information than what had been available without a specific mission simulation. It indicates that if there is a priority on being acoustically undetected, employing a lower number of UAVs is preferable. In terms of radar detectability, though, there was no difference between the eight variants.

In terms of overall aggregated MoE, variants 8 and 6 are judged to be the most effective (in both cases of considering either the median or the arithmetic mean over the replications). The efficiency consideration, however, can change the ranking considerably. Therefore, it is of utmost importance for stakeholders to clarify which effort and/or cost items should be factored in.

Method-Related Findings

Overall, the enhanced metrics testbed delivers more accurate and realistic agent behavior than before, with higher model fidelity. It allows for looping over variants and replications, and automatically applying various MCDA methods to the extensively logged data. Logged data and results can be displayed graphically and sensitivity analysis can be conducted.

The metrics application to the IMINT vignette confirms that many, sometimes contradictory, criteria are necessary to accurately judge the success of a mission. The criteria can also be fused to overall mission MoE in a still traceable and comprehensible way. Furthermore, vignette definition and variation over mission plans significantly impact the result and are required for assessing force compositions' capabilities.

5.3 Metamodel Derivation

The attempt at deriving a useful metamodel for predicting mission effectiveness for the IMINT vignette was not quite successful. A linear or linear-quadratic regression model is obviously too simplistic for directly relating force composition input to mission-level MoE output, due to the inherent complexity of a cooperative mission, which cannot really be expected to observe this type of behavior.

The given model considers the number of UAVs, but not the sensor resolution nor the planned cruise altitude, even though it clearly can be seen from Fig. 6 and Fig. 7 that these indeed play a noticeable role. Thus, the regression model cannot be considered adequate.

A flight altitude term should yield some tipping point related to GSD and the operational ceiling of the aircraft, whereas a sensor resolution term should have diminishing returns.

6 Conclusion and Future Research

Our proposed concept of metric derivation, simulation, and measurement for the assessment of mission effectiveness and efficiency of multiple cooperating aircraft was validated for single mission executions of different multi-UAV systems by SMEs, who previously had derived mission criteria and corresponding weights. Overall, it was evaluated as suitable and more complete than previous approaches of assessing aerial systems-of-systems. Improvements were implemented according to SME feedback.

Then, the full concept was executed for an expert-approved IMINT mission vignette and preselected force compositions over a range of variants and replications. Meaningful metrics for detectability (as part of survivability) and electrooptical detection were implemented. This enabled comparison between the force compositions in terms of mission effectiveness and efficiency, indicating which of the considered force packages and mission plans should be chosen.

SME evaluation (like in Sects. 3.1 and 4.1) of the results of this simulation experiment shall be conducted. Additionally, a more in-depth design of experiments concept is still necessary to ensure all relevant measures are picked up, while computation time is minimized. Also, a refined concept to ensure the validity of mission execution plans/CONOPS might be required. In future research, further vignettes and different and more complex force packages will be investigated with the aim of making predictions about future cooperative airborne systems. More relevant criteria and MoE for the assessment of cooperating aerial vehicles must be derived for these missions. Their sensitivity and full value range will be explored. One interesting additional criterion that must be discussed is planning effort. It is expected to increase dramatically when using larger and/or heterogenous force packages, at one point perhaps leading to smaller or less complex force compositions being more effective or efficient than larger ones.

Regarding regression analysis, better metamodel derivation might not only require more data points, but rather a more involved approach, meaning different types of functions. Also, machine learning, e.g. training a neural network, could be a feasible approach.

References

1. Morawietz, S., Strohal, M., Stütz, P.: A decision support system for the mission-based evaluation of aerial platforms: advancements and final validation results. In: 18th AIAA Aviation Technology, Integration, and Operations Conference 2018, Atlanta, Georgia, USA, 25–29 June 2018: Held at the AIAA Aviation Forum 2018. Curran Associates Inc., Red Hook (2018). https://doi.org/10.2514/6.2018-3975

2. Morawietz, S., Strohal, M., Stütz, P.: A mission-based approach for the holistic evaluation of aerial platforms: implementation and proof of concept. In: 18th AIAA/ISSMO Multidisciplinary Analysis and Optimization Conference 2017, Denver, Colorado, USA, 5–9 June 2017: Held at the AIAA Aviation Forum 2017. Curran Associates Inc, Red Hook (2017). https://doi.org/10.2514/6.2017-4152

3. Youngling, E.W., Levine, S.H., Mocharnuk, J.B., Weston, L.M.: Feasibility study to predict combat effectiveness for selected military roles: fighter pilot effectiveness, Accession Number: ADA041650. McDonnell Douglas Astronautics, St. Louis (1977). https://apps.dtic.mil/sti/citations/ADA041650. Accessed 20 Apr 2021

4. Portrey, A.M., Schreiber, B.T., Bennett Jr, W.: The pairwise escape-G metric: a measure of air combat maneuvering performance. In: 2005 Proceedings of the Winter Simulation Conference, Orlando, FL, USA, 4 December 2005, pp. 1101–1108. IEEE (2005). https://doi.org/10.1109/WSC.2005.1574365

5. Schreiber, B.T., Stock, W.A., Bennett Jr, W.: Distributed mission operations within-simulator training effectiveness baseline study: metric development and objectively quantifying the

degree of learning. AFRL-HE-AZ-TR-2006–0015-Vol II. Air Force Research Laboratory, Mesa, Arizona (2006)

6. Kelly, M.J.: Performance measurement during simulated air-to-air combat. Hum. Factors **30**(4), 495–506 (1988)

7. Seethaler, J., Strohal, M., Stütz, P.: Finding metrics for combat aircraft mission efficiency: an AHP-based approach. In: Deutscher Luft-und Raumfahrtkongress 2020. Deutsche Gesellschaft für Luft-und Raumfahrt - Lilienthal-Oberth e.V., Bonn (2020). https://doi.org/10.25967/530013

8. Pettersen, S.S., Fagerholt, K., Asbjørnslett, B.E.: Evaluating fleet effectiveness in tactical emergency response missions using a maximal covering formulation. Nav. Eng. J. **131**, 65–82 (2019)

9. Habayeb, A.R.: Systems Effectiveness. Elsevier Science, Burlington (1987)

10. Buscher, U., Wels, A., Franke, R.: Kritische Analyse der Eignung des Fuzzy-AHP zur Lieferantenauswahl. In: Bogaschewsky, R., Eßig, M., Lasch, R., Stölzle, W. (eds.) Supply Management Research. Aktuelle Forschungsergebnisse 2010, [Tagungsband des wissenschaftlichen Symposiums Supply Management; Advanced studies in supply management, Bd. 3], 1st edn, vol. 140, pp. 27–60. Gabler, Wiesbaden (2010)

11. Seethaler, J., Strohal, M., Stütz, P.: Multi-UAV mission efficiency: first results in an agent-based simulation. In: Mazal, J., Fagiolini, A., Vasik, P., Turi, M. (eds.) MESAS 2020. LNCS, vol. 12619, pp. 169–188. Springer, Cham (2021). https://doi.org/10.1007/978-3-030-70740-8_11

12. Morawietz, S.: Konzipierung und Umsetzung eines Unterstützungssystems zur vergleichenden Bewertung von Luftfahrzeugen. Dissertation, Universität der Bundeswehr München (2018)

13. Ruhland, A.: Entscheidungsunterstützung zur Auswahl von Verfahren der Trinkwasseraufbereitung an den Beispielen Arsenentfernung und zentrale Enthärtung. Dissertation, Technische Universität Berlin (2004)

14. Whitcomb, C.A.: Naval ship design philosophy implementation. Nav. Eng. J. **110**(1), 49–63 (1998). https://doi.org/10.1111/j.1559-3584.1998.tb02385.x

15. Jain, V., Sangaiah, A.K., Sakhuja, S., Thoduka, N., Aggarwal, R.: Supplier selection using fuzzy AHP and TOPSIS: a case study in the Indian automotive industry. Neural Comput. Appl. **29**(7), 555–564 (2016). https://doi.org/10.1007/s00521-016-2533-z

16. Brans, J.P., Vincke, P., Mareschal, B.: How to select and how to rank projects: the PROMETHEE method. Eur. J. Oper. Res. **24**, 228–238 (1986)

17. Zwicky, F.: The morphological approach to discovery, invention, research and construction. In: Zwicky, F., Wilson, A.G. (eds.) New Methods of Thought and Procedure: Contributions to the Symposium on Methodologies, pp. 273–297. Springer, Heidelberg (1967). https://doi.org/10.1007/978-3-642-87617-2_14

18. Ritchey, T.: Fritz zwicky, morphologie and policy analysis (1998)

19. Likert, R.: A technique for the measurement of attitudes. Arch. Psychol. **22**(140), 55 (1932)

20. Office of the Secretary of Defense: Unmanned Aircraft Systems Roadmap 2005–2030, Washington, D.C (2005). https://fas.org/irp/program/collect/uav_roadmap2005.pdf. Accessed 18 May 2021

21. Stodola, P., Drozd, J., Mazal, J., Hodický, J., Procházka, D.: Cooperative unmanned aerial system reconnaissance in a complex urban environment and uneven terrain. Sensors **19**(17), 3754 (2019). https://doi.org/10.3390/s19173754

22. Stodola, P., Drozd, J., Nohel, J., Hodický, J., Procházka, D.: Trajectory optimization in a cooperative aerial reconnaissance model. Sensors **19**(12), 2823 (2019). https://doi.org/10.3390/s19122823

23. Ögren, P.: Increasing modularity of UAV control systems using computer game behavior trees. In: Guidance, Navigation, and Control and Co-Located Conferences. AIAA Guidance, Navigation, and Control Conference, Minneapolis, Minnesota (2012). https://doi.org/10.2514/6. 2012-4458

24. Faconti, D.: BehaviorTree/Groot (2018–2021). https://github.com/BehaviorTree/Groot. Accessed 29 July 2021

25. The Mathworks Inc.: MATLAB. Version 2020a, Natick, Massachusetts (2020–2021)

26. Quigley, M., et al.: ROS: an open-source robot operating system. In: ICRA Workshop on Open Source Software, vol. 3 (2009)

27. Raia, F.: Tacview - The Universal Flight Data Analysis Tool (2021). https://www.tacview. net/. Accessed 29 July 2021

28. DIN Deutsches Institut für Normung e.V.: Acoustics—normal equal-loudness-level contours (ISO 226:2003). Beuth, Berlin (2006). ICS 13.140 (DIN ISO 226:2006-04). https://www.iso. org/standard/34222.html

29. Harney, R.C.: Combat Systems. Volume 1. Sensor Elements. Part I. Sensor Functional Characteristics, Monterey, California (2004)

30. Ball, R.E.: Fundamentals of Aircraft Combat Survivability. Analysis and Design. American Institute of Aeronautics and Astronautics, Reston (1985)

31. Li, H., Zuo, H., Su, Y., Xu, J., Yin, Y.: Study on segmented distribution for reliability evaluation. Chin. J. Aeronaut. **30**(1), 310–329 (2017). https://doi.org/10.1016/j.cja.2016. 12.008

Modelling and Simulation for Behavioral Codes of Robotics and Autonomous Systems

Salvatore De Mattia[✉]

NATO Modelling and Simulation Centre of Excellence, Piazza R. Villoresi 1, 00143 Rome, RM, Italy

mscoe.cde04@smd.difesa.it

http://www.mscoe.org

Abstract. The employment of Robotics and Autonomous Systems (RAS) and robotic swarm in the future military operational environment is going to be one of the main challenges to modern warfare. Modelling and Simulation (M&S) technology allows to support Concept Development and Experimentation (CD&E) activities to develop new robotic autonomous capabilities, like robotic platforms, human-machine teaming interactions, Command and Control of robots, Artificial Intelligence for mission planning, decision support and self-training of robots and robot swarms. NATO M&S Centre of Excellence (M&S CoE) supports in this field NATO, national initiatives, and technical activities with particular engagements with the Allied Command for Transformation and the Science and Technology Organization. M&S COE has been developing the Research on Robotics for Concepts and Capability Development (R2CD2) project since 2016 to deliver in three annual phases an open, scalable, modular, standard-based prototypical architecture of M&S tools for experimentation on RAS and Robotic Swarms.

The latest upgrade of this project is called R2CD2-EVO, which allows studying, analyzing and counter-measuring RAS systems and swarms, highlighting a behavioral script describing the process reacting to external stimulus from modelled sensors, according to mission tasks and dynamics attitude of the simulated system. One of the evolution of this platform could be the creation of a synthetic environment in order to validate and test attitude algorithms developed for a RAS system, in a military and complex scenario. This capability extends the wargaming concepts to RAS, acting as a training test arena, not only for standard training of personnel employing RAS systems during a military mission, but also as attitudinal training algorithms projected to capability development. Thanks to the boundary conditions, modelling it is possible to perform a multi-domain analysis, such as electromagnetic, cyber and machine decisional effects. As final results of these research and development activities, the R2CD2 EVO prototype is now proposed as the cornerstone for the development of a RAS Synthetic Environment Architecture to support Concept Development, Experimentation, Training and Exercise activities on Unmanned Autonomous multi domain Systems (UAxS) both for NATO and Nations partnering the M&S Centre of Excellence.

Keywords: Behavioral codes · RAS

© Springer Nature Switzerland AG 2022
J. Mazal et al. (Eds.): MESAS 2021, LNCS 13207, pp. 180–190, 2022.
https://doi.org/10.1007/978-3-030-98260-7_11

1   Synthetic Environment for Robotic Autonomous System (RAS)

1.1   Introduction

The use of Robotics and Autonomous Systems (RAS) and robotic swarms will be one of the main challenges in the future operating environment [5].

Modeling and Simulation (M&S) allows supporting Concept Development and Experimentation (CD&E) activities, in order to investigate new technologies, such as autonomous robotic capabilities and define human-machine interactions linked to Command and Control (C2) of robots [6]. In this field, it is also important to study the artificial intelligence for planning and execution of mission, in addition to the decision support and the self-learning of robotic systems. NATO M&S COE has been developing the Research on Robotics for Concepts and Capability Development (R2CD2) project since 2016 to provide an open, scalable, modular and standards-based prototype architecture in three annual phases, for the testing of robotic platforms in military operations with related countermeasures [9].

This capability was also designed to contribute to the testing of the C2SIM [8] interoperability language within the MSG-145 (Modelling and Simulation Group). In detail, the Center focused on the aspects concerning the extension of the C2SIM standard to Unmanned Autonomous Systems (UAxS), in order to obtain an exchange of orders and reports between C2 and simulation systems. R2CD2 is an open M&S platform initially designed for CD&E activities for the C2 of simulated UAxS. The prototype is continuously updated to extend its possible applications [5].

R2CD2 is based on a federated simulation, allowing implementing interoperability between different tools, using the HLA (High Level Architecture – IEEE 1516) protocol, in order to combine the functional features of each simulator [1].

In the Fig. 1, an overall picture of the R2CD2-EVO simulation is reported, as an open, modular, scalable and standard-based platform, where both tools and RTI (Runtime Infrastructure) connection involved are highlighted.

The evolution of this project (R2CD2-EVO) was addressed in a study aimed at the capacitive development of RAS platforms, proposing a synthetic environment to carry out experimentation on autonomous robotic systems to be used in military operations [4], performing different simulations of the scenario, with the aim of adopting analytical considerations in relation to the stochastic results obtained. The complete architecture includes, in addition to the M&S tools, the integration of cyber effects to contextualize both the use of autonomous robotic systems in the cyber layer and the wargaming capability applied in the robotic field with specific applications (MASA - SWORD) for the analysis of the Course of Action (CoA). These capabilities are going to be implemented in order to enhance the application and versatility of this platform, regarding concept and capacity development.

In this regard, M&S was able to allow the development of the behavioral codes of the unmanned platforms modeled within the simulators, in order to structure specific attitudes linked to particular mission execution characteristics. Behavioral codes provide autonomy to unmanned systems, which must be able to adapt, with different reactive actions, to the surrounding conditions of the scenario. The triggers for the attitude engine consist of data detected by sensors modeled and placed on board each platform, with

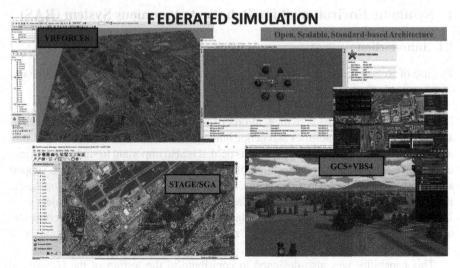

Fig. 1. Simulation of the R2CD2-EVO project.

different functional performances, to contextualize the individual operational actions at the technological level defined for each entity within the synthetic environment.

The project conceived as part of the capacitive development of autonomous robotic systems aims to verify and validate the attitude models implemented on the platforms, through the aid of M&S, creating a cyclical synergy between the simulated and the real world, would allow to outline the peculiar and doctrinal aspects related to the use of this and to adapt the autonomous actions of the platforms in diversified technical, operational and information contexts, allowing to quantify the impact of autonomous systems on military operations and, subsequently, to study and implement respective countermeasures technological and tactical-procedural.

NATO M&S COE organized a practical demonstration of this synthetic environment for the Italian Army General Staff, in the context of RAS experimentation campaign, highlighting the aspects strictly connected to the development of behavioral codes for autonomous robotic platforms inside simulators, without prejudice to the additional applications of this architecture and its versatility of use.

1.2 Network Architecture

The synthetic environment developed for the RAS experimentation is based on the use of different federated simulators in connection with a specific communication protocol.

A HLA federation takes place through the activation of a middleware, called RTI (Run Time Infrastructure) which essentially acts as a communication bus to each simulator connected, able to share the defined attributes. In order to configure the simulators for connection to RTI [1], it is necessary to set the main parameters characterizing the network within configuration files defining HLA standard.

For the architecture of synthetic architecture for RAS project, the following federated simulators are used, as reported in Fig. 2:

- VT MAK - VR Forces (constructive simulator);
- C2SIM client-server communication emulating communication with a SISO-compliant protocol;
- Presagis – STAGE 16 (constructive simulator);
- Bohemia Interactive Simulations - VBS4 (constructive/virtual simulator).

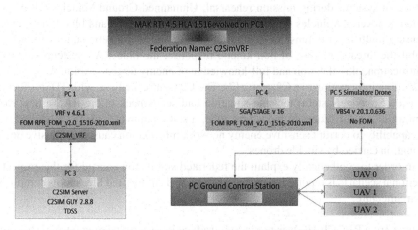

Fig. 2. HLA federated simulation in the R2CD2-EVO project.

The simulators publish and subscribe entities and events according to the HLA standard: the sharing of information, in addition to a correspondence of FOM (Federate Object Model) to define the structure of the incoming and outgoing information on each simulator, also requires a mapping activity to allow a consistent representation of the events that occur on all the entities of the federation.

In this operational scenario, different RAS entities with different levels of autonomy have been described. The attitudinal autonomy modeled within the synthetic environment is implemented in the form of a functional algorithm, developing a code (LUA), aimed at describing the processes constituting the behavioral mission of each system.

1.3 R2CD2-EVO Simulation

This project is mainly based on a federated simulation, which was build using different modelled platforms, both human and robotic. This aspect firstly demonstrated the interoperability between different simulators (constructive and virtual) using HLA1516 evolved standard for data exchanging and compatibility with C2SIM, being some units directly driven using C2SIM protocol. The upgrade of the R2CD2-EVO project implements virtual simulation (VBS 4) and a Ground Control Station (GCS) for real and simulated Unmanned Aerial Systems (UAS), demonstrating also hardware/software in the loop.

The platform data-setting scenario was developed implementing an urban environment based on a future (2035) mega-city model (Archaria 5 × 5 km synthetic terrain) [2].

New swarms behaviors were also implemented in the constructive simulation, through LUA scripting language. The friendly and enemy forces were modeled as UAS with different types of sensors (e.g. optics, infrared and radio frequency transceiver), in order to manage the external inputs given by the environment and triggering specific behavioral scripts developed inside the decisional engine of simulated platforms. In particular, different levels of autonomy were implemented in order to verify the reacting attitude of systems during mission rehearsal. Unmanned Ground Vehicles (UGV) and Unmanned Aerial Vehicles (UAV) were employed in either red and blue forces, experimenting multi-layer defense or exploiting kinetic and non-kinetic actions in order to inhibit the threats. In case of non-kinetic response, the blue RAS systems (after the identification, classification and full-knowledge of enemy targets) exploit sensor fusion forwarding information to a C2 center [7]. The C2 Centre analyzes with TDSS (Tactical Decision Support System) tool the situation and sends back to the RAS a new route pointing to a safety zone. In the safety zone, a static jammer device is modelled with the capability to corrupt sensitive enemy network information causing explosive device detonation carried by suicide drones.

In order to synthetically explain the federated simulation created for this project, a functional scenario is reported, focusing on modelled entities and M&S tools composing the architecture:

- Stage 16: a RECCE UGV is going to introduce in the critical area (airport), in order to gather information for triggering next attack;
- VRForces: UGV introduced inside airport area recalls 3 RAS reconnaissance drones on a specific route for the engagement and killing of enemy aircraft systems;
- Stage 16: an ASIO UAV is setting for a reconnaissance mission inside the critical area (airport). Once it has identified a potential threat (unauthorized UGV), it begins to follow it at a certain distance, escorting it. UAV ASIO communicates to the C2 station to report the identification of the potential threat;
- VBS4: C2 simulated to provide an order to 2 drones (a decoy drone – DJI Phantom – and a streamer drone – Black Hornet), which fly according to an established route according to waypoints, programmed in the pre-mission phase through the simulated real GCS. The area to be covered is along the access point to the critical area (airport). At the same time, an order is given to infantry team (quick response force), that engages the terrestrial robotic threat (UGV) with kinetic activity;
- C2SIM: C2 send an order (using C2SIM protocol) to a swarm composed of 3 RAS UAV, modelled with a medium level of autonomy (LoA = 3). These drones carry out a reconnaissance throughout the airport area to identify additional enemy platforms (the optimized routes for the swarm are calculated using TDSS tool interfaced with C2SIM Gui);
- VBS4: the drone that is streaming a real video is flying along the same path of the decoy one, following it from a higher altitude. This is useful to resume the interaction of the enemy drones with a friendly platform. The video is used to study the enemy TTPs put in place by the attacked an to plan further countermeasures;

- C2SIM: RAS UAV swarm constantly sends reports (C2SIM) to provide the C2 with situational awareness and send the positions and status of enemy platforms detected during the reconnaissance mission;
- VRForces: following the data provided by the sensors on board UAVs and the analysis carried out through the streaming video, it is shown that the enemy RAS UAVs have an autonomous suicide behavior. It means that they are able to automatically direct to the potential threats acquired by sensors and activate an on board IED (Improvised Explosive Device) through a proximity threshold. Once the blue force RAS UAVs recognized threats (red forces RAS suicide UAVs), they automatically switch to another behavior, with respect to default one (reconnaissance mission), in order to react and to countermeasure the threat. The behavior adopted consists in pointing to a specific area where a static jammer tower is modelled, which transmits a radio frequency signal designed to inhibit the sensors of the enemy RAS. Once in the jamming area, the red RAS UAVs are programmed in order to create much damage as possible, trying to hit the target drones, according to last collected position by sensor and adjusts it through trigonometric formulas. For this reason, the suicide drones will self-destruct trying to involve blue RAS UAVs in the explosion.

The results of this simulation are important in order to understand how to use the Modelling & Simulation as a development arena to design, verify and validate the autonomous missions of different RAS platforms in the military operations. The stochastic outcomes provided by R2CD2-EVO show that, before focusing on autonomous mission of the RAS platforms, it is meaningful to fully characterize the physical features of the robotic entities, especially concern the sensors and the dynamics. The LUA scripting language allows creating different behavioral codes for RAS entities and enhancing the autonomy capabilities performances through multiple simulations.

1.4 Future Considerations and Proposals

The proposed platform represents a synthetic environment able to verify, validate and test the use of RAS in military operations of different types. The RAS M&S would allow, not only developing innovative concepts of high technological depth, but also to determine the peculiar doctrinal elements concerning the use of RAS systems within the military missions.

From a technical point of view, a RAS wargaming could be carried out where different teams, for example operating on different platform simulators, could design attitudinal algorithms to be implemented on specific modelled RAS entities, in order to develop and analyze different courses of action in the use of autonomous robotic systems and to improve the attitudinal logic within the systems (Robotic Wargaming). The multiple simulation would allow acquiring multiple operational cases to "train" the RAS platform modeled on a wide spectrum of causality, thus reducing the behavioral error differentials that can be verified in a process algorithm.

The engineering skills applied to the M&S field, (e.g. the characterization and design behavioral codes) would provide a valid and essential aid for a real prototype development. M&S activity for developing technological capabilities is a significantly important process in the context of the transformation of military forces within increasingly complex and multi-factorial scenarios.

The implemented platform could allow developing the concepts and relevant elements of a real RAS system. The development of a real physical object (prototype) could follow a process based on the following steps:

1. Identification of an unmanned system based on an "open" platform in order to connect different sensors (e.g. optical, infrared, radio frequency, radar, acoustic, lidar) to provide greater versatility using the platform;
2. Modelling the unmanned platform in terms of the main physical parameters to derive the complete dynamics of the simulated entity and achieve maximum correspondence with the real platform;
3. Modelling all sensors on unmanned platform to verify their performance effectiveness in case studies built inside the constructive environment;
4. Design behavioral algorithms (e.g. via LUA programming language) with definition of execution priorities, referring to level of autonomy to be achieved on the platform. The designed algorithms are gradually tested within multiple simulations, carrying out a technical validation of what is present in the technical-operational requirements;
5. Transposition of the algorithms within an electronic board (e.g. Field Programmable Gate Array and/or microcontroller) adaptable and consistent with the unmanned platform, in order to provide different RAS capabilities, in terms of attitude processes, to the robotic system. In this context, what is modelled and simulated in a synthetic environment is tested under realistic conditions by carrying out an operational validation of the system.

The multiple passages from the real world to the simulated one are essential for creating a reliable RAS system operating in diversified operating contexts, with the use of feedback for the development, verification and testing of the autonomous behavioral engine of the system.

Sensors are the components that allow the autonomous platform to make the changes from one behavioral state to another, following the finite state machine created inside the autonomous engine. Therefore, their modelling is a fundamental and necessary operation to perform a test on attitudinal algorithms of the autonomous system.

The diagram expressing the proposed process using the M&S tools for capability development is reported in Fig. 3, highlighting the main processes concerning each phase and the related transitions. In this schematic diagram is possible to underline the technical validation and the operational validation of autonomous attitude in a cyclic implementation.

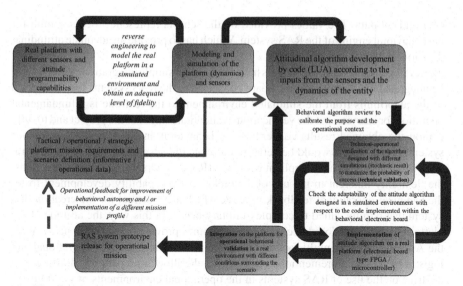

Fig. 3. Schematic diagram of RAS capability development process using M&S tools.

2 Technical Development of RAS Capability

The platform presented and developed in the RAS context allows adoption and endorsement of fundamental procedures for a complex capacity development, where M&S is able to draw up the primary design cornerstones and technical-operational verification/validation.

Through the scenario developed to show the use of RAS systems in military operations, it is possible to define the main points of interest, highlighted on a technological capability that can be used in the complex military scenarios of the future operations. The elements of interest and the perspectives of the RAS project are listed below:

- the modelling of the complex scenario of the future (e.g. Archaria) could facilitate the formulation of adequate technical-operational requirements for the development of a RAS platform. The scenario must be expressed under multiple and diversified layers (tactical, operational, strategic, electromagnetic, cybernetic, social, etc…), with the aim of developing a synthetic environment, where the simulation of the RAS platforms would allow to analyze more realistic behavioral responses;
- a training platform for RAS systems would allow the concept of wargaming to be extended to autonomous systems, in order to study and analyze different courses of action;
- the modelling and simulation of dynamics and physical components of the RAS platform is a complex operation from a technical point of view, due to the high engineering eclecticism required, at different levels of abstraction, in the knowledge of the subsystems suitable for the accomplishment of missions. In particular, the fundamental components to be modelled and simulated include the platform sensors, which will act as the main element of generation and processing of the external

environment data. The values provided by the sensors represent process stimuli for the behavioral engine of the RAS system. Which has to provide an adequate attitudinal decision-making;

- the developed behavioral algorithms have to represent simple integration operations, to be integrated inside an electronic board connected to the RAS system. The translation of the algorithms from the simulated environment to the real one is a fundamental operation for carrying out the verifications/validations stages of the project and to fully exploit the obtained results concerning the behavioral analysis of the autonomous system. In this step, it would be better to correlate the development of the behavioral code within an algorithm compliant with ROS (Robotics Operating System) functions;
- the flexibility inherent in the use of a simulation for capability development foster the technical-operational feedback process, which are essential for the creation of a system that is fully suited to complex military needs. In this way, the attitude of the platform could be tailored according to guidelines provided by operational team of the Armed Forces, as multi-factorial parameters in the information, operational and logistic field. The aforementioned activity would allow a preparatory drafting of a doctrine on the use of RAS systems in the operational environments of the future to support military missions;
- the development of code within the simulators to structure a behavioral algorithm of the RAS system constitutes the basic technical activity of the autonomous platform. This step is closely related to the previous ones and should be contextualized with the level of autonomy to be given to the platform and the type of mission to be summarized within the decision-making processes;
- through artificial intelligence, the attitudinal algorithms would reach a greater level of versatility and functionality, giving a significant and greater impact to the M&S tool to train autonomous platforms within a synthetic environment. In this way, the RAS platform, depending on the different conditions surrounding the scenario, could improve the behavioral responses in multiple external causalities and minimize the percentage of error, with the progressive mitigation of any decision gaps not calculated in the design phase;
- the platform for the simulation of RAS systems would allow to test and adopt counter-measures for robotic systems [3], through the implementation of complex models for the adoption of kinetic and not kinetic activities, able to inhibit the target systems. In this context, the decision-making implemented inside the platforms would require the integration of countermeasure systems within certain attitudinal processes, in order to obtain maximum effectiveness and allow the capabilities of autonomous systems to be extended in a remote and automated way.

In the Fig. 4, a graphical representation of cyclic simulated and real world is reported, starting from a real RAS platform in order to create a synthetic entity of high fidelity and use it in the synthetic environment as a developed arena of physical and behavioral characteristics of autonomous platforms in the military operations. The basic idea to implement this cyclic process lies on an electronic board able to allow the simulated and the real world translation, in terms of feasible synthetic algorithms tested within the constructive environment.

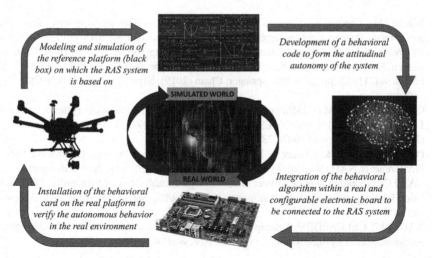

Fig. 4. Graphical representation of cyclic simulated and real world process for RAS development.

3 Conclusion

In conclusion, through behavioral programming in a simulated environment, the starting point for a completely innovative and technological project is provided. The design of an electronic board integrating the behavioral algorithms defined, tested and validated in a simulated environment, interconnected to the logic functions of the autonomous system, would provide a turning point in the sector of robotic autonomous systems. The interconnection and interoperability with the logic processes of the platform, would allow to increase the Level of Autonomy (LoA) of an unmanned reference platform, also COTS (Commercial Off The Shelf) type, with a low/null Level of Autonomy (LoA ≈ 0), providing it some autonomy behaviors. This would allow the creation of an electronic behavioral card, whose code is tested, verified and validated in a synthetic environment, operationally usable and adaptable according to the scenario, operational requirements and TTPs. In this way, the board is able to define a predefined and versatile LoA, according to the mission, to face complex scenarios and domains, where threat is dynamically changing.

References

1. IEEE: 1516-2010-IEEE Standard for Modeling and Simulation (M&S) High Level Architecture. IEEE Standard Association (2010)
2. NATO ACT: NATO Urbanization Project (2015). http://www.act.nato.int/activities/natourban isation-project. Accessed Oct 2019
3. Biagini, M., Corona, F.: Modelling and simulation architecture supporting NATO counter unmanned autonomous system concept development. In: Hodicky, J. (ed.) MESAS 2016. LNCS, vol. 9991, pp. 118–127. Springer, Cham (2016). https://doi.org/10.1007/978-3-319-47605-6_9

4. Biagini, M., Corona, F., Casar, J.: Operational scenario modelling supporting unmanned autonomous systems concept development. In: Mazal, J. (ed.) MESAS 2017. LNCS, vol. 10756, pp. 253–267. Springer, Cham (2018). https://doi.org/10.1007/978-3-319-76072-8_18

5. Biagini, M., Corona, F.: M&S based robot swarms prototype. In: Mazal, J. (ed.) MESAS 2018. LNCS, vol. 11472, pp. 283–301. Springer, Cham (2019). https://doi.org/10.1007/978-3-030-14984-0_22

6. Corona, F., Biagini, M.: C2SIM operationalization extended to autonomous systems. In: Mazal, J., Fagiolini, A., Vasik, P. (eds.) MESAS 2019. LNCS, vol. 11995, pp. 389–408. Springer, Cham (2020). https://doi.org/10.1007/978-3-030-43890-6_32

7. Pullen, J.M., Patel, B., Khimeche, L.: C2-Simulation Interoperability for Operational Hybrid Environments. NATO Modelling and Simulation Symposium 2016. Bucharest, Romania (2016)

8. NATO Collaboration Support Office: MSG 145 Operationalization of Standardized C2 Simulation Interoperability (2019). https://www.sto.nato.int/Pages/activitieslisting.aspx. Accessed Oct 2019

9. NATO ACT CEI CAPDEV: Autonomous Systems Countermeasures (2016). https://www.innovationhub-act.org/project/counter-measures. Accessed Oct 2019

Intersection Focused Situation Coverage-Based Verification and Validation Framework for Autonomous Vehicles Implemented in CARLA

Zaid Tahir[1,2](✉) and Rob Alexander[2]

[1] Assuring Autonomy International Programme, Department of Computer Science, University of York, York, UK
zaidt@bu.edu, zaid.butt.tahir@gmail.com
[2] ECE Department, Boston University, Boston, USA
rob.alexander@york.ac.uk

Abstract. Autonomous Vehicles (AVs) i.e., self-driving cars, operate in a safety-critical domain, since errors in the autonomous driving software can lead to huge losses. Statistically, road intersections which are a part of the AVs operational design domain (ODD), have some of the highest accident rates. Hence, testing AVs to the limits on road intersections and assuring their safety on road intersections is pertinent, and thus the focus of this paper. We present a situation coverage-based (SitCov) AV-testing framework for the verification and validation (V&V) and safety assurance of AVs, developed in an open-source AV simulator named CARLA. The SitCov AV-testing framework focuses on vehicle-to-vehicle (V2V) interaction on a road intersection under different environmental conditions and intersection configuration situations (start/goal locations), using situation coverage criteria for automatic test suite generation for safety assurance of AVs. We have developed an ontology for intersection situations, and used it to generate a situation hyperspace i.e., the space of all possible situations arising from that ontology. For the evaluation of our SitCov AV-testing framework, we have seeded multiple faults in our ego AV, and compared situation coverage-based and random situation generation. We have found that both generation methodologies trigger around the same number of seeded faults, but the situation coverage-based generation tells us a lot more about the weaknesses of the autonomous driving algorithm of our ego AV, especially in edge-cases. Our code is publicly available online and since the simulation software (CARLA) is open-source, anyone can use our SitCov AV-testing framework and use it or build further on top of it. This paper aims to contribute to the domain of V&V and development of AVs, not only from a theoretical point of view, but also from the viewpoint of an open-source software contribution and releasing a flexible/effective tool for V&V and development of AVs.

Keywords: Autonomous driving · Autonomous vehicle · Self-driving car · CARLA · Verification and validation · Safety assurance · Automatic test case generation · Situation coverage · Coverage criteria

© Springer Nature Switzerland AG 2022
J. Mazal et al. (Eds.): MESAS 2021, LNCS 13207, pp. 191–212, 2022.
https://doi.org/10.1007/978-3-030-98260-7_12

1   Introduction

Autonomous vehicles (AVs) are no longer an idea of science fiction, they are being tested and even being used in some cities around the world. As seen in the past that with the adoption of new technologies, new kinds of hazards arise, similarly AVs have brought up a whole new concoction of hazards [1]. Since human lives are dependent when dealing with AVs, this makes AVs a safety-critical system. Hence, the safety of AVs cannot be taken callously.

The Problem of AV Safety Assurance. The safety standards applied to road vehicles currently, such as the ISO/PAS 21448:2019 [2] and ISO 26262 [3], do not translate well to AVs (SAE level 3 or above) [9], due to the fact that a human driver is required to take over in case of an emergency. An emerging standard UL-4600 [4] does not assume human drivers, but it only gives some guidelines to build a valid safety cases for autonomous systems, it is not at all prescriptive and can not be. Recently a new safety standard for automated vehicles has been published, the ISO 22737 [21], but it is quite limited. The ISO 22737 assumes low-speed automated driving, where routes are pre-defined within restricted operational design domains (ODDs).

In order to tackle this issue of safety of AVs, researchers have been employing various strategies, mostly using simulation softwares as a baseline since on-road testing of AVs is quite risky and costly. In [5] the authors attempt to model the distribution of disturbances over failures in a vehicle-to-vehicle (V2) interaction as a safety validation approach, this approach relies on selecting a particular set of disturbances to be injected. Moritz K. et al. [6] proposes automatic critical scenario generation based on minimization of solution space, but this method relies quite a lot on discretization of solution space and constraining the behaviour of the vehicles. Greg C. et al. [7] have proposed an agency-directed approach to test-suite generation for testing AVs, the scenario considered in their experiments is quite simple along with fidelity of simulation being quite low.

The papers mentioned above are quite recent but one thing lacking in these and most of the papers related to safety assurance of AVs is that the fidelity of simulators used for their experimentations, is quite low. Since the fidelity of the simulations is low, realistic camera images from the dash cam of the ego AVs can not be used since that functionality is not available in low fidelity simulators, and doing online-testing [8] of the perception system of AVs is really important while assuring the safety of the AV automated driving algorithm, since many AVs on roads these days rely on camera only (e.g., Tesla AVs). Along with no option of 3D rendering and live feed of dash-cam images being obtained by ego AVs in low fidelity simulators, the physics engine of low fidelity simulators is quite weak as well. The physics engine of such low fidelity simulators usually models just the basic equations of motions while neglecting some basic principles such a road friction, vehicles' tire friction etc.

Exploration & Selection of AV Simulators. With such considerations of limitations of AV simulators in mind, before moving on with the designs of our experiments and developing and testing our situation coverage-based (SitCov) AV-testing framework, we tried and tested a few AV simulators first before selecting CARLA, these are as follows: (1) MATLAB Automated Driving (AD) Toolbox [10]: This tool box has the upside of flexible scenario designing but the downside is that customizing experiments with AV Autonomous Emergency Braking (AEB) activated is not easily doable and there is no AV dash-cam option; (2) CarMaker [11]: This simulator has customizable scenarios with AEBs activated but it uses TCL script which is an outdated programming language and its interfacing with third party softwares is quite tedious and is not opensource; (3) CARLA [12]: Scenario customization is really easy, it has a high fidelity physics engine along with realistic 3D rendering and a long list of available sensors including the AV dash-cam. It uses Python language which is the goto programming language for researchers in various fields of Artificial Intelligence (AI). Also, CARLA is opensource which was one of the main aims of our research, to provide the public an effective AV-testing tool. Hence with all these upsides in mind, we selected CARLA as our simulator. Our code is publicly available here [20].

AV Coverage-Based Testing Methods. Since AVs are a complex integration of systems of systems (SOSs), they face countless hazards due to the huge search space of inputs to the AVs. One methodology from software testing which is employed when testing a huge input space is called coverage criteria [13], which suits testing of AVs as well. Researchers have used the following coverage-based testing approaches for AVs recently [1]: (1) Scenario Coverage [15]; (2) Situation Coverage [14, 33]; (3) Requirements Coverage [14].

Our Papers' Contribution. In this paper we have developed a novel situation coverage-based (SitCov) AV-testing framework for the V&V and safety assurance of AVs and have used situation coverage as the coverage criterion for our automatic test suite generation, and we have come up with a unique/novel derivative of an ontology that we named situation hyperspace for our situation coverage-based situation generation, which we will elaborate in detail in the coming sections.

Situation Coverage. We define our situation coverage methodology as follows: *Situation coverage is a coverage criterion which takes into account the external and internal situations of the autonomous system (AV in our case), the automatic test suite generator (SitCov AV-testing framework in our case) uses the situation coverage metrics to know which situations have already covered and how many times each situation has been generated, based on this information the next batch of situations would be generated by the automatic test suite generator so that more of the situation space is covered and close to uniform distribution is achieved in case of repetitions of situations if all situations have been covered once.*

We would also like to define the term *situation* as *the initial configuration of the input space before the start of the simulation run (the temporal development of scenes).*

Research Questions. In order to quantify the situation space around our ego AV we have developed an ontology and used it to derive the situation hyperspace [16] which we will elaborate further in the next section. In this paper we investigate the following research questions (RQs): (1) RQ1: *How does situation coverage-based test suite generation perform against random generation from a viewpoint of revealing seeded faults?* (2) RQ2: *Does the situation coverage-based test suite generation provide any additional value in terms of the confidence in the safety metrics outputs of our SitCov AV-testing framework?*

We will look to answer these RQs in the subsequent sections. The rest of the paper is divided into the following sections. Section 2 presents the situation hyperspace that is used to quantify the situations around the ego AVs for our SitCov AV-testing framework. Section 3 breaks down the methodology and the development of the SitCov AV-testing framework in CARLA. Section 4 lays out the experimentation results and its in-depth analysis. Section 5 concludes the paper with some ideas for future work to build further on top of this proposed framework.

2 Situation Hyperspace

The literal meaning of hyperspace is "space of more than three dimensions/axis". By situation hyperspace [16], we refer to the multi-dimensional external world around the ego AV. This situation hyperspace has been constructed in a methodical way so that our SitCov AV-testing framework can systematically navigate through it to generate interesting and challenging situations for our ego AV using situation coverage-based generation to make sure good coverage of the situation hyperspace is executed by the SitCov AV-testing framework.

In order to come up with the ontology to derive the situation hyperspace for our framework, we have examined various AV world ontologies and ODDs. These include works by Krzysztof C. [17] in which the author has designed the operational world model (OWM) of the AV and AV ODD model presented by National Highway Traffic Safety Administration (NHTSA) [18]. We also studied the PhD thesis of Philippe N. [19] in which it is highlighted that road intersections are one of the highest risk areas among all road structures for road users with 30% of all road accidents occurring at intersections, with 14% of road accidents on intersections resulting in death.

After dissecting the literature on ontologies regarding AV ODDs we have come up with the situation hyperspace for our SitCov AV-testing framework as shown in Fig. 1 (left). In the block diagram of the situation hyperspace, we have the top layer as the situation hyperspace axis, it is further divided into environmental conditions axis and intersection axis. Other axis can be added to the situation hyperspace but right now since we were implementing this in our SitCov AV-testing framework in CARLA, we are using just two axes since we needed to keep the number low, of situation elements in the situation hyperspace and their combinations, which was practical when implementing it in our code. Nonetheless, these two axes are high priority axis as per literature that is another reason we have added and tested them first. The SitCov AV-testing framework selects the situation elements from both environmental conditions and intersection axes and combines them to generate the discrete situation where the AV simulation actually runs in CARLA.

The Environmental Conditions Axis. For the environmental conditions axis, we further subdivided it into the following situation elements, after studying a variety of AV ODD and world ontologies [17, 18]: (1) Friction (road friction); (2) Fog Density; (3)

Precipitation; (4) Precipitation Deposits; (5) Cloudiness; (6) Wind Intensity; (7) Wet-ness; (8) Fog Distance. These situation elements of the environmental conditions axis can be seen in Fig. 2, each element has been discretized into 6 discrete bins and our SitCov AV-testing framework chooses from those.

The Intersection Axis. In the intersection axis, T intersection has been chosen as the axis to do AV-testing on since it has one of the highest accident rates [19]. We have also highlighted other axis types that could be added to the intersection axis, such as 4-legged intersection, offset 4-legged intersection, circular 4-legged intersection, different skew angles of intersection legs in all types. But for the sake of simplicity and implementation we have chosen T-intersection as the only element of the intersection axis.

In our SitCov AV-testing framework, we are essentially doing pairwise testing of our ego AV with an adversarial other vehicle which we will now refer to as other vehicle (OV). The reason for selecting pairwise encounters of the ego AV with the OV is that V2V accident rates on intersections are way above all other types of accidents on intersections as mentioned here [19]. The pairwise testing between the ego AV and OB is done on a T-intersection in different environmental conditions selected from the environmental conditions axis by our framework, and in different intersection configurations i.e., dif-ferent start and goal locations for the ego AV and OV, selected from the intersection axis, as seen in Table 1. For the different intersection configurations, we have come up with collision-points c_1, c_2, c_3, c_4 as seen in the center of the T-intersection in Fig. 1(right), these are the closest point of approach (CPA) for our ego AV and OV for their pairwise interaction during each simulation run. This CPA approach has been inspired from the work of [22]. The four collision-points we have come up with are one of the highest risk areas on an intersection and are similar to the conflict-points on intersections mentioned in [17]. Hence, testing our ego AV on one of the most dangerous road structures (inter-section) on the highest risk areas (intersection conflict-points) would be an effective way to stress the ego AV and test it thoroughly, which our SitCov AV-testing framework is doing.

For the pairwise AV-testing on the T-intersection, we will consider right-hand traffic, the main reason being that many road ontologies that we have followed, used right-hand traffic. There will be three possible start and goal locations for the vehicles (AV on the T-intersection. As we would only consider the vehicles to be moving inwards towards the T-intersection after spawning at their start locations, we are considering cross-collisions [23] at the moment, we will incorporate diverging and merging type collisions [23] in the later stages also. The SitCov AV-testing framework generates the OV such that it comes in conflict with our ego AV at the CPA i.e., one of the collision-points (c_1, c_2, c_3, c_4).

Notations for Intersection Pairwise Conflict-Point Interactions. Notations for the intersection pairwise conflict-point interactions between ego AV and the OV (as seen in Table 1) are elaborated as follows: (1) L, R, and B, define left, right and base legs of the T-intersection; (2) S implies start location variable; (3) S_x, x subscript defines the type of vehicle. It can either be AV or OV in the pairwise encounter; (4) S_{xy}, y subscript defines the location of the start point S of the vehicle; (5) G implies goal location variable; (6) G^z, z superscript indicates the final destination/goal location of the vehicle; (7) The multiply operator "x", is the intersection pairwise conflict-point interactions operator between the variables as seen in the "Conflict Point Interaction" column of Table 1, the notations in that column essentially tells what were the start/goals locations of the ego AV, the OV and the particular conflict-point where the ego AV and OV met during a simulation run by the SitCov AV-testing framework.

Intersection pairwise conflict-point interactions when ego AV starts at the base, have been shown in Fig. 1(right) along with their notations and can be compared with Table 1. The complex notations shown as conflict-point interaction column in Table 1 have been simplified in the last column of the Table 1, i.e., intersection situation labels, and these simplified labels will be used to represent these complex conflict-point interactions between our ego AV and the OV. These intersection situation labels are the situation elements of our intersection axis of the situation hyperspace. We encourage the readers to see how these complex conflict-point interactions of the intersection axis have been modelled in our code [20] using dictionaries and lists in Python language.

The next section elaborates how does the SitCov AV-testing framework selects the situation elements from environmental conditions and intersection axis of the situation hyperspace, to generate the discrete situation in which the V2V interaction between the ego AV and OV takes place in the CARLA simulator. The next section also highlights key elements in the development of the SitCov AV-testing framework in CARLA.

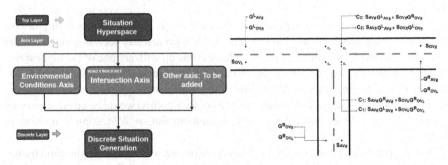

Fig. 1. Situation hyperspace block diagram (left). T-intersection with notations (right).

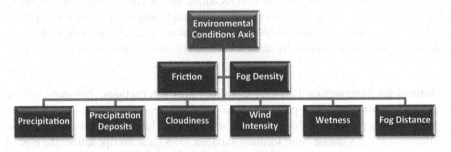

Fig. 2. Environmental conditions axis subdivision elements.

Table 1. Intersection situations & notations

Sr	Start loc of AV	Goal loc of AV	Conflict Point Interaction	Intersection Situation Label
1	S_{AVB}	G^R_{AVB}	$C_1: S_{AVB}\, G^R_{AVB} \times S_{OVL}\, G^R_{OVL}$	IntSit-8
		G^L_{AVB}	$C_1: S_{AVB}\, G^L_{AVB} \times S_{OVL}\, G^R_{OVL}$	IntSit-5
			$C_2: S_{AVB}\, G^L_{AVB} \times S_{OVR}\, G^L_{OVR}$	IntSit-6
			$C_2: S_{AVB}\, G^L_{AVB} \times S_{OVR}\, G^B_{OVR}$	IntSit-7
2	S_{AVR}	G^B_{AVR}	$C_2: S_{AVR}\, G^B_{AVR} \times S_{OVB}\, G^L_{OVB}$	IntSit-10
			$C_4: S_{AVR}\, G^B_{AVR} \times S_{OVL}\, G^R_{OVL}$	IntSit-11
			$C_4: S_{AVR}\, G^B_{AVR} \times S_{OVL}\, G^B_{OVL}$	IntSit-12
		G^L_{AVR}	$C_2: S_{AVR}\, G^L_{AVR} \times S_{OVB}\, G^L_{OVB}$	IntSit-9
3	S_{AVL}	G^B_{AVL}	$C_4: S_{AVL}\, G^B_{AVL} \times S_{OVR}\, G^B_{OVR}$	IntSit-1
		G^R_{AVL}	$C_4: S_{AVL}\, G^R_{AVL} \times S_{OVR}\, G^B_{OVR}$	IntSit-2
			$C_1: S_{AVL}\, G^R_{AVL} \times S_{OVB}\, G^R_{OVB}$	IntSit-3
			$C_1: S_{AVL}\, G^R_{AVL} \times S_{OVB}\, G^L_{OVB}$	IntSit-4

3   The SitCov AV-Testing Framework

This section lays out the development of the proposed SitCov AV-testing framework from top to bottom, from the theory to its implementation in CARLA. Below we start with explaining the theoretical mechanics of our SitCov AV-testing framework along with its connection with test adequacy criteria from software testing [25].

3.1   Situation Coverage-Based AV-Testing Test Suite Generation Methodology

In this subsection we will briefly explain the methodology behind the situation coverage-based situation generation for AV-testing by our SitCov AV-testing framework.

Test Adequacy Criteria for AV-Testing. The inspiration of the SitCov AV-testing Framework has been taken from the concept of test adequacy criterion from software testing [25], which essentially states that test adequacy criterion provides a stopping rule and/or a measure of test quality. A test adequacy criterion highlights the testing requirements and the test suits needed to satisfy those requirements, and it determines the observations required during the process of testing.

In this paper, we have made a connection between this concept of test adequacy criterion from software testing with AV testing, by applying it in our SitCov AV-testing framework as seen in Fig. 3 (right). Our SitCov AV-testing framework utilizes situation hyperspace for situation coverage metrics along with fault injection. This will provide us with reliable test adequacy criteria-based metrics. Figure 3 (right) shows the test adequacy criteria highlighted in blue being employed by our SitCov AV-testing framework, further detailed definitions of these test adequacy criteria along with their connection with our framework are out of the scope of this paper, the test adequacy criteria definitions can be found here though [25].

Handling the AV Nominal vs Functional Safety Problem. The safety assurance of AVs is particularly a challenging task due to AVs being a cyber-physical, we not only have to make sure that the functional safety of the AV is addressed i.e., software/hardware of the AV is bug-free and all the functions are being executed correctly, but also that, we have to make sure that we have addressed the nominal safety [26] aspect of the AV as well i.e., making sure the AV is making safe and logical decisions, assuming that the software/hardware of the AV are operating error-free.

To address both the issues of AV functional and nominal safety, we will use the fundamentals provided by software testing as described above in conjunction with agent-based simulation in CARLA.

Automatic Test Suite Generation of our SitCov AV-testing Framework. The situation coverage-based generation of the discrete situations from the situation hyperspace is done by counting how many times each bin of the situation elements of each of the environmental conditions and intersection axis, has been generated. Then the counts (number of times that bin of the situation element was used for a discrete situation generation simulation run) of each bin of a particular situation element of an axis of the situation hyperspace, let's say the precipitation element from the environmental conditions axis, or the intersection situation labels from the intersection axis as mentioned in Table 1, are fed into a SoftMax function to generate a normalized probability distribution over those bins, then those probability values are inverted and normalized and act as weights to a weighted random number generator, higher the weight, higher the probability of that bin to be selected. In simple terms, those bins of a situation element, let's say the precipitation element of the environmental conditions axis or the intersection situation labels from the intersection axis, that have been generated fewer times than the other bins (for discrete situation generation), have the highest probability to be selected by the SitCov AV-testing framework for the discrete situation generation in the next simulation run. This way we aren't setting hard rules that all situations have to be generated exact equal number of times, we are leaving some room for exploration along with exploitation by using weighted random number generation while still making sure that with a very high probability good coverage of the situation hyperspace is being achieved. This whole process of situation coverage-based generation from the situation hyperspace can be seen in our code here at [20], this process can also be seen in the block diagram of the SitCov AV-testing framework in Fig. 3 (left), along with other processes that are occurring such as keeping track of collisions happening between ego AV and OV, and increasing the counter of those situation element bins that were used to generate the last discrete situation for the simulation run, etc.

The AV-IP Problem & Our Solution. A really important but seldom talked about problem for the practical V&V and safety assurance of AVs is the IP (Intellectual Property) problem. As top tech (Waymo, Apple, etc.) and car manufacturing (Tesla, BMW, etc.) companies compete with each other in the AV domain, with projected trillions of dollars of revenue on the line, these companies are investing millions and billions of dollars in the AV industry, and to say the least, they would not be very pleased to share their entire AV IPs with third party safety experts for the sole purpose of V&V and safety assurance of AVs.

Keeping in view, this IP problem of AVs, we have designed our SitCov AV-testing framework to treat AVs as a black-box or in some instances as a grey-box, which will

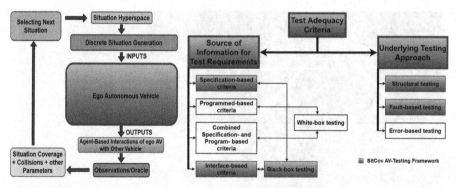

Fig. 3. SitCov AV-testing framework block diagram (left). Test adequacy criteria used (right).

enable our research to be industry ready and hopefully contribute directly to saving lives in the process. The users of our SitCov AV-testing framework will just have to insert their AV autonomous driving stack as a black-box and run the SitCov AV-testing framework, the framework will output situation coverage metrics of all situation elements of the situation hyperspace and those situations that the AV autonomous driving stack cannot deal with will be highlighted along with the percentages of the situation covered in the situation hyperspace and their repetitions. This process will be further elaborated in the experiments results and analysis section.

3.2   Developing the SitCov AV-Testing Framework in CARLA

This subsection presents the main softwares used for the development and experimentation of the SitCov AV-testing framework. Following were the baseline software packages used: (1) CARLA [12]; (2) Scenario Runner [27]; (3) Tensorflow Object Detection API [28]. All of these software packages are opensource and Python is the programming language used to run them. A few screenshots of the development of the SitCov AV-testing framework can be seen in Fig. 4. Further elaboration of development of the framework is below.

Using CARLA & Scenario Runner for Simulations.   As stated in the previous sections that CARLA was chosen due to its upsides as compared to other AV simulators that includes it being opensource, regularly updated with previous versions maintained on GitHub website, 3D rendering and AV dashcam video feed are available along with other sensors. We have used Scenario Runner [27] on top of CARLA for the development of our SitCov AV-testing framework. Scenario Runner is just an API built on top of CARLA software and it is used to generate certain scenarios. The intersection scenarios generation from Scenario Runner was of our interest and we edited that intersection class to run our AV-testing test suites on a T-intersection by generating situation coverage-based situations from the situation hyperspace that we designed in Python using dictionaries and lists [20].

Fig. 4. Development of our SitCov AV-testing framework in CARLA and Scenario Runner. The tools provided in CARLA such as waypoints highlighting and their interconnection with each other and other road structures such as intersections etc., were really helpful in speedy development of our SitCov AV-testing framework.

Developing Automated Driving Algorithm for Ego AV using Tensorflow Object Detection API in CARLA. In order to evaluate our SitCov AV-testing framework we needed an autonomous driving (AD) algorithm for the ego AV in our simulation runs, and we also need an algorithm to drive the OV as well since our SitCov AV-testing framework ensures that the ego AV and OV meeting in the intersection at the collision-points shown in the previous section, under different situations generated from the situation hyperspace using situation coverage-based generation. So, for the ego AV autonomous driving, ideally, we would have liked to get that algorithm from an OEM such as Tesla, then our SitCov AV-testing framework would have been properly evaluated, since we would have had high confidence that there was a very low probability of the ego AV crashing in the OV due to faults in the autonomous driving algorithm. Even though our SitCov AV-testing framework would have treated the ego AV autonomous driving algorithm as a black-box, we still wouldn't have been able to get such an algorithm from an OEM easily. So, we had to develop it on our own autonomous driving algorithm for the ego AV.

Types of Autonomous Driving Algorithms. We had the option to choose from 3 overarching pipeline of autonomous driving as mentioned in [12]: (1) Modular pipeline; (2) Imitation learning; (3) Reinforcement learning. Imitation learning and reinforcement learning pipeline of autonomous driving isn't very effective when it comes to rule-based driving. Imitation learning does perform a lot better than reinforcement learning in rule-based driving [12] but modular pipeline is the best option for rule-based driving and it is what actual driving is like in the real world as well, traffic-rules-based. Hence, we chose modular pipeline for our ego AV autonomous driving.

Our Modular Autonomous Driving Pipeline. Our modular autonomous driving pipeline has following main stages: (1) Perception; (2) Local Planner; (3) Continuous Controller.

The local planner stage was already implemented in CARLA examples, which used A-star algorithm [29] to compute the optimal path from one point of the map to the other using the roads and following traffic rules. We used this directly and our assumption here is that our ego AV has an ideal GPS sensor which tells it the exact location because the local planner we are using is accessing the ground truth for accessing the location of the location of the ego vehicle which is the ego AV or OV, since we are using this local planner in both.

The continuous controller stage is a PID controller for lateral and longitudinal control of the ego vehicle, and this was also already implemented in CARLA that we used directly for the ego AV and OV controllers.

The Perception Stage (Using Deep Neural Network). The ego AV uses one more stage for its autonomous driving that the OV does not, it is the perception stage. Just like Tesla AVs, we are using only camera as the perception sensor for our ego AV. We are using a very deep Single-Shot multi-box Detection (SSD) Convolutional Neural Network (CNN) pre-trained on 350,000 images of the MS COCO dataset, the SSD mobilenet [28, 30, 31], to detect the incoming OV and locate its position in the dash-cam feed (images) of the ego AV and to apply emergency brakes (AEBs) if the OV is too close. We are using the Tensorflow Object Detection API to implement this pre-trained SDD mobilenet CNN.

Parameters of Object Detection. The perception stage of our ego AV has three main parameters that can be set for efficient object detection using the pre-trained SDD mobilenet CNN that is taking images from the dash-cam of the ego AV in CARLA. Following are the three parameters:

1. **Threshold of distance for object detection:** To detect the distance of the OV from the image we are using a computer vision technique [32] to check how much percentage of the image feed is the detected OV taking up, the closer is the OV to the ego AV, the more percentage of the image from the ego AV would contain the OV and at a certain threshold value, our ego AV will apply the AEBs to avoid collision. We will refer to this threshold value as the "threshold of distance for object detection" in the later sections.
2. **Probability of detection for object detection threshold:** This is the threshold value for probability output from the SDD mobilenet CNN, which tells us the probability if there is a car detected in the images received from the dash-cam of our ego AV in CARLA. Since we are doing V2V pairwise testing with an OV, we only need to care about the probability of detecting a car in the received images, though we could also check for detection of buses, trucks, people, animals, etc., we recommend including other dynamic agents in future work. Setting this parameter for probability of detection is really important, i.e., the minimum probability value that our ego AV would consider as a detection of an OV in the image. Since too high a value of this parameter would result in missing detection and collisions happening in extreme weathers such as heavy rain and fog etc., where probability of detection of OV would be lower even if it was in front of our ego AV.
3. **Centering limits parameters:** These parameters look at if the OV detected by our SDD mobilenet CNN is close to the center of the received image. If the detected OV is too far away from the center, even if the OV is really close, AEBs are not activated.

So, this makes up the autonomous driving pipeline of our ego AV, provides a useable driving pipeline to test the main focus of this paper, the SitCov AV-testing framework and its implementation in CARLA. A bit of effort was also put into implementing the autonomous driving modular pipeline in CARLA as elaborated above and we would like to recommend the readers to have a look at its implementation in the code if they're looking to get a working implementation of autonomous driving [20]. Next, we will show the screenshots of a few experiments, i.e., the discrete situations generated by our SitCov AV-testing framework, just so that the readers get a gist of how these experiments are happening practically in CARLA software with the test suites generated by our SitCov AV-testing framework.

Running Test Suites on the SitCov AV-Testing Framework. In Fig. 5 we can see experiment #1 generated by the SitCov AV-testing framework. The temporal progression of the simulation run starts from top left image to the top right, then the bottom left and the last timestamp is the bottom right picture in Fig. 5. The SitCov AV-testing framework selected a sunny day with clear weather and generated the ego AV on the left side of the intersection and the OV was generated on the top side of the intersection as seen in the first picture of the temporal sequence of experiment #1, all these situations were selected from the situation hyperspace using situation coverage-criterion by our framework. In all the pictures we can see another window opened up in the top right corner of each picture, this window is showing the images received by the dash-cam of the ego AV and it is also showing the object detection probability values and detection boxes if an OV is detected. As seen in the third picture of the temporal sequence, the ego AV has detected the OV really close to it and close to its centers as well (the path of progression of the AV) and the probability of detection is above the threshold value we have set, hence AEBs are activated and the ego AV comes to a full stop and a collision is avoided and the experiment is a success. This OV can be seen going away in the last picture of the temporal sequence.

Figure 6 shows experiment #2, in which the SitCov AV-testing framework has generated a heavy rain and a foggy discrete situation from the situation hyperspace. The OV is coming in from the top of the intersection and the ego AV is coming in from the right side of the intersection. Due to the heavy rain and fog, the ego AV was not able to detected the OV in time for AEB and a collision happened between the two. This accident was avoidable if the ego AV had detected the OV just a few moments earlier.

Figure 7 show experiment #3, in which there is just a little bit of rain but heavy fog and a lot of precipitation deposits on the road, selected by our SitCov AV-testing framework. The OV is coming in the intersection from the left side and the ego AV is coming in from the right side and the OV suddenly turns towards the ego AV and collides. There was much the ego AV could do as both vehicles were following the traffic rules before the OV suddenly turned towards the ego AV and collided. We will refer to such accidents/failures as unavoidable accidents/failures.

4   Experimentation Results and Analysis

We provide the details of our experimentation setup and a detailed analysis and evaluation of our SitCov AV-testing framework in this section. We have carried out experiments

Fig. 5. Experiment #1 Sunny Day Accident Avoided: Success.

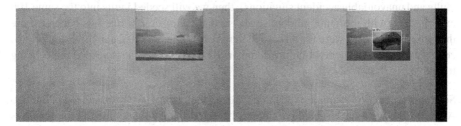

Fig. 6. Experiment #2 Heavy Rainy & Foggy Day & Avoidable Accident Collision: Fail.

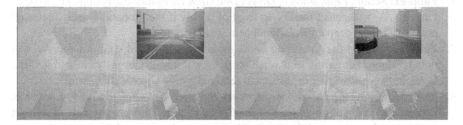

Fig. 7. Experiment #3 Rainy Foggy Day & Unavoidable Accident Collision: Fail.

to evaluate our SitCov AV-testing framework and our goal is to answer our research question RQ1 and RQ2, that how well does our SitCov AV-testing framework perform compared to random situation generation, and what additional useful information does our SitCov AV-testing framework gives us that random situation generation cannot.

For the experimentations we used a core-i7 PC with a 4GB GPU and 32 GB RAM. We ran experiments in sets of 20, and repeated it 5 times with a different starting random seed, to make it a total of 100 experiments. Each set of 20 experiments took almost 30 min to run, so one set of 100 experiments took about 2.5 h.

Evaluation Metrics and Seeding Faults. We use seeded faults to evaluate our SitCov AV-testing frame. The process starts with us seeding a few faults in the ego AV software, and when the ego AV crashes in the OV, we say that the *seeded fault* has been *triggered*. But there is a caveat, as we mentioned in the previous section of our autonomous driving algorithm development, we have used a fairly simple algorithm and just one RGB-camera sensor on our ego AV, hence there will be a lot of non-seeded background fault causing failures (crashes of ego AV in OV) and apart from that there will be some unavoidable accident situations as shown in Fig. 7 that when the OV suddenly turns towards the ego AV.

So, we will denote crashing of ego AV into the OV as a failure, whether it is due to a seeded fault being triggered or it is due to non-seeded background faults or simply unavoidable failures as mentioned above. Then we will first conduct experimentations without seeding any faults and note how many failures we get per 100 experiments which we will call as *general failures* (which include both non-seeded background faults and unavoidable accident situations), then we will seeded faults into the same 100 experiments (with the same random seeds, so that the discrete situations generated are the same as what they were for no seeded faults experimentations), and the extra failures counts exceeding the general failure counts will be labelled as those seeded faults being triggered.

Seeded Faults. Following are the three types of faults that have been seeded in our ego AV software, we have targeted parameters of object detection, detailed in the previous section, for seeding our software faults:

1. **Fault #1:** Setting *Probability of Object Detection Threshold* really high, i.e., to 0.95. Our ego AV will detect the OV in front of it only if the probability of its detected outputted by the deep convolutional neural network is 0.95.
2. **Fault #2:** Setting the *Centering limits parameters* too rigid, i.e., our ego AV will only label the OV in front of it as a hazard, even if it is dangerously close to it, if the OV is in the exact center of the image received by the ego AV dash-cam.
3. **Fault #3:** Setting the *Threshold of distance for object detection* to a really low value, i.e., our ego AV will only consider the OV as a hazard if it is extremely close to it because we have lowered the threshold of distance for object detection parameter.

4.1 Intersection Situation Generation Results

In this subsection we are highlighting the performance of our SitCov AV-testing framework vs random situation generation w.r.t the intersection situations being generated from the intersection situation axis as mentioned in Table 1. These experiments will explore which intersection situations are dangerous for our ego AV, having the highest failure rates.

No Faults Seeded Intersection Situations Experiments. Our first set of experiments compares intersection situation generation results of our SitCov AV-testing framework vs random generation, when no faults have been seeded.

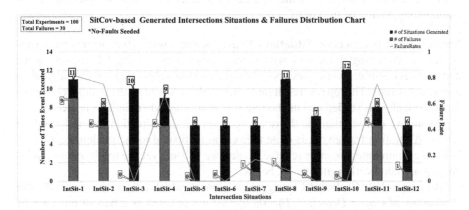

Fig. 8. SitCov-based intersection situation generation & failure distributions, no faults seeded.

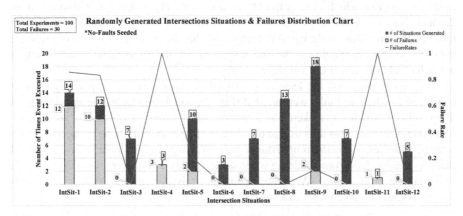

Fig. 9. Random intersection situation generation & failure distributions, no faults seeded.

In Figs. 8, 9 the labels mentioned on the x-axis are the intersection situation labels that have been shown in detail in Table 1, which mention particular intersection situations as derived from the notations used and explained in the previous sections. The labels will be used in the coming figures as well.

As seen in Fig. 9, random intersection situation generation has quite erratic failure rates values vs the situation coverage-based (SitCov-based) intersection situation generation from our SitCov AV-testing framework in Fig. 8, even though their total number of failures are equal. This is due to the uneven distribution of situations produces by the random generation, which is why we can see for some intersection situations the failure rate is 100% for random generation in Fig. 9, where as this is not the case for situation coverage-based intersection situation generation in Fig. 8, because each intersection situation was tested somewhat evenly and hence we can have more confidence in the failure rates of the intersection situations provided by our SitCov AV-testing framework and then we can in-turn look at the ego AV autonomous driving algorithm and see why does it have high failure rates in these particular intersection situations and how can the autonomous driving be improved.

Faults Seeded Intersection Situations Experiments. Figure 10 compares the results of SitCov AV-testing framework with random generation. On the left colored in red and black is our framework's results and on the right in yellow and blue is the random generation results. The top row results are when fault #1 is seeded, the middle row results are when fault#2 is seeded and the bottom row results are when fault #3 is seeded.

Again, we can see that the distribution of intersection situation generation of our SitCov AV-testing framework is quite uniform as compared to the random generation, hence the failure rates are more reliable. Hence, we will analyze the results of our SitCov AV-testing framework from Fig. 10, below.

As seen in Fig. 10, fault #1 is triggered the most, which tells us that the parameter *Probability of Object Detection Threshold*, that we tinkered as a part of our fault #1 seeding, is extremely important for autonomous driving.

After fault #1, fault #2 is triggered the most (Fig. 10), which refers to the *Centering limits parameters*, which detects the OV as a hazard only when it is within a certain distance from the center of the line of sight of ego AV, i.e., from the center of the image taken by the dash-cam on the ego AV. This means that the developers need to be wary of not only the vehicles coming in the line of sight of the ego AV, but also vehicles further out in the field of view (FOV) of the ego AV.

Fault #3 is triggered the least (Fig. 10) but its failure rate is still really high as compared to when no faults were seeded (Fig. 8), which highlights the importance of the parameter which was tinkered as a part of fault #3, i.e., the *Threshold of distance for object detection*. This tells us that developers of AVs should not be aggressive while setting the distance *Threshold of distance for object detection* to a really low value and that AVs need to have some padding while setting safety distance from the OVs in front of it so that even if the ego AV or the OV makes a mistake, the ego AV has some extra safety distance to make up for that mistake.

Table 2 two shows the number of triggered faults from the experiments shown in Fig. 10, by subtracting the total failures of these faults seeded experiments from Fig. 10 by the failures caused when no faults were seeded in experiments shown in Fig. 8, 9.

Figure 11 shows the faults triggered by our SitCov AV-testing framework vs random generation from the experiments shown in Fig. 10, and as seen in Fig. 11 the faults being triggered by SitCov-based generation and random generation are somewhat similar in number but the additional benefit that our SitCov-based generation gives us is that the failures rates/fault triggering rates are much more reliable since each situation is thoroughly tested by the SitCov AV-testing framework as compared to random generation where one out of the 12 possible intersection situations (as mentioned in Table 1) could have been tested only one time whereas the other 11 would have been tested 99 times in total, incase of our 100 simulation runs experiments.

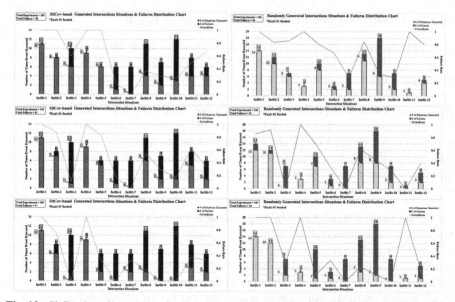

Fig. 10. SitCov-based vs Random intersection situation generation & failure distributions, fault-1 seeded.

Table 2. Faults triggered

Intersection Situation Generation Method	No-Faults Seeded Failures (f_{NF}) (100 Runs)	Fault-1 Seeded Failures (f_{F1}) (100 Runs)	Fault-2 Seeded Failures (f_{F2}) (100 Runs)	Fault-3 Seeded Failures (f_{F3}) (100 Runs)	# Times Fault-1 Triggered $= f_{F1} - f_{NF}$	# Times Fault-2 Triggered $= f_{F2} - f_{NF}$	# Times Fault-3 Triggered $= f_{F3} - f_{NF}$
Random Generation	30	69	54	40	39	24	10
Situation Coverage-based Generation	30	58	51	37	28	21	7

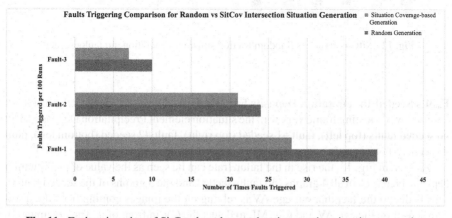

Fig. 11. Faults triggering of SitCov-based vs random intersection situation generation

4.2 Environmental Conditions Situation Generation Results

We will focus on the environmental conditions situations from the environmental conditions axis (Fig. 2) when comparing our SitCov AV-testing framework with random generation. It is to be noted that actually all of these situation elements from environmental conditions axis and intersection axis are interlinked and correlated as when the discrete situation is generated, it selects some values of all the situation elements from all the axis (environmental conditions axis and intersection axis) from the situation hyperspace, but we are looking at the situation elements separately during our evaluation since analyzing the correlations between the situation elements and their effect on the success/failure of ego AV would be a research project on its own and we definitely recommend it for future work.

No Faults Seeded Road Friction Experiments. Figure 12 shows the results of our SitCov AV-testing frame (in the red and black graph) vs random generation (in the blue and yellow graph) w.r.t road friction, which is a situation element from the environmental conditions axis. Friction values have been divided into 6 bins between 0 and 1 with lowest value being 0.1. The experiments show that random generation as expected has generated highly uneven distribution of friction bins whereas our SitCov AV-testing framework has generated more of an even generation of friction bins. The results from both the graphs in Fig. 12 show us that the failure rate does not get higher as the road friction gets lower, rather the failure rate across all the friction bins is somewhat evenly distributed. This means that this situation element, friction, doesn't affect the success of our ego AV directly that much otherwise it would have had a high failure rate for lower friction values.

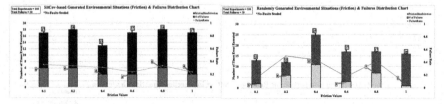

Fig. 12. SitCov-based vs Random friction situation generation, no faults seeded.

Faults Seeded Precipitation Deposits Experiments. Figure 12 displays the results of our SitCov AV-testing framework w.r.t the situation element Precipitation Deposits when no seeded faults (top left), fault #1 seeded (top right), fault #2 seeded (bottom left), fault #3 seeded (bottom right).

As seen in Fig. 13, the rise in the failure rate can be seen as the value of precipitation deposits increases in all 4 graphs (no faults seeded and the 3 graphs of the seeded faults). This is due to the fact that our ego AV is relying on the images coming from the dashcam sensor on the ego AV and uses ML to detect the OV in front of it. The precipitation deposits produce a lot of noise in the incoming images which leads to errors in detecting the OV which causes the ego AV to miss detecting the hazard in front of it, leading to a crash/failure. We can see a big jump in failure rate when fault #1 is seeded, this tells us as we mentioned before that the parameter which fault #1 corresponds to is *Probability of Object Detection Threshold*, and when the image received by the dash-cam on the ego AV is already noisy due to the precipitation deposits, and we introduce the fault #1 which is having a very high detection threshold, our ego AV simply fails to detect the OV in front of it many times and hence has a high failure rate when the precipitation deposits value is high especially when the fault #1 is seeded.

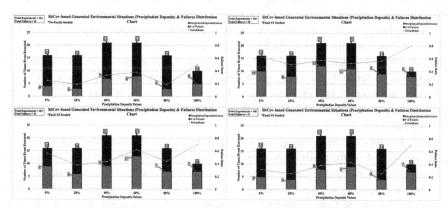

Fig. 13. SitCov-based precipitation deposits situation generation & failure distributions, no faults seeded.

4.3 Summary of Our Results

With respect to RQ1, our experiments suggest that the fault revealing capabilities of our SitCov AV-testing framework vs random generation is more or less the same.

With respect to RQ2, the results produced by the SitCov AV-testing framework are much more convincing than random situation generation, because the results produced by the SitCov AV-testing framework provide additional confidence in the failure rates as the situation coverage-based situation generation of our SitCov AV-testing framework make sure that the situation hyperspace is efficiently covered while also making sure that in case of repetitions of situations, we have an even distribution of repetitions rather than highly varying distributions of situations as seen for random situation generation.

5 Conclusion and Future Work

In this paper we have developed a novel situation coverage based (SitCov) AV-testing framework which uses situation hyperspace, a derivative of an ontology that we developed, to automatically generate test suites according to a situation coverage criterion to test AVs for their V&V and safety assurance.

This paper contributes not only theoretically but we have also released our code publicly so researchers can get a head start if they are looking to work in this direction. Our framework has shown that it can find seeded faults and highlight situations with high failure rates and this framework also provides us with confidence in the failure rates of situations it generates due to the even distributions of generating situations, due to the novel mechanism of situation coverage-based situation generation mechanism we have developed for our SitCov AV-testing framework. The developed SitCov AV-testing framework treats AV autonomous driving algorithm as a black-box, hence battling against the AV-IP problem, while testing the AV thoroughly and evenly in all kinds of situations, so that we would have high confidence in the results provided by our SitCov AV-testing framework.

For the future work, we have made our SitCov AV-testing framework really expandable such that many more axis can be added in the situation hyperspace and a lot more dynamic obstacles can be added in the simulation runs instead of just V2V interactions that is currently being done. So, we recommend that the situation hyperspace be expanded more with additional axis, e.g., adding different kinds of roads, adding diverging and merging collisions instead of only cross-collisions with OV, adding pedestrians and cyclists, etc.

We also recommend making use of the failure rate of situations and using that information to skew the generation of next batch of situations in the direction of situation elements that were causing more failure (high failure rate), which can be done using the same mechanism we have developed for doing situation coverage-based situation generation for our SitCov AV-testing framework. Similarly, machine learning can also be added in our SitCov AV-testing framework to generate more interesting situations, i.e., edge-cases, by learning the patterns of combinations of environmental conditions and intersection situations that cause the highest failure rates in the results of our SitCov AV-testing framework test-cases.

Acknowledgements. The research presented in this paper has been funded by European Union's EU Framework Programme for Research and Innovation Horizon 2020 under Grant Agreement No. 812.788.

References

1. Tahir, Z., Alexander, R.: Coverage based testing for V&V and safety assurance of self-driving autonomous vehicles: a systematic literature review. In: 2020 IEEE International Conference On Artificial Intelligence Testing (AITest), pp. 23–30 (2020). https://doi.org/10.1109/AITEST49225.2020.00011
2. International Organization for Standardization, ISO/PAS 21448:2019 Road vehicles—Safety of the intended functionality (2019). https://www.iso.org/standard/70939
3. International Organization for Standardization, ISO 26262-1:2018 Road vehicles—Functional safety (2018). https://www.iso.org/standard/68383
4. Underwriters Laboratories, Presenting the Standard for Safety for the Evaluation of Autonomous Vehicles and Other Products. https://ul.org/UL4600
5. Anthony, C., Lee, R., Kochenderfer, M.J.: Scalable autonomous vehicle safety validation through dynamic programming and scene decomposition. In: 2020 IEEE 23rd International Conference on Intelligent Transportation Systems (ITSC). IEEE (2020)

6. Klischat, M., Althoff, M.: Generating critical test scenarios for automated vehicles with evolutionary algorithms. In: 2019 IEEE Intelligent Vehicles Symposium (IV). IEEE (2019)
7. Chance, G., Ghobrial, A., Lemaignan, S., Pipe, T., Eder, K.: An agency-directed approach to test generation for simulation-based autonomous vehicle verification. In: 2020 IEEE International Conference on Artificial Intelligence Testing (AITest). IEEE (2020)
8. Haq, F.U., Shin, D., Nejati, S., Briand, L.: Comparing offline and online testing of deep neural networks: An autonomous car case study. In: 2020 IEEE 13th International Conference on Software Testing, Validation and Verification (ICST). IEEE (2020)
9. Society of Automotive Engineers, SAE J-3016 international report at https://www.sae.org
10. MathWorks Automated Driving Toolbox. https://uk.mathworks.com/products/automated-dri ving.html
11. CarMaker: Virtual testing of automobiles and light-duty vehicles. https://ipg-automotive.com/ products-services/simulation-software/carmaker/
12. Dosovitskiy, A., Ros, G., Codevilla, F., Lopez, A., Koltun, V.: CARLA: an open urban driving simulator. In: Conference on Robot Learning. PMLR (2017)
13. Kitchenham, B.A., et al.: Refining the systematic literature review process—two participant-observer case studies. Empir. Softw. Eng. **15**(6), 618–653 (2010). https://doi.org/10.1007/s10 664-010-9134-8
14. Alexander, R., Hawkins, H., Rae, D.: Situation coverage – a coverage criterion for testing autonomous robots, pp. 1–20 (2015)
15. Ulbrich, S., Menzel, T., Reschka, A., Schuldt, F., Maurer, M.: Defining and substantiating the terms scene, situation, and scenario for automated driving. In: IEEE Conference Intelligent Transportation System Proceedings, ITSC, vol. 2015-Octob, pp. 982–988 (2015)
16. Tahir, Z.: Situation hyperspace—using a simulated world to obtain situation coverage for AV safety assurance. https://assuringautonomy.medium.com/situation-hyperspace-using-a-simulated-world-to-obtain-situation-coverage-for-av-safety-assurance-39fa5ea203cd
17. Krzysztof, C.: Operational world model ontology for automated driving systems - part 1: road structure. https://doi.org/10.13140/RG.2.2.15521.30568
18. Thorn, E., Kimmel, S., Chaka, M.: A framework for automated driving system testable cases and scenarios. National Highway Traffic Safety Administration USA (2018)
19. Philippe, N.: Safety-critical scenarios and virtual testing procedures for automated cars at road intersections. Diss. Loughborough University (2018)
20. Tahir, Z.: Situation Coverage-based AV-Testing Framework in Carla. https://github.com/zai dtahirbutt/Situation-Coverage-based-AV-Testing-Framework-in-CARLA
21. International Organization for Standardization, ISO 22737:2021, Intelligent transport systems—Low-speed automated driving (LSAD) systems for predefined routes—Performance requirements, system requirements and performance test procedures. https://www.iso.org/sta ndard/73767
22. Xueyi, Z., Alexander, R., McDermid, J.: Testing method for multi-UAV conflict resolution using agent-based simulation and multi-objective search. J. Aerosp. Inf. Syst. **13**(5), 191–203 (2016)
23. Krzysztof, C.: Operational world model ontology for automated driving systems - part 2: road users, animals, other obstacles, and environmental conditions. https://doi.org/10.13140/RG. 2.2.11327.00165
24. Krzysztof, C.: Operational design domain for automated driving systems - taxonomy of basic terms. https://doi.org/10.13140/RG.2.2.18037.88803
25. Zhu, H., Hall, P.A.V., May, J.H.R.: Software unit test coverage and adequacy. ACM Comput. Surv. **29**(4), 366–427 (1997). https://doi.org/10.1145/267580.267590
26. Shwartz, S.S., Shammah, S., Shashua, A.: On a formal model of safe and scalable self-driving cars. https://arxiv.org/abs/1708.06374

27. ScenarioRunner for CARLA. https://github.com/carla-simulator/scenario_runner
28. Jonathan, H., et al.: Speed/accuracy trade-offs for modern convolutional object detectors. In: Proceedings of the IEEE Conference on Computer Vision and Pattern Recognition (2017)
29. LaValle, S.M.: Planning Algorithms. Cambridge University Press, Cambridge (2006)
30. COCO dataset. https://cocodataset.org/
31. Howard, A.G., et al.: Mobilenets: efficient convolutional neural networks for mobile vision applications. ArXiv preprint arXiv:1704.04861 (2017)
32. Kinsley, H.: Object detection with Tensorflow - Self Driving Cars p. 17. https://www.youtube.com/watch?v=UAXulqzn5Ps
33. Babikian, A.A.: Automated generation of test scenario models for the system-level safety assurance of autonomous vehicles. In: Proceedings of the 23rd ACM/IEEE MODELS (2020)

AxS/AI in Context of Future Warfare and Security Environment

Search for Similarity Transformation Between Image Point Clouds Using Geometric Algebra for Conics

Anna Derevianko$^{(\boxtimes)}$ and Pavel Loučka

Brno University of Technology, Brno, Czech Republic
{Anna.Derevianko,Pavel.Loucka}@vutbr.cz

Abstract. We introduce a novel way of searching for the similarity transformation of 2D point clouds using Geometric Algebra for Conics (GAC). In our approach, we do not represent the image objects by their contour but, instead, by the ellipse fitted into the contour points. Such representation makes the consequent similarity search fast and memory-saving. Examples of application on the real object images are also included.

Keywords: Geometric algebra · Clifford algebra · Transformation · Point cloud · Image processing

1 Introduction

The paper considers the search for a transformation, consisting of translation, rotation and scaling, which allows one object to be as close as possible to another in order to simplify their further comparison. However, existing methods are computationally demanding: neural networks or iterative closest point search require plenty of time and memory [7]. Therefore, we have presented a relatively fast method. As a fundamental contour of our object, we consider ellipses inscribed into the extracted contour points in a specific way, namely, using GAC-based Iterative Conic Fitting Algorithm. This will simplify the search for the required transformation.

2 Geometric Algebra for Conics

Let us briefly describe the basic concepts of GAC.

By geometric algebra we mean a Clifford algebra with a specific embedding of a Euclidean space in such a way that the intrinsic geometric primitives as well as their transformations are viewed as its elements, precisely multivectors. For more details see [3] and [2,6].

For our goals we will use the algebra for conics, proposed by C. Perwass to generalize the concept of (two–dimensional) conformal geometric algebra $\mathbb{G}_{3,1}$,

© Springer Nature Switzerland AG 2022
J. Mazal et al. (Eds.): MESAS 2021, LNCS 13207, pp. 215–226, 2022.
https://doi.org/10.1007/978-3-030-98260-7_13

[11] with the notation of [4]. In the usual basis $\bar{n}, e_1, e_2, n$, embedding of a plane in $\mathbb{G}_{3,1}$ is given by

$$(x, y) \mapsto \bar{n} + xe_1 + ye_2 + \frac{1}{2}(x^2 + y^2)n,$$

where e_1, e_2 form Euclidean basis and $\bar{n}$ and n, defined by specific linear combination of additional basis vectors e_3, e_4 with $e_3^2 = 1$ and $e_4^2 = -1$, are the coordinate origin and infinity, respectively, [11]. Hence the objects representable by vectors in $\mathbb{G}_{3,1}$ are linear combinations of $1, x, y, x^2 + y^2$, i.e. circles, lines, point pairs and points. For the general conics, we need to add two terms: $\frac{1}{2}(x^2 - y^2)$ and xy. It turns out that we need two new infinities for that and also their two corresponding counterparts (Witt pairs), [9]. Thus the resulting dimension of the space generating the appropriate geometric algebra is eight.

Let $\mathbb{R}^{5,3}$ denote the eight–dimensional real coordinate space $\mathbb{R}^8$ equipped with a non–degenerate symmetric bilinear form of signature $(5, 3)$. The form defines Clifford algebra $\mathbb{G}_{5,3}$ and this is the Geometric Algebra for Conics in the algebraic sense. To add the geometric meaning we have to describe an embedding of the plane into $\mathbb{R}^{5,3}$. To do so, let us choose a basis of $\mathbb{R}^{5,3}$ such that the corresponding bilinear form is

$$B = \begin{pmatrix} 0 & 0 & -1_{3\times 3} \\ 0 & 1_{2\times 2} & 0 \\ -1_{3\times 3} & 0 & 0 \end{pmatrix}, \tag{1}$$

where $1_{2\times 2}$ and $1_{3\times 3}$ denote the unit matrices. Analogously to CGA and to the notation in [11], the corresponding basis elements are denoted as follows

$$\bar{n}_+, \bar{n}_-, \bar{n}_\times, e_1, e_2, n_+, n_-, n_\times.$$

Note that there are three orthogonal 'origins' $\bar{n}$ and three corresponding orthogonal 'infinities' n. In terms of this basis, a point of the plane $\mathbf{x} \in \mathbb{R}^2$ defined by $\mathbf{x} = xe_1 + ye_2$ is embedded using the operator $\mathcal{C} : \mathbb{R}^2 \to \mathcal{C}one \subset \mathbb{R}^{5,3}$, which is defined by

$$\mathcal{C}(x, y) = \bar{n}_+ + xe_1 + ye_2 + \frac{1}{2}(x^2 + y^2)n_+ + \frac{1}{2}(x^2 - y^2)n_- + xyn_\times. \tag{2}$$

The image $\mathcal{C}one$ of the plane in $\mathbb{R}^{5,3}$ is an analogue of the conformal cone. In fact, it is a two–dimensional real projective variety determined by five homogeneous polynomials of degree one and two.

Definition 1. *Geometric Algebra for Conics (GAC) is the Clifford algebra $\mathbb{G}_{5,3}$ together with the embedding $\mathbb{R}^2 \to \mathbb{R}^{5,3}$ given by (2) in the basis determined by matrix (1).*

Note that, up to the last two terms, the embedding (2) is the embedding of the plane into the two–dimensional conformal geometric algebra $\mathbb{G}_{3,1}$. In particular, it is evident that the scalar product of two embedded points is the same as in $\mathbb{G}_{3,1}$, i.e. for two points $\mathbf{x}, \mathbf{y} \in \mathbb{R}^2$ we have

$$\mathcal{C}(\mathbf{x}) \cdot \mathcal{C}(\mathbf{y}) = -\frac{1}{2} \|\mathbf{x} - \mathbf{y}\|^2, \tag{3}$$

where the standard Euclidean norm is considered on the right hand side. This demonstrates linearisation of distance problems. In particular, each point is represented by a null vector. Let us recall that the invertible algebra elements are called versors and they form a group, the Clifford group, and that conjugations with versors give transformations intrinsic to the algebra. Namely, if the conjugation with a $\mathbb{G}_{5,3}$ versor R preserves the set Cone, i.e. for each $\mathbf{x} \in \mathbb{R}^2$ there exists such a point $\bar{\mathbf{x}} \in \mathbb{R}^2$ that

$$R\mathcal{C}(\mathbf{x})\tilde{R} = \mathcal{C}(\bar{\mathbf{x}}), \tag{4}$$

where $\tilde{R}$ is the reverse of R, then $\mathbf{x} \to \bar{\mathbf{x}}$ induces a transformation $\mathbb{R}^2 \to \mathbb{R}^2$ which is intrinsic to GAC. See [4] to find that the conformal transformations are intrinsic to GAC.

Let us also recall the outer (wedge) product, inner product and the duality

$$A^* = AI^{-1}. \tag{5}$$

However we use the definitions as in [11]. Note that in GAC the pseudoscalar is given by $I = \bar{n}_+\bar{n}_-\bar{n}_\times e_1 e_2 n_+ n_- n_\times$. For our purposes, we stress that these operations correspond to sums and products only. Indeed, the wedge product is calculated as the outer product of vectors on each vector space of the same grade blades, while the inner product acts on these spaces as the scalar product. The extension of both operations to general multivectors adds no computational complexity due to linearity of both operations. Let us also recall that if a conic C is seen as a wedge of five different points (which determines a conic uniquely), we call the appropriate 5–vector E^* an outer product null space representation (OPNS) and its dual E, indeed a 1–vector, the inner product null space (IPNS) representation.

Let us recall the definition of inner product representation. An element $A_I \in \mathbb{G}_{5,3}$ is the inner product representation of a geometric entity A in the plane if and only if $A = \{\mathbf{x} \in \mathbb{R}^2 : \mathcal{C}(\mathbf{x}) \cdot A_I = 0\}$. Hence, given a fixed geometric algebra, the representable objects can be found by examining the inner product of a vector and an embedded point. A general vector in the conic space $\mathbb{R}^{5,3}$ in terms of our basis is of the form

$$v = \bar{v}^+\bar{n}_+ + \bar{v}^-\bar{n}_- + \bar{v}^\times\bar{n}_\times + v^1 e_1 + v^2 e_2 + v^+ n_+ + v^- n_- + v^\times n_\times$$

Hence a conic is uniquely represented (in a homogeneous sense) by a vector in $\mathbb{R}^{5,3}$ modulo this subspace. This gives the desired dimension six. In other words, the inner representation of a conic in GAC can be defined as a vector

$$Q_I = \bar{v}^+\bar{n}_+ + \bar{v}^-\bar{n}_- + \bar{v}^\times\bar{n}_\times + v^1 e_1 + v^2 e_2 + v^+ n_+. \tag{6}$$

The classification of conics is well known. Among the non–degenerate conics there are three types, the ellipse, hyperbola, and parabola. Now, we present the vector form (6) appropriate to the simplest case, i.e. an axes–aligned ellipse E_I with its centre in the origin and semi–axes a, b. Correctness may be verified easily by multiplying its vector by an embedded point which means the application of (1) and (2). The corresponding GAC vector is of the form

$$E_I = (a^2 + b^2)\bar{n}_+ + (a^2 - b^2)\bar{n}_- - a^2 b^2 n_+. \tag{7}$$

More generally, an ellipse E with the semi–axes a, b centred in $(u, v) \in \mathbb{R}^2$ rotated by angle θ is in the GAC inner product null space (IPNS) representation given by

$$
\begin{aligned}
E = {} & \bar{n}_+ - (\alpha \cos 2\theta)\bar{n}_- - (\alpha \sin 2\theta)\bar{n}_\times \\
& + (u + u\alpha \cos 2\theta - v\alpha \sin 2\theta)e_1 + (v + v\alpha \cos 2\theta - u\alpha \sin 2\theta)e_2 \\
& + \tfrac{1}{2}\left(u^2 + v^2 - \beta - (u^2 - v^2)\alpha \cos 2\theta - 2uv\alpha \sin 2\theta\right) n_+.
\end{aligned}
\tag{8}
$$

For proofs and further details about other conics see [4].

2.1 Parameter Extraction

It is well known that the type of a given unknown conic can be read off its matrix representation, which in our case for a conic given by

$$
Q = \begin{pmatrix}
-\tfrac{1}{2}(\bar{v}^+ + \bar{v}^-) & -\tfrac{1}{2}\bar{v}^\times & \tfrac{1}{2}v^1 \\
-\tfrac{1}{2}\bar{v}^\times & -\tfrac{1}{2}(\bar{v}^+ - \bar{v}^-) & \tfrac{1}{2}v^2 \\
\tfrac{1}{2}v^1 & \tfrac{1}{2}v^2 & -v^+
\end{pmatrix}. \tag{9}
$$

The entries of (9) can be easily computed by means of the inner product:

$$
\begin{aligned}
q_{11} &= Q_I \cdot \tfrac{1}{2}(n_+ + n_-), \\
q_{22} &= Q_I \cdot \tfrac{1}{2}(n_+ - n_-), \\
q_{33} &= Q_I \cdot \bar{n}_+, \\
q_{12} &= q_{21} = Q_I \cdot \tfrac{1}{2}n_\times, \\
q_{13} &= q_{31} = Q_I \cdot \tfrac{1}{2}e_1, \\
q_{23} &= q_{32} = Q_I \cdot \tfrac{1}{2}e_2.
\end{aligned}
$$

It is also well known how to determine the internal parameters of an unknown conic and its position and the orientation in the plane from the matrix (9). Hence all this can be determined from the GAC vector Q_I by means of the inner product.

The parameters of a conic can be obtained from the matrix (9) of its IPNS representation, for example:

– center of an ellipse or hyperbola:

$$
x_c = \frac{q_{12}q_{23} - 2q_{22}q_{13}}{4q_{11}q_{22} - q_{12}^2}, \quad y_c = \frac{q_{13}q_{12} - 2q_{11}q_{23}}{4q_{11}q_{22} - q_{12}^2} \tag{10}
$$

– semiaxis of an ellipse:

$$a, b = \frac{\sqrt{(2A(q_{11} + q_{22} \pm \sqrt{(q_{11} - q_{22})^2 + q_{12}^2})}}{(4q_{11}q_{22} - q_{12}^2))}, \tag{11}$$

where $A = q_{11}q_{23}^2 + q_{22}q_{13}^2 - q_{12}q_{13}q_{23} + (q_{12}^2 - 4q_{11}q_{22})q_{33}$
– angle of rotation

$$\theta = \begin{cases} -\arctan \frac{q_{22} - q_{11} - \sqrt{(q_{11} - q_{22})^2 + q_{12}^2}}{q_{12}}, & q_{12} \neq 0 \\ 0, & q_{12} = 0, \quad q_{11} < q_{22} \\ \frac{\pi}{2}, & q_{12} = 0, \quad q_{11} > q_{22} \end{cases} \tag{12}$$

Other parameters can be derived with the help of eigenvalues of the quadratic form matrix. For more details see [8].

2.2 Transformations

The main advantage of GAC compared to other models (for instance, $\mathbb{G}_6$) is that it is fully operational in the sense that it allows all Euclidean transformations, i.e. rotations and translations. But not just that, it also allows scaling in the sense of (4). Hence, like in the case of CGA (or $\mathbb{G}_{3,1}$), one obtains all conformal transformations. The exact form of GAC versor for rotation (rotor), translation (translator), and scaling (scalor) is given as follows.

The rotor for a rotation around the origin by the angle φ is given by $R = R_+(R_1 \wedge R_2)$, where

$$R_+ = \cos(\tfrac{\varphi}{2}) + \sin(\tfrac{\varphi}{2})e_1 \wedge e_2, \tag{13}$$
$$R_1 = \cos(\varphi) + \sin(\varphi)\bar{n}_\times \wedge n_-, \tag{14}$$
$$R_2 = \cos(\varphi) - \sin(\varphi)\bar{n}_- \wedge n_\times. \tag{15}$$

The translator is given by $T = T_+T_-T_\times$, where

$$T_+ = 1 - \tfrac{1}{2}ue_1 \wedge n_+ \tag{16}$$
$$T_- = 1 - \tfrac{1}{2}ue_1 \wedge n_- + \tfrac{1}{4}u^2 n_+ \wedge n_- \tag{17}$$
$$T_\times = 1 - \tfrac{1}{2}ue_2 \wedge n_\times \tag{18}$$

for a translation in the direction e_1 around u. Similarly, for a translation in the direction e_2 around v one has

$$T_+ = 1 - \tfrac{1}{2}ve_2 \wedge n_+ \tag{19}$$
$$T_- = 1 + \tfrac{1}{2}ve_2 \wedge n_- - \tfrac{1}{4}v^2 n_+ \wedge n_- \tag{20}$$
$$T_\times = 1 - \tfrac{1}{2}ve_1 \wedge n_\times \tag{21}$$

The scalor for a scaling by $\alpha \in \mathbb{R}^+$ is given by $S = S_+ S_- S_\times$, where

$$S_+ = \frac{\alpha+1}{2\sqrt{\alpha}} + \frac{\alpha-1}{2\sqrt{\alpha}} \bar{n}_+ \wedge n_+, \tag{22}$$

$$S_- = \frac{\alpha+1}{2\sqrt{\alpha}} + \frac{\alpha-1}{2\sqrt{\alpha}} \bar{n}_- \wedge n_-, \tag{23}$$

$$S_\times = \frac{\alpha+1}{2\sqrt{\alpha}} + \frac{\alpha-1}{2\sqrt{\alpha}} \bar{n}_\times \wedge n_\times. \tag{24}$$

For proof see [4]. All transformations apply on a vector in GAC by conjugation (4) of the appropriate versor formal exponential. This holds also for translations and rotations, for their precise form see [4]. Consider python implementation of transformations. Now let us demonstrate transformations by visualization on example.

Example 1. Consider axis-aligned ellipse with the semi-axes $a = 2, b = 4$ centred in $(u, v) = (0, 2)$.

The result of applying the rotation by angle $\phi = \frac{\pi}{6}$, scaling by 2 and translations by vectors $(0, 2), (2, 0), (-3, 2)$ respectively is shown in Fig. 1 [1].

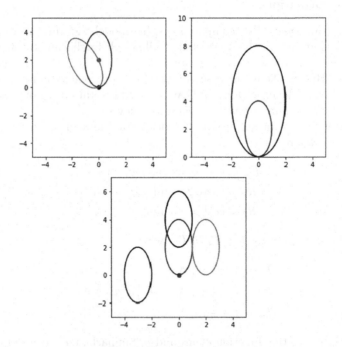

Fig. 1. Transformations

3   Conic Fitting Algorithm

The iterative conic fitting algorithm employed in the article uses a direct GAC-based conic fitting algorithm described in [5]. Even though the direct fitting algorithm is computationally fast and possesses invariance w.r.t. rotations and isotropic scaling, it is not invariant w.r.t. translations, and thus not suitable for our purposes. Fortunately, we can largely remedy this flaw by using the iterative algorithm (Algorithm 1), which is "almost invariant" (i.e. invariant up to the prescribed precision) w.r.t. translations, rotations and scaling as well.

Comparison of direct and iterative algorithm regarding the translational (non-)invariance can be seen in Fig. 2.

The main idea of the iterative algorithm can be summed up in two steps:

1. Translate the initial point set in such way that the conic fitted to the translated point set using the direct fitting algorithm has the centre at the origin of coordinate system.
2. Apply the inverse translation of the fitted origin-centred conic.

Alas, because it is nearly impossible to analytically calculate the translation necessary for step 1., it cannot be performed directly in terms of the direct algorithm. Therefore, instead of step 1., we iteratively move the point set closer and closer to the origin of coordinates as long as the fitted conic has its centre sufficiently close to the origin.

Algorithm 1. Iterative Conic Fitting Algorithm

    Inputs: point set p, precision ε, maximum number of iterations k_{max}
    Output: conic C_f

1: Using the direct conic fitting algorithm from [5], fit the point set p with the initial conic C
2: Find the centre $S_0 = (x_{S0}, y_{S0})$ of the initial conic C
3: $k \leftarrow 0$ ▷ number of iterations
4: $d \leftarrow \|S_0\|$ ▷ distance between the fitted conic's centre and the coordinate system origin
5: $S \leftarrow S_0$ ▷ centre of a conic C
6: **repeat**
7:      $p \leftarrow$ point set p translated by vector $-S$
8:      Using the direct conic fitting algorithm from [5], fit the translated point set with a conic C
9:      Find the centre $S = (x_S, y_S)$ of the conic C
10:      $d \leftarrow \|S\|$
11:      $k \leftarrow k + 1$
12: **until** $d < \varepsilon$ or $k = k_{max}$
13: $C_f \leftarrow$ conic C translated by vector $S_0 - S$

---

4 Algorithm for Finding the Transformation

Scaling parameter for our transformation can be found by Algorithm 2.

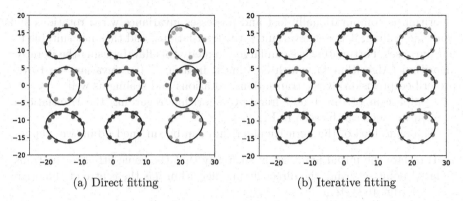

(a) Direct fitting   (b) Iterative fitting

Fig. 2. Comparison of conic fitting algorithms applied on a differently translated sample point set. Precision prescribed for the iterative algorithm was $\varepsilon = 10^{-6}$ and, in all cases, 3 or 4 iterations sufficed to achieve the precision.

Algorithm 2

    Inputs: two co–centric ellipses, where one is the scaled copy of another
    Output: scaling parameter SP

1: Construct a line l passing through the points $\bar{n}^+$ and e_2:

$$l = e_2 \wedge n_+ \wedge \bar{n}_+ \wedge n_- \wedge n_\times.$$

2: Find the intersection points $C = E_1 \cap l$, $B = E_2 \cap l$ of the line and both ellipses, i.e. solve a quadratic equation in a Euclidean space.
3: The scale parameter between ellipses is

$$SP = \frac{|\bar{n}_+ \cdot \mathcal{C}(B)|}{|\bar{n}_+ \cdot \mathcal{C}(C)|}.$$

For more details, see [1]. Therefore, we may find the desired transformation, using the following algorithm.

Algorithm 3

    Inputs: $Image_1$, $Image_2$
    Output: Transformation Tr

1: **for** $i = 1, 2$ **do**
2:     Upload $Image_i$ and create a binary thresholded $ImageT_i$
3:     Find the contours in the thresholded $ImageT_i$
4:     Find the contour ($Contour_i$) enclosing the biggest area
5:     Apply **Algorithm 1** to $Contour_i$
6:     Extract parameters $x_c^i, y_c^i, a_i, b_i, \theta_i$ from the fitted conic C_i
7:     Apply translation T of conic C_i by vector $(-x_c^i, -y_c^i)$
8: Find the transformation $Tr = SR$ for the conic C_2, where
    R is rotation of a conic around the origin by the angle $\theta_1 - \theta_2$;
    S is scaling of a conic by factor s; which is found in **Algorithm 2**

By applying transformation Tr to the conic C_2 in a way $TrC_2\widetilde{Tr}$ we get the conic, aligned with the C_1. By applying that transformation to the all points in cloud set it is possible to get aligned point cloud sets. Perceptual hash algorithms describe a class of functions for generating comparable hashes. Image characteristics are used to generate an individual (but not unique) fingerprint, and these fingerprints can be compared with each other.

For our research we use perceptual hashes. In cryptography, every hash is random. The data used to generate the hash acts as a source of random numbers, so that the same data will give the same result, but different data will give a different result. If the hashes are different, then the data is different. If the hashes match, then the data is most likely the same (since there is a possibility of collisions, the same hashes do not guarantee that the data will match). In contrast, perceptual hashes can be compared with each other and inferred about the degree of difference between the two datasets [10].

5   Simulation

Let us demonstrate the operation of the algorithm using the following example.

First let us compare two images using perceptual hashes. The parameter shows the degree of similarity of the images. For similar images the output parameter is up to 15, for different images it is more than 15. We will try to apply the algorithm for the same image, but with a rotation. as a result, we get a parameter equal to 27, that shows pictures as different ones.

```
hash1   1111111111111111111111111111111111111000111000000110000000
10000001
hash2   1111111110011111100001111000011111000101100001111110001111
11111111
result:  27
```

Now we will find the transformation and apply it to the image (rotate and translate).

Consider the image in Fig. 3. After highlighting the contours, the maximal one was selected, it is shown in Fig. 3c. After fitting the ellipse into the given contour, an ellipse was obtained with the following parameters:

```
S = [209.28270157371304, 219.21051222645468]
a = 143.03570493523614, b = 39.44706773871411
theta = 179.01303550495453°
```

Therefore, the necessary translation for moving the ellipse center to the origin is $T = T(-209.28270157371304, -219.21051222645468)$ (Fig. 4).

We carry out a similar procedure with another image and, as a result, we obtain an ellipse with the following parameters:

```
S = [78.46521295752942, 208.2824206859099]
a = 144.4629567013477
b = 39.379824424229525
theta = 88.9105709714248°
```

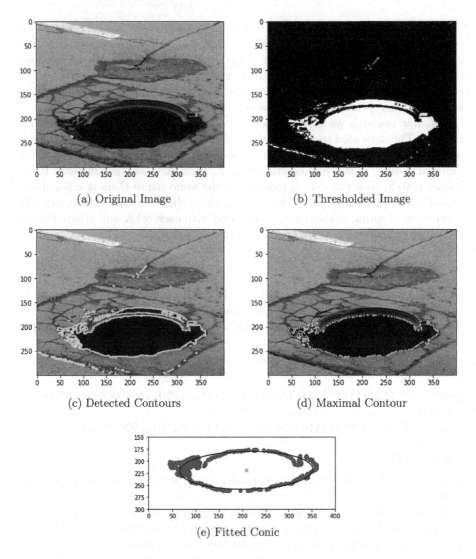

(a) Original Image

(b) Thresholded Image

(c) Detected Contours

(d) Maximal Contour

(e) Fitted Conic

Fig. 3. Simulation

Note that transferring conics to the origin was necessary for the scaling procedure to work correctly. After transferring both ellipses to the origin rotate them by the angle difference $R = R(90.1024645335°)$. In our particular case we get scaling parameter equal to 1 due to the fact that sample images were just rotated. So the needed transformation consists only of translation and rotation. Now the parameter is already equal to 3, so images are found to be similar.

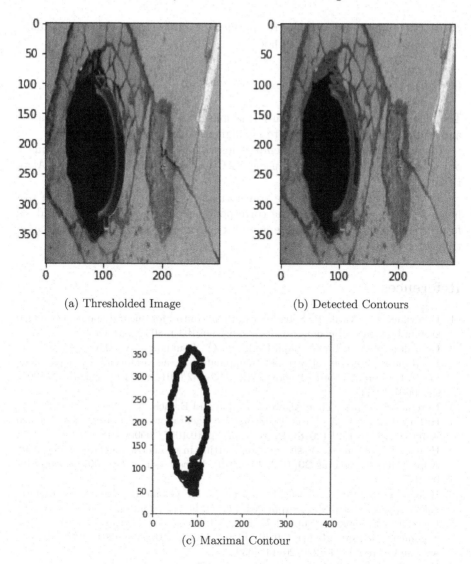

(a) Thresholded Image   (b) Detected Contours

(c) Maximal Contour

Fig. 4. Simulation

```
hash1  11111111111111111111111111111111111110001110000000110000000
10000001
hash2  11111111111111111111111111111111111110000110000000011000000011000001
result:  3
```

Perceptual hashes method works well in the case when the pictures have different sizes, and it also works when there is a little noise in the image (the noise disappears when the picture is scaled). Using our transformation (rotation, translation and scaling) we solve the problem of rotated images, that are usually found to be different by this method.

6   Conclusion

The algorithm, searching for transformation, consisting of translation, rotation and scaling, allowing one object to be aligned to another, was presented.

It was proposed to consider ellipses inscribed into the contour of an image object in a specific way, namely, using GAC-based Iterative Conic Fitting Algorithm.

The example of applying the transformation search algorithm on the real image was demonstrated and the corresponding transformation was found. In future it is planned to use more complex images and extend the list of used transformations.

References

1. Derevianko, A., Vašík, P.: Solver-free optimal control for linear dynamical switched system by means of geometric algebra. arXiv:2103.13803 [math.OC]
2. Gonzalez-Jimenez, L., Carbajal-Espinosa, O., Loukianov, A., Bayro-Corrochano, E.: Robust pose control of robot manipulators using conformal geometric algebra. Adv. Appl. Clifford Algebras **24**(2), 533–552 (2014). https://doi.org/10.1007/s00006-014-0448-2
3. Hestenes, D.: Space-Time Algebra. Gordon and Breach, New York (1966)
4. Hrdina, J., Návrat, A., Vašík, P.: Geometric algebra for conics. Adv. Appl. Clifford Algebras **28**(3), 1–21 (2018). https://doi.org/10.1007/s00006-018-0879-2
5. Hrdina, J., Návrat, A., Vašík, P.: Conic fitting in geometric algebra setting. Adv. Appl. Clifford Algebras **29**(4), 1–13 (2019). https://doi.org/10.1007/s00006-019-0989-5
6. Hrdina, J., Návrat, A., Vašík, P., Matoušek, R.: Geometric algebras for uniform colour spaces. Math. Meth. Appl. Sci. (2017). https://doi.org/10.1002/mma.4489
7. Khazari, A.E., Que, Y., Sung, T.L., Lee, H.J.: Deep global features for point cloud alignment. Sensors **20**(14) (2020). https://doi.org/10.3390/s20144032. https://www.mdpi.com/1424-8220/20/14/4032
8. Korn, G., Korn, T.: Mathematical Handbook for Scientists and Engineers: Definitions, Theorems, and Formulas for Reference and Review. Dover Civil and Mechanical Engineering Series, Dover Publications (2000). https://books.google.cz/books?id=xHNd5zCXt-EC
9. Lounesto, P.: Clifford Algebra and Spinors, 2nd edn. CUP, Cambridge (2006)
10. Niu, X.M., Jiao, Y.H.: Overview of perceptual hashing, **36**, 1405–1411 (2008)
11. Perwass, C.: Geometric Algebra with Applications in Engineering. Springer, Heidelberg (2009). https://doi.org/10.1007/978-3-540-89068-3

Using SUMO to Construct Dynamic Urban Modeling Scenarios for Military Transport

Mark Anthony Cowan[✉]

United States Army Corps of Engineers ERDC ITL CAB, Vicksburg, MS, USA
Mark.A.Cowan@usace.army.mil

Abstract. SUMO (Simulation of Urban MObility) is a microscopic simulator of continuous multimodal vehicular and pedestrian traffic along large urban road networks. First released in 2001 by Berlin's Institute of Transportation Systems, SUMO has been used to investigate the effects of vehicle pollution and noise, to generate traffic forecasts during large athletic events, and to model in-vehicle telephony to assess performance of traffic surveillance devices. A SUMO scenario can quickly be initiated by importing street network topologies from OpenStreetMap (OSM), a freely-available, constantly updated, crowd-sourced and -tagged mapping service. With the OSM network as a foundation, one can easily add, delete, and modify traffic lanes and the timing of traffic lights to explore the effects upon local traffic over time as the populace attempts to route between its origin and destination pairs, making adjustments to their routes as necessary. The tags within OSM add the possibility of choosing origin-destination regions for the motorists and pedestrians based upon city zoning, from which we can likely infer some features of the demographic layout and thereby add more realism to the traffic simulation.

In this paper, we will focus on building these scenarios for military transport across large urban areas and collecting the results across many runs of the SUMO software, varied by the random seeding of the model and changing the open/close times of some important lanes. While limited visualization tools exist for single runs, we will fortify these with a more global view and summarize the analytic results for military decision-makers, enabling them to anticipate potential traffic bottlenecks that could interfere with their mission and to choose optimally among alternate routes on-the-fly as new information arrives.

Keywords: Military transport · Scenario · Mobility · Simulation · SUMO · Urban · Data analysis

1 Simulation of Urban Mobility (SUMO)

1.1 SUMO's Purpose

Recognizing that graduate students were perennially implementing their own complex simulators of widely-varying quality as the initial steps to study traffic phenomena, developers at Berlin's Institute of Transportation Systems released SUMO (Simulation

This is a U.S. government work and not under copyright protection in the U.S.; foreign copyright protection may apply 2022
J. Mazal et al. (Eds.): MESAS 2021, LNCS 13207, pp. 227–248, 2022.
https://doi.org/10.1007/978-3-030-98260-7_14

of Urban MObility) [1] in 2001 as a common trusted technical sandbox to explore these types of questions. This saved the students' time by re-directing their focus to the more pertinent research questions at-hand, rather than verifying and validating the inner workings of home-grown traffic simulation engines. Since that first code release, SUMO has continued development and grown into a mature microscopic simulator of multimodal vehicular and pedestrian traffic within large urban road networks. Over the years since its inception, preliminary development, and initial public release, SUMO has continued to be used in academic circles to research different aspects of traffic behavior, but it has also been widely adopted as a tool-of-choice by urban planners and municipalities to project how infrastructural alternatives could affect traffic behavior and mobility of their urban populace, both during and after construction. Suddenly policy-makers have a tool enabling them to analyze thoroughly the options they face, quantify the benefits and disadvantages for their citizenry and the commercial entities within their region, and justify the difficult and costly decisions they must take. Examples include: simulating traffic in Berlin, carbon reduction studies, real-time road management solutions, traffic monitoring by cell phones, traffic forecasting for the FIFA World Cup tournament in 2006, optical sensor studies for the optimization of traffic lights, evaluation of the effects of C2X communication on traffic, and others [2].

1.2 Programmatic Description of SUMO

Programmatically, SUMO is an open source, freely-available package running on both the Windows and the Linux operating systems. (For the Debian and Ubuntu variants of Linux, SUMO is part of the regular distribution packages and can generally be installed worry-free). Coded in C++, SUMO's functionality can be accessed via the Traffic Control Interface [3] API (usually denoted TraCI) by the Python, C++, .NET, Matlab, and Java programming languages with varying degrees of completeness. Python presently supports all TraCI commands. As open source, all SUMO code can be downloaded and examined, for instance, to see various algorithm implementations, to explore the hooks provided by the API, to walk through the data structures, or simply to debug a newly-discovered problem. No matter the underlying purpose, SUMO's code is transparent and hence, does not suffer from weaknesses often inherent in proprietary code. As such, it lowers the barriers to the rapid gaining of expertise by its new users and benefits immensely in a security/performance/future expansion context from the additional eyes searching through the code. Its only legal demand upon the user is the acceptance of the Eclipse Public License (version 2.0) [4].

SUMO comes in two flavors: GUI-driven and command line-driven. Early exploration of traffic behavior upon a network is often conducted within the confines of the GUI binary to harvest all the benefits of visualization for debugging purposes. Within the GUI, the model is simpler to configure, especially in its earliest stages. There the network can be confirmed to be connected properly, the lanes and signage (or traffic lights) correctly placed, and the general distribution of the origin-destination pairs for vehicles and pedestrians verified. Visualization of the environment and the movement of the traffic is key in the discovery and eventual rectification of anomalous results. Pedestrians and vehicles can be highlighted and tagged to draw user attention to behaviors under investigation. Various internal GUI tools permit customization of the environment, vehicle properties, and pedestrian actions. Any variance from expectations can easily be teased out, the conditions under which they occurred noted, and modifications implemented to pull outliers back into line. However, once all of these conditions meet the standard, it is suggested that the researcher fall back to the command line-driven binary for computational speed efficiency. All desired options can be called from the command line, and the resulting model output will be the same as for the GUI, with a noticeable improvement in time-to-completion. As will be described more completely later in this paper, the command line-driven binary can be used in parallel among multiple computer cores and driven by a different random seed. From the command line, the user can also specify the level of detail required as model output from a simple and very general summary to very highly-detailed information about each vehicle and each pedestrian within the simulation at each instant in time.

1.3 Near-Term Future Extensions of SUMO and Its Ancillary Tools

SUMO remains a viable candidate for recent and near-future research involving traffic modeling. For example, SUMO is being used to plan city infrastructural changes due to the introduction of autonomous cars. SUMO package extensions and integration with other communication-centered models have offered insight into message-passing between cars with the ultimate aim of smoothing out traffic flow. As autonomous cars begin to interact with traditional driver-managed cars in a hybrid traffic mix, SUMO is helping to determine placement of electric vehicle chargers to optimize the power grid load. Recent research and some forward-leaning future studies are presented at the annual SUMO conference in Berlin. For additional examples of SUMO's wide-spread and growing use, the conference proceeding are freely-available online for download and review [5].

1.4 USACE Research Needs for Dynamic Elements in an Urban Environment

For many years the US Army Corps of Engineers Geotechnical and Structures Laboratory (GSL) at the Engineer Research and Development Center (ERDC) has researched the mobility of military transport convoys within an urban environment with an eye toward anticipating blockages or slow-downs in the route ahead, modifying the present route (if necessary), and completing the mission in a time-efficient manner. Foreseeing these traffic jams and providing enough time to choose alternative routes is assisted by use of unmanned aerial systems (UAS) containing diverse sensor platforms feeding data back to the command center and/or to the transport teams on the ground. Presently a virtual environment is being created inside the ANVEL software suite [6] to account for the mobility characteristics of various military vehicles, environmental challenges, and sensor platform noise. However, lacking until now was the dynamic interaction of the convoy with other vehicles and pedestrians, all of whom may exhibit unexpected, mission-affecting behaviors. SUMO will provide this missing link for the Mobility in Complex Urban Environments System Model and Integration (MCUE SMI) program and give the user an opportunity to test widely across individual driver behaviors (say, to be more or less aggressive) and pedestrian clustering parameters (for instance, to be more or less prone to cluster in or near intersections to obstruct the flow of traffic).

2 Creation of a Road Network in SUMO

2.1 Capture OpenStreetMap Topologies with osmWebWizard

Vehicles and pedestrians require a road and sidewalk network (respectively) to move upon. These networks can be created *ex nihilo* by hand within the SUMO GUI binary (or from ASCII text files). For very small scenarios, where some very basic functionality is being tested, a hand-rolled network may suffice, and SUMO tutorials [7] provide ample insight into how to accomplish this. However, for more realistic simulation environments (such as needed for military transport within an urban area), SUMO offers the opportunity to import real-world-generated road (and sidewalk) data. Among the many data sources for this purpose is OpenStreetMap (OSM), a freely-available crowd-sourced map that collects and incorporates modifications to the traffic networks constantly from volunteers uploading new data as the ground situation changes (perhaps due to lane removal, widening, or other construction efforts). Likewise, there are several venues on the internet for downloading OSM datasets; however, many, if not most, of them will require special handling or additional conversion to push the data into a form acceptable for SUMO ingestion. To sidestep this issue completely, one should use the Python script osmWebWizard.py, usually found in the $SUMO_HOME/tools directory. Kickoff of the script within a terminal (or within a Python engine) initiates a web browser that is configured initially to show a portion of metropolitan Berlin, as shown in Fig. 1.

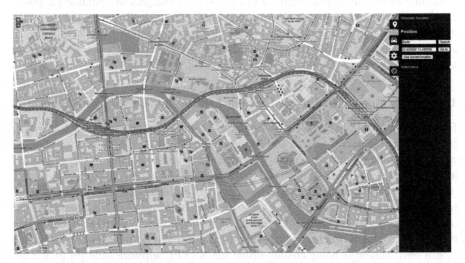

Fig. 1. Opening screen for Python tool OSMWebWizard, centered on Berlin, Germany

To the right of the map, four buttons are displayed, corresponding to (1) position map, (2) configure vehicles, (3) setting simulation options, and (4) displaying icon copyright information, the latter of which may safely be ignored. Clicking on any of these buttons will expand or contract the configuration pane from the browser's right-most side.

Pressing the top button (position map in Fig. 2) prompts the user to select a scenario position, either by searching on the name of a geographical entity, by specifying a latitude/longitude pair, by accepting the current location, or by moving the map to the desired location and selecting its area by defining a bounding box. The map within the browser reacts to mouse actions similarly to Google (TM) Maps and other online map servers— i.e., the area displayed can be changed by sliding the map to the left or right, and the

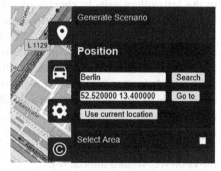

Fig. 2. The position map tool

map can be zoomed into or away from by use of the mouse scroll wheel. Of course, the more complicated the urban area (i.e., the more streets it has highlighted by the user), the bigger the resulting file will be. At times, the OSM server may simply refuse to deliver the product, if the chosen area is too large. The user should chose a minimally-acceptable urban area over which to conduct the simulation. This is also beneficial later from a SUMO performance perspective. The author had little luck with the Search (by name) function, but that may be due to some nonstandard nomenclature, instead of functional deficiency in the tool. The latitude/longitude lookup, on the other hand, worked well. When the "Select Area" box is checked, a default region placed upon the map may not be suitable for the desired scenario. If needed, one can move the mouse pointer to any corner of the default highlighted region and drag the corners to the best position. A similar result can be obtained by realigning the edges instead, as shown in Fig. 3.

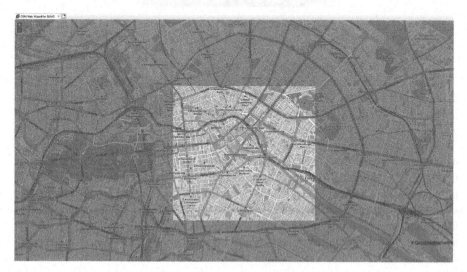

Fig. 3. The select area tool, centered in Berlin.

The second button from the top (as shown in Fig. 4 to the left) offers vehicle choices of cars, trucks, buses, motorcycles, bicycles, pedestrians, trams, urban trains, trains, and ships. Of these, this paper will exclusively focus on cars and pedestrians. Any combination of these may be included within the scenario, simply by checking the box to the right of the desired vehicle and entering acceptable values in the numerical boxes defining the "through factor" and the overall "count". These parameters are precisely defined under the "Demand Generation" section of the OSMWebWizard tutorial [8].

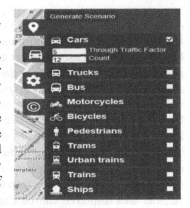

Fig. 4. The configure vehicles tool

Fig. 5. Setting simulation options tool

By default SUMO's time step is in unit seconds. This may be modified on the command line with the step-length switch; however for the purposes of this paper, a unit of one second will be assumed. The third button from the top of the side panel (displayed in Fig. 5) shows a default simulation duration of 3600 s (i.e., one hour), but the user may configure this to any value needed.

When all configuration elements have been specified, pressing the "Generate Scenario" button at the very top of the panel will write out all configuration files meeting the required conditions to the user's tools directory (or a preconfigured directory, if provided) within a datestamped and timestamped directory having format yyyy-mm-dd-hh-mm-ss. Inside this directory are the xml files defining Dijkstra routes for all of the chosen vehicles types from randomly-selected origins and destinations, along with alternate routes, and a couple of other SUMO configuration files related to the network construction.

2.2 Configure the Traffic Lights System with NETEDIT

Although the OpenStreetMap file contains the position of every tagged traffic light, it lacks the logic underpinning their behavior. This is supplemented by NETEDIT, which is callable directly from the SUMO-GUI binary (or it may be executed independently as the need arises). Within the SUMO-GUI, NETEDIT can be accessed from the bottom choice ("Open in Netedit...") in the "Edit" pulldown menu. Within the NETEDIT program, one can add logical behaviors to the junctions by entering the "Traffic Light Mode", specifying the Junction ID by left clicking on the desired intersection, and issuing the command to "Create TLS". Below this, one can see the traffic light cycle is timed to 90 s (by default), and a particularly adventuresome user can modify the traffic light's behavior in detail by changing the Phases. Assigning a logical behavior to an intersection's lights [9] can be repeated for as many lights as desired. To incorporate the final results into the SUMO-GUI simulation run, the TLS must be saved in XML format and the underlying *.net.xml file (representing the network) must be saved as well. Then the vehicles' response to the newly-created traffic lights can be tested by returning to the SUMO-GUI program and reloading the affected configuration files. See Fig. 6 for an example.

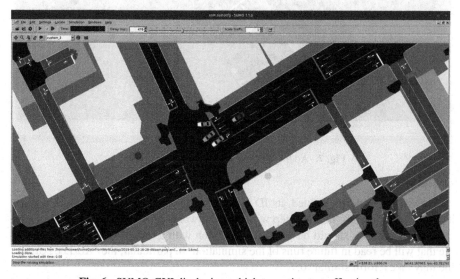

Fig. 6. SUMO-GUI displaying vehicles reacting to traffic signals

Perhaps a simulationist is not satisfied with the network as downloaded from Open-StreetMap, but rather requires changes (say, due to future construction) and would like to simulate traffic under these future conditions. Within NETEDIT, the "Edit" functionality permits one to add edges, make connections, move or delete various elements of the network, and inspect the results of the changes. The entire topology is open to modification—from substantial to mundane, but care must be shown to maintain reasonable connectivity at the intersection junctions. In-depth discussion of all the available editing tools is well outside the scope of this paper, but detailed tutorials on these network editing activities [10] can be found at the SUMO website.

2.3 Embed Rerouters to Divert Traffic Temporarily

Suppose that, during the course of running a military scenario, a lane (or group of lanes) should not be available for public use, say, from 8 AM until 6 PM. Outside of these hours, normal traffic is allowed. The first step in rerouting traffic is to define a rerouter along an edge within the network (immediately before the edge to isolate from traffic) using SUMO-GUI. This simply entails right clicking on a lane (not an edge) and choosing to "Add rerouter". Figure 7 below gives an example of setting rerouters across six lanes of traffic.

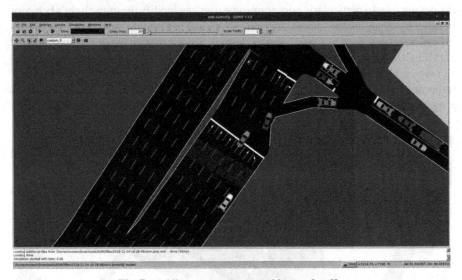

Fig. 7. Adding rerouters across 6 lanes of traffic

Once the rerouter is defined, its ID is available by right clicking upon the rerouter icon positioned during the last step within the network. That ID and any additionally-created rerouter IDs should be saved for inclusion with a text-based "additional" file that will be read in future runs of the simulation. In particular, the "additional" file may take a form similar to:

```
<rerouter id="rerouter_0" edges="-17389805#4">
        <interval begin="300" end="1300">
                <destProbReroute id="181447605"/>
        </interval>
</rerouter>
```

for each rerouter created. This excerpt shows the rerouter ID (here, "rerouter_0") placed upon an edge denoted "-17389805#4". The time during which the neighboring edge will be unavailable begins at 300 s into the simulation and ends at 1300 s into the run, for a total of 1000 s of lane closure. Any vehicle originally routed along this edge is rerouted to "181447605" for the duration. Then the information for the interval and the rerouter is closed, and additional rerouter information can be placed after this to ensure complete closure across all lanes leading to the isolated lane that is being temporarily removed from service. There are many more switches that can be activated in regards to traffic rerouting than are hinted at here, including user-defined probability of taking different lanes, rather than always taking a single predefined lane (as shown above). Many more details about coding the rerouter parameters are presented in the online documentation [11].

To ensure inclusion of all rerouter activities within the following simulation runs, the file "osm.sumocfg" must be amended. Typically, near the top of the XML file, there is a section devoted to model input, where the network file is defined, followed by a list of route files. "Additional" files come immediately below that, often including a reference to "poly" files (showing the polygonal footprint of buildings within the SUMO-GUI visualization). Since all files in the "osm.sumocfg" are comma-separated, one may simply place a comma following the poly file name and add on the rerouter's file name, for example:

```
<additional-files value="osm.poly.xml,myRerouters.add.xml"/>
```

Once this has been saved and all configuration files reloaded within SUMO-GUI, the user should see the effects of the rerouters during the shutdown time period as defined above by the interval begin and end time (in simulator seconds).

3 Initialization of Vehicles on Network

3.1 Populate SUMO's World with Vehicles Using RandomTrips.py

SUMO's network connectivity and traffic signals are of prime importance for the successful running of a military transport simulation. However, without the presence of vehicles, the simulation has no interest for military decision-makers. There are at least nine methods to populate the SUMO environment with vehicles [12], but this paper will describe only one of the simplest methods, namely, use of the Python script randomTrips.py, which will uniformly distribute the origin and destination points for all vehicles and pedestrians throughout the network. This script is found under the $SUMO_HOME/tools directory and can be initiated as:

```
<SUMO_HOME>/tools/randomTrips.py -n osm.net.xml -e 50
```

Here the script is called with a network defined by the file "osm.net.xml" and a trip end time defined at 50 s into the simulation (and the start time is defined as 0 by default [though user-changeable]). Obviously a simulationist may desire that the vehicles be active in the simulation for much longer than these 50 s or that the start time begin later than the defaulted 0 s. These parameters and many more switches to this program are defined and detailed thoroughly in the SUMO documentation [13]. The instructions here, however, should be sufficient to get started and begin to see vehicles moving through the SUMO-GUI visualization. As output, the `randomTrips.py` script generates files that have the form `<networkName>.<vehicleType>.trips.xml`, across all vehicle types previously defined in the OSMWebWizard. For example, an output file may take the name "`osm.truck.trips.xml`" or "`osm.bus.trips.xml`".

Although the topic will not be explored deeply here, randomly-placed origin and destination pairs may not serve the needs of the simulationist well. Among the eight other methods of generating trip input information for SUMO is a tool called ACTIVITYGEN. If demographics data for the network under consideration can be obtained, say, from a census source, or be inferred from other available geographically-informed population data, it is possible to introduce more "ground truth" into the model by having vehicles originating near homes, driving to industrial or commercial areas for work during standard work hours, and returning to the residential areas at the end of the day. Other trips may involve shopping at retail outlets during traditional store hours. If the underlying demographics data is good and care is taken in the generation of these trips, the results of the SUMO model should improve markedly and approach real-world conditions.

3.2   Define Vehicle Characteristics

Vehicle characteristics are generally defined through vehicle class. At a low-level of resolution, one could classify all Trucks as having a specific fixed length, acceleration and deceleration rates, maximum speeds, and some level of driver imperfection. However, for the purposes of a military simulation where there exist several classes of vehicle, it is possible to distinguish vehicle characteristics at a much higher-level of resolution, even down to individual vehicles, if necessary. These classes are often defined in the `*.rou.xml` files near the top, and the instantiations of the vehicles are referenced by a type pointing back to these classes. This may be particularly important if the military vehicles being tested have unusual shape or performance characteristics, or if they are involved in "follow the leader"-style tests. Without the user making an effort to change the vehicle characteristics to conform to their real-world situation, SUMO will generally assume common default values, some of which are dynamic (i.e., varying over the course of the simulation run), while others are hard-coded to be static. A full list of vehicle characteristics and their descriptions can be found in the SUMO documentation [14].

To ensure the vehicles are performing as intended, within the SUMO-GUI binary, one can change the default settings vehicle visualization to use raster images where the color is determined by any number of user-defined characteristics that may change over time. Highlighting the extreme behaviors of vehicles will permit a simulationist to pause the simulation when a critical value is seen (via the vehicle's color) and right click on the vehicle to view its parameters. There one can see the precise value for any desired vehicle parameter at that point in time. Judicious use of these visualization tools should serve as an aid to configuration debugging for many of the vehicle characteristics.

4 Configuration of Pedestrians on Networks

4.1 SUMO Offers Multimodal Transport Opportunities for Pedestrians

In real life, pedestrians walk (sometimes briskly) along urban sidewalks, loiter near retail store fronts, congregate in parks and other urban green zones, generally obey traffic laws by crossing at intersections, and occasionally obstruct traffic by lingering in the crosswalk too long. Pedestrians may be atomic, walking along as a singleton minding their own business, or they may converse with others, sauntering along side-by-side. Some walk fast with purpose; others walk a bit more leisurely, less constrained by time. Sometime a pedestrian walks to a bus stop, awaits her bus's arrival, hops on and rides to her stop, exits the bus, and continues the remainder of her journey, perhaps on-foot, perhaps using other modes of urban transportation (taxi, train, tram, etc.)

Pedestrian-vehicle interaction is key to constructing a reliable, credible urban traffic simulation. To remove pedestrians from consideration or to marginalize their effects would vastly undermine any simulator's claim to veracity. SUMO even manages multimodal transport opportunities for urban dwellers, perhaps offering many modes along a route.

4.2 SUMO Randomizes Pedestrian Behaviors

SUMO adds some randomization effects into its pedestrian behavior by default, but simulationists can overwrite these defaults by recourse to the source code if it is deemed absolutely necessary. Pedestrians naturally are configured to have a credible speed, although as is the case for vehicles, particular lanes may have maximum speeds that are not exceeded. Generally, these characteristics should not be changed by researchers, without strong justification.

Early pedestrian models were more atomic and non-interacting, but later releases of SUMO incorporated striping [15] into the pedestrian behaviors on sidewalks, thereby introducing more realism. Loitering in certain areas can be encouraged by parameter changes, and pedestrians can be made more (or less) risk averse in their interactions with the environment, especially with cars and trucks at intersections. In much the same way as vehicles can be enticed to go off-road in SUMO (with a bit of reconfiguration), pedestrians too can be encouraged to step off the sidewalks and explore parks and other green zones without the obligation strictly to follow predetermined sidewalks. Apart from SUMO's original focus on urban traffic modeling, some researchers [16] have stretched it to cover pedestrian movement within and emergency exits from public buildings and sports complexes.

4.3 Populate SUMO's World with Pedestrians Using RandomTrips.py with a Switch

The pedestrian demand in the SUMO model can be generated by the same script used for vehicle demand generation, namely randomTrips.py, by enabling people with the switch "--pedestrians". As before, the agents (this time, pedestrians) are randomly-assigned from a uniform distribution the origin and destination pairs within the modeling space. By default, Dijkstra's algorithm is used to find the shortest routes between these locations, with a cost associated with the effort. As seen within the file "`osm.pedestrian.trips.xml`", for each walker, the script defines an instance of class pedestrian and gives it a distinct person ID (of the form "`ped12`"). It pins down a departure time from its origin node and defines its destination node. The files "`osm.pedestrian.rou.xml`" and "`osm.pedestrian.rou.alt.xml`" include this same general information but appends to it the complete list of edges (as determined by Dijkstra's algorithm) from origin to destination, with the "`alt`" file exclusively associating a cost for the suggested route for comparison with any later on-the-fly calculated alternatives. Again, if demographic information is available or can be inferred with some accuracy, pedestrians can be distributed throughout the modeling space in accordance with the real-world data to provide more model realism via the ACTIVITYGEN tool [17].

4.4 Visualize Pedestrian Behaviors Within SUMO-GUI

Within the SUMO-GUI program, the pedestrians can easily be identified, labeled for recognition from afar, and their output monitored as simulation model time progresses. Out of the box, SUMO pedestrians can be hard to see. To alleviate this difficulty, initiate a scenario, allow it run for a few moments (enough time for pedestrians to enter the scene), pause, and point the "Locate" menu to "Locate Persons". Inside the popup entitled "Person Chooser", the simulationist can choose a pedestrian (here denoted by its distinct person ID) to center his efforts upon. Having made that decision and closed out the "Person Chooser" menu, one can then "Edit" the visualization, choose to "Show person name", and enlarge the font point size. Changing the color of the text is also recommended, if the default does not stand out enough from the background. When the researcher accepts these changes, zooms out from the pedestrian currently under the microscope, and continues the simulation, data will then be available for any pedestrian per time step. Pausing the simulation grants the researcher time to investigate available data by right clicking the pedestrian and choosing "Show Parameters" associated with that agent. Information provided in the resulting popup includes the origin and destination edge IDs, the edge currently occupied and physical position upon that edge, the pedestrian's instantaneous speed in meters per second, the speed factor, the angle relative to the model's frame of reference, any accrued waiting time (in seconds), and the desired departure time from the origin node.

More will be reported about the full XML output of the pedestrian model later in this report.

4.5 SUMO has Some Limitations in its Pedestrian Model

There are many similarities between the vehicle and the pedestrian models in SUMO; however, the pedestrian model lacks presently in one key aspect: vehicles can easily be rerouted (as described in Sect. 2.3 above), but currently this is **not** the case for pedestrians. For example, a lane reserved for vehicular traffic may be closed off for a few hours during the day, making cars and trucks take an alternate route around the closure. This functionality does not exist for pedestrians on sidewalks, so temporary closures are not possible. The sidewalk can be permanently removed via the NETEDIT tool, if desired, but simply turning the sidewalk on or off to traffic for some portion of the day is not presently an option. Near-future releases reportedly will address this deficiency.

This concludes the paper's treatment of the basic SUMO configuration elements useful for military simulationists conducting urban traffic modeling. The online documentation and tutorials [10] provide a vast treasure trove of information at much higher resolution, detailing the intricacies of all switches and options available for both the SUMO-GUI and command-line SUMO executables. Readers serious about exploring SUMO's full functionality are strongly urged to explore these developer-composed assets, following up closely with published research articles and queries in online fora.

5   Full Output XML of SUMO Run

5.1   Choose "floating Car Data" Output

SUMO outputs its model results in many different formats, the best choice being highly context-sensitive to the research scope. One can focus upon vehicles in general, particular vehicles, lanes/edges within the network, general summaries aggregating across time steps, raw data dumps of all entities (static or dynamic) over all time steps, and other variants [18]. This paper will restrict its attention to "floating car data" (FCD) output, described by the online documentation [19] as "behav[ing] somewhat like a super-accurate high-frequency GPS device for each vehicle".

In Fig. 8, a screen capture from the online SUMO documentation shows the data SUMO outputs resulting from calling the FCD switch "--fcd-output <OUTPUT_FILENAME>":

Name	Type	Description
timestep	(simulation) seconds	The time step described by the values within this timestep-element
id	id	The id of the vehicle
type	id	The name of the vehicle type
speed	m/s	The speed of the vehicle
angle	degree	The angle of the vehicle in navigational standard (0-360 degrees, going clockwise with 0 at the 12'o clock position)
x	m or longitude	The absolute X coordinate of the vehicle (center of front bumper). The value depends on the given geographic projection
y	m or lattitude	The absolute Y coordinate of the vehicle (center of front bumper). The value depends on the given geographic projection
z	m	The z value of the vehicle (center of front bumper). **Note:** This value is only present if the network contains elevation data
pos	m	The running position of the vehicle measured from the start of the current lane.
lane	id	The id of the current lane.
slope	degree	The slope of the vehicle in degrees (equals the slope of the road at the current position)
signals	bitset	The signal state information (blinkers, etc). Only present when option --fcd-output.signals is set.

Fig. 8. A screen capture from the online SUMO documentation showing fields from FCDOutput

5.2   Restrict Output to Bare Minimum to Avoid Handling Issues

This information is available for *each vehicle* at *each time step* of the simulation run. Note that x and y are by default reported in meters from the origin point of the network frame of reference, but they can easily be converted into the more useful longitude and latitude, respectively, with a single function call. Given the precision of the data, the frequency of the data, and the number of vehicles involved, the resulting output file can reach an overwhelming size. Care should be taken to restrict the output to the minimum necessary information required to meet the needs of the research; file transfer, editing, and parsing may become difficult with exceedingly large files, enough so that at times, large files must be partitioned and handled in pieces, rather than as a whole, due to memory constraints. For example, it is possible for Python's cElementTree function to run out of memory when parsing a large XML file from SUMO's FCD output [20], resulting in a need to call.clear() on each element after it has been processed.

5.3   Create Database to Manage Results from SUMO Batch Runs

Previous discussion in this paper has focused almost exclusively on use of the SUMO-GUI binary, rather than the command-line (non-graphical) SUMO executable. Once configured with the help of SUMO-GUI and debugged with the assistance of its visualization, the SUMO model can easily and profitably (from a performance perspective) be called from the command line and its output be pushed to a unique set of filenames. It is recommended that a database system be created to receive the parsed output data for each simulation run. This promotes a clean separation of model runs from the results analysis stage and lends itself to later visualization by means of popular tools. In the discussion that follows, it is assumed that the open-source PostGRES database [21] has been configured, and tables created to receive the parsed data from the "floating car data" upon completion of the simulation runs. Retention of the data can enhance data analysis efforts and assist in narrowing-in on optimal visualization schema that will effectively convey the message of the analysis to decision-makers. These advantages, however, come at the price of the additional administrative overhead associated with the creation and maintenance of the database plus the cost of the disk space to retain what could eventually turn into hundreds (or thousands) of simulation runs.

Rather than parse the SUMO output results and insert them into the respective Post-GRES table line-by-line, it was decided to complete the run, parse the output into comma-separated values in an ASCII text file, and copy the the text file into the appropriate table.

5.4   Define Database Tables for PERSON and VEHICLE

To match the "floating car data" shown in the screen capture above (Fig. 8), two Post-GRES tables were created, one dedicated to pedestrian movements within the network, the other to vehicles. The tables were defined as follows:

```
create table PERSON_FCD
(
simruntimestamp REAL,
time REAL,
id VARCHAR,
lon REAL,
lat REAL,
angle REAL,
speed REAL,
pos REAL,
edge VARCHAR,
slope REAL
);


create table VEHICLE_FCD
(
simruntimestamp REAL,
time REAL,
id VARCHAR,
lon REAL,
lat REAL,
angle REAL,
type VARCHAR,
speed REAL,
pos REAL,
lane VARCHAR,
slope REAL
);
```

All data, except for the simruntimestamp, are parsed directly from the SUMO XML output. The column simruntimestamp simply designates a unique timestamp associated with the running of the model (usually seconds since midnight 01 January 1970) to keep datasets from distinct SUMO runs separable. The time column is related to the seconds timestamp of the model run, while the id column uniquely identifies the pedestrians or vehicles. By default, instead of longitude or latitude, SUMO reports the x or y position of the pedestrians or vehicles relative to the model's locally-defined origin, but this can be overridden by inclusion of the "--fcd-output.geo" switch in the SUMO binary command-line call. The VEHICLE_FCD table includes a type column missing from the PERSON_FCD table; this may describe the vehicle as passenger, hov, taxi, bus, coach, delivery, truck, or one of many other designations [22]. The remaining columns are similar and relate the speed, the position along the edge (or lane), and the slope of the edge (or lane) being traversed by the pedestrian or vehicle at that timestep. Taken together, these fields can entirely reconstruct all elements of the SUMO model run. With careful construction of database queries, deeper insights into dynamic traffic conditions may be gained.

6 Construction of SUMO Wrapper for Monte Carlo Parameter Studies

6.1 Randomize SUMO Input

A single run of SUMO is for all practical purposes meaningless. In models like this, a slight perturbation in initial conditions can have vast effects on the output. As powerful a tool as SUMO is, limiting it to one run cannot possibly afford the deep insights into traffic volume over time that would justify exploration of alternate routing strategies within an urban environment. How might one resolve this issue and still go forward with the hard-won SUMO configurations this paper has articulated?

The SUMO developers have built randomness into the DNA of the tool. The documentation reports use of the Mersenne Twister algorithm to generate random numbers as seeds for the simulation. By default, the seed value is fixed, which is a "best practice" standard, since it ensures repeatable behavior during those early stages when the user tries to establish her configuration to specification, then test, and debug. However, once correctness of the configuration parameters has been demonstrated, randomness reduces the model results' dependency upon initial conditions. Varying only the universal seed for SUMO ensures the decoupling of times to load vehicles, probabilistic flows, the vehicle driving dynamics, and any on-board vehicle devices [23]. Many more variables are available for change, such as altering the probabilities of taking different routes, modifying the vehicle speed distribution functions, and varying the vehicle departure times. However, a more radical (and recommended) source of variation within the SUMO model runs would be to return to the Python script randomTrips.py and generate many different sets of origin-destination pairs for the vehicles and pedestrians. (Recall that, as given "out-of-the-box", SUMO uses a uniform distribution to choose these origin-destination pairs, but this can be substantially changed by use of the ACTIVITYGEN tool, incisively informed by demographics information).

6.2 Build Wrapper with TraCI API to Explore Parameter Space

Regardless of the method ultimately chosen to induce variation into SUMO input, a researcher can use the TraCI API to execute numerous model runs for sequential or (embarrassingly) parallel processing, depending upon available computational resources. An off-the-shelf laptop should suffice for handling hundreds, if not thousands, of SUMO simulation runs; the determinant would likely be adequate disk space to hold and manipulate the output results, but the continued miniaturization of solid state devices, their accompanying cost reductions, and increasing file transfer rates may make that issue moot. Careful wrapping of the output file names and the use of metadata will enable future users reviewing the model output results to distinguish between simulation runs performed with parameter changes. A call to PostGRES's internal psql commands to copy from the parsed output directory will quickly load all SUMO output data into the appropriate tables for data analysis and visualization efforts.

6.3 Generate Queries to Discover Alternatives

When the simulation data have been dumped into the two PostGRES tables, the simulationist can reproduce any aspects of the SUMO runs by issuing carefully-crafted SQL queries against the tables. A small sample of queries include:

- What urban traffic lane was most used during the second hour of the simulation run?
- What were the top five sidewalks where pedestrians walked the slowest?
- What was the longest average wait time for a traffic signal? When and where did it occur?
- What was the minimum/median/average/maximum distance a pedestrian walked during the simulation? How does this change with the introduction of bus lines along the most popular routes?
- How many intersections did a pedestrian cross on average along her route?
- What was the average (versus median) vehicle queue length due to round-robin behavior at stop signs?
- How does turning off a particular traffic lane affect the volume of traffic on nearby roads? How long does it take to bring the traffic volume on those roads back down to normal levels once access to that lane is restored?

7 Analysis of SUMO Results

7.1 Focus Statistical Output to Address Mission Alternatives

As seen by the small list of queries above, the knowledge that can be extracted from these tables is really limited only by the creativity of the researcher and the number of the substantially different SUMO runs made. Some fundamental meta-questions are: of the queries available, what statistics would be most profitable to military decision-makers tasked with transport through an urban environment? Where should they focus their attention for maximal benefit? What feasibly can be ignored? What concerns originally motivated the simulation? What traffic factors will the military transport effort likely affect? Conversely, what traffic factors will have the most effect upon the military convoy? What potential ground events most worry the military leader planners? How could such worries be assuaged by some upfront planning for alternative actions? What equipment limitations (such as vehicle turning radius or steep slopes) constrain their routing options?

7.2 Avoid Pitfalls Leading to Spurious Interpretations

It should be reiterated that there is no substitute for many realistic, carefully-crafted SUMO input decks with adequate variation in the key parameters under study. Without these, the simulation results would be highly dubious, reflecting the old computer adage of "garbage in, garbage out". However, supposing these conditions are met, a military decision-maker may, for example, want to know the hour-by-hour traffic volume, both vehicular and pedestrian, on the full set of edges along three or four potential routes under consideration. Such queries would not be made for a single SUMO run, but across numerous representative runs, and their results would be normalized so as not to overstate expected counts. Perhaps an hour is not a good way to divide the time, and a quarter-hour is better. Minor adjustments to the queries can be made to explore the most fruitful way to bin the time durations under review. The results of this type of query may permit the transport team leaders to rank order their routing options to assure minimal interference with current traffic patterns and for the transport to be interrupted minimally in turn.

7.3 Confirm Statistical Distributions of Data Before Using Common Techniques

Another useful query may give insight into how much local traffic would be affected by the short temporary closure of some highway edges to facilitate the passage of a sizable military convoy. As input, this would require that SUMO be run many times with several variants of road closures at different times during the day. To quantify the effect, comparison of model output data may be approached by using built-in SQL aggregate statistical functions; however, more formal methods such as designing experiments to run ANOVA and compare F-ratio values may not apply since these assume samples drawn independently from normally-distributed populations, all having a common variance—features not necessarily assured in traffic data. As such, this precludes the use of some well-known multiple comparison techniques until the foundational assumptions of the statistical tools can be confirmed.

7.4 Distinguish Locally-Important Versus Globally-Important Network Edges

A more complex query could attempt the following: For a chosen edge in the network, show the routing from the origin to the destination and all intermediate edges in between for any vehicles passing through that edge; all components edges should then be weighted by vehicular traffic volume as vehicles enter and exit the simulation. From a vehicle's perspective this query addresses these questions: For any vehicle that passes along the chosen edge, where did it come from, what edges did it pass through on its way toward the chosen edge, what edges will it take upon exit from the chosen edge, and where is its final destination? From an edge perspective, aggregating the results of these questions across all vehicles that pass through the chosen edge should give a sense of how important the chosen edge is locally versus globally. That is, does the majority of its users come from neighboring edges, or are they more widely-dispersed? Similar queries could address concerns about pedestrians, rather than vehicular traffic.

7.5 Visualize Traffic Flow as a Heatmap over Time

Generally the output of the database queries are reports that can be passed along to military decision-makers, advising them of how best the suggested alternatives may meet their mission requirements. Other uses may be to emphasize the unexpected first- and second-order effects of considered decisions. Although statistical aggregates of the simulation output values may offer some action-ready insights, the geographical element of the network makes tabular reports somewhat cumbersome. With just a little more effort, more outstanding and visually-engaging products can be generated that "tease out" relationships among simulation components that would be difficult, if not impossible, to discover otherwise. The spatial element, when properly displayed, especially in conjunction with the added dimension of time, can effectively tell a story that may be missed when limited to tabular reports. Of particular interest in this regard is the construction of a heatmap, highlighting those areas of the map under traffic flow. High volume areas are colored from the red/orange part of the spectrum, while lower trafficked zones may be represented in the blue/purple range. A snapshot over the whole simulation run could be helpful, but of particular interest would be a series of heatmaps displaying changes in the traffic flow over time. There is a full crop of tools that will generate heatmap visualizations, including several new entrants to this field such as FOLIUM [24].

8   Future Work

8.1   Introduce Ego Car

Future work in this topic can neatly be divided into near-term and longer-term. SUMO as described above is a mobility simulation. Initial conditions are set, vehicle and pedestrians are positioned at their origin nodes at their start times, the simulation clock is initiated, and the simulation is executed—no user interaction required or desired. Among the first additions to be made is the introduction of a user-controlled "ego car". The user can then drive within the SUMO network, interact with pedestrians and vehicles, queue up as necessary for traffic signals, and generally participate in all aspects of the traffic flow. Since all vehicles, prior to their virtual introduction within SUMO, are pre-routed from their origin to destination nodes, a driven car in this space must continuously "reroute" from its computer-generated optimal route (since its destination is unknown to the model). As any avid video gamer will testify, computer keyboards are particularly atrocious as a manner to steer a vehicle, so a video game controller, such as used in the Xbox series, can be attached via USB connector slot and, with the proper hardware drivers installed, can serve to steer, accelerate/decelerate, and brake the vehicle at the user's command.

8.2   Take the Vehicle Off-Road

As established in the earlier discussion, all of SUMO's simulation activities occur on a network, so it will be necessary to investigate the changes needed to permit vehicles to move off-road, since military transport may occasionally require this capability. In a similar vein, future research will explore pedestrian movements and congregation in green spaces, such as parks or rural areas, where sidewalks are not defined and free movement is the norm, rather than the exception.

8.3   Reroute Pedestrians for a Fixed Time Period

SUMO developers are reportedly working on the ability to reroute pedestrians from chosen sidewalks, much like the functionality currently available to restrict vehicles from particular lanes at fixed time intervals. Upon release of this new option, an investigation will ensue to determine how the closure of lanes periodically affects pedestrian movements and what effect that may have upon vehicular transport globally. Most importantly, what effect may closures of vehicular and pedestrian routes have upon the choice of military transport routes and how alternatives are ranked?

8.4 Tie SUMO Behavior into the ANVEL Software Suite

The SUMO-GUI handles a two-dimensional visualization of the simulation fairly well, but other ERDC-sponsored mobility research has requirements to marry the dynamic elements (vehicles and pedestrians) with a first-person view of the lead vehicle in a military convoy in a three-dimensional landscape. The real-time SUMO simulation results will be integrated with ANVEL software via APIs to bring the dynamic elements in the driver's near-field to life within the three-dimensional space. Since drones with platforms of full-spectrum sensors will be tasked by the military planners to preview road conditions along the potential routes of the convoy, additional requirements specify a much wider drone's-eye point of view for which SUMO will serve to populate the dynamic elements.

8.5 Explore New Pedestrian Behaviors

SUMO's pedestrian behavior is at present limited. Future research will entail enhancing the pedestrian behavior in accordance with recent pedestrian models, such as Mississippi State University's Intermodal Simulator for the Analysis of Pedestrian Traffic (ISAPT) program [25], accepting the tradeoff of more realism for higher computational cost. The pedestrian's new-found realism may encourage more clustering, congregating effects that hostile forces would rely upon to obstruct the forward movement of military transport vehicles. Studies in this area will seek to discover apparent factors that anticipate any such obstruction efforts.

8.6 Introduce Multiple Drivers Simultaneously into SUMO

In the longer-term, user control of a single vehicle may be extended to control of a lead vehicle in a convoy where the other trucks play "follow the leader". Eventually the simulation may include users controlling four or five vehicles, while the rest of the vehicles in the model are SUMO-controlled. To compensate for any lack of verisimilitude in pedestrian behaviors, direct user manipulation of pedestrians should soon follow.

9 Conclusion

This paper has shown some relatively easy techniques to configure a SUMO simulation (mostly) via its GUI tool for the purpose of constructing dynamic environmental elements for use by military planners to assess and choose routes among alternates most likely leading to mission success. SUMO has a long history of use by urban planners, and this history is reflected in some arcane, rather complex, configuration choices and a very comprehensive documentation set. This paper proposes sidestepping some of the more complicated switches for a simple original configuration, debugging by recourse to the GUI's visualization, and, once satisfied there, collecting simulation runs into a couple of PostGRES database tables for the generation of statistical analysis, heatmaps of traffic flow, and other visualization products. In this way, SUMO can serve as a framework for the genesis and movement of dynamic elements in the urban space that can be incorporated into physics-based 3D tools such as ANVEL or as a stand-alone driving/mission training simulator for multiple concurrent users. Wrapped inside a script that varies key parameters in the simulation, SUMO can act as the engine for a Monte Carlo exploration of a vast urban space, the results of which can help military planners foresee obstacles to their mission plans and construct resilient and robust solutions that can evolve as situations change on the ground.

Acknowledgements. The author offers sincere thanks to Joshua R. Fairley (ERDC GSL), Dr. John (Gabe) Monroe (ERDC GSL), Dr. Jerrell Ballard (ERDC ITL), Dr. Jeff Hensley (ERDC ITL), Dr. Alicia Ruvinsky (ERDC ITL), and Kevin Winters (ERDC ITL).

References

1. GitHub repository of Eclipse SUMO for download. https://github.com/eclipse/sumo/. Accessed 11 Jan 2020
2. Projects – SUMO Documentation. https://sumo.dlr.de/docs/Other/Projects.html. Accessed 11 Jan 2020
3. TraCI – SUMO Documentation. https://sumo.dlr.de/docs/TraCI.html. Accessed 11 Jan 2020
4. Libraries Licenses – SUMO Documentation. https://sumo.dlr.de/docs/Libraries_Licenses. html#code_in_the_repository and Eclipse Public License 2.0 | The Eclipse Foundation. https:// www.eclipse.org/legal/epl-2.0/. Accessed 11 Jan 2020
5. Deutsches Zentrum für Luft- und Raumfahrt e.V. (DLR) – Institute of Transportation Systems – Eclipse SUMO – Simulation of Urban Mobility. https://www.dlr.de/ts/en/desktopde fault.aspx/tabid-9883/16931_read-41000/. Accessed 11 Jan 2020
6. AnvelSim – Home | Facebook. https://www.facebook.com/pages/category/Robotics-Com pany/AnvelSim-250058572374411/. Accessed 11 Jan 2020
7. Tutorials – SUMO Documentation. https://sumo.dlr.de/docs/Tutorials/index.html
8. OSMWebWizard – SUMO Documentation. https://sumo.dlr.de/docs/Tutorials/OSMWebWiz ard.html#demand_generation. Accessed 11 Jan 2020
9. Traffic Lights – SUMO Documentation. https://sumo.dlr.de/docs/Simulation/Traffic_Lights. html. Accessed 11 Jan 2020
10. Tutorials – SUMO Documentation. https://sumo.dlr.de/docs/Tutorials/index.html. Accessed 11 Jan 2020
11. Rerouter – SUMO Documentation. https://sumo.dlr.de/docs/Simulation/Rerouter.html. Accessed 11 Jan 2020
12. Introduction to demand modelling in SUMO – SUMO Documentation. https://sumo.dlr.de/ docs/Demand/Introduction_to_demand_modelling_in_SUMO.html. Accessed 11 Jan 2020
13. Trip – SUMO Documentation. https://sumo.dlr.de/docs/Tools/Trip.html#randomtripspy. Accessed 11 Jan 2020
14. Definition of Vehicles, Vehicle Types, and Routes – SUMO Documentation. https://sumo. dlr.de/docs/Definition_of_Vehicles,_Vehicle_Types,_and_Routes.html#available_vtype_att ributes. Accessed 11 Jan 2020
15. Pedestrians – SUMO Documentation. https://sumo.dlr.de/docs/Simulation/Pedestrians.html# model_striping. Accessed 11 Jan 2020

16. Erdmann, J., Krajzewicz, D.: Modelling pedestrian dynamics in SUMO. In: SUMO 2015 – Intermodel Simulation for Intermodal Transport Proceedings, pp. 103–117. Deutsches Zentrum für Luft- und Raumfahrt e.V., Berlin, May 2015
17. Persons – SUMO Documentation. https://sumo.dlr.de/docs/Specification/Persons.html. Accessed 11 Jan 2020
18. Output – SUMO Documentation. https://sumo.dlr.de/docs/Simulation/Output/index.html#available_output_files. Accessed 11 Jan 2020
19. FCDOutput – SUMO Documentation. https://sumo.dlr.de/docs/Simulation/Output/FCDOutput.html. Accessed 11 Jan 2020
20. Personal website of Bosco Ho. https://boscoh.com/programming/reading-xml-serially.html. Accessed 11 Jan 2020
21. PostgreSQL: The world's most advanced open source database. https://www.postgresql.org. Accessed 11 Jan 2020
22. Definition of Vehicles, Vehicle Types, and Routes – SUMO Documentation. https://sumo.dlr.de/docs/Definition_of_Vehicles,_Vehicle_Types,_and_Routes.html#abstract_vehicle_class. Accessed 11 Jan 2020
23. Randomness – SUMO Documentation. https://sumo.dlr.de/docs/Simulation/Randomness.html#random_number_generation_rng. Accessed 11 Jan 2020
24. GitHub repository of folium examples. https://github.com/python-visualization/folium/tree/main/examples. Accessed 11 Jan 2020
25. Usher, J., McCool, R., Strawderman, L., Carruth, D.W., Bethel, C.L., May, D.: Simulation modeling of pedestrian behavior in the presence of unmanned mobile robots. Simul. Modell. Pract. Theory **75**, 96–112 (2017)

Utility as a Key Criterion of a Decision-Making on Structure of the Ground Based Air Defence

Vlastimil Hujer[1] , Vlastimil Slouf[2], and Jan Farlik[3]([⊠])

[1] Joint Operations Command, Armed Forces, Prague, Czech Republic
[2] Retia, a.s., Pardubice, Czech Republic
[3] University of Defence, Brno, Czech Republic
jan.farlik@unob.cz

Abstract. The article devotes to a rational approach to the development of the Ground Based Air Defense (GBAD). The rational approach is influenced by the operational requirements to fulfill a given task on one hand and by a set of economic indicators on the other hand. The aim of this article is a specification of a marginal utility of GBAD elements within the operational needs and within an economic efficiency as a criterion of a decision-making on GBAD structure. As to the operational requirements, an expected efficiency of GBAD is a fundamental indicator which displays a ratio of fulfillment of the given task. At political-military level, the economic efficiency of GBAD structure is a significant indicator.

The contribution of GBAD elements to the expected efficiency of the entire GBAD structure depends on joint activities. For instance, a ratio of destroyed air targets during joint activities of a radar and an effector is higher than bare sum of ratios of destroyed air targets by the radar and the effector separately. A proper contribution of GBAD elements to the expected efficiency and the expenses can be determined by the instruments of the game theory.

A presented method rationalizes process of the decision-making on the structure of GBAD by defining the utility in the operational domain and in the economic domain. The method is applicable in the other branches.

Keywords: Efficiency · Utility · Demand · Ground based air defense · Decision-making · Rational approach · Modeling · Simulation · Game theory

1 Introduction

The aim of this article is to provide differing approach to evaluation of an efficiency of the ground-based air defense (GBAD). As of these days, the efficiency is measured by number of destroyed targets without considerations of economic model of demand of GBAD elements and without consideration of a total and a marginal utility in accordance with real needs defined by valuable targets (defended objects) and by anticipated enemy air threat.

An integral part of the state's defense is GBAD. At a time of lasting peace in Central Europe, where air strikes and reconnaissance are not a major threat, GBAD is not a critical part of the state's defense system. It is this aspect in peacetime and in the presence of a market economy that adds an economic dimension to decisions on the development of GBAD.

© Springer Nature Switzerland AG 2022
J. Mazal et al. (Eds.): MESAS 2021, LNCS 13207, pp. 249–260, 2022.
https://doi.org/10.1007/978-3-030-98260-7_15

2   GBAD as Pure Public Good

GBAD as a set of measures that contributes to the defense of the country is a pure public good (good means a physical or non-physical benefit), so it is an indivisible good, no citizen of the country can be excluded from its consumption and citizens are not rivals in its consumption as stated by Varian (2010, p. 695) [8]: "People can't purchase different amounts of public defense; somehow they have to decide on a common amount." Constructing the demand curve for public goods in general, including, of course, the country's defenses and air defense, is an extremely complex problem, because markets (consumers) such as defended objects, airspace altitude, destroyed enemy targets cannot be clearly formulated. In addition, it is clear that the expenditures in the assessed chapter of the state budget are of public interest. Citizens are certainly interested in whether defense is ensured effectively and efficiently [6], whether the government behaves transparently, whether it is not possible to acquire individual weapons systems cheaper abroad, and whether it is at all possible to provide methods for assessing the rational level of costs.

Stiglitz states (1997, p. 367) [1]: "Defense spending cannot simply be determined by asking individuals how much they are willing to pay". Responsible actors are faced with a set of conflicting influences that cannot be cardinally expressed, such as:

- public election and the resulting mandate of elected bodies;
- the purpose, which is again given by an expert assessment of the risks determined by information about the plans and intentions of potential enemies;
- expert assessment of the capabilities of domestic and foreign industry;
- confidence on possible scenarios for the future development of macroeconomic indicators, both the country, the Union, the Alliance and potentially hostile countries;
- confidence on the capabilities and will of a potentially enemy country to launch an attack;
- formulation of one's own rational attitude with regard to the above-mentioned influences;
- an assessment of the will of citizens to defend themselves and to agree on the allocation of funds for defense and defense capabilities, including the readiness to accept possible losses.

For the public with limited resources to provide all services, the ability to rationally assess needs is vital. The need can be assessed from two opposing perspectives: operational and economic [4–7].

3   Combat Efficiency of GBAD

A suitable indicator for assessing the operational needs of GBAD is the combat efficiency of the structure of GBAD and battle formations of the GBAD units. The quantification of the combat efficiency of the combat unit in this article represents the degree of preservation of the function of the defended object, which is based on the assumed task of

GBAD units, which are to defend the defended objects. The degree of preservation of the function of the defended object can be quantified by the evaluating a rate of products being produced, services being provided etc. In general, task to GBAD may be defined as "to destroy all enemy air threat", etc. For the purposes of this article, the structure of GBAD is limited to a battle formation. Combat efficiency is affected by many influences, including:

- tasks and structure of ground-based air defense;
- terrain and weather;
- the number and character of the defended objects and the size of the area of operation;
- type, activity, maneuvering scheme and intensity of action of air threat.

Often neglected partial influences include malfunctions, jamming or for example level of training (knowledge, skills, experience, moral and mental state) of crews of their own and hostile means. The following figure shows a possible and for the needs of the article simplified battle formation of GBAD unit.

The efficiency of the increasing number of elements of the GBAD battle formation can be illustratively defined by the curve in the following graph in Fig. 2, where the number of elements of the battle formation increases with the number of missile launchers (ML). With the gradual increase in the number of missile launchers increases the combat efficiency of a such battle formation until the situation of saturation, when the number of missile launchers no longer brings an increase in combat efficiency with respect to some influences such as the intensity of air attacks in the area of operation.

The combat efficiency thus defined corresponds to the definition of the term utility used in economic theory, which defines utility as a degree of satisfaction resulting from the consumption of goods. In economic theory, a rational consumer seeks to maximize his benefit. From this point of view, rational behavior also occurs in the operational area, where commanders and chiefs, as responsible persons, strive to achieve maximum combat efficiency and thus benefit (Fig. 1).

In the example shown, saturation starts occurring at number of 16 missile launchers. For the completeness, it can be stated that with a low number of missile launchers, the action of air attack means on the missile launchers plays a role. Regarding the increasing number of missile launchers and the intensity and nature of the action of air attack means, the effect of this action decreases. The course of the curve must be verified experimentally, however, its course is not important in clarifying the importance of the benefits of individual elements of the GBAD unit. The situation is similar for the overall structure of GBAD itself, which in this simplified case corresponds to the defined GBAD set-up.

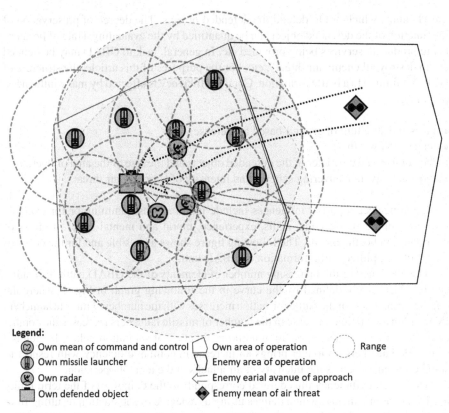

Legend:

(C2) Own mean of command and control	⬠ Own area of operation		⦵ Range
⊞ Own missile launcher	⬡ Enemy area of operation		
⊛ Own radar	⬻ Enemy earial avanue of approach		
⬛ Own defended object	◈ Enemy mean of air threat		

Fig. 1. Simplified possible battle formation of GBAD with enemy air avenues of approach.

It is possible to obtain reliable inputs for determining the course of the development curve of the combat efficiency of battle formations (troop structures) of GBAD in basically several ways. These methods include, in particular, expert estimation (see the example and its extension in this article), experiments or the processing of statistical data. Statistical data from conflicts are insufficient regarding a size of the set of accessible data, especially for newly emerging own and enemy means. Mathematical modeling and software simulations with the application of operations research methods such as the Monte Carlo method and game theory methods seem to be the most suitable. Modeling and simulation will enable the quantification of the combat efficiency of GBAD, taking into account all the above-mentioned influences.

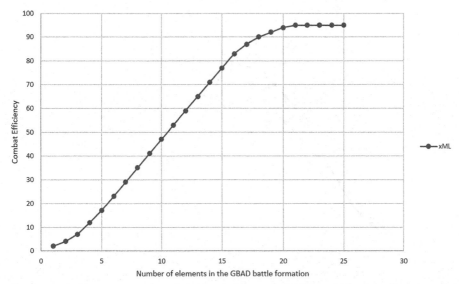

Fig. 2. Development of combat efficiency of the GBAD battle formation including only one element.

The graph in Fig. 2, as the expert estimation, shows that the increase in combat efficiency of the GBAD formation is negligible from the number of 18 missile launchers. The question is to what extent the acquisition of additional firepower is rational in view of the rising financial costs and the limited increase in the combat efficiency of the GBAD formation. An increase in the combat efficiency of the battle formation can be considered a marginal benefit, which expresses how much the total benefit will increase if the amount of consumed goods (elements of the battle formation: missile launchers, command and control (C2) nodes, radars, etc.) increases by one at a time. As the number of elements in the battle formation increases total utility increases as well, however marginal utility decreases.

The simplified example above can be extended by adding other elements, namely C2 nodes and surveillance and reconnaissance means, such as radars, to the GBAD formation. This would on the one hand significantly increase the efficiency of the GBAD formation, but on the other will reduce the number of missile launcher within the same number of all elements in the GBAD formation. The following figure shows the situation.

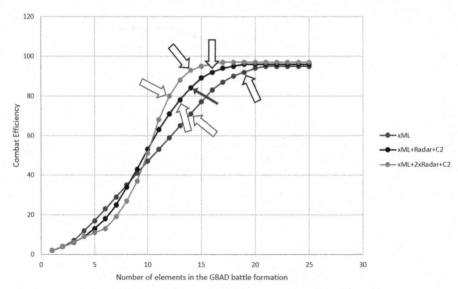

Fig. 3. Development of combat efficiency of the GBAD battle formation including various elements. (Color figure online)

The graph in Fig. 3 above shows the possible development of the combat efficiency of the GBAD battle formation depending on the number of different elements. Battle formation of GBAD composed only of missile launchers (blue color) was analyzed earlier and is extended by two more illustrative cases. The first extension is a battle formation extended by a reconnaissance device (radar) and by C2 node (orange). Similarly, the second extension is a battle formation supplemented by two radars and one by command and control node (gray). It can be deduced from the graph that with a low number of all elements, the combat efficiency of the GBAD battle formation composed exclusively of missile launchers (without radar and C2 node) is more effective than battle formation supplemented by radars and C2 nodes. Only with the increasing number of all elements in the battle formation, the advantage begins with radars and C2 nodes into the battle formation. Again, the intensity of the action of air attack and other influences plays a role.

Deciding on the structure of the GBAD battle formation in the example and on its extensions is simple. Figure 3 shows that the same approximate combat efficiency of about 92% (see red arrows) can be achieved by 19 elements composed exclusively of missile launchers or 16 elements, one of which is the radar, one C2 node and 14 missile launchers. The same combat efficiency can also be achieved by the battle formation consisting of 14 elements, which include two radars, one C2 node and 11 missile launchers.

4 Formulation of Demand of GBAD as the Public Goods

GBAD is demanded by the citizens of the country. Each of them has a certain priority, each of them has a certain ability and will to pay for the provided public goods (here the military means) in the form of taxes. The amount is determined by public choice, because each of the political parties also represents a strategic documents containing information to what extent GBAD must be provided. Determining an effective amount of public goods is difficult. Schiller (2004, p. 81) [2] states: "Additional public sector activity is desirable only if the benefits of that activity outweigh the opportunity costs".

However, in the case of GBAD, the assessment of profit in terms of comparison with the sacrificed opportunity to consume other private goods is highly debatable. Most citizens only become aware of certain security risks when there is a threat in their vicinity. Moreover, a citizen of a country bordering on allies may not be aware of it at all, and this psychological-knowledge influence may be reflected in the public choice of populist entities. One of the ways to solve the problem is to set standards. Linhartová (2017, p. 205) [3] also discusses the problem: "A public authority that is competent to define the standard (volume and quality) of public service must decide even without knowing the individual demand curves (graph curves) of individual consumers in the market. It will do so on the basis of expert judgment or empirical investigation and sets a standard Q_C level (author's note - amount of public service provided)".

The following equation is a general expression of the author's proposal, which forms the basic structure for expert assessment and other empirical investigations to determine not only the standard, but also the specification of demand and effective amount of public goods (services).

$$D_{PPVO} = f(VP; PMV; IN; HR) \tag{1}$$

Where the individual symbols mean:

- D_{PPVO} - demand for public goods (service) that is represented as air defense;
- VP - public demand - defined by various surveys of individual demand and public opinion polls on a selected sample of the population structure. It is expressed by the degree of belief of the evaluator about the level of public demand in the interval <0; 1>, while the value 0 expresses a state where no demand for air defense exists and a value of 1 indicates a situation where all respondents demand a higher level of air defense;
- PMV - prediction of macroeconomic development - expressing that the state will be able in terms of its fiscal and monetary situation to implement appropriate development projects to ensure public goods, expressed by the evaluator's belief in predicting macroeconomic development in the interval <0; 1>, where a value of 0 indicates a state in which the country will not be able of it (high inflation, unsustainable public debt, high interest rates, etc.) and a value of 1 indicates a state of a positive expectation of further economic development;
- IN - infrastructure - as a result of expert assessment, which expresses whether it is important to implement air defense at all, if there is anything to defend. The degree of the evaluator's belief in the level of infrastructure in the interval <0; 1>, where the value 0 indicates a state where the state has nothing to defend and air defense is an unnecessary service and a value of 1 indicates a state that indicates a high level of infrastructure with a high impact on society;

- HR - expected threat - as a result of expert assessment, which expresses the existence of a real threat of air attack of a potential enemy, expressed by the degree of belief of the evaluator about the threat level in the interval <0; 1>, while the value 0 expresses the state when the state is not endangered and a value of 1 indicates a condition that indicates the existence of an imminent threat.

Economic costs also play a role in the acquisition and operation of GBAD equipment. Costs can include [5] the cost of acquisition, operation, maintenance and other parts of the life cycle (calibration, certification, modernization, etc.). The economic costs also include the cost of training, the acquisition of funds for training, etc. A possible illustrative development of costs depending on the number of elements of the GBAD unit is shown in the following figure.

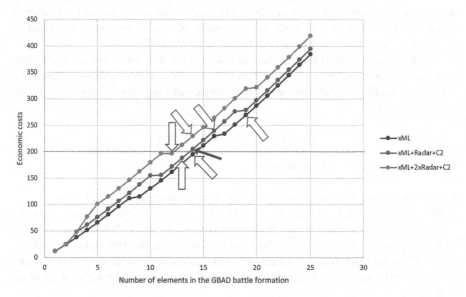

Fig. 4. Development of economic costs. (Color figure online)

The Fig. 4 shows that the most economically expensive is to extension of the example including two radars and C2 node. Individual shifts on individual cost curves represent possible discounts. The data given are illustrative.

When comparing the graph with the development of combat efficiency, it can be read, for example, that in the above selection of battle formations (19, 16 and 14 elements, see red arrows), from the point of view of the lowest economic costs, the most advantageous battle formations consists of 14 elements.

5   Relationship of Elements

The situation will change significantly if, with regard to the assessment of supply and demand, see the text above, a limit is set on the economic costs of acquiring elements of GBAD. This limit can be, for instance, 200 units, see red dashed line in the graph above. In this case, you can get battle formations with 14, 13 and 12 elements (see blue arrows in the graphs above). Selected battle formations, however, represent battle formations with lower combat efficiency, where a variant of 12 elements including two radars and one C2 node represents a combat efficiency of approximately 80% (see blue arrows in Fig. 3).

Further analysis of the situation leads to the conclusion that a slight increase in economic costs allows addition of the battle formation with 14 elements (see green arrow in the figures above), which contains one radar and one C2 node, and achieve higher combat efficiency of approximately 84%.

This simplified example and its extensions show that a situation of choosing a rational structure of the GBAD battle formation, taking into account economic costs, can and often does lead to a complex decision-making task.

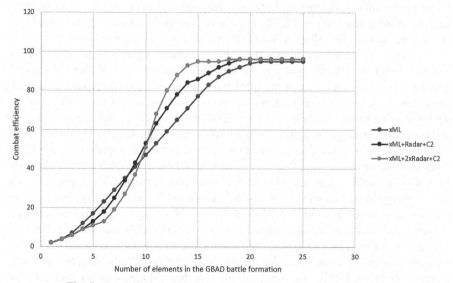

Fig. 5. Expected influence on the development of combat efficiency.

In real practice, various situations can arise that further complicate rational choice. This situation can be, for instance, a limited number of missile launchers that can be controlled by one C2 node, see Fig. 5. The figure shows that with 15 elements of the battle formation including one radar and C2 node, or 16 elements of the battle formation, respectively involving two radars and one C2 node, the combat efficiency of the structure is growing more slowly. This is due to the fact that missile launchers above 12 are not effectively controlled by C2 node.

From the graph of Fig. 3 can be read that the increase in the combat efficiency of the battle formation consisting only of missile launchers (blue curve) from 14 to 15 elements is approximately 6%. This increase represents the marginal benefit of the combat efficiency for the numerically increased battle formation, however, it may not represent the contribution of added element to the overall combat efficiency of the battle formation. This can be illustrated by the following situation, where it can be assumed that if only one element is in the GBAD battle formation, namely radar, the resulting combat efficiency of a such battle formation will be less than if the battle formation consists only of a missile launcher. Suppose the combat efficiencies of the elements themselves are 5% for radar and 15% for missile launcher. It can also be assumed that the combat efficiency of a battle formation consisting of two same elements will be higher than the sum of the combat efficiency of the individual elements of the battle formation, for example 25%, with respect to mutual cooperation and early detection of air enemy threat. Using the game theory and its part of the cooperating games, it is easy to determine the share of the individual elements in the resulting combat efficiency. In the situation shown, this is the average value of the sum of the combat efficiency of one element (5% or 15%) and the share of the same element in the total combat efficiency (25%–15% or 25%–5%). Thus, it can be calculated that the contribution of the radar to the total combat efficiency is 7.5% and the contribution of the missile launcher to the total combat efficiency is 17.5%. The above procedure is based on the formula for calculating the so-called Shapley Value within the cooperative game theory [9].

As mentioned in the text above, the battle formation of GBAD units does not and will not comprise of purely one type of the GBAD element and the battle formation will always include elements of the three basic types – means of reconnaissance and surveillance, the command and control means and lastly means of fires. In the following demonstrative example, see Fig. 6, a set "C2SR" comprises of complements: command and control nodes and surveillance and reconnaissance radars (C2SR). User must decide what a combination of complements to choose. For the purpose of this article, one C2SR is capable to control actions of four missile launchers. Lower or higher number of missile launchers are inefficient in regard to tactical principles in battles. The higher utility is achieved only by adding one C2SR together with four missile launchers. The marginal utility is influenced by number of C2SR and of missile launchers. In this demonstrative example, user is allowed to acquire all necessary sets of C2SR and missile launchers. Figure 6 shows increase of the total utility of the five missile launchers and C2SR with a various probability of air target destruction. In case of probability of 0.26, total utility increases up to five C2SR and twenty missile launchers. Marginal utility decreases and is zero at the higher number of C2SR and missile launchers. One C2SR with the four missile launchers and probability 0.9 leads to comparable the utility as five C2SR with twenty missile launchers and probability 0.26. Acquisition of additional GBAD elements relates to further tasks of GBAD like defense of vital national assets.

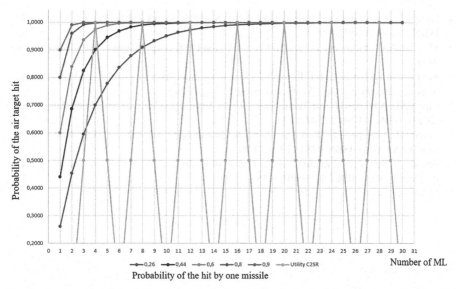

Fig. 6. Assumption of the influence of limited number of missile launchers controlled by C2 nodes.

6 Conclusion

Contribution of the element added to the GBAD battle formation can be obtained by the mathematical modeling and software simulation and by the application of the game theory. The economical cost of the same element is given. The economic efficiency can be quantified by comparing both variables.

The presented example and its extensions show that based on the economic limitations the chosen structure of the GBAD battle formation may not be of the top combat efficiency but can be considered as effective within economic context.

The mathematical modeling and software simulations allow the determination of the expected real operational need by calculating the combat efficiency as the utility and together with definition of the demand of GBAD as the public good provides inputs for the rational decision-making in the development of GBAD in the context of economics of the country whose people do not consider the air threat as the essential. The apparatus is also applicable to the other industries and countries.

References

1. Stiglitz, J.E.: Economy in Public Sector (In Czech). Grada Publishing, Prague (1997). ISBN 80-716-9454-1
2. Schiller, B.R.: Microeconomy Today (In Czech). Computer Press, Brno (2004). ISBN 80-251-0109-6
3. Linhartova, V.: Standardization in public sector (In Czech). Wolters Kluwer, Prague (2017). ISBN 978-807552-726-4

4. Hampl, P.: Creativity and Honesty (In Czech). Naštvané Matky Publishing, Olomouc (2019). ISBN 978-80-906573-5-9
5. Mankiw, N.G.: Economy Basics (In Czech). Grada Publishing, Prague (2009). ISBN 80-716-9891-1
6. Ministry of Defense, Czech Republic. http://www.mocr.army.cz/finance-a-zakazky/resortni-rozpocet/resortni-rozpocet-5146/. Accessed 19 July 2021
7. Chvoj, M.: Advanced Game Theory in the World Around (In Czech). Grada Publishing, Prague (2013). ISBN 978-80-247-4620-3
8. Varian, R.: Intermediate Microeconomics: A Modern Approach, 8th edn. W.W. Norton & Co., New York (2010). ISBN 978-0-393-93424-3
9. Shapley, L.S.: In Contributions to the Theory of Games, vol. 2. Princeton University Press, Princeton (1953). ISBN 9780691079356

Human Detection in the Depth Map Created from Point Cloud Data

Adam Ligocki[✉][iD] and Ludek Zalud[iD]

Faculty of Electrical Engineering and Communication, Brno University of Technology,
Technicka 12, Brno, Czechia
adam.ligocki@vutbr.cz

Abstract. This paper deals with human detection in the LiDAR data using the
YOLO object detection neural network architecture. RGB-based object detection
is the most studied topic in the field of neural networks and autonomous agents.
However, these models are very sensitive to even minor changes in the weather
or light conditions if the training data do not cover these situations. This paper
proposes to use the LiDAR data as a redundant, and more condition invariant
source of object detections around the autonomous agent. We used the publically
available real-traffic dataset that simultaneously captures data from RGB camera
and 3D LiDAR sensors during the clear-sky day and rainy night time and we
aggregate the LiDAR data for a short period to increase the density of the point
cloud. Later we projected these point cloud by several projection models, like
pinhole camera model, cylindrical projection, and bird-view projection, into the
2D image frame, and we annotated all the images. As the main experiment, we
trained the several YOLOv5 neural networks on the data captured during the day
and validate the models on the mixed day and night data to study the robustness
and information gain during the condition changes of the input data. The results
show that the LiDAR-based models provide significantly better performance during the changed weather conditions than the RGB-based models.

Keywords: LiDAR data · RGB camera · Point cloud · Projection · YOLO ·
Object detection · Neural network · DCNN

1 Introduction

These days, we can see the growing number of various applications of autonomous
agents in Advanced Driving Assistant Systems (ADAS) and in many fields of industrial automation or even in our homes. These systems have very high demands on the
security of the people that interact with them. For example, in autonomous cars, we talk
about vehicles' crew that is permanently at risk as the cars are moving with very high
velocities or the very poorly protected pedestrians in the moment of collision with the
vehicle. On the other hand, in the industry, we see various systems and robotic manipulators that use large forces to handle very heavy cargoes. All these systems have to
be designed with the security systems in mind, that would prevent them from harming
living beings, even if they break the security rules.

© Springer Nature Switzerland AG 2022
J. Mazal et al. (Eds.): MESAS 2021, LNCS 13207, pp. 261–272, 2022.
https://doi.org/10.1007/978-3-030-98260-7_16

Many different approaches allow detecting humans' presence and make it possible to avoid the collision. For autonomous cars, the most common is the usage of RGB cameras, LiDARs, radars, etc. In the industry sector, there are cameras, mechanical and optical branches in use. Overall, if we want to detect the presence of a human in some area, RGB cameras combined with the neural network object detectors are these days the easiest and the most reliable method to do so.

Fig. 1. During the night, the RGB camera was covered by the water from rain and partially blinded. It significantly affects the inference of the neural network object detectors. At the very same time, the LiDAR sensor worked without any significant degradation of the data quality.

In many edge cases, the RGB camera is not a source of reliable data. Situations, like light condition changes, bad weather, rain, fog, etc., make the RGB camera nearly useless (see the Fig. 1). Therefore, we decided to study the LiDAR sensor and its usage in detecting humans as the laser time of flight (TOF) sensor provides significantly higher robustness regarding unfavorable conditions. In the end, the LiDAR-based human detector could help handle edge situations when the common RGB-based detection methods fail.

2   Related Works

The primary motivation for our work is that LiDAR sensors are more robust during light or weather conditions changes than RGB cameras. For example, the standard camera optics are blinded even during the soft rain when the LiDAR sensors can handle these conditions up to rain intensity of several mm per hour [4,5,8].

The field of object detection in LiDAR data is quite well covered, not only by classic point cloud processing method, like clustering [2,19], which could also be applied on pedestrian detection [12,21]. On the other hand, these days the application of the neural network is the most frequently used approach. Let us mention the well known PointNet++ [14], Complex-Yolo [18], or the BirdNet [1].

All the papers mentioned above deal with the point cloud data. We handle the problem differently in this work, and we process point cloud data like an image. We based it, on the idea of transfer learning [20]. Neural networks require an enormous amount of training data, which is very difficult to collect. However, we can use the model trained on a similar problem, finetune it, and adjust it to the problem we want it to deal with.

Our idea is to use the pre-trained RGB neural network and apply it on depth map data generated from the LiDAR point cloud data. In this way, we can cover not only human detection in the field of autonomous cars but also in the indoor and other security applications, where the RGB-Depth cameras are used [22,23].

Object detection in depth maps images using neural networks is not a new idea. There are many papers, like [6,17] or [3], but they all focus on a fusion of the depth map and common RGB camera data. Only very few papers study the raw depth map and compare the different projection types and points of view [13], but non of those dataset compares the RGB and point cloud data in the way, we do.

3   Dataset

To train and evaluate all the proposed models, we created our own dataset based on the open source Brno Urban Dataset (BUD) [10]. The Brno Urban Dataset is a publically available set of real-life road traffic records that contains data from four RGB cameras, a single thermal camera, three 3D LiDAR sensors, IMU, and an RTK GNSS receiver. The dataset also provides the calibration data for the physical layout of the sensors as homogeneous transformations w.r.t. the IMU in the center of the sensory framework and internal calibration parameters for all cameras.

3.1   Depth Image Generation

We used four recording sessions from the BUD in total. It comprises of about half an hour of the data recorded during both day and night time and we processed it using the open-source Atlas Fusion framework [9]. It allows us to aggregate point cloud data from the Livox Horizon LiDAR sensor, pair them with the corresponding image from the very front RGB camera, and project this point cloud data into the RGB camera plane. For projecting point clouds into the camera frame, we used the pinhole camera model. All the generated images are 1920×1200 px, and we projected each point as a 11×11 px square. This way, we created the depth map-like image for every single RGB frame. The grayscale intensity of the projected point was estimated based on the point's distance from the camera's optical center as linear regression, 255-pixel value for zero distance, and the 0 value for points in the distance of 50m, or more (Fig. 2).

In total, we created about 20 000 RGB-depth map pairs. For those, we extract only the pairs that contained pedestrians in the RGB image. Also, as the cameras capture images at 10 Hz, we removed the following nine pairs after every single selected RGB-depth pair containing pedestrians. This way, we removed from the dataset the RGB-depth map pairs that cover the same scene, only captured from a slightly different perspective.

In total, we selected four hundred pairs of RGB images and corresponding depth maps. All containing at least one pedestrian.

Fig. 2. The example of the RGB images (left) with the corresponding depth map image (right) generated by projecting point cloud to camera frame, using the pinhole camera model.

3.2 Different Projections

As a part of the paper, we studied not only the simple pinhole camera projections to the physical camera, but we also tested the other types of projection of the point clouds to compare the possible information gain in the pedestrian detection task.

First we proposed creating the depth maps for a virtual camera that would take place three meters above the physical one, and the virtual camera would be tiled by 20 degrees below the horizon. Then, we tried to validate the hypothesis that the neural network would deal better if it would combine the visual information of the pedestrian's presence with the shadow in the point cloud data behind him. The depth map images for the virtual camera have the same parameters as the depth map created by projecting point cloud to the physical camera, 1920×1200 px, and points were projected as 11×11 px squares. We set the pixel intensity in the same way as the projection to the physical camera.

Next, we used the spherical projection into the physical camera's optics center. We created the 1800×600 px blank image, and for each point in the 90×30 deg field of view, we projected it into the image as a 11×11 px square. This way, each row and column in the image corresponds to 0.05 deg in the spherical coordinates. Finally, the pixel intensity was estimated in the same way as the previous two projection methods.

The very last projection type is the bird view projection, where we took the 50×30 m area in front of the camera, where each pixel represented the 0.1×0.1 cell on the ground. This results in a 500×300 px image. For every point in the point cloud, we used its X and Y position coordinates to estimate the corresponding image pixel, and we set this pixel to the value given by the linear regression between the -2 m (0) and the 3 m (255) of the Z coordinate of the point. If several points occupied the same cell, we projected the highest one.

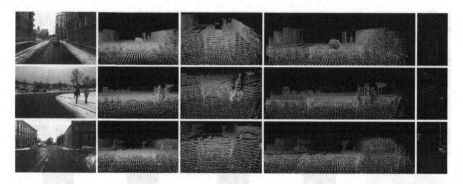

Fig. 3. Visualization of different projection types of the same scenes. From left to right: real RGB, pinhole projection to physical camera, pinhole projection to virtual camera, spherical projection to physical camera and bird-view projection.

This way, we created four different depth maps by projecting the aggregated point cloud for each captured RGB image.

3.3 Annotation

We selected the dataset, which contains 400 RGB images, 300 captured during the day and 100 images captured during the rainy night. For each RGB image, we generated four different projections of the point clouds from the corresponding time. In total, it gives us 2000 images. We annotated them all by hand using the LabelImg tool https://github.com/tzutalin/labelImg (Fig. 4).

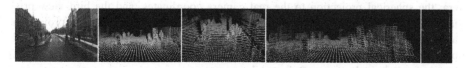

Fig. 4. Example of annotations for all used types of projection. The layout of different projections is same, like in Fig. 3.

3.4 Morphological Modifications

We aggregate the captured point cloud data for one second. Still, the images created by projecting the point cloud data to the camera frame are pretty sparse. Thus we tried to extend the area of each point by applying the morphological operation of dilatation on each depth map image. We applied the dilatation using the kernels of 3×3, 5×5, 7×7, and 9×9 px size (Fig. 5).

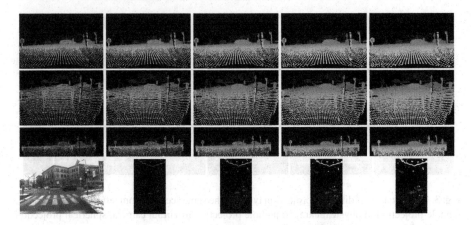

Fig. 5. Visualization of different projections (in rows) and the application of morphological oper-ation of dilatation. First image in every column is always the original image, followed by the dilatation by kernel of size 3, 5, 7, 9. In last row the RGB image is appended. For bird-view projection only kernels sizes of 3, 5 and 7 were applied.

Annotations for images modified by the dilatation stay the same as the annotations for the original image, which serves as an input to the morphological operation.

3.5 Summary

In total, we made 400 RGB images, 300 captured in the day, 100 captured in the night, during the rain. We generated the depth map for each RGB image by projecting point cloud to the physical camera, to the virtual camera placed 3m above the physical cam-era, the spherical projection to the real camera coordinates, and the bird view pro-jections. Additionally for each physical camera projection, virtual camera projection and spherical projection image, we extended the dataset by applying the morphological operation of dilatation with a kernel size of 3, 5, 7, and 9. For the bird view projection, we used only kernels of sizes 3, 5, and 7.

Summing it all together, we prepared 8000 different images split into the 20 small training datasets, each of 400 images.

4   Models Training and Evaluation

To train the neural network models, we used the YOLOv5 framework [7]. It imple-ments the entire YOLO architecture, based on the YOLOv3 [16] for four models of different complexities, S, M, L, and X. The framework also allows using training data augmentation, which allows models to reach better results even for small datasets.

The YOLO architecture is an end-to-end neural network model that takes the raw image on the input and proposes a tensor representing the set of proposed bounding boxes on the output. The output tensor is a 3D data structure, where the X and Y coordinates correspond to the grid cell that divides the input image. The depth dimension of the tensor encapsulates the parameters of the proposed bounding boxes. See the Fig. 6.

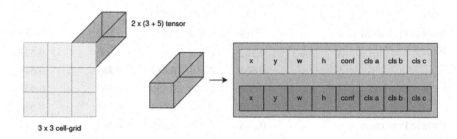

Fig. 6. The YOLO [15] is the full end-to-end neural network. On the input, it takes raw N-channel images and split them to the SxS cell grid. On the output of the neural network, there is a tensor of SxSxM size. The M represents the length of the vector that holds the numbers that represent parameters of proposed bounding boxes, like position, dimension, class score, and bounding box confidence.

To evaluate the performance of the trained models, we used the metrics commonly used in the object detection tasks, the mAP(0.5) and mAP(0.5:0.95). Additionally, we also added the F1 score, which expresses the relation between the recall and the precision of the model.

Our entire dataset is divided into 20 smaller subsets - one RGB subset, five subsets of depth maps created by projecting point clouds by pinhole camera model into the camera frame (one original subset and four derived subsets generated using the morphological dilatation), five subsets of depth map by projecting point clouds to the virtual camera above the physical one, five subsets for spherical projection, and four subsets of bird view projections.

Each subset comprises 400 images, 300 captured in the day and 100 captured during the night. Thus, we used 240 day-time images to train the neural network model, 60 daytime images to validate the model during the training process, and the 100-night images we use to estimate the robustness of the model as the light conditions change.

We trained three YOLOv5 X models for every one of 20 subsets. We used the transfer learning technique, using the pre-trained YOLOv5 weights trained on the RGB data and trained to detect bounding boxes of 80 COCO classes [11]. To adjust the neural network for our task, we modified the last layer of the model so it detects only one class on the output, the pedestrians. After training three models on each subset, we evaluated each of them and measured their performance on the day-time validation data using the mAP(0.5), mAP(0.5:0.95), the F1 score metrics and calculated the average for each of those criteria. See the results in the Table 1.

Table 1. The table shows the performance of trained neural networks, measured on validation data captured during the day. The number "Kernel Size" gives the NxN dimensions of the kernel used to dilate the depth map before the training process.

Daytime Dataset	Dil. Kern. Size	Precision	Recall	mAP(0.5)	mAP(0.5:0.95)	F1
RGB	–	0.674	0.843	0.830	0.457	0.749
Pinhole Projection (Physical Cam)	–	0.514	0.672	0.635	0.339	0.583
	3	0.562	0.707	0.667	0.360	0.627
	5	0.604	0.691	**0.687**	**0.363**	0.644
	7	0.606	0.674	0.656	0.362	0.638
	9	0.614	0.687	0.664	0.356	0.648
Pinhole Projection (Virtual Cam)	–	0.548	0.645	0.616	0.341	0.592
	3	0.594	0.657	0.644	0.367	0.624
	5	0.602	0.696	**0.681**	**0.391**	0.645
	7	0.583	0.660	0.647	0.368	0.645
	9	0.576	0.669	0.656	0.385	0.619
Spherical projection (Physical Cam)	–	0.524	0.647	0.607	0.282	0.579
	3	0.521	0.657	0.605	0.284	0.581
	5	0.618	0.673	**0.622**	0.299	0.645
	7	0.573	0.646	0.615	**0.306**	0.608
	9	0.474	0.701	0.616	0.305	0.566
Bird View	–	0.482	0.775	**0.694**	0.291	0.595
	3	0.504	0.718	0.673	**0.307**	0.592
	5	0.456	0.699	0.631	0.297	0.552
	7	0.431	0.716	0.637	0.295	0.538

Later, for each projection type we selected the models trained on the morphologically modified dataset with the best average mAP(0.5) score above others. We assume that the selected models are trained on those datasets, where the morphological dilatation upscaled projected points to the size, meaning that the neural network can detect the objects with the best performance. We later evaluated these selected models on the night-part of the same projection type and morphological modification. The results are shown in the Table 2 below.

5   Discussion

The results of the models evaluation are shown in the Table 1 (day data) and in the Table 2 (night data).

We trained three neural networks on a day data for each subset and calculated the average over their results. For each projection type, we chose the best average results concerning the kernel size. These results are shown in Table 1. Here the performance

Table 2. Performance of trained models on the data captured during the rainy night. Compared to the "Day Data", LiDAR data-focused models outperform the RGB object detector during the night.

Night Dataset	Dil. Kern. Size	Precision	Recall	mAP(0.5)	mAP(0.5:0.95)	F1
RGB	–	0.643	0.580	0.582	0.259	0.609
Pinhole Projection (Physical Cam)	5	0.689	0.900	**0.879**	**0.482**	0.780
Pinhole Projection (Virtual Cam)	5	0.712	0.872	0.871	0.420	**0.784**
Spherical projection (Physical Cam)	5	0.699	0.856	0.814	0.300	0.770
Bird View	–	0.475	0.816	0.716	0.270	0.600

on the RGB data is mentioned as a reference to compare object detection in the depth map with the RGB-based state-of-the-art. The depth map-based approaches give worse results compared to the RGB. However, if we compare only the models trained on the depth map subsets, the best results are reached by the simple pinhole camera projection into the physical camera, placed on the roof of the car, and by the same projection to the virtual camera that is placed three meters above the physical one. On the other hand, the spherical projection and bird-view projection did not show any advantage, and their performance is significantly worse.

Considering the kernel size, Table 1 shows that in most cases, we get the best results for the dilatation with a kernel size of 5×5. If we consider that the point cloud data were projected to the image frame as 11×11 squares, applying the 5×5 dilatation, we can assume that for the 1920×1200 image, the best results were reached using the 19×19 px size of projected points. For bird-view projection, the dilatation did not help to get any better performance. It corresponds with the smaller size of the bird-view image and the smaller size of the objects in the image. However, this paper studies the different sizes of the projected points very briefly. In the future, this topic could be extended.

Later, we used these best models selected by the performance on the day data, and we evaluated them on the corresponding data that we recorded during the night. The results are presented in Table 2. Here we see, that all the depth-based models outperform the RGB object detector.

Figure 7 visualizes several detections for each projection type.

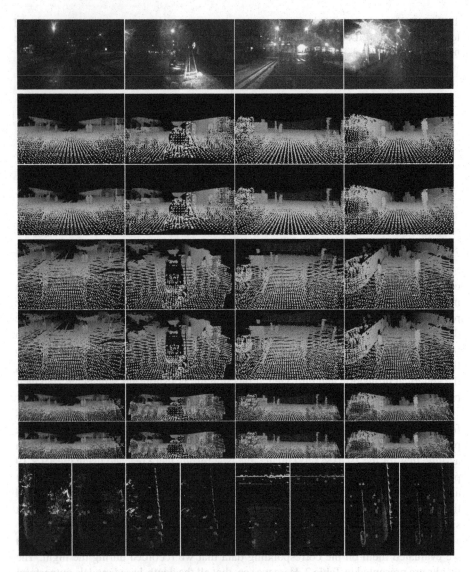

Fig. 7. Each column shows the data captured at the exact same moment. On the top, the RGB image illustrates the environment in front of the car. Then follow by the row the simple pinhole projection to the physical camera, pinhole projection to the virtual camera, spherical projection to the physical camera, and bird view projection. Always the image with neural network's detections is above the ground truth annotation. For bird-view projection, the image with detections is on the left and ground truth on the right side of the column.

6   Conclusion

This paper proposes an experiment that evaluates different projection methods of LiDAR's point cloud data into the 2D depth map images and tests the YOLO neural network's capability to detect objects in these images. In total, we created 20 different 400-images datasets and trained the YOLOv5 X neural network model on the data captured during the day. Later we select the best model for each projection type, and we evaluate them on the data recorded during the rain and night.

The results confirmed our original hypothesis that object detection on depth map images shows significantly worse results during good weather conditions (0.830 for RGB vs. 0.687 mAP(0.5) for depth map), but as the light and weather worsen, the LiDAR-based depth map object detection outperforms the object detection on RGB data (0.582 for RGB vs. 0.879 mAP(0.5)).

Acknowledgement. The work has been performed in the project ArchitectECA2030: Trustable architectures with acceptable residual risk for the electric, connected and automated cars, under grant agreement No 877539/8A20002. The work was co-funded by grants of Ministry of Education, Youth and Sports of the Czech Republic and Electronic Component Systems for European Leadership Joint Undertaking (ECSEL JU). The work was supported by the infrastructure of RICAIP that has received funding from the European Union's Horizon 2020 research and innovation programme under grant agreement No 857306 and from Ministry of Education, Youth and Sports under OP RDE grant agreement No CZ.02.1.01/0.0/0.0/17_043/0010085.

References

1. Beltrán, J., Guindel, C., Moreno, F.M., Cruzado, D., Garcia, F., De La Escalera, A.: BirdnNet: a 3D object detection framework from lidar information. In: 2018 21st International Conference on Intelligent Transportation Systems (ITSC), pp. 3517–3523. IEEE (2018)
2. Börcs, A., Nagy, B., Benedek, C.: Instant object detection in lidar point clouds. IEEE Geosci. Remote Sens. Lett. **14**(7), 992–996 (2017)
3. Chen, X., Ma, H., Wan, J., Li, B., Xia, T.: Multi-view 3D object detection network for autonomous driving. In: Proceedings of the IEEE conference on Computer Vision and Pattern Recognition, pp. 1907–1915 (2017)
4. Filgueira, A., González-Jorge, H., Lagüela, S., Díaz-Vilariño, L., Arias, P.: Quantifying the influence of rain in lidar performance. Measurement **95**, 143–148 (2017)
5. Goodin, C., Carruth, D., Doude, M., Hudson, C.: Predicting the influence of rain on LIDAR in ADAS. Electronics **8**(1), 89 (2019)
6. Guo, Z., Liao, W., Xiao, Y., Veelaert, P., Philips, W.: Deep learning fusion of RGB and depth images for pedestrian detection. In: 30th British Machine Vision Conference, pp. 1–13 (2019)
7. Jocher, G., et al.: ultralytics/yolov5: v3.1 - Bug Fixes and Performance Improvements (2020). https://doi.org/10.5281/zenodo.4154370
8. Kutila, M., Pyykönen, P., Holzhüter, H., Colomb, M., Duthon, P.: Automotive lidar performance verification in fog and rain. In: 2018 21st International Conference on Intelligent Transportation Systems (ITSC), pp. 1695–1701. IEEE (2018)
9. Ligocki, A., Jelinek, A., Zalud, L.: Atlas fusion-modern framework for autonomous agent sensor data fusion. arXiv preprint arXiv:2010.11991 (2020)

10. Ligocki, A., Jelinek, A., Zalud, L.: Brno urban dataset-the new data for self-driving agents and mapping tasks. In: 2020 IEEE International Conference on Robotics and Automation (ICRA), pp. 3284–3290. IEEE (2020)
11. Lin, T.-Y., et al.: Microsoft COCO: common objects in context. In: Fleet, D., Pajdla, T., Schiele, B., Tuytelaars, T. (eds.) ECCV 2014. LNCS, vol. 8693, pp. 740–755. Springer, Cham (2014). https://doi.org/10.1007/978-3-319-10602-1_48
12. Liu, K., Wang, W., Wang, J.: Pedestrian detection with lidar point clouds based on single template matching. Electronics 8(7), 780 (2019)
13. Luo, Y., Zhang, C., Zhao, M., Zhou, H., Sun, J.: Where, what, whether: multi-modal learning meets pedestrian detection. In: Proceedings of the IEEE/CVF Conference on Computer Vision and Pattern Recognition, pp. 14065–14073 (2020)
14. Qi, C.R., Yi, L., Su, H., Guibas, L.J.: PointNet++: deep hierarchical feature learning on point sets in a metric space. arXiv preprint arXiv:1706.02413 (2017)
15. Redmon, J., Divvala, S., Girshick, R., Farhadi, A.: You only look once: unified, real-time object detection. In: Proceedings of the IEEE Conference on Computer Vision and Pattern Recognition, pp. 779–788 (2016)
16. Redmon, J., Farhadi, A.: YOLOv3: an incremental improvement. arXiv preprint arXiv:1804.02767 (2018)
17. Seikavandi, M.J., Nasrollahi, K., Moeslund, T.B.: Deep car detection by fusing grayscale image and weighted upsampled lidar depth. In: Thirteenth International Conference on Machine Vision, vol. 11605, p. 1160524. International Society for Optics and Photonics (2021)
18. Simony, M., Milzy, S., Amendey, K., Gross, H.M.: Complex-YOLO: an Euler-region-proposal for real-time 3D object detection on point clouds. In: Proceedings of the European Conference on Computer Vision (ECCV) Workshops (2018)
19. Sualeh, M., Kim, G.W.: Dynamic multi-lidar based multiple object detection and tracking. Sensors 19(6), 1474 (2019)
20. Tan, C., Sun, F., Kong, T., Zhang, W., Yang, C., Liu, C.: A survey on deep transfer learning. In: Kůrková, V., Manolopoulos, Y., Hammer, B., Iliadis, L., Maglogiannis, I. (eds.) ICANN 2018. LNCS, vol. 11141, pp. 270–279. Springer, Cham (2018). https://doi.org/10.1007/978-3-030-01424-7_27
21. Wang, H., Wang, B., Liu, B., Meng, X., Yang, G.: Pedestrian recognition and tracking using 3D LiDAR for autonomous vehicle. Robot. Auton. Syst. 88, 71–78 (2017)
22. Wang, P., Li, W., Ogunbona, P., Wan, J., Escalera, S.: RGB-D-based human motion recognition with deep learning: a survey. Comput. Vis. Image Underst. 171, 118–139 (2018)
23. Zhou, K., Paiement, A., Mirmehdi, M.: Detecting humans in RGB-D data with CNNs. In: 2017 Fifteenth IAPR International Conference on Machine Vision Applications (MVA), pp. 306–309. IEEE (2017)

Robust Decision Making via Cooperative Estimation: Creating Data Saturated, Autonomously Generated, Simulation Environments in Near Real-Time

Israel Toledo-Lopez[1,2], Dylan Pasley[1,2](✉) ⓘ, Raul Ortiz[1,3],
and Ahmet Soylemezoglu[1,2]

[1] Construction Engineering Research Lab, Champaign, IL 61801, USA
Dylan.a.pasley@erdc.dren.mil
[2] University of Illinois, Champaign, IL 61801, USA
[3] University of Puerto Rico – Mayaguez Campus, Boulevard Alfonso Valdes, 00680 Mayaguez, Puerto Rico

Abstract. Every branch of the U.S Military, as well as foreign military agents, have a vested interest in the broad applications and development of robotic systems. Advancements in data collection and storage capabilities has exposed an opportunity to increase the utility of simulated environments. At the most basic level, operations that involve robotic systems require detailed simulation environments to test algorithms and edge cases. The wealth of information collected from robotic platforms can be utilized to autonomously generate simulation environments, which can provide a robust platform for enhanced decision-making capabilities.

Current industry standards depend on labor intensive post processing methods which generate static simulation environments. These simulation environments lack much utility beyond controlled testing. To address this gap, we introduce the foundational research for an intelligent simulation module, a system that utilizes sensory data, collected from semi-autonomous robotic mapping platforms, to generate in near-real time high fidelity digital twin simulation environments of real-world locations. With this system, end-users will be provided with the details they need to make operational decisions without the delay of post processing.

Our system bridges the ROS platform with Unity3D game engine to achieve the generation of its simulated environments. Combat Engineer operations that rely on autonomous robotic platforms will benefit from having a system that can generate high fidelity digital twin simulation environments to aid testing research, mission planning, and robotics control. In general, the intelligent simulation system will allow for robust decision making in autonomous mobile robots, by improving navigation, path planning, coordination between agents, and task planning.

This research has the potential of being utilized in hardware in the loop scenarios where multi-agent control and coordination is required to complete a mission thus advancing the field of cooperative estimation. Further, with the use of virtual reality technology, an operator could potentially be inserted into an operation site virtually; the virtual environment and agents operating within it would be parallel to the physical site, and the operator can then possibly supervise, control, and coordinate both virtual and real hardware robotic systems remotely.

© Springer Nature Switzerland AG 2022
J. Mazal et al. (Eds.): MESAS 2021, LNCS 13207, pp. 273–289, 2022.
https://doi.org/10.1007/978-3-030-98260-7_17

Keywords: Near real time · Autonomous platforms · Digital twin · No post processing · Unity3D · ROS · Multi-agent control and monitoring · Virtual reality

1 Introduction

This paper introduces the research, including the challenges and successes that we have encountered while attempting to automate the process of generating a 3D simulated environment which can be extended and further utilized with some additional research.

1.1 The Problem Space

Military engineers are presented with a broad scope of responsibilities that have a direct impact on the built environment. One factor that is unique to the military engineer and sets them apart from their civilian peers is the expectation to operate within austere environments. The austere nature of these environments presents a significant amount of risk to the field engineers assigned to analyze the potential usability of specified areas of interest. To better carry out their mission, military engineers need to be provided with the appropriate tools as well as accurate and robust sets of data. Current methods of data collection provide a significant increase, both in the volume, and fidelity of data, that engineering units have at their disposal. However, the process of converting this data into a usable information platform is still a time consuming and resource intensive process. The shortcoming of the current process sets the foundation of this research effort. A concrete example is as follows; a military engineer is tasked with overseeing a construction effort in an unknown environment. Data is collected on the unknown environment, to provide the engineer with enough information for the planning and execution of the mission. The system this paper proposes takes this data and automatically generates a simulated environment; this simulated environment provides, to the engineer, the collected information in a manner that is now easy to digest and actionable. The engineer can interact with this simulation and plan out all the individual tasks for the mission. The system this paper proposes also has the ability to display simulated vehicles that can be synced to their physical counterparts, this means that engineers can also monitor the whole operation in real time utilizing our system.

1.2 Interdisciplinary Topics

The sections below outline the approaches we took in analyzing modern processes of collecting, storing, processing, and displaying data which include the use of robotic and autonomous platforms. Our decision to explore Unity 3d was informed by related research which conducted a comparative analysis of Gazebo and Unity [5, 16, 17] and found that it offers a degree of extensibility that can be utilized in robotics applications. Additionally, there has been a growing interest in connecting Unity to ROS [3, 4, 6] in recent years. Even with recent developments in the field it is important to note that the generation of a 3d simulation environments is inherently complicated, time consuming, and resource intensive task.

1.3 Primary Contributions

The primary contribution of this research is to create an enhanced 3D simulation environment which can be generated, updated, and displayed without relying on post processing. An additional requirement of this research mandates that this environment should be generated with minimal input from the end users. By addressing the issue stated above with an automated system we would enable the rapid utilization of data collected from the field in near real time. The development of such a system carries significant implications in the development and application of multi robot systems [6, 7], training machine learning algorithms [5], human machine interfacing via virtual and augmented reality [8–10] and cooperative estimation. Our own internal research confirms that there is a growing need for the capability to control and monitor multiple platforms in a safe and efficient manner. Although the research detailed in this passage does not fully meet this need, it begins to create a foundation that is robust and flexible enough to incorporate these capabilities in future research efforts.

By the end of this effort, we have been able to generate a 3D simulated environment without the extensive hassle of post processing. We were able to cut down the total time and effort required to create a simulated environment while ensuring that the environment accurately represents the real-world landscape where the data was collected. The findings of this research are promising, however there remains a significant amount of work to be done in order to optimize and further automate this process. This will be elaborated on in the results section of this paper.

1.4 Paper Outline

Section 2 will give a brief overview of the technology and research that provide the foundation for the efforts discussed in this paper. Section 3 will elaborate further on related research before we delve into the technical details that address the primary scope of this paper. Section 4 will provide the technical details that outline how we were able to create our simulation environment. The related research section (Sect. 3) will make use of the terms defined in the background section (Sect. 1) and will prepare the reader to better understand the methods section (Sect. 4). Finally, Sect. 5 concludes this paper by providing the results, limitations, and future efforts of our research.

2 Background and Rational

This section will briefly introduce the Robotics for Engineer Operations (REO) research project that preceded the efforts outlined in this paper. By the end of this section there should be a clear understanding of where our data sets come from, how they are processed, and where they are stored prior to map creation. To do this we briefly introduce the platform hardware, as well as the software dependencies that exist in our system. Additionally, this section should provide an explanation of our objectives which will later tie into our justification and the impact of applying this research.

2.1 Project Background

The effort to explore autonomously generated simulation environments is an expansion of the REO research effort being conducted at the Construction Engineering Research Center. REO seeks to extend the capabilities of military engineer units into the modern era by providing a semi-autonomous robotic platform which is capable of surveying and accurately depicting real-world operating environments to expose any challenges or obstacles that the warfighter might encounter. One outcome of this research to date is the development of a large data storage system, which we refer to as the Site Model Database. Proper utilization of the Site Model Database has allowed us to create a data visualization system that is generated in near real time, without the need of post processing, that provides the end user with a robust information system that can be used to guide them in their decision-making process.

An in-depth description of the REO system is outside of the scope of this manuscript. However, in order to explain our data acquisition strategy, it is necessary to provide a high-level overview of the REO system. This REO overview can be found in the system hardware section below. Additionally, the software section will cover the typical workflow that is followed during data collection and processing.

2.2 System Hardware

The data collection portion of this research is carried out on a semi-autonomous robotic platform which is designed to function in a completely offline environment by utilizing standard simultaneous localization and mapping algorithms. The robotic platform that we utilize can be retrofitted with sensor payloads that are specific to its individual mission. For the purpose of this paper the platform was equipped with a 16 channel Velodyne sensor which provides us with our primary point cloud dataset. In order to collect enough information to enable the user to make robust decisions we also use stereo and mono cameras to collect colorized image data. This data is combined with the lidar data in order to create a depth image of the operating environment. In order to facilitate on board processing the payload includes at least one Karbon 700 computers. In order to conduct our tests we ran the program on Razor's Blade 15 studio edition laptop.

The configuration presented above provides the general hardware pieces that are needed to gather and process the data that we use to create our simulated environment. To explain the software components of this research the next section will provide a brief workflow and address the individual software pieces that are responsible for each section of the workflow.

2.3 Software

Initially, our platform is introduced to the environment where it will begin collecting data and building a map using iterative closest point (ICP). In an effort to increase the ability for the platform to localize, our system ingests an initial set of a-priori data. This a-prioi data is data which has already been previously collected and made available for use. In our case the data sets we utilize as a-priori data are 2D GeoTiffs (Fig. 1).

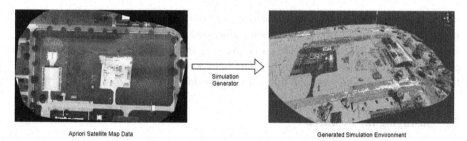

Apriori Satellite Map Data   Generated Simulation Environment

Fig. 1. A-prioi input and 3D mapped output

In order to facilitate the interchange of data between internal subsystems of our platform we rely on ROS. ROS is a set of open-source software libraries and tools which facilitate many common robotic operations [13]. Our data is sent through a series of ROS nodes where it is translated and voxelated before it finally reaches our internal database which we refer to as the Site Model Database. The voxel becomes a core data structure, which is addressed below in the methods section. In order to work with the large quantity of data that is collected in the map building process we chose to use a non-relational database scheme which is currently facilitated by MongoDB. Once the data has gone through some initial processing and is stored in the Site Model Database it can be used to generate display outputs for 2D, 3D, and AR/VR systems. At this point, the initial processing of data is complete, and it is now stored in a format that we can work with to create a simulated environment inside of Unity 3D.

It is important to highlight some of the significant software dependencies that influenced our design decisions of this system's architecture. Perhaps the most significant requirement belongs to ROS and Unity. ROS is compatible with Linux operating systems, whereas Unity 3D is currently most stable on Windows. This set of requirements presents a significant challenge in the attempt to create a bridge between the two operating systems. Fortunately, this issue has been addressed and has seen significant development in previous years [1]. Some related research suggests that in the coming years and the further development of ROS 2 will eliminate this dependency on a ROS-Unity bridge. For now, this connection is made using previously explored socket technology and ROS # [11]. For the purpose of comparability, it is important to note that this project used the HTC Vive system for our virtual reality display. This system operates Steam VR which is the software component for the HTC Vive. Any other dependencies and limitations of this research will be elaborated on in the results and conclusion section of this paper.

2.4 Objectives

After conducting this research, it has become clear that there are multiple objectives that can be obtained through continued advancement of this foundational research. First and foremost, the objective of this research is to establish an applied framework for automating the creation of simulated environments which are identical to real world locations. In essence, we aim to create digital twins of real-world operating environments in an autonomous fashion. This research establishes a proof of concept that carries significant implications to future work which will be addressed below. With these implications in mind, it is also an objective of this research to create an environment using software components that are flexible and extensible. The intention of this environment is to enhance the end user's ability to view and interact with the data collected.

2.5 Justifying the Research

By creating a 3dimensional simulation environment that is accurate and reflects the real-world operating environment we provide the end user with a data rich, responsive, and integrated environment that is created for, tasks that involve the use of cooperative estimation. There are two functional scales at which we operate at for this research. The first is at the sensor-to-sensor level, where a suite of sensor work together to accurately locate objects as they appear in the real world. Once this research is expanded the capability to use and track multiple machines in the same operating environment can be utilized to perform cooperative estimation as a means of confirmation of the systems accuracy.

3 State of the Art

Past work in the area of modeling and simulation in the fields of robotics has focused mostly on generating an accurate representation of a robotic agent that can be observed and utilized in a simulated environment. This paper focuses on the simulated environment itself; generating a simulated environment near-real time, that is as close to a real location as possible so that robotic agents can be studied in this simulated space. In this section, we will discuss some of the past works that are related to this research.

Babaians et al. make two contributions with their work. First, they propose a method to interface ROS from the Unity engine. Second, they simulate various robotic sensors such as LIDAR, RGBD and Monocular cameras in Unity. Although our work does not share their objectives, it does shed some light on the advantages and shortcomings of ROS# as ROS to Unity interface (which is the interface we use in our research). Their work also highlights the tools and techniques available in Unity to simulate robotic sensors, which is something that will be added to our system in future efforts. This paper also highlights that: "Simulators cannot guarantee the final result for industrial or mobile robotic applications since the success of off-line programming depends on how similar the real environment of the robot is to the simulated environment" [1]. This statement supports the need for a system like ours, that generates a simulated environment that is analogous to the real environment the robot operates in.

Codd-Downey et al. develop in their work a virtual reality-based teleoperation interface for autonomous systems by bridging ROS and Unity [2]. Utilizing their system, a user can take control and teleoperate a robotic ground vehicle through a virtual reality platform. The simulated environment that the user can drive the robotic vehicle through is mapped by the robot utilizing a standard SLAM algorithm and afterwards it is pre-stored and loaded into Unity as a particle system. Thus, the simulated environment in this system is static and offers little details to the user; in comparison, the system proposed in our work generates dynamic maps that offer detailed information of the robotic vehicle's surroundings.

Hu et al. present in their work a real-time three-dimensional simulation system, ROSUnitySim, for local planning by miniature unmanned aerial vehicles (UAVs) in cluttered environments [3]. In this work the authors focus on sensor modeling (mainly LIDAR), the interface between ROS and Unity and multiplatform control. This last feature is something that we would like to add to our system in future efforts; the ability to control and observer several robotic vehicles at the same time. The simulated environment that was used to fly the simulated UAVs in this work was static. handcrafted and did not necessarily resemble any real-life location. Handcrafting a map requires time and effort from developers and it is not guaranteed to resemble the target real-life location, thus here we see an example where our system would've been a better environment choice for the simulated UAV flights.

4   Methodology

The following sections present the architecture of the Environment Simulation System developed in this work. This research was conducted in order to generate a near-real time simulated environment that is dynamic, immersive, and interactive. This section concludes with a discussion of the separate components that comprise our system. Using the processes outlined below we were able to successfully establish an automated approach to displaying large quantities of data into an interactive simulated environment. The foundational platform that was established with these methods can be used to improve the level of detail provided to the end user and ultimately better inform their decision-making process.

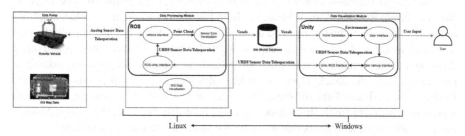

Fig. 2. Architecture for the environment simulation system

4.1   Architecture

An overview of the proposed architecture can be seen in Fig. 2. As shown in the figure, there are four main components that make up this system: the data pump, data processing module, database, and data visualization module. The general workflow of this system is as follows; first, the data pump component introduces real-world data into the system. This data is then used to generate the simulated environment. Next, the data processing module takes in the real-world data, provided by the data pump, generates voxel objects, and stores them in the database; it also handles one end of the interface between the simulated environment and the robotic vehicle.

The Site Model database houses the voxel objects generated by the data processing module. This database is where the voxel objects will be stored until they are called upon for visualization. Finally, the data visualization module retrieves the voxels from the database, generates the simulated environment, and renders the environment for the user. Additionally, the data visualization module handles user input and facilitates interactions between the simulated environment and the simulated vehicle. it also takes care of user input and the other end of the interface between the simulated environment and the robotic vehicle.

4.2 Data Storage and Utilization Components

This section will present an in-depth description of the backend of the developed system. The components discussed in this section are responsible for gathering real-world data, generating voxel objects, and storing said voxels.

Data Pump. The Data Pump gathers and introduces real-world data into the system through its two main methods: the GIS maps and the robotic vehicle. In the first method, we utilize GIS maps, which can be collected from either satellite imagery, or unmanned aerial vehicle (UAV) imagery, to provide an accurate geographic description of the area to be modeled by the system. All the pixels in these maps are geographic points tied to real coordinates, therefore they introduce to the system a list of points with latitude, longitude, altitude and RGBA color information that is then used to generate the voxels.

The second method of data collection utilizes our robotic platform. The J8 Atlas Xtreme Terrain Robot robotic vehicle, is a ROS based electric, eight-wheeled, amphibious, and all-terrain mobile unmanned ground vehicle (UGV) [14]. Once this platform is fitted with an array of sensors it can capture a detailed map of its surroundings and inject this map into the simulation generator. Our J8 load-out is capable of gathering lidar data, providing camera feedback, reporting its live GPS location, as well as other important feedback data. Although the different data collected by the J8 are important, this system is primarily interested in the lidar data. Using the lidar data, the Data Processing Module can generate the voxel objects that make up the simulated world environment.

Data Processing Module. The Data Processing Module is responsible for managing the interface between the system and the robotic vehicle. This module is also responsible for generating and storing voxels, as well as managing the Linux end of the interface between the robotic vehicle and the simulated environment. This module is comprised of four components: the vehicle interface, the sensor data voxelization service, the ROS-Unity interface, and the GIS data voxelization service.

Due to the current limitations of ROS that are outlined in the related research section above, each of these components operate within the Linux operating system. In order to interact with the robotic platform directly, we placed the vehicle interface, ROS-Unity interface, and the Sensor Data Voxelization service inside the ROS platform. This design decision can be seen in Fig. 2. The Vehicle Interface is a series of ROS scripts that enable communication and data transfer between the on-board computer(s) in the robotic vehicle and the rest of the system. The ROS-Unity Interface handles communication between the robotic vehicle and the simulated environment on the Linux side of things. This component is essentially a bridge between Linux and Windows, specifically between ROS and Unity. The ROS-Unity Interface mainly makes use of ROS#, a set of open-source software libraries and tools in C# for communicating with ROS from.NET applications, in particular Unity [12].

Within the ROS# libraries, the ROS-Unity Interface utilizes the ROS package *file_server*, which allows the system to transfer files between platforms. With this package the system can transfer the robotic vehicle URDF files (XML format that represents a robot model) to Unity where the model is rendered for the user. Another important ROS package used by the ROS-Unity Interface is the *rosbridge_server*, this package creates a WebSocket connection that allows outside platforms to interact with ROS through JSON messages. The *rosbridge_server* package can receive JSON strings and convert them into ROS calls, as well as convert ROS responses into JSON string. The ROS-Unity Interface utilizes this package to transfer sensor data and teleoperation commands between the two platforms. The GIS Data Voxelization service is a series of scripts that take GIS map data and generate voxel objects with it. The service iterates through the pixels of a GIS map and retrieves the following information from each pixel: geographical coordinates (latitude, longitude, altitude), local pose (xyz coordinates based on the origin of the map and x y z w orientation), and RGBA color information (red, green, blue, alpha).

With this information the service generates a voxel object that corresponds to each pixel and stores it in the Site Model Database. The voxel object contains an id, the type of material, its geographical coordinates, and its local pose. The type of material is determined by the service utilizing a simple classifier based on the color information of the pixel. Figure 3 shows the structure of a voxel object with details on the information it holds. Once all the voxels are generated, they are inserted into the Site Model database. The generation of our simulated environment happens in two phases. First the Voxels that are created from GIS map data form are added into the environment. Once the GIS generated voxels are added, the system ingests the data that was collected from the J8's sensors. After the J8 data is added, our simulated environment is complete.

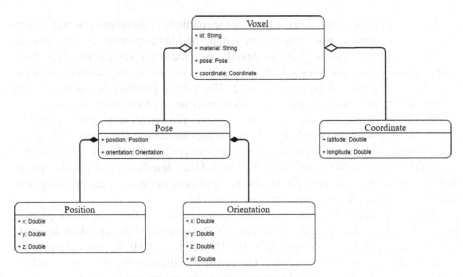

Fig. 3. Voxel object UML diagram

The last component of the Data Processing Module, the Sensor Data Voxelization service, is a series of scripts that take in colorized point cloud data from the Vehicle Interface and generate voxel objects with it. The service iterates through each point in the point cloud data, extracts relevant information (abovementioned in the GIS Data Voxelization service description), generates a voxel object for each point and inserts them into the Site Model Database. This service is responsible for consistently updating the simulated environment, thus any changes in the real-life environment captured by the robotic vehicle will be reflected in the simulated one. Also, this service fills the information gaps in the environment left by the GIS map data; for example, a GIS map may have information on the roof of a building, but no information on any of its walls, with the Sensor Data Voxelization service, it is possible to generate that missing information and introduce it to the simulated environment.

Site Model Database. The Site Model Database component contains the data elements required to create the simulated environment as a list of voxel objects. These voxel objects are inserted by the Data Processing module and retrieved over the network by the Data Visualization module. This database was set up utilizing MongoDB, a document database that stores data in JSON-like documents. The advantages of using MongoDB are its data flexibility (data structure can be changed over time), ad hoc queries, it is a distributed database, and it is free to use. Figure 4 shows an example of a voxel object stored in the Site Model database.

```
{ ⊟
    "id" : "4445492.980|391627.700|234.75|16",
    "material" : "grass",
    "Pose" : { ⊟
        "position" : { ⊟
            "x" : 78.15000915527344,
            "y" : 0.15000000596046448,
            "z" : 234.75
        },
        "orientation" : { ⊟
            "x" : 0,
            "y" : 0,
            "z" : 0,
            "w" : 1
        }
    },
    "Coordinates" : { ⊟
        "latitude" : 40.152799063537344,
        "longitude" : -88.27242206639227
    }
}
```

Fig. 4. Voxel object example in the database

4.3 Data Visualization and Interaction

This section presents a detailed description for the frontend of the developed system. The component discussed in this section is responsible for generating and rendering the simulated environment, as well as managing the user interface.

Data Visualization Module. The Data Visualization module is responsible of managing the user interface, managing the simulated vehicle interface, generating and rendering the simulated environment with voxel objects, and managing the Windows end of the interface between the robotic vehicle and the simulated environment. This module is made up of four components: the World Generation service, the Unity-ROS interface, User Interface, and Sim Vehicle Interface. These components operate in the Windows operating system and within the Unity game engine as seen in Fig. 2.

The World Generation service is a series of scripts that generate and render the simulated environment based on the voxel objects available in the Site Model database. The simulated environment is made up of quads (Unity primitive plane that has edges of one unit long and its surface is oriented in the XY plane of the local coordinate space [15]) that are meshed to form blocks (simple cubes with one-unit long sides), these blocks are then meshed to form chunks and the chunks are lined up to form the world or simulated environment. Figure 5 shows the structure of the simulated environment.

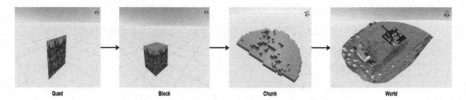

Fig. 5. Structure of the simulated environment

The steps taken to achieve the simulated environment generation and rendering can be seen in the flowchart in Fig. 6 and 7. The Unity-ROS Interface makes use of ROS#. Specifically, the following ROS# plugins are utilized: *RosBridgeClient* and *UrdfImporter*. The *RosBridgeClient* plugin is the .NET API that interacts with ROS through the *rosbridge_server* package and allows communication between the platforms from the Unity side. The *UrdfImporter* plugin imports the robot's URDF files to Unity, parses them, and generates a model of the robot in the Unity environment (Fig. 8).

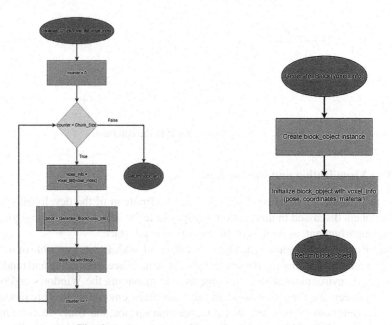

Fig. 6. Left to right: generate chunk, generate block

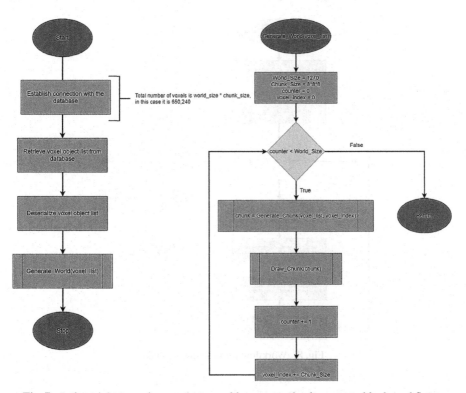

Fig. 7. Left to right: overview, generate world, generate chunk, generate block workflows.

The Sim Vehicle Interface is a series of scripts tied to the robotic vehicle simulated model that allows the user to interact with the simulated robotic vehicle and with the actual robotic vehicle. With this interface, and using the ROS# plugins, the simulated robot could receive sensor feedback from the physical platform and display it for the user. Additionally, it could receive user input (like teleoperation commands from Unity) and execute them in the physical robotic vehicle. Thus, this interface permits a synchronization between the simulated robotic model and the physical platform; it also gives the user control over the simulated robotic vehicle in the simulated environment.

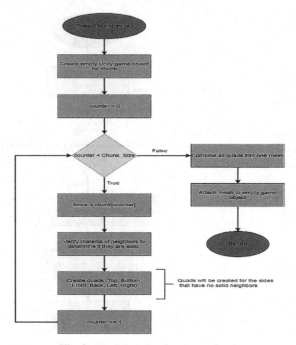

Fig. 8. Workflow for drawing a chunk

The final component in the Data Visualization Module is the User Interface, which is a series of scripts that allow the user input commands to interact with the simulated environment and simulated robotic vehicle. The User Interface gives the user a Virtual Reality (VR) and a First-Person control over the whole system. The VR controls were achieved utilizing Unity's XR Interaction Toolkit plugin. This plugin tracks the user's VR headset movement (in this case the HTC Vive Pro) to control the in-game camera and the hand controllers to move around the simulated environment. The First-Person controls were achieved with a series of Unity scripts, and they allow the user to control the simulated robotic vehicle and move around the simulated environment with the use of regular joystick controllers or with a mouse and keyboard.

5   Conclusion

5.1   Results

Through the efforts of this research, we have been able to connect the Site Model Database system hosted in Linux to the Unity 3d platform hosted in Windows. The large technical challenge here is presented by bridging these separate systems and allowing them to properly communicate with each other without introducing significant lag and without relying on postprocessing. This has been an issue that was thoroughly explored in previous research. Once this was resolved we were able to pull large amounts of voxelated data (around 650,00 voxels that cover a radius of 250 m around the GIS map's origin) into the unity environment in approximately 6 min and 26 s. Figure 9 shows the render time of the environment in relation to the number of voxels.

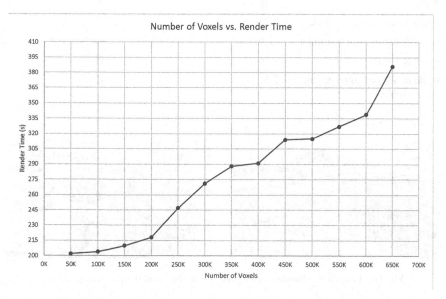

Fig. 9. Graph of the number of voxels vs. the render time (in seconds)

This time expands in an inconsistent manner once we integrated the virtual reality interface into the system. When conducting this exercise, we have experienced rendering times of around 7 to 7:30 min. Table 1 shows the render time for the system in different scenarios and with different combinations of elements integrated into the system, like for example the virtual reality controller. As for fps (frames per seconds) performance, we did not observe any change in performance in relation to the number of voxels, it maintained a steady rate between 900 and 1000 fps. We did observe fps drops when we integrated other elements into the system, for example by integrating the virtual reality controller we observed the frames per second drop to around 42 fps. Table 1 shows the frames per second observed in the different test scenarios. At this point the ability to automate the production of digital environments has been established.

With the current state of the research, we have been able to establish Unity 3d as a flexible environment. Using Unity, we have been able to establish control of a camera avatar allowing a user to navigate the world freely. Additionally, we have been able to establish rudimentary control of a simulated robotic platform and drive it around the simulated world. Additional work is needed to improve this system to add more features and increase its utility. These improvements are discussed in the future work section. Screenshots of the product can be seen in Fig. 10.

Table 1. Scenario based performance comparisons based on render time

Sim objects rendered	Render time (minutes/seconds)	Frames per seconds
World	6:26	900
World & sim. robotic vehicle	6:27	700
World & first-person controller	6:26	84
World & VR controller	7:30	45
World & VR controller & sim robotic vehicle	7:31	42

Fig. 10. Left: J8 rendered in simulation. Right: Simulated environment output

5.2 Future Work

The previous section established that we have successfully created an automated work-flow for generating a 3D Simulated environment inside of Unity. Additionally, we have established the ability to navigate a viewport around the environment as well as control a simulated robot. This research confirmed that Unity is collaborative and flexible enough to handle additional capabilities with further research.

Short term work will focus on optimizing the display and generation of the world environment. The goal of this effort will be to establish a consistent and reliable load time while increasing the resolution of displayed data. Additional work efforts will focus on utilizing the connection between the simulated platform and the real world platform, to simulate several different robotic sensors, increase the level of control the user can exhibit within Unity, add user interface options that take provide better situational awareness to the end user, increase the level of automation in this process, add the ability to send world modification updates back to the site model database, and explore this platform's potential to control multiple real world robots. Once this final effort is complete this system would be able to utilize cooperative estimation algorithms as a means of redundant verification of real-world object placement.

References

1. Babaians, E., Tamiz, M., Sarfi, Y., Mogoei, A., Mehrabi, E.: ROS2Unity3d; High-performance plugin to interface ROS with Unity3D engine. In: 9th Conference on Artificial Intelligence and Robotics and 2nd Asia-Pacific International Symposium 2018, pp. 59–64. IEEE (2019)
2. Codd-Downey, R., Mojiri, P., Speers, A., Wang, H., Jenkin, M.: From ROS to unity: leveraging robot and virtual environment middleware for immersive teleoperation. In: Proceeding of the IEEE International Conference on Information and Automation, pp. 932–936. IEEE, China (2014)
3. Hu, Y., Meng, W.: ROSUnitySim: development and experimentation of a real-time simulator for multi-unmanned aerial vehicle local planning. Simulation **92**(10), 931–944 (2016)
4. Hussein, A., García, F., Olaverri-Monreal, C.: ROS and unity based framework for intelligent vehicles control and simulation. In: IEEE International Conference on Vehicular Electronics and Safety (ICVES), pp. 1–6. IEEE, Spain (2018)
5. Konrad, A.: Simulation of mobile robots with unity and ROS- a case-study and a comparison with Gazebo. In the Department of Engineering Science University West (2019). diva2:1334348
6. Meng, W., Hu, Y., Lin, J., Lin, F., Teo, R.: ROS+Unity: an efficient high-fidelity 3D multi UAV navigation and control simulator in GPS-denied environments. In: IECON 2015 - 41st Annual Conference of the IEEE Industrial Electronics Society, pp. 002562–002567. IEEE, Japan (2015)
7. Reid, R., Cann, A., Meiklejohn, C., Poli, L., Boeing, A., Braunl, T.: Cooperative multi-robot navigation, exploration, mapping and object detection with ROS. In: IEEE Intelligent Vehicles Symposium (IV), pp. 1083–1088. IEEE, Australia (2013)
8. Roldán, J.J., et al.: Multi-robot systems, virtual reality and ROS: developing a new generation of operator interfaces. In: Koubaa, A. (ed.) Robot Operating System (ROS). SCI, vol. 778, pp. 29–64. Springer, Cham (2019). https://doi.org/10.1007/978-3-319-91590-6_2
9. Rosen, E., Whitney, D., Phillips, E., Ullman, D.: Testing robot teleoperation using a virtual reality interface with ROS reality. In: Proceedings of the 1st International Workshop on Virtual, Augmented, and Mixed Reality for HRI, Illinois, USA (2018)
10. Sidaoui, A., Elhajj, I., Asmar, D.: Human-in-the-loop augmented mapping. In: IEEE/RSJ International Conference on Intelligent Robots and Systems (IROS), pp. 3190–3195. IEEE, Spain (2018)
11. Whitney, D., Rosen, E., Ullman, D., Phillips, E., Tellex, S.: ROS reality: a virtual reality framework using consumer-grade hardware for ROS-enabled robots. In: IEEE/RSJ International Conference on Intelligent Robots and Systems (IROS), Spain (2018)
12. ROS-Sharp GitHub. https://github.com/siemens/ros-sharp/wiki/. Accessed 25 July 2021
13. ROS Home Webpage. https://www.ros.org/. Accessed 25 July 2021
14. Army Technology Webpage. https://www.army-technology.com/projects/j8-atlas-xtreme-terrain-robot-xtr/. Accessed 14 July 2021
15. Esri Webpage. https://www.esri.com/en-us/what-is-gis/overview. Accessed 27 July 2021
16. Marian, M., Stîngă, F., Georgescu, M.-T., Roibu, H., Popescu, D., Manta, F.: A ROS-based control application for a robotic platform using the gazebo 3D simulator. In: 21st International Carpathian Control Conference (ICCC), Slovakia (2020)
17. de Melo, M.S.P., da Silva Neto, J.G., da Silva, P.J.L., Natario Teixeira, J.M.X., Teichrieb, V.: Analysis and Comparison of Robotics 3D Simulators. In: 21st Symposium on Virtual and Augmented Reality (SVR), Brazil (2019)

Algorithm Development of the Decision-Making Process of an Engineer Specialization Officer

Ota Rolenec[✉] , Karel Šilinger , and Martin Sedláček

University of Defence, Kounicova 65, 66210 Brno, Czech Republic
ota.rolenec@unob.cz

Abstract. Military operations represent a complex involvement of various types of forces, requiring as far as possible their flawless alignment during the maneuver to meet the objectives of the operation. In order to use the forces effectively, it is necessary to successfully carry out the planning process of the task force to create an operational order and to be able to effectively command and control subordinate units during the conduct of the operation based on the development of the situation in the combat zone. The article describes the algorithm development of individual steps of the decision-making process of the brigade combat team staff officer of the engineer specialization when designing the mobility and counter-mobility support of task force during the planning of the operation. The proposed individual steps of the decision-making process were designed based on the analysis of documents related to the engineer support of the troops and the authors' own experience from practical exercises and exercises on simulators. Each step of the algorithm requires input data from the user, based on which it adds values to the designed engineer modular elements. The output and the main benefit of the algorithm are the values of modular elements and engineer-based recommendations for their use. Based on the algorithm, software was programmed to support the decision-making process of the engineer officer, leading to a reduction in the probability of human error during the planning process given the different levels of knowledge.

Keywords: Algorithm · Counter-mobility support · Decision-making process · Military engineering · Mobility support

1 Introduction

Current and future armed conflicts will be characterized by a high degree of dynamics [1]. The gradual introduction of robotic assets [2] into the equipment of armed forces [3, 4] and the control of automated activities will place high demands on staff members in the planning and management of combat and combat support. Preparing staff members in the ability to correctly assess the battlefield situation and adapt to evolving situations is crucial to enable sound decision making and effective management of all subordinate units in an operation. To support decision making, it is necessary to have well-developed standard operating procedures and costings for unit-level activities at the staff and unit levels. Furthermore, it is advisable to have software tools that enable the unification of some parts of the decision making process [5].

© Springer Nature Switzerland AG 2022
J. Mazal et al. (Eds.): MESAS 2021, LNCS 13207, pp. 290–307, 2022.
https://doi.org/10.1007/978-3-030-98260-7_18

This approach also corresponds with Environment 4.0, elsewhere also Industry 4.0 or "Revolution 4.0", which is an environment whose characteristics seek to capture the various initiatives responding to what is called the fourth industrial revolution. Examples include the German initiative of 2013 called Industrie 4.0, the Industrial Internet Consortium or the Smart Manufacturing Leadership Coalition in the USA, or similar projects and programmes in Japan and China. All of them accentuate the "new" philosophy of systemic use, integration and interconnection of various technologies with the dominant role of information and communication technologies, considering their continuous and very rapid development. This "new" philosophy, in many of its supporting characteristics, corresponds to the issue of implementation of the NATO Network Enabled Capability (NNEC) concept, i.e. the concept of using modern information and communication technologies for effective performance [6, 7] of activities and functions in the security environment.

Although information on command and control systems is generally classified, there are articles on the general approach to this issue. In [8] is proposed a decision-making system to solve dynamic, multi-objective and unequal-area construction site layout planning problem to reduce construction costs. The system is using mathematical optimization models max-min ant system, modified Pareto-based ant colony optimization algorithm and the intuitionistic fuzzy TOPSIS method. Article [9] aims to elaborate an intelligent decision support system that provides relevant assistance to urban planners in urban projects by using machine learning classifiers, naive Bayes classifier and agglomerative clustering. The paper [10] presents a decision support system to help Command and Control operators to help the decision-making process in case of interdiction operations so that the success rate increases. Article [11] presents military tactical planning systems using computational system called fuzzy-genetic decision optimization which combines two soft computing methods, genetic optimization and fuzzy ordinal preference, and a traditional hard computing method, stochastic system simulation, to tackle the difficult task of generating battle plans for military tactical forces.

The shortcomings and deficiencies of the current command and control system modeling and proposed a command-and-control system modeling based on the entity-relationship is presented in [12]. Combined with the Lanchester model considering efficiency of command this paper takes command and control system models for scenario analysis. Paper [13] describes a new way to combine NATO partners' command and control systems and simulation systems into a system of systems, that can support military training, course of action analysis, and mission rehearsal for a coalition.

This article deals with the support of the decision making process of the engineer specialisation staff at the brigade and battalion level in the creation of engineer groups (modules) for engineer mobility support and counter-mobility support. When assessing the tasks performed in the above mentioned engineer roles, the survivability of the engineer role must also be taken into account, especially to assess protective structures [14] for their overcoming. The algorithm will allow the user to encompass all the factors needed to select appropriate engineer assets and units capable of effectively performing the required tasks. Thus, this article presents a different approach to developing a command and control system than that mentioned in the aforementioned papers, where the decision-making activities of members of a single specialty are optimized under predefined steps of a well-defined algorithm.

2 Algorithm Development

Determination of algorithm steps and ranges of values of individual coefficients is based on the study of national regulations and study texts of the University of Defence, calculation standards for individual military engineering works and tactical-technical data of engineering technology. The output of the algorithm are proposals of engineer modular elements related to the current structures of the engineer units of the Czech Army Corps of Engineers as well as to the engineer technology of the armies of NATO countries [15]. The algorithm described below is based on the previous version of the algorithm on the basis of which the APOSŽPP program was developed [16]. The new version of the algorithm includes:

– added steps in the mobility support part of the calculating engineer elements;
– modification of the original steps (changes in the descriptions of some steps) and modifications in the proposed number of engineering recommendations;
– extension of the whole section used for the creation of mine-laying means.

The algorithm should be used with the accompanying excel file "DATA", which contains tabs with information about the individual designed engineer modular elements intended to support movement and to limit enemy movement. Each tab contains the name, organizational structure and capabilities of the modular element in the area of engineer support of troop movement or movement restriction. In addition, the "DATA" file contains engineer recommendations related to engineer equipment or conditions for accomplishing engineer support tasks.

The algorithm requires the input of specific values or information about the occurrence of certain obstacles or the performance of some task of engineering support of troops. The user usually obtains this information from the combat order related to the operation or from studying the map documentation of the area of operation, or from the conducted reconnaissance in the areas of interest.

The algorithm is divided into three basic parts. The first part concerns the design of modules for engineer mobility support. The second part deals with calculations to design suitable engineer modules for counter-mobility support. The last part of the algorithm is intended to allocate the design of modules for engineer support activities between the light and heavy brigade combat team type.

2.1 Mobility Support Part

Mobility support part of the algorithm can be divided into several sub-parts. The first one is dealing with the occurrence of explosive roadblocks and their possible overcoming (see Fig. 1). This part of the algorithm deals with tabs 1 to 8 containing engineer modular elements based on engineer teams and engineer demining teams containing transport vehicles (T-815 medium off-road vehicle, BVP-2 infantry fighting vehicle and Pandur wheeled armoured personnel carrier), explosive mine clearing devices (towed explosive ordnance deminer Python and portable explosive ordnance deminer SAPLIC), mechanical mine clearing devices (bulldozer and demining plough) and surface clearance devices (additional minesweeping device POOZ).

Step 'A' assesses the assumed depth of anti-tank minefields to be overcome, which the enemy will establish on threatened lines of advance of adversary forces, particularly in calculated mobility corridors. Depending on the evolution of the situation, anti-tank minefields may also be established by means for establishing minefields remotely. The ability to overcome anti-tank minefields is very important because they can cause casualties, disrupt the battle formation, divert the direction of advance of own forces to areas ready for destruction (Kill Zones) and negatively affect the psyche of soldiers.

Larger input values or the performance of a certain task usually means an increase in the number of individual tabs. In the next steps, however, this increase is linked to the previous possible value of the bookmarks, so that their value is not increased beyond an unrealistic limit.

Step "B" assesses the assumed depth of antipersonnel minefields that own troops will have to overcome. Anti-personnel mines may be part of anti-tank minefields or non-explosive barricades, or they may be established separately. The fastest way to overcome them is to use explosive demining devices. The intervals used for the algorithm steps are related to the tactical-technical capabilities of the proposed modules.

Step "C" assesses preparedness for the possible occurrence of surface mines on paved roads. This step of the algorithm primarily forces the user to assess possible enemy measures depending on the evolution of the situation on the battlefield. Especially in an offensive operation, it is necessary to calculate the possible deployment of means establishing minefields at a distance.

Another sub-part of the algorithm (see Fig. 2) is dealing with overcoming man-made objects or non-explosive roadblocks and landslides and performing tasks in urbanized areas, representing a very common task in current conflicts. In step "D" and "E", tabs 1 to 13, 26 to 29 are modified to include, in addition to the aforementioned engineer units and demining assets, engineer engineered modular elements equipped with engineer earthmoving machines (universal finishing machine UDS, tractor-trailer JCB 4CX, wheel loader KN-251, excavator-loader MPEV and armoured amphibious bulldozer AACE) which are very effective for clearing non-explosive roadblocks in urbanized environments. In addition, the tabs contain engineer reconnaissance and demarcation modular elements (light armoured reconnaissance vehicle) that are essential for early identification and assessment of adversary roadblocks. The last type of recommended engineer elements is transportation equipment that allows for the transport of soil to backfill roadway embankments or the transport of necessary engineer tools (T-815 S3 dump truck in both unarmored and armored versions). This part of the algorithm also includes recommendations for increased consumption of explosives when performing engineer tasks in urbanized areas [17]. This is because underground and aboveground spaces are also evaluated in urbanized spaces.

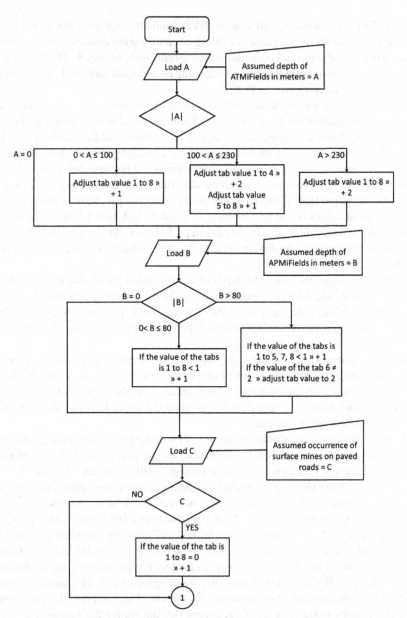

Fig. 1. Sub-part of the algorithm with steps "A"–"C".

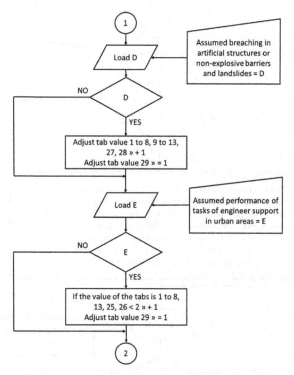

Fig. 2. Sub-part of the algorithm with steps "D"–"E".

The third sub-part of the algorithm (see Fig. 3) is dealing with the width and depth of dry and wet gap crossing. Step "F" calculates the engineered modular elements containing assault and accompanying bridges used in the armies of NATO countries (2nd echelon bridge layers AM-50/70, SISU Leguan, PAR-70, MS-20 and BR 90 ABLE; modular assault bridge PTA MAB; armoured vehicle launched bridge Mowag Piranha 3 and assault bridges Leopard 2 Leguan, Wolverine HAB and MT-55). The values of each tab are adjusted based on the ability of each asset to bridge a given gap width. Step 'G' assesses bridge assets whose design requires the placement of bridge piers in the gap.

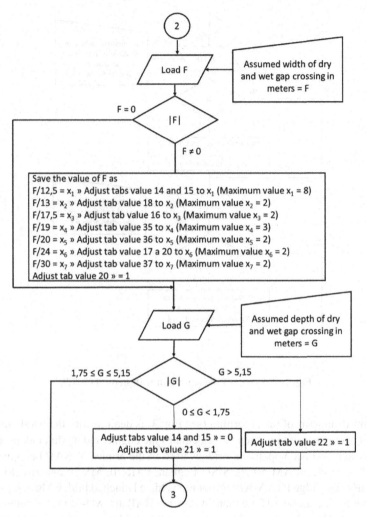

Fig. 3. Sub-part of the algorithm with steps "F"–"G".

The next part of the algorithm deals with wet gap crossing and the earthworks in overcoming all types of obstacles (see Fig. 4). In step "H" it is necessary to assess whether the water velocity allows the construction of bridge crossing site requiring the use of bridge supports. If the watercourse velocity is too high, only single span bridges may be used and neither bridge ends nor bridge abutments may be placed in the watercourse.

Steps "I" and "J" assess the implementation of bank modifications when crossing an obstacle out of contact with the enemy or under enemy fire. This has implications for the selection of types of engineered earth moving equipment. While in the first case, ground vehicles with basic ballistic protection should be sufficient to accomplish this task, in the second case, it is necessary to use engineer amphibious ground vehicles with a high degree of protection against enemy fire. Amphibious armored bulldozers are suitable for this task.

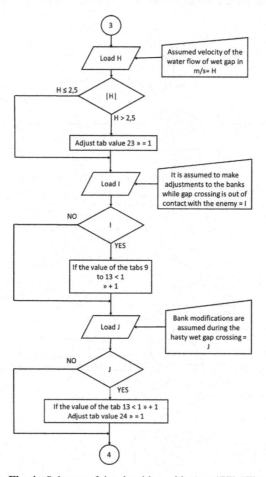

Fig. 4. Sub-part of the algorithm with steps "H"–"J".

The last part of the algorithm, which deals with engineer support for movement, is concerned with terrain modification to enable the movement of own forces (see Fig. 5). Step "K" specifies the number of longitudinal and tranverse paths to monitor and maintain in the area of operations. Their number affects the number of engineer earth moving equipment, engineer reconnaissance elements and technical modular elements representing a combination of transport equipment, displacement equipment and some types of engineer earth moving equipment. The selection in steps 'L' and 'M' allows increasing values for engineering modular elements and transport modular elements. These steps evaluate the possibility of muddying the roads due to meteorological conditions, overcoming natural obstacles and take into account the task of building and maintaining forward landing areas. In step 'N', the existence of explosive threats is assessed to calcu-late the clearance of paths and areas out of contact with the enemy. A positive selection under this step indicates a possible increase in the proposed number of engineer earth moving vehicles and mine clearance vehicles and a recommendation for deployment of EOD units.

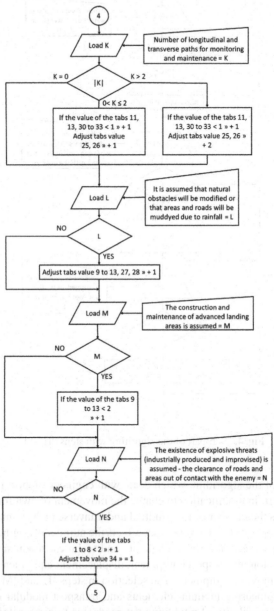

Fig. 5. Sub-part of the algorithm with steps "K"–"N".

2.2 Counter - Mobility Support Part

The second main part of the algorithm deals with the design of engineer mine-laying means. The algorithm's designs are particularly useful for obstacle detachments performing the task of establishing anti-tank minefields during a manoeuvre. As a first step in the first sub-part is the selection of available mine-laying means used in the armies of NATO countries (trucks with mines and slides, MV-3 mine thrower, minesweepers Minotaur, Skorpion, SUM Kalina and Volcano). This information is important for the calculations in the next steps of the algorithm.

In step "P" there is the possibility to define a custom minelayer. For further needs of the algorithm the name of the means, the amount of mines carried, the range value and the number of means are stored. Step "Q" allows to repeat this option based on the user's need (see Fig. 6).

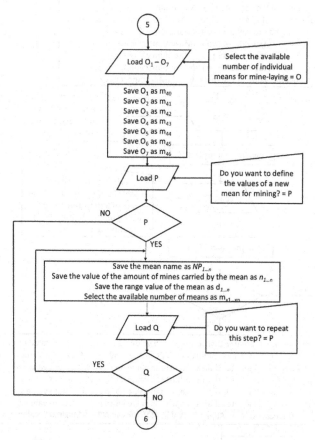

Fig. 6. Sub-part of the algorithm with steps "O"–"Q".

In the next sub-part of the algorithm (see Fig. 7), the quantity values for each obstacle are calculated based on the input data. In step "R", the expected maximum number of mines in the minefields to be established during the conduct of the operation is entered. This data is used to determine the maximum number of individual mine-laying devices with respect to their available quantity and their capabilities regarding the quantity of mines carried. In the 'S' and 'T' steps, the expected maximum length of the anti-tank minefields to be established and their calculated density are entered, considering for all proposed assets the use of only anti-tank mine types acting on the entire vehicle profile.

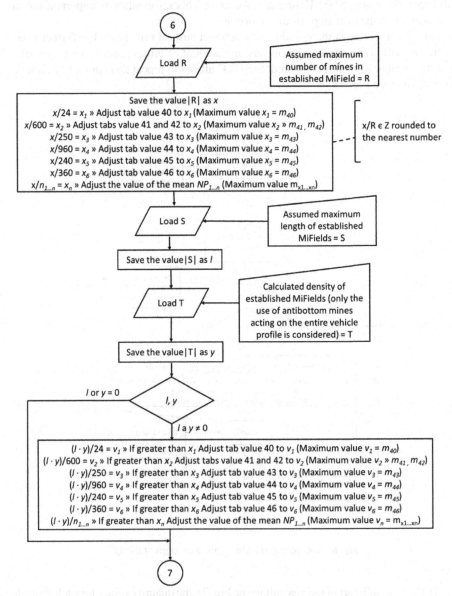

Fig. 7. Sub-part of the algorithm with steps "R"–"T".

These data are used to determine the number of individual mine clearance devices based on the calculation of the quantity of mines to be laid in each minefield. If the determined values are higher than in the previous step, the total number of devices is increased.

The last part of the algorithm concerning the design of mine countermeasures deals with the possible supply of ammunition to the formed blocking units during combat and the range of the mine countermeasures. In step "U" the user enters the total number of mines calculated for a given phase of the operation. If this value is higher than the amount of mines carried by each type of calculated minelayer, a warning is displayed regarding cooperation with logistics units in the area of resupplying units with ammunition vehicles or possible mine supply during the conduct of combat.

In step 'V', the requirement for a minimum spacing to lay a minefield is assessed. This step of the algorithm relates to the tactical situation and the range of the minelayer. A longer distance from the edge of the minefield when establishing a passage means more protection for the asset, as it must be considered that in combat operations, obstacles will

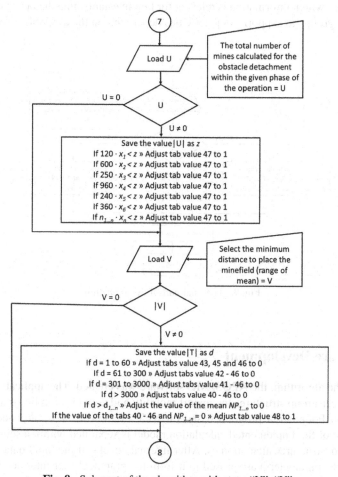

Fig. 8. Sub-part of the algorithm with steps "U"–"V".

always be under enemy fire to increase their effectiveness (see Fig. 8). If the required range exceeds the range of the calculated resource, the value of 0 is set for that resource and it will not be present in the proposed resources.

2.3 Resulting Part

The last main part of the algorithm (see Fig. 9) is used to display the resulting values of the engineering modular elements and engineering recommendations. In the last step, the user chooses whether to display values for the engineer support of a light task force (in the conditions of the Czech Armed Forces, this is a task force based on the 4th Rapid Deployment Brigade) or for the engineer support of a heavy task force (in the conditions of the Czech Armed Forces, this is a task force based on the 7th Mechanised Brigade) in the area of engineer support of movement. Engineer mine-laying assets are displayed identically for both types of task groups.

A member of the engineer specialisation staff chooses, according to his level (battalion, brigade), which information is relevant for him in establishing the optimal structure of engineer groups to support troop activities depending on the available resources.

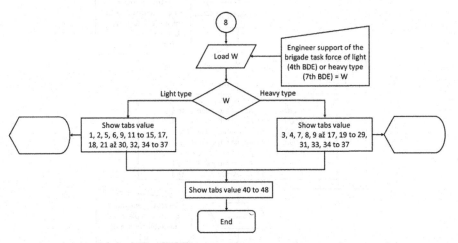

Fig. 9. The final part of the algorithm.

3 Software Development

Based on the algorithm, the APOSŽMC software was created. The application to optimize engineer group structures to support mobility and counter-mobility is a Windows Presentation Foundation program based on NET Framework 4.5.2 architecture. The source code of the implemented calculation model is separated within a self-contained class named StructureCalculation.cs. After internal checks of the input data, data variables (inputs, parameters) are passed to it from the graphical user interface for further processing.

The software has been improved over the previous version to provide a more user-friendly environment. The findings were based on experimental validation conducted with members of the Engineer Specialization Staff. The operation of the application is intuitive and ToolTip help is available if needed (by hovering the mouse cursor over the appropriate application control). The individual controls are hidden from the user until they can be used. The gradual uncovering of the application controls allows the user to be clearly directed to the next steps. The number of inputs required from the user is eliminated as much as possible in the program. When the application is started, the user is presented with an introductory window (see Fig. 10).

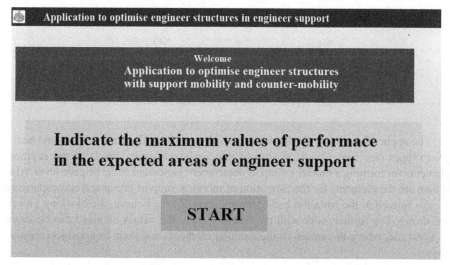

Fig. 10. Home window of APOSŽMC application in Czech language.

In the initial window, the user can get acquainted with the basic characteristics of the program by clicking on the option "About the application". Pressing the START button will start a series of steps that require the user to enter the necessary values to determine the corresponding engineered modular elements (see Fig. 11).

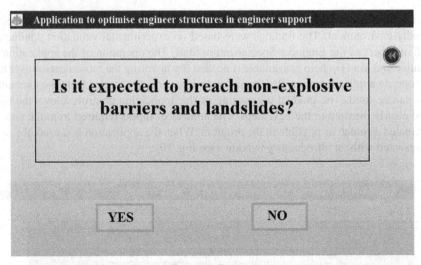

Fig. 11. One of the steps of the APOSŽMC application in Czech language.

The application allows the display of engineer modular elements for light and heavy group types (see Fig. 12). At the top of the displayed results are listed the engineer elements for forming a mobility support detachment generated at the brigade level. Also shown are the elements for the formation of mobility support group and accompanying groups formed at the battalion and company level. At the bottom, the blocking groups are shown. For further work with the resulting data, the values obtained can be saved in a text file, where the names of the modular elements and their composition are also shown.

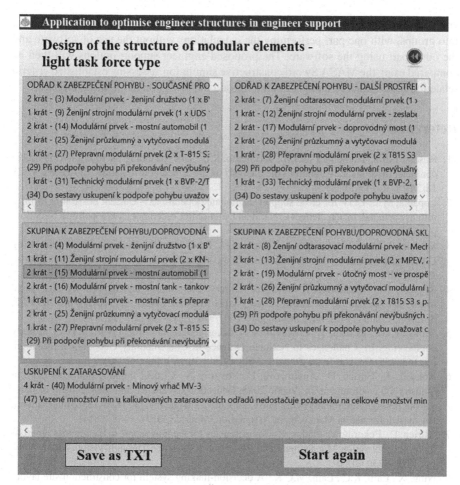

Fig. 12. Results of APOSŽMC application in Czech language.

4   Conclusion

The developed algorithm allowed us to program a software tool that is intuitive and requires minimal user training. However, when using this and similar programs, the user must be aware that he/she must always be able to perform the specified tasks without using them. When assessing the results of a program based on the above algorithm, one must think critically about the values obtained. The displayed values of the engineering modular elements are of a recommendatory nature and serve to support the decision-making of the officer of engineering expertise. The decision support can have benefits in reducing the time consuming nature of the planning process, but also in the effectiveness of the designs for individual engineer support tasks. This is a benefit that is not only required within the Czech Army Corps of Engineers.

In the next part of the research it is planned to conduct staff drills at engineer units. The aim of the staff exercises will be to perform calculations of the deployment of engineer forces and assets in combat operations. The staff members will be divided into groups, with one part performing the calculations without the use of software and the other part using the software. The proposed engineer support options will then be compared using the MASA Sword constructive simulation, assessing established criteria related to the conduct of combat operations and the provision of engineer support.

References

1. Rosero, L.F.T., Hernandez, A.J.B., Sarmiento, R.B.: Understanding the multiple Colombian conflicts: theoretical evolution in the analysis of the armed confrontation. CO-HERENCIA **18**(34), 119–155 (2021)
2. Lopatka, M.J.: UGV for close support dismounted operations – Current possibility to fulfil military demand. In: Challenges to National Defence in Contemporary Geopolitical Situation, p. 16 (2020)
3. Stodola, P., Drozd, J., Šilinger, K., Hodický, J., Procházka, D.: Collective perception using UAVs: autonomous aerial reconnaissance in a complex urban environment. Sensors **20**(10), 2926 (2020)
4. Kopuletý, M., Palasiewicz, T.: Advanced military robots supporting engineer reconnaissance in military operations. In: Mazal, J. (ed.) MESAS 2017. LNCS, vol. 10756, pp. 285–302. Springer, Cham (2018). https://doi.org/10.1007/978-3-319-76072-8_20
5. Sedláček, M., Dohnal, F.: Possibilities of using geographic products in tasks of military engineering. In: Challenges to National Defence in Contemporary Geopolitical Situation, pp. 145–155. General Jonas Žemaitis Military Academy, Vilnius (2020)
6. Koleňák, J., Ullrich, D., Ambrozová, E., Pokorný, V.: Critical thinking and leadership in industry 4.0 environment. In: Vision 2025: Education Excellence and Management of Innovations through Sustainable Economic Competitive Advantage, pp. 1826–1836. International Business Information Management Association, Madrid (2019)
7. Ullrich, D., Pokorný, V., Sládek, P.: Competencies for leading people in the security environment. In: Innovation Management and Education Excellence Through Vision, Milan, Italy, pp. 1722–1730 (2020)
8. Ning, X., Lam, K.C., Lam, M.C.K.: A decision-making system for construction site layout planning. Autom. Constr. **20**(4), 459–473 (2011)
9. Khediri, A., Laouar, M.R., Eom, S.B.: Improving intelligent decision making in urban planning: using machine learning algorithms. Int. J. Bus. Anal. **8**(3), 40–58 (2021)
10. Loeches, J., Vicen-Bueno, R., Mentaschi, L.: METOC-driven vessel interdiction system (MVIS): supporting decision making in command and control (C2) systems. In: Oceans 2015, Genova: Ctr Congressi Genova, Genova (2015)
11. Kewley, R.H., Embrechts, M.J.: Computational military tactical planning system. IEEE Trans. Syst. Man Cybern. Part C-Appl. Rev. **32**(2), 161–171 (2002)
12. Meng, H., Song, X.: The modeling and simulation of command and control system based on capability characteristics. In: Xiao, T., Zhang, L., Ma, S. (eds.) ICSC 2012. CCIS, vol. 327, pp. 255–261. Springer, Heidelberg (2012). https://doi.org/10.1007/978-3-642-34396-4_31
13. Pullen, J.M., Mevassvik, O.M.: Coalition command and control - simulation interoperation as a system of systems. In: 11th Systems of System Engineering Conference, Kongsberg: Norway. IEEE (2016)
14. Štoller, J., Zezulová, E.: The basic properties of materials suitable for protective structures and critical infrastructure. In: Kravcov, A., Cherepetskaya, E.B., Pospichal, V. (eds.) Durability of Critical Infrastructure, Monitoring and Testing. LNME, pp. 211–221. Springer, Singapore (2017). https://doi.org/10.1007/978-981-10-3247-9_24

15. Cibulová, K., Rolenec, O., Garba, V.: A selection of mobility support engineering devices of NATO armies usable in the Czech armed forces combat operations. In: Proceedings of International Conference of Military Technologies Brno 2019, p. 8870016. Institute of Electrical and Electronics Engineers Inc., Brno (2019)
16. Rolenec, O., Šilinger, K., Žižka, P., Palasiewicz, T.: Supporting the decision-making process in the planning and controlling of engineer task teams to support mobility in a combat operation. Int. J. Educ. Inf. Technol. **2019**(13), 33–40 (2019)
17. Öğünç, G.İ: The effectiveness of armoured vehicles in urban warfare conditions. Defence Sci. J. **71**(1), 25–33 (2021)

Reconnaissance in Complex Environment with No-Fly Zones Using a Swarm of Unmanned Aerial Vehicles

Petr Stodola$^{(\boxtimes)}$ ⓘ and Jan Nohel ⓘ

Department of Intelligence Support, University of Defence, Brno, Czech Republic
petr.stodola@unob.cz

Abstract. This paper presents the model of reconnaissance of the area of interest performed by a swarm of cooperative Unmanned Aerial Vehicles. The reconnaissance operation of the area may be conducted in a complex environment in which obstacles and/or terrain may block the line of sight from sensors and, thus, occlude some parts of the area. Also, the no-fly zones can be defined where no vehicle is permitted at any time of the operation. The model focuses on planning trajectories of all the vehicles in the swarm in such a way that the vehicles avoid the no-fly zones, the sensors of the vehicles explore as large area as possible during the operation and, at the same time, the operation is performed as quickly as possible. Exact and stochastic (metaheuristic) algorithms are proposed to find a solution to the task at hand. A set of experimental tasks, based on the real military reconnaissance scenarios, is proposed to verify the presented model and algorithms.

Keywords: Reconnaissance · Unmanned aerial vehicles · No-fly zones · Complex environment · Metaheuristic algorithm · Experiments

1 Introduction

The use of Unmanned Aerial Vehicles (UAVs) for tasks such as monitoring, exploration, surveillance or reconnaissance for information gathering has become commonplace in many domains – civil as well as military. The reasons are the precise and fast results, easy control, and technological and financial availability.

This paper presents the use of UAVs for military purposes – in a reconnaissance operation. This research extends the previous research of the authors in which the model of aerial reconnaissance in complex environment was proposed [1]. The objective is to plan trajectories of available UAVs in the swarm to explore as large area of interest as possible as quickly as possible. The exploration is done via sensorial systems of UAVs from a set of waypoints (electro-optical sensors are primarily considered in this paper).

The planning of trajectories is conducted in two phases: (1) deployment of waypoints from which sensors perform monitoring; (2) planning routes between waypoints. The goal of the first phase is to deploy a number of waypoints in such a way that as large area as possible is explored; the goal of the second phase is to plan routes of individual

© Springer Nature Switzerland AG 2022
J. Mazal et al. (Eds.): MESAS 2021, LNCS 13207, pp. 308–321, 2022.
https://doi.org/10.1007/978-3-030-98260-7_19

UAVs so that the whole reconnaissance operation is performed in mutual cooperation as fast as possible.

In this paper, the original approach is extended in several areas, connected especially with the no-fly zones:

- No-fly zones. These are areas in which no UAV can be located at any time of the reconnaissance operation; i.e. trajectories must be planned so that the vehicles avoid these zones during the operation. The objective of the task remains the same: the complete and fast reconnaissance.
- Exact algorithm for initial waypoints deployment. This algorithm was proposed to increase the efficiency of the metaheuristic algorithm used for deployment of waypoints.
- Extension of the algorithm for waypoints deployment. The algorithm was modified so that the waypoints are not to be deployed in no-fly zones.
- Extension of the algorithm for planning routes. The algorithm was modified so that the routes (when UAVs move between waypoints) avoid the no-fly zones.

The idea of no-fly zones appears in some other publications and research papers. In [2], the authors proposed an algorithm for visual reconnaissance by a single UAV. It plans a route for this UAV using a genetic algorithm while taking no-fly zones into account. The authors deploy waypoints solving integer linear programming problem. The no-fly zones change the objective function – penalties are introduced for points laying inside these zones.

The authors of [3] discuss the methodology of forming an air defence plan to counteract a group of drones based on multiagent modelling. The model presented plans trajectories for UAVs taking into account a set of circular no-fly zones or obstacles. In [4], the authors address the dynamic formation reconstruction and trajectory replanning problem in the air patrol task using multiple fixed-wing unmanned aerial vehicle formations. The possibility that some of the vehicles may break down during operation is considered as well as no-fly zone avoidance and intervehicle collision avoidance.

The problem of controlling a hypersonic cruise vehicle with waypoint and no-fly zone constraints is examined in [5]. The authors proposed a non-uniform control approach by transforming waypoints into variable optimization parameters of pseudospectral method which can satisfy the constraints completely. In [6], an online estimation algorithm based on the drag acceleration-energy profile for generation of entry landing footprints for an entry hypersonic vehicle while satisfying the no-fly zone constraint is examined. Based on the constraint model of the no-fly zone, flying around strategies are proposed for different conditions, and a reachable area algorithm is designed.

A bio-inspired algorithm based on the memetic whale optimization integrating the Gauss pseudo-spectral methods is proposed in [7] for the hypersonic vehicle entry trajectory optimization problem with no-fly zones. The whale optimization works as an initializer in the first searching stage of the entire searching due to its strong global search ability and non-sensitivity to the initial values. The authors of [8] combines the convex optimization and multi-resolution technique for the spacecraft close-range trajectory planning; affine approximation of no-fly zone is derived by surface tangent-plane formula.

The motivation of the authors of this paper is to draw the model nearer to the real conditions which are common in the military operations and to provide the complete solution. No-fly zones could be defined either on the strategic level or on the tactical level by commander. The model is part of the software tool Tactical Decision Support System (TDSS) designed for commanders to support them in their decision making process on the battlefield [9]. It is composed of a number of models of military tactics for the various tactical tasks such as reconnaissance [1], surveillance [10], manoeuvre of units [11], observation post planning [12] and others [13–18].

The rest of the paper is organized as follow. Section 2 presents the complete model of the reconnaissance problem. Section 3 proposes a solution to the problem using a set of exact and metaheuristic algorithms. In Sect. 4, a number of experiments, based on the real military reconnaissance scenarios, is proposed to verify the presented model and algorithms. Section 5 concludes the paper.

2 Model

Let A_0 be an area of interest to be explored. This area is composed of an infinite number of points laying inside some bordered shape (that is a polygon in this model) defined in the area of operations and copying the relief of the terrain.

Let $U = \{U_1, U_2, \ldots, U_m\}$ be a finite set of available UAVs ready for the operation $(m > 0)$. Let $W = \{W_1, W_2, \ldots, W_n\}$ be a finite set of waypoints from which sensors of UAVs explore the area of interest $(n > 0)$. The number of waypoints and their positions are not known at the beginning of the operation. The position of any waypoint is defined as $W_i = (x_i, y_i, h_i)$ – the two-dimensional Cartesian coordinates and the height above the ground level.

UAVs in the swarm may be heterogeneous from the point of view of their motion and control parameters but must be equipped with the same types of sensors. From each waypoint, any UAV in the swarm $U_i \in U$ may explore some part of the area of interest $a_i \subseteq A_0$. The size of a_i depends on the location of the waypoint, parameters of the sensor and terrain and/or obstacles in the area which may block the visual line of sight (VLOS) and occlude some parts of A_0 (see [1] for more details). The total explored area a_{total} when the reconnaissance operation is finished is expressed in formula (1).

$$a_{total} = \bigcup_{i=1}^{n} a_i \tag{1}$$

Let $Z = \{Z_1, Z_2, \ldots, Z_z\}$ be a finite set of no-fly zones in which no UAV is permitted during the reconnaissance operation $(z \geq 0)$. Constraint (2) must be satisfied; i.e. the area of interest must not be entirely overlapped with no-fly zones.

$$\bigcup_{i=1}^{z} Z_i \subset A_0 \tag{2}$$

The objective of the first optimization problem is to determine the number of waypoints and their positions in such a way that (a) there is as low number of waypoints n as possible whereas (b) the total explored area a_{total} is as large as possible. As these two conditions go against one another, the minimum portion of the explored area ω_{min} needs to be specified ($0 < \omega_{min} \leq 1$). Formulae (3) and (4) shows the objectives subject to condition (5).

$$Minimize(n) \tag{3}$$

$$Maximize(a_{total}) \tag{4}$$

$$|a_{total}| \geq \omega_{min} \cdot |A_0| \tag{5}$$

In the second optimization problem, the trajectories of all available UAVs are planned so that all the waypoints are visited by at least one UAV and the reconnaissance operation is performed as fast as possible. R_i denotes a route for a UAV $U_i \in U$. Every route is composed of an infinite number of points through which the UAV flies in the correct order during the operation: $R_i = \{R_i^1, R_i^2, \ldots\}$. T_i is the time necessary for a single UAV $U_i \in U$ to fly along its route and then return back to its home position. Then, the objective is in formula (6).

$$Minimize(max(T_1, T_2, \ldots, T_m)) \tag{6}$$

The no-fly zones limit both the positioning of waypoints and planning the routes. In the former, condition (7) must be satisfied; that means no waypoint can be positioned inside any no-fly zone. In the latter, condition (8) must be satisfied; no point laying on the route of any UAV can go through no-fly zones.

$$W_i \notin Z_j \quad \text{for all} \quad W_i \in W \quad \text{and} \quad Z_j \in Z \tag{7}$$

$$R_i^k \notin Z_j \quad \text{for all} \quad U_i \in U, R_i^k \in R_i \quad \text{and} \quad Z_j \in Z \tag{8}$$

3 Solution

In this section, the solution to the problem formulated in the previous text is presented. First, the approach used to optimize the waypoints deployment is examined (Sect. 3.1). Then, the algorithm to optimize trajectories of UAVs is discussed (Sect. 3.2). The emphasis is put on the novel features.

3.1 Waypoints Deployment

The objective of this optimization problem is to find the number and positions of a set of waypoints according to optimization criteria in formulae (3) and (4) subject to conditions (5) and (7).

Figure 1 shows an example. On the left, the blue lines border the area of interest, red areas are no-fly zones. The goal was to explore at least 97% of the area of interest ($\omega_{min} = 0.97$). To do this, a number waypoints (green dots) were deployed ($n = 35$). On the right, the green colour represents the explored area; i.e. the area covered by sensors

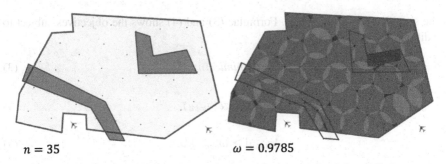

$n = 35$ $\omega = 0.9785$

Fig. 1. Example situation for waypoints deployment (Color figure online)

from waypoints which is 97.85%. There are two UAVs available for the reconnaissance operation (labelled A and B).

The optimization problem is divided into the two subproblems. The first is the estimation of the number of waypoints necessary to cover the specified portion of the area of interest.

For this first subproblem, an exact algorithm is proposed – see Algorithm 1. The inputs of the algorithm is the area of interest A_0, no-fly zones Z, the portion of the area needed to be explored ω_{min} and sensor parameters. First, the algorithm calculates best possible circular coverage d_{range} on the flat surface based on the sensor parameters (point 2). Then, this distance is modified by the minimum necessary coverage coefficient ω_{min} and the safety coefficient $c_{reserve}$ which ensures the full coverage (point 3). The latter takes into account the influence of terrain and obstacles in the area of operations. Empirically, from various experiments based on real scenarios, this value is set to $c_{reserve} = \sqrt{2}$.

Estimate_Number_of_Waypoints (A_0, Z, ω_{min}, sensor parameters)

1. $W = \emptyset$

2. **calculate** d_{range} based on sensor parameters

3. $d_{range} = \frac{c_{reserve}}{\omega_{min}} \cdot d_{range}$

4. **for each** edge E_i of A_0 and $Z_j \in Z$

5.   **calculate** a set of points X at distance $d_{range}/2$ from the edge E_i
    and with distance d_{range} between nearest neighbours in this set

6.   **for each** point $X_k \in X$

7.   **if** X_k is inside A_0, outside all $Z_j \in Z$ and distance to the nearest
        waypoint in W is not smaller than d_{range}

8. $W = W + \{X_k\}$

9. **return** $|W|$, W

Algorithm 1. Algorithm to estimate the number of waypoints

The algorithm inserts waypoints in the area of interest uniformly from individual edges of the area of interest A_0 and all no-fly zones $Z_j \in Z$ (points 4 to 8). The points are inserted in regular rows from each edge until crossing another edge (point 5). The waypoints are inserted into set W when they lie inside the area of interest but outside no-fly zones and the distance to other waypoints in W is smaller than d_{range} (points 7 and 8). The algorithm returns the estimated number of waypoints $n = |W|$ as well as the initial deployment of waypoints W (point 9).

The result of the algorithm on the situation from Fig. 1 is shown in Fig. 2. On the left, positions of the 35 waypoints are shown; on the right, the blue colour represents the coverage from these waypoints which is 85.17%.

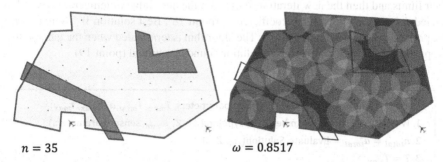

$n = 35$ $\omega = 0.8517$

Fig. 2. Estimation of the number of waypoints and initial waypoints deployment (Color figure online)

The second subproblem is the optimization of the positions of n waypoints using the optimization criterion in formula (4). Each waypoint is defined by 3 independent variables: $W_i = \{x_i, y_i, h_i\}$. This means that, in total, the problem is composed of $3n$ optimization varibles.

For solution, the metaheuristic algorithm based on the simulated annealing principle was proposed (see Algorithm 2). The inputs are the area of interest, no-fly zones, minimum necessary coverage, sensor parameters and control parameters of the algorithm. The control parameters are as follows:

- T_{min}: the minimum limit of temperature.
- T_{max}: the initial setting of temperature.
- α: cooling coefficient ($0 < \alpha < 1$).
- k_{max}: maximum number of transformations per iteration.
- r_{max}: maximum number of replacements per iteration.

At the beginning, the algorithm Estimate_Number_of_Waypoints (Algorithm 1) is used to estimate the number of waypoints and to determine their initial positions (point 1). The solution is evaluated using function Evaluate_Solution by computing the total coverage a_{total} (point 2).

The algorithm works in iterations (points 4 to 18). In each iteration, the constant value of temperature is used. A number of solution transformations (limited by coefficient k_{max}) using function Transform_Solution is performed and evaluated during each iteration (points 7 and 8). This transformed solution W' replaces the original with probability $p(W' \rightarrow W)$ given by the Metropolis criterion (points 9 to 13) – see formula

(9). This probability depends on the difference in qualities of the original and transformed solutions (better solution always replaces the original) and the current value of temperature. The number of replacements per iteration is limited by coefficient r_{max}.

$$p(W' \rightarrow W) = \begin{cases} 1 & \text{for } a'_{total} \geq a_{total} \\ e^{-\frac{a_{total} - a'_{total}}{T}} & \text{otherwise} \end{cases} \tag{9}$$

The iteration ends when the number of transformations or replacements exceeds their limits and then the new iteration starts with the new value of temperature which is cooled down using the cooling coefficient α (point 18). Best solution W^* found during the process is saved (points 14 to 16). The algorithm is terminated when the temperature drops below its lower limit and best solution found is returned (point 19).

Deploy_Waypoints $(A_0, Z, \omega_{min},$ sensor parameters, $T_{max}, T_{min}, \alpha, k_{max}, r_{max})$
1. $W = W^* = $ Estimate_Number_of_Waypoints$(A_0, Z, \omega_{min},$ sensor parameters$)$
2. $a_{total} = a^*_{total} = $ Evaluate_Solution(A_0, Z, W)
3. $T = T_{max}$
4. **while** $T > T_{min}$ **do**
5. $\quad$ $k = r = 0$
6. $\quad$ **while** $k < k_{max}$ **and** $r < r_{max}$ **do**
7. $\quad\quad$ $W' = $ Transform_Solution(T, W)
8. $\quad\quad$ $a'_{total} = $ Evaluate_Solutionn(A_0, Z, W')
9. $\quad\quad$ **calculate** $p(W' \rightarrow W)$
10. $\quad\quad$ **with probability** $p(W' \rightarrow W)$ **do**
11. $\quad\quad\quad$ $W = W'$
12. $\quad\quad\quad$ $a_{total} = a'_{total}$
13. $\quad\quad\quad$ $r = r + 1$
14. $\quad\quad$ **if** $a'_{total} > a^*_{total}$ **then do**
15. $\quad\quad\quad$ $W^* = W'$
16. $\quad\quad\quad$ $a^*_{total} = a'_{total}$
17. $\quad\quad$ $k = k + 1$
18. $\quad$ $T = \alpha \cdot T$
19. **return** W^*

Algorithm 2. Algorithm to deploy waypoints

The solution evaluation is performed using Algorithm 3. The explored areas from individual waypoints are computed gradually using function Explored_Area but only for waypoints which do not lie inside any no-fly zone; otherwise such waypoint is marked as invalid and the corresponding explored area is set to zero. The total explored area is calculated by unification of individual areas a_i – see formula (1).

Evaluate_Solution (A_0, Z, W)

1. **for each** $W_i \in W$
2. ┊ **if** $W_i \in Z$ **then do**
4. ┊ ┊ ┊ $a_i = \emptyset$
5. ┊ **else do**
6. ┊ ┊ ┊ $a_i = $ Explored_Area(A_0, W_i)
7. $a_{total} = \bigcup a_i$
8. **return** a_{total}

Algorithm 3. Algorithm to evaluate a solution

The process of the solution transformation is presented in Algorithm 4. The transformation is performed on a single optimization variable $w_i \in W_j \in W$. The variable is selected using a random number generator with uniform distribution (point 1). The selected variable is then changed using a random number generator with normal distribution (points 2 and 3). The size of the change depends on the current value of temperature and the range of variable w_i. This range corresponds to the size of the area of interest in case of variables representing Cartesian coordinates (i.e. when $w_i = x_i$ or $w_i = y_i$), or the difference between the maximum and minimum limits of the permitted height in case of variables representing height above the ground level (i.e. when $w_i = h_i$).

Transform_Solution (T, W)

1. $w_i = $ Random_Variable(W)
2. $\sigma = \dfrac{(T - T_{min}) \cdot \text{Range}(w_i)}{T_{max} - T_{min}}$
3. $w_i = w_i + RandN(0, \sigma^2)$
4. **return** W

Algorithm 4. Algorithm to transform a solution

3.2   Planning Trajectories

The objective of the second optimization problem is to find a set of routes for all available UAVs in the swarm using the optimization criterion in formula (6) subject to condition (8). The problem can be transformed into the problem related to the well-known Multi-Depot Vehicle Routing Problem (MDVRP) [1]. The goal is to visit all waypoints by available UAVs so that the operation is as fast as possible.

The route R_i of each UAV $U_i \in U$ is composed of a set of waypoints which are visited one after another it the correct order; formula (10) must be valid (no waypoint can be avoided). A set of routes $R = \{R_1, R_2, \ldots, R_m\}$ is the solution to this optimization problem.

$$W = \bigcup_{i=1}^{m} R_i \tag{10}$$

The existence of no-fly zones means that the direct connection between arbitrary pair of waypoints may not exist. In that case, the shortest route avoiding the no-fly zones is calculated as shown on examples in Fig. 3. Violet lines show the connections between three pairs of waypoints. Whereas there is a direct connection between waypoints W_1 and W_2, the route around no-fly zones must be found in case of connections between waypoints W_3 and W_4, and W_5 and W_6.

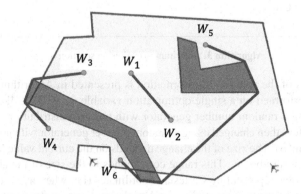

Fig. 3. Connection of waypoints around no-fly zones (Color figure online)

The Algorithm 5 plans the routes for UAVs in the swarm. In the first phase (points 1 to 10), a graph $G = (V, E)$ is created from base UAV positions, waypoints and polygon points of no-fly zones. Edges E_{ij} between each pair of vertices V_i and V_j of the graph exist only if there is a direct connection between the vertices; i.e. an edge does not cross any no-fly zone. In the next phase (points 11 to 15), a distance matrix D is created; D_{ij} denotes distance between each pair of UAV positions and waypoints. If there is no direct connection between vertices, a shortest path algorithm in point 15 finds the shortest route in the graph G between these vertices (Dijkstra's algorithm is used).

Finally, the algorithm which optimizes the routes for UAVs is executed. This algorithm is metaheuristic using Ant Colony Optimization (ACO) principle. This task is solved as the MDVRP problem. Distance matrix D created in the previous phase is used by this algorithm to express the distances between waypoints. More information about the principles of algorithm itself can be found in [1].

Plan_Routes (U, W, Z)

1. $V = U \cup W$
2. **for each** $Z_k \in Z$ **do**
3.   **for each** point of polygon $P_i \in Z_k$ **do**
4. $V = V + \{P_i\}$
5. **for every** pair of vertices $V_i \in V$ and $V_j \in V$ **do**
6.   **if** line segment from V_i to V_j does not cross any $Z_k \in Z$ **then do**
7. $E_{ij} = E_{ji} = |V_i - V_j|$
8.   **else do**
9. $E_{ij} = E_{ji} = -1$
10. $G = (V, E)$
11. **for every** pair of vertices $V_i \in U \cup W$ and $V_j \in U \cup W$ **do**
12.   **if** $E_{ij} \geq 0$ **then do**
13. $D_{ij} = D_{ji} = E_{ij}$
14.   **else do**
15. $D_{ij} = D_{ji} = \text{FindShortestPath}(V_i, V_j, G)$
16. $R = \text{OptimizeRoutes}(U, W, D)$
17. **return** R

Algorithm 5. Algorithm to plan routes

Figure 4 shows the result for an example based on the situation from Fig. 1. On the left, the optimal solution is shown for the situation when there are no no-fly zones; the reconnaissance operation takes 8:50 min. On the right, the optimal solution is shown where UAVs avoid no-fly zones; the operation takes 9:27 min.

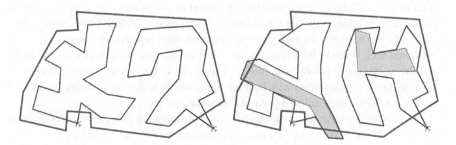

Fig. 4. Optimal routes for UAVs available for the reconnaissance operation

4 Experiments

A set of experiments was designed to verify the proposed algorithms. The experiments are composed of five benchmark instances which are based on the reconnaissance scenarios typical in the real reconnaissance operations. The parameters of the scenarios are summarized in Table 1.

Table 1. Benchmark scenarios

Scenario	Area of interest	No-fly zones	UAVs available	Min. coverage ω_{min}
s01	2.539 km^2	0.414 km^2	2	97%
s02	2.539 km^2	0.617 km^2	3	90%
s03	13.492 km^2	1.489 km^2	4	95%
s04	13.492 km^2	1.008 km^2	5	94%
s05	7.713 km^2	2.026 km^2	8	96%

All computations were executed on the computer with parameters as follows: Intel Core i9-9900K CPU @ 3.60 GHz, 32 GB RAM. The results presented below achieved via stochastic algorithms were averaged from 50 independent optimization trials.

The first set of experiments are aimed at the estimation of the number of waypoints needed to cover the required portion of the area of interest and the influence of the initial waypoints deployment via Algorithm 1. The algorithm was executed with control parameters as follows: $T_{max} = 10$, $T_{min} = 0.01$, $\alpha = 0.9$, $t_{max} = 1000$, $r_{max} = 100$.

The results are in Table 2. Column n shows the estimated number of waypoints and ω_0 is the portion of the area explored when using the initial waypoints deployment (Algorithm 1). The algorithm optimizing the positions of waypoints (Algorithm 2) was executed under the two different conditions: (a) the initial solution in generated randomly (each variable is generated in the corresponding range using the random number generator with uniform distribution); (b) the initial waypoints deployment provided by the Algorithm 1 is used as the first solution entering the algorithm. Thus, the benefits of the initial deployment could be examined. The size of the explored area under these conditions are denoted as ω_{rand} (the former condition) and ω_{init} (the latter condition).

The results in Table 2 show that the estimation of the number of waypoints based on the required portion of the area of interest is very precise. With a single exception of scenario s05, the explored area exceeds the required limit just slightly. In scenario s05, the explored area is lower than that required. But in this case, it is caused by the huge size of the no-fly zones (more than 26% of A_0) where no waypoints can be positioned, and also by the complex terrain with built-up areas which make the scanning by sensors more difficult. Thus, the explored portion cannot be any bigger.

The initial deployment of waypoints by the exact algorithm provides rough positions of waypoints with the portion of the explored area between 80% and 90% (see column ω_0 in Table 2). However, the difference between the final quality in solutions in optimization via the metaheuristic algorithm when using a randomly generated initial solution (see column ω_{rand} in Table 2) or initial solution provided by the exact algorithm as an input into the simulated annealing (see column ω_{init} in Table 2) is insignificant. The reason is that, if given enough optimization time, the simulated annealing manages to find a solution close to the global optimum even for complex problems (e.g. in case of scenario s05, there are 308 waypoints which is 924 optimization variables).

The second set of experiments are aimed at planning the trajectories of available UAVs in the swarm using Algorithm 5. The objective is to minimize the longest time

Table 2. Deployment of waypoints

Scenario	n	ω_0	ω_{rand}	ω_{init}
s01	35	85.17%	97.33%	97.64%
s02	79	80.07%	90.15%	90.22%
s03	125	87.73%	95.57%	95.61%
s04	192	88.91%	94.13%	94.15%
s05	308	81.51%	93.42%	93.41%

T needed by one of the UAVs with the longest route to fly along it (see optimization criterion in formula (6)). The total distance D travelled by all UAVs is also recorded (but this parameter was not used as the optimization criterion).

Table 3 presents the results. Column T_{free} shows the total time of the operation with waypoints positioned as presented in Table 2 when there were no no-fly zones; column D_{free} is the total distance travelled by all the vehicles. Columns T_{no-fly} and D_{no-fly} express the same variables but this time with no-fly zones as recorded in Table 1. The time of the reconnaissance operation is logically longer when UAVs must avoid the no-fly zones (it ranges from 3% to 13% in case of scenarios s01 to s05). It depends of course on the situation at hand, especially on the shapes and size of no-fly zones. Figure 4 shows the result for scenario s01 (on the left the situation without no-fly zones; on the right the situation with no-fly zones).

Table 3. Planning trajectories for UAVs

Scenario	T_{free}	D_{free}	T_{no-fly}	D_{no-fly}
s01	8:50	10.51 km	9:27	11.18 km
s02	12:14	14.66 km	13:06	15.61 km
s03	20:27	48.76 km	21:10	50.58 km
s04	20:14	22.31 km	20:56	62.45 km
s05	11:25	55.06 km	12:54	62.17 km

5 Conclusions

This paper deals with the model of reconnaissance of the area of interest performed by a swarm of cooperative UAVs. The original research of the authors was extended in several ways:

- The new tactical constraint in the form of no-fly zones was inserted into the model and its impact on the two optimization problems (deployment of waypoints and trajectories optimization) was examined.

- The new exact algorithm for the estimation of the number of waypoints needed was proposed and verified.
- The stochastic algorithms used for the both optimization problems were extended to reflect the new constraint.

The proposed principles were examined on a set of scenarios that were designed to reflect the standard and typical military reconnaissance operations. The new features proposed in this manuscript were implemented into the Tactical Decision Support System and are available now to be used for commanders on the battlefield to support them in their decision-making process.

References

1. Stodola, P., Drozd, J., Šilinger, K., Hodický, J., Procházka, D.: Collective perception using UAVs: autonomous aerial reconnaissance in a complex urban environment. Sensors **20**(10), 2926 (2020)
2. Shang, B., Wu, C., Hu, Y., Yang, J.: An algorithm of visual reconnaissance path planning for UAVs in complex spaces. J. Comput. Inf. Syst. **10**(19), 1–8 (2014)
3. Savchenko, V., Shchypanskyi, P., Martyniuk, O.R.: Air defense planning from an impact of a group of unmanned aerial vehicles based on multi-agent modeling. Int. J. Emerg. Trends Eng. Res. **8**(4), 1302–1308 (2020)
4. Wang, Y., Yue, Y., Shan, M., He, L., Wang, D.: Formation reconstruction and trajectory replanning for multi-UAV patrol. IEEE/ASME Trans. Mechatron. **26**(2), 719–729 (2021)
5. Lv, L., et al.: A novel non-uniform optimal control approach for hypersonic cruise vehicle with waypoint and no-fly zone constraints. Int. J. Syst. Sci. **52**(13), 2704–2724 (2021)
6. Fu, S., Lu, T., Yin, J., Xia, Q.: Rapid algorithm for generating entry landing footprints satisfying the no-fly zone constraint. Int. J. Aerosp. Eng. **2021**, 1–16 (2021). Article ID: 8827377
7. Zhang, H., Wang, H., Li, N., Yu, Y., Su, Z., Liu, Y.: Time-optimal memetic whale optimization algorithm for hypersonic vehicle reentry trajectory optimization with no-fly zones. Neural Comput. Appl. **32**(7), 2735–2749 (2018). https://doi.org/10.1007/s00521-018-3764-y
8. Li, B., Zhang, H., Zheng, W., Wang, L.: Spacecraft close-range trajectory planning via convex optimization and multi-resolution technique. Acta Astronaut. **175**, 421–437 (2020)
9. Stodola, P., Mazal, J.: Tactical decision support system to aid commanders in their decision-making. In: Hodicky, J. (ed.) MESAS 2016. LNCS, vol. 9991, pp. 396–406. Springer, Cham (2016). https://doi.org/10.1007/978-3-319-47605-6_32
10. Stodola, P.: Unmanned surveillance problem: mathematical formulation, solution algorithms and experiments. Mil. Oper. Res. **25**(2), 31–47 (2020)
11. Stodola, P., Nohel, J., Mazal, J.: Model of optimal maneuver used in tactical decision support system. In: International Conference on Methods and Models in Automation and Robotics 2016, Miedzyzdroje, pp. 1240–1245 (2016)
12. Stodola, P., Drozd, J., Nohel, J., Michenka, K.: Model of observation posts deployment in tactical decision support system. In: Mazal, J., Fagiolini, A., Vasik, P. (eds.) MESAS 2019. LNCS, vol. 11995, pp. 231–243. Springer, Cham (2020). https://doi.org/10.1007/978-3-030-43890-6_18
13. Bruzzone, A.G., Procházka, J., Kutěj, L., Procházka, D., Kozůbek, J., Scurek, R.: Modelling and optimization of the air operational manoeuvre. In: Mazal, J. (ed.) MESAS 2018. LNCS, vol. 11472, pp. 43–53. Springer, Cham (2019). https://doi.org/10.1007/978-3-030-14984-0_4

14. Mazal, J., et al.: Modelling of the microrelief impact to the cross country movement. In: International Conference on Harbor, Maritime and Multimodal Logistics Modelling and Simulation 2020, pp. 66–70 (2020)
15. Blaha, M., Silinger, K., Potuzak, L.: Data binding issue in fire control application for technical control of artillery fire. In: World Multi-Conference on Systemics, Cybernetics and Informatics 2017, vol. 1, pp. 4–8 (2017)
16. Hoskova-Mayerova, S., Talhofer, V., Otrisal, P., Rybansky, M.: Influence of weights of geographical factors on the results of multicriteria analysis in solving spatial analyses. ISPRS Int. J. Geo Inf. **9**(8), 489 (2020)
17. Sekelova, M., Korba, P., Hovanec, M., Mrekaj, B., Oravec, M., Szabo, S.: Options of measuring the work performance of the air traffic controller. In: Transport Means 2018, Trakai, Lithuania, pp. 1476–1481 (2018)
18. Drozd, J., Neubauer, J.: Use of an aerial reconnaissance model during the movement of oversized loads. J. Defense Model. Simul. **17**(4), 447–456 (2020)

Simulation Environment for Neural Network Dataset Generation

Aleš Vysocký[(✉)] [iD], Stefan Grushko[iD], Robert Pastor[iD], and Petr Novák[iD]

Department of Robotics, Faculty of Mechanical Engineering,
VSB-Technical University of Ostrava, 70800 Ostrava, Czech Republic
ales.vysocky@vsb.cz

Abstract. We present a simulation setup in the robot simulation software CoppeliaSim which is used for a synthetic dataset generation for training the neural network. In the simulator we can generate either color and depth images which can be tuned according to the real cameras mounted to the robot or robotic workplace. Vision sensors capture the simulated scene which contains different environment features, obstacles and objects of interest which can be labeled automatically with another filtering vision sensor. Except static environment which can be imported in case of known setup or generated based on height-field or simple objects. We can simulate randomly or with a specific pose oriented and positioned objects which may appear in the field of view of the robot. As an output the system produce RGB or depth information which is stored as a RGB or a gray-scale image or a combined RGBA image including the RGB data extended by depth data stored in the alpha channel. Second product of the system is a label describing different detectable classes for the neural network. The simulator is able to generate large datasets in a short period of time and produce a highly customized learning base for the neural network.

Keywords: Simulation · Neural network · Synthetic dataset

1 Introduction

Synthetic dataset is a learning set of data for training a neural network which is created in an artificial simulated environment. Neural network based image recognition systems used in a cluttered environment require a large and highly diverse dataset. Manual creation of a such dataset is a time consuming and arduous work. In the first step the scene has to be set up according to the situation we want to include in the training base of the neural network next we need to capture this situation and label the regions of interest.

In a comparative study by Dandekar et al. [1] are described different techniques of synthetic dataset acquisition. There are already available datasets made by different research groups including RGB, depth and labeled ground truth images. SceneNet [2] is a collection of data of indoor scenes. Dataset SYNTHIA

© Springer Nature Switzerland AG 2022
J. Mazal et al. (Eds.): MESAS 2021, LNCS 13207, pp. 322–332, 2022.
https://doi.org/10.1007/978-3-030-98260-7_20

[3] includes millions of data samples of the urban environment simulated in an environment including roads, buildings, cars and thousands of different objects. Except rendering different objects in the scene under specific conditions, there might be incorporated also influencing factors extending the basic situations. This can be the weather or light conditions as is described in the work of Khan et al. [4] or in TartanAir dataset [5]. There is a difference between synthetic data generated in the simulation and real world data, in the case of image recognition also depending on the capturing device. Domain adaption [6] considers the differences between simulated and a real data.

In our research we focus on detection and localisation of hands and arms of the operator working with the collaborative robot. For the hand recognition there are datasets available based on real RGB-D data [7,8], special type is dataset with the RGB-D source but labeled automatically with the additional sensor [9], in this study is a comparison between different datasets and labeling methods. Another way is a synthetic dataset generated for RGB tracking [10] or systems using 3D data to recognize the pose of the hand [11]. These systems usually use a simplified estimation of the position of the hand, it uses either other detector to localise the specific region with the hand or methods based on specific shape or color detection. Use of depth image instead of RGB helps to focus more to the shape of the hand than the color which can be in the industrial environment more various in connection with using protective equipment such as gloves. Using RGB and depth together may bring a synergistic effect. There are also datasets including not only the hands but also tools and other objects which may be used during working operation [12] or during interaction with objects [13].

According to Pasieka [14] there is a significant evolution of building synthetic datasets. It can be stochastic, rule-based or generated by the artificial intelligence systems. For the generation we can use simulation environment like CoppeliaSim, game engines like Unity or Unreal Engine or custom software tools usually built on computer vision libraries. The simplest methods are based on combination of 2D images and different augmentations of its parts, more advanced methods are based on placing a 3D object to the 3D environment including the influence of other objects and environment like the occlusion, distortion or lights and ray casting if we need also the color image and not only the depth information.

In the Sect. 2 we describe the simulation environment and conditions for dataset generating. Description of the dataset image is provided. In the Sect. 3 we show the process of unification and tuning of the dataset with real conditions captured with an RGB-D camera. In the Conclusion section we depict further steps and usecase of the dataset.

2   Experiment Setup

In this section we describe conditions and the environment where we want to use the image recognition with the neural network based system. The simulation environment used in this research is CoppeliaSim software. CoppeliaSim provides simulation environment for script-controlled multi-object scenarios as well as simulation of vision sensors which are used in this research. There is also powerful remote API available for interaction with the simulator from a custom software tools. We specify the vision sensor - simulated in the simulation environment and the real camera which will be used with the neural network based recognizer, then we describe the setup for the environment and the desired output.

2.1   Virtual Vision Sensor

In this experiment setup we use an Intel RealSense D435i RGB-D camera. This camera provides up to 1920×1080 pixel resolution 16 bit color image at maximal frame rate 30 FPS or 60 FPS for lower resolutions. Depth is calculated based on active stereo vision technology. Camera provides either the 1280×720 raw image from both infrared cameras used for stereo vision computation of the distance or the camera provides already calculated depth stream which is even refined with pattern emitted with laser projector. Resolution of the depth stream is 1280×720 at 30 FPS or even 90 FPS for lower resolutions. In the presented approach we try to simulate this depth stream. It is important to pay attention to the specific characteristics and limitations of the stereo vision technology. There is a shadow and invalid data band on one side of the image which are caused by unavailability of the data from both sensors (Fig. 1).

In the CoppeliaSim environment there is an object called vision sensor available which can provide either RGB image and depth. According to extrinsic parameters of the real camera we can use 2 vision sensors, one for RGB and second for depth which will be shifted because of the distance between sensors on the real camera. Setting the perspective angle and resolution unites the field of view of the real and virtual vision sensor. To display depth on the image we use intensity clipped to the minimum and maximum distance which is set equally with the real camera. The product is a gray-scale image (Fig. 1) which can be saved and used later for training of the neural network. In this study we do not use the RGB data, but we use a second vision sensor which also captures the depth. This sensor does not capture the whole scene but only a specific objects of interest, output of this image is transformed to binary producing the mask which is described later in the paper.

The main aim for a synthetic dataset generator is to produce depth images corresponding to the output of the depth camera scaled to the gray-scale image.

2.2   Working Environment

By specifying the obstacles directly according the environment where the system will be used we can set objects which should be detected, which should be recognized or ignored. There are three types of obstacles which are used in the generator:

– **Static environment** A static environment in the industrial applications may be the working table, frame of the workplace, several component supply

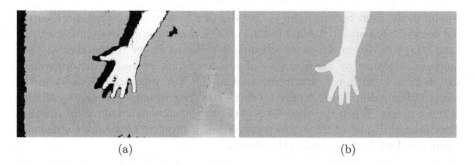

(a)   (b)

Fig. 1. Depth image colorized in gray-scale in a specific clipping distance: (a) Image captured by the depth camera with visible shadow which is a characteristic for the stereo vision method, (b) image captured in simulation.

devices, clamping units etc. This type of obstacles are usually available as a 3D model of the workstation. This model can be imported to the simulation scene as a mesh. We want to teach the neural network to ignore the static environment.

- **Objects of interest** Objects of interest are things or body parts which we want to detect, distinguish in the scene and localize. The goal is to label these objects in the dataset and train the neural network to recognize these objects.
- **Other objects** Object in the scene which may be similar to the objects of interest but we do not want to detect those objects and we need to teach the neural network to ignore those objects as well as the static environment. This may be the handled/manufactured parts, tools, free wires etc.

In our experiments we want to detect a hand and arm of the operator so we use a 3D model of the right hand in the simulation. It is not necessary to include left hand to the dataset generation because during training phase of the neural dataset we can add augmentations to the input image such as flips and rotations. In different setups we use hand with open palm gesture and pointing with index finger. During dataset generation the hand position is changing. This may be done randomly but we use a spatial linear pattern, where the fingertip of the index finger is shifted in every iteration in one direction, at the end of the area which is captured by the camera the fingertip is shifted in a perpendicular direction, when the finger finishes the layer it is shifted by the selected increment in the z-direction closer to the camera. With this pattern we ensure to capture the object of interest in all positions of the workplace. Moreover we set a semi-random orientation of the hand. Roll, pitch and yaw are limited in order to make the hand visible in obtainable positions according to the real environment. In our experiment we use limits $\pm 90°$ for yaw, $\pm 60°$ for pitch and $\pm 30°$ for roll of the hand. The combination of the position and the random orientation is checked during two tests (Fig. 2). First test is a check if the majority of the hand is visible within the camera field of view. Images with only fragment of the hand may confuse the neural network. We check if the fingertip of the index finger and a point in the center of the palm are present in a truncated pyramid representing the camera work-space. Second test is a check if an obstacle does not cover the hand, this may occur when the obstacle position is closer to the camera than the hand. Because we use obstacles of simple shapes, we test if the centroid of the obstacle is within a certain distance in the X-Y plane to the point in the center of the palm, if this occurs we check the Z coordinate (the depth) of both of those points. When the obstacle is closer to the camera we move the obstacle more further. If any of those two tests records a problem situation, the scene is regenerated without capturing the image.

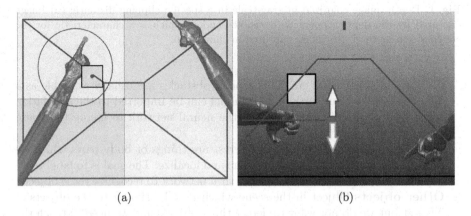

(a)   (b)

Fig. 2. Tests for omission of invalid situations: (a) top view indicating the obstacle within a monitored area of the hand on the left and the fingertip of the right hand located outside the field of view of the camera, (b) side view indicating the position of the obstacle closer to the camera than is the hand.

Tests are demonstrated in the Fig. 2. There are two situations in the scene, hand on the left pass the first test because both fingertip and center of the palm are within the camera field of view represented with blue color. The hand on the right fails the first test because the fingertip is outside the truncated pyramid. The hand on the left fails during the second test because the centroid of the yellow cube representing the obstacle is within the observed distance and the cube is closer to the camera than the hand.

2.3   Dataset Image

Output of this generator is a set of image couples (Fig. 3). Depth image of the scene is saved as gray-scale image. 8 bit format allows to save the depth in resolution of 256 values, this is approximately 4 mm of depth resolution if a clipping distance is 1 m. This resolution may be sufficient for recognition of bigger objects, such as the hand. If we want to detect more details, the depth map may be stored in different format, we used 8 bit resolution in order to keep the size of the file small. Depth image is stored as a single channel 8 bit depth gray-scale image.

Second image is a mask or ground truth for neural network training process. The mask is binary and it is captured with a vision sensor which can detect only objects of interest and the result of the depth sensor is transformed to binary image. The image may be stored as single channel 1 bit depth gray-scale image. Mask corresponds to the desired output of the neural network based recognizer.

Both images are saved directly from the CoppeliaSim software, further post-processing is done with custom software tools based on OpenCV library. Images are saved to separate folders with the same name for easy pairing in the training process.

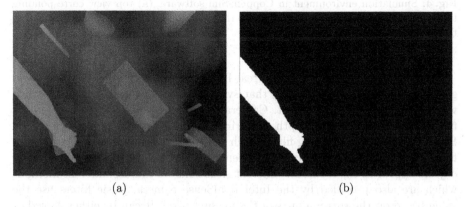

(a)   (b)

Fig. 3. Output of the generator is a couple of depth image (a) and a binary label (b) where white are pixels of the object of interest and black is the rest.

3   Synthetic Dataset Generation

In the first simulation scenario (Fig. 4), there is no static environment and the camera is placed one meter above the plane surface. We simulate the floor or the surface below the camera with a height-field which represents uneven surface. In the real camera image there are surface irregularities but also some reflections which can cause false irregularities on the flat surface. The height-field is randomly generated. In our scenario we use a height-field which is wider than the camera field of view and by changing position of the height-field below the camera we make the background random. The height-field is generated from the gray-scale image based on the Perlin noise. Perlin noise is a gradient type noise which creates a very natural pseudo-random appearance. It is used in virtual landscape simulation.

Obstacles in this test are simple shape primitives such as cuboid boxes and cylinders. Long cylinders with a small diameter represent tools or pencils which may occur in the scene and the shape is close to fingers. Moving the obstacles to random positions and orientations creates a highly variable cluttered environment.

Fig. 4. Simulation environment in CoppeliaSim software: (a) top view corresponding to the view from the camera, (b) view from the side with visible height-field illustrating the background noise.

Comparing the image made by real RGB-D camera (Fig. 5) to the dataset generator output (Fig. 3) shows, that synthetic and real data is similar which was intended in the experiment. Camera image was post-processed to remove the shadow caused by the depth capturing technology which was shown on the Fig. 1. We used a hole filling filter which fills in the missing values. Our custom filter use a static background environment which is captured before the moving objects appear in the scene, for different situations we can use different methods which are also provided by the Intel RealSense camera. Basic filters use the value based on the surroundings of the missing pixel. It can be either copied or calculated to best fit to the missing region.

It is not necessary, that the hand, or the object of interest in general, is closer to the camera than other objects in the background. Some recognizers use this feature to separate the area with the hand from the background. In our system we don't take the object of interest as the object in the foreground but objects of interest are labeled with the labeling vision sensor. This makes the system more versatile but we need to handle the possible occlusions which may disrupt the learning process.

If it is necessary to use higher resolution of the output image, there may be problem with sharper edges in the simulation than in the real camera image, where reflections on the surface and other disruptive effect cause that the shape of the object is not as sharp as in the simulation. Further post processing, such as a blur filter or another noise added to the image helps with unification of the image.

In the second experiment (Fig. 7) we set the environment according to a real workplace which is available in our laboratory. There is a robot Universal Robots UR3 mounted to the table which is intended for human-robot cooperation during assembly. The robot is capable of handling the maximal payload of 3 kg and the operation radius is 500 mm. This robot is intended for collaborative operation next to the human operator and it is certified as safe for collisions under a specific circumstances (regulated velocity, safe tool and other safety precautions essential

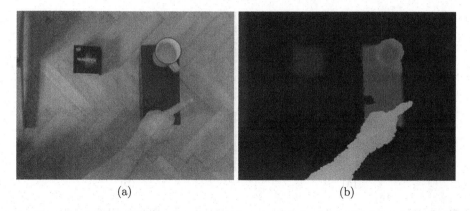

(a)   (b)

Fig. 5. Image captured with a real RGB-D camera mounted 1 m above the ground, scene includes objects of basic shapes and the hand: (a) RGB image of the scene, (b) depth image with applied hole filling filter.

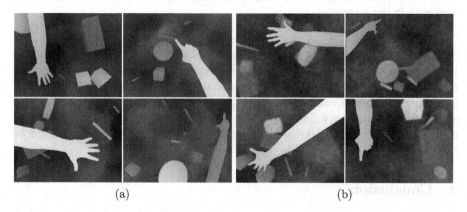

(a)   (b)

Fig. 6. Augmentations used in the dataset: (a) different gestures may be used in the simulation according to imported model of the hand, (b) selective blur filter applied to the image may help to blunt the edges of the simulated objects and additional noise represents different reflections and surface irregularities.

for the safe human-machine cooperation). The robot has six degrees of freedom and the body of the robot is made of aluminium cylinders and plastic covers. The plane of the working table is approximately one meter above the ground which means that the standing operator can freely operate with his hands on the table. One meter above the working plane is a platform where an RGB-D camera is mounted same as in the first experiment. 3D model of the construction of the workplace is imported to the simulation software and the virtual sensor is placed to the same place as the real camera is mounted.

(a)   (b)

Fig. 7. Experimental workplace with the robot Universal Robots UR3: (a) Simulation environment CoppeliaSim with imported workplace including robot UR3. Figurine has a hand with pointing gesture mounted to the right arm. Output of the generator are depth (upper) and label (bottom) images, (b) image captured with real camera at the workplace in the laboratory.

Object of interest is a part of the figurine, hand and arm are visible for the labelling sensor. The hand and arm are moving in the same pattern as in the first experiment. Instead of random basic shape obstacles there is a robot which is moving to random valid positions. Lightweight industrial robot as UR3 may appear in the image very similar to the human arm. In this experiment we intend to teach the neural network based recognizer to recognize the arm but to ignore the robot during operation (Fig. 6).

4   Conclusion

Proposed synthetic dataset generator is able to generate large sets of labeled images in a short period of time. Moreover the CoppeliaSim simulation environment may be operated in multiple instances on a single computer or it can be run on more machines with different augmentations of the scene. In our experiment we store the images on the hard drive of the computer, so for better operation we try to minimize the size of the file. Size of the image 320×240 pixels which was used in the experiment was sufficient for the given scene and objects of interest. We used a post-processing for dataset split into train/validation and test parts. In this simulation environment this could be done by additional random process or by random save to different locations without additional post-processing of the dataset. By increasing the size of the dataset we can capture more details of the scene, this may be necessary if we want to detect screw heads or small parts, on the other hand bigger file takes up more space on hard drive which may be significant in datasets including millions of images. Secondly the training phase of the neural network gets longer with larger images. This also applies to the inference phase where we need the time for recognition as short as possible to ensure the operation without delays.

The scene based on the imported workplace and a 3D scan of the hand in the simulation is a source of images very similar to the images taken by the depth camera in the real workplace. Simple post-processing of the simulated image or the real camera image may further increase the similarity between those data sources. Simulation image can be blurred for refinement of sharp edges or we can add some additional noise to prepare the network for the real data. The noise and unclear segments are cleaned from the real camera image. The setup of the real camera and the properties of the post-processing of the generated images may be tuned to get the best combination for reaching the best similarity between generated images and real camera images in the specific environment.

With this generator we can simulate environment with some specific shapes and characteristics typical in the industrial area. The dataset is more specific and the neural network recognizer may be trained with stronger emphasis to the specific workplace. 3D model of the workplace is usually available from the design stage and is a part of documentation of the workplace. Therefore the process of setup of the environment in the simulation software is fast. First tests with the training of the neural network have very positive results and specification of the training phase and comparing to the existing recognition ways will be done in the further research. For industrial purposes we need to localise the hand and recognise the gesture for a natural human robot interaction, there is also a second purpose to use the localisation for safety reasons. In this case the system must be very robust for predicting the collision between the operator and a moving part of the machine and the operation must be without delays and if possible close to the real-time operation. Basic dataset we intend to use in semantic segmentation with hand area extraction. This may be used as an input to the OpenPose network which requires color image with specified hands locations. From the CoppeliaSim environment we can extract hands and fingers joints positions and gesture types and train network for detection.

Acknowledgment. This work was supported by the Research Platform focused on Industry 4.0 and Robotics in Ostrava Agglomeration project, project number CZ.02.1.01/0.0/0.0/17_049/0008425 within the Operational Programme Research, Development and Education.

References

1. Dandekar, A., Zen, R.A.M., Bressan, S.: A comparative study of synthetic dataset generation techniques. In: Hartmann, S., Ma, H., Hameurlain, A., Pernul, G., Wagner, R.R. (eds.) DEXA 2018. LNCS, vol. 11030, pp. 387–395. Springer, Cham (2018). https://doi.org/10.1007/978-3-319-98812-2_35
2. McCormac, J., Handa, A., Leutenegger, S., Davison, A. J.: SceneNet RGB-D: 5M photorealistic images of synthetic indoor trajectories with ground truth. arXiv (2016)
3. Ros, G., Sellart, L., Materzynska, J., Vazquez, D., Lopez, A.M.: The SYNTHIA dataset: a large collection of synthetic images for semantic segmentation of urban scenes. In: Proceedings of the IEEE Conference on Computer Vision and Pattern Recognition (CVPR), June 2016

4. Khan, S., Phan, B., Salay, R., Czarnecki, K.: ProcSy: procedural synthetic dataset generation towards influence factor studies of semantic segmentation networks. In: Proceedings of the IEEE/CVF Conference on Computer Vision and Pattern Recognition (CVPR) Workshops, June 2019

5. Wang, W., et al.: TartanAir: a dataset to push the limits of visual SLAM. In: 2020 IEEE/RSJ International Conference on Intelligent Robots and Systems (IROS) (2020)

6. Loghmani, M.R., Robbiano, L., Planamente, M., Park, K., Caputo, B., Vincze, M.: Unsupervised domain adaptation through inter-modal rotation for RGB-D object recognition. arXiv preprint arXiv:2004.10016 (2020)

7. Tompson, J., Stein, M., Lecun, Y., Perlin, K.: Real-time continuous pose recovery of human hands using convolutional networks. ACM Trans. Graph. **33**, 1–10 (2014)

8. Qian, C., Sun, X., Wei, Y., Tang, X., Sun, J.: Realtime and robust hand tracking from depth. In: 2014 IEEE Conference on Computer Vision and Pattern Recognition, pp. 1106–1113 (2014). https://doi.org/10.1109/CVPR.2014.145

9. Yuan, S., Ye, Q., Stenger, B., Jain, S., Kim, T.: BigHand2.2M benchmark: hand pose dataset and state of the art analysis. In: Proceedings of the IEEE Conference on Computer Vision and Pattern Recognition (CVPR), July 2017

10. Yang, D., Moon, B., Kim, H., Choi, Y.: Synthetic hands generator for RGB hand tracking. In: TENCON 2018–2018 IEEE Region 10 Conference (2018)

11. Malik, J., et al.: DeepHPS: end-to-end estimation of 3D hand pose and shape by learning from synthetic depth. arXiv (2018)

12. Shilkrot, R., Narasimhaswamy, S., Vazir, S., Hoai, M.: WorkingHands: a hand-tool assembly dataset for image segmentation and activity mining. In: BMVC (2019)

13. Hasson, Y., et al.: Learning joint reconstruction of hands and manipulated objects. In: Proceedings of IEEE Computer Society Conference on Computer Vision and Pattern Recognition, CVPR (2019)

14. Pasieka, M.: The evolution of synthetic data: a comparison of three data generation methods. Mostly AI. https://mostly.ai/2020/10/28/comparison-of-synthetic-data-types/. Accessed 21 Aug 2021

Traversability Transfer Learning Between Robots with Different Cost Assessment Policies

Josef Zelinka$^{(\boxtimes)}$, Miloš Prágr , Rudolf Szadkowski , Jan Bayer ,
and Jan Faigl

Faculty of Electrical Engineering, Czech Technical University in Prague,
Technická 2, 166 27 Prague, Czech Republic
{zelinjo1,pragrmi1,szadkrud,bayerja1,faiglj}@fel.cvut.cz
https://comrob.fel.cvut.cz

Abstract. Predicting mobile robots' traversability over terrains is crucial to select safe and efficient paths through rough and unstructured environments. In multi-robot missions, knowledge transfer techniques can enable learning terrain traversability assessment the robots did not experience individually. The knowledge can be incrementally aggregated for homogeneous robots since they can treat foreign knowledge as their own. However, robots with different perceptions might experience the same terrain differently, so it is impossible to aggregate the shared knowledge directly. In this paper, we show how to learn a model that transfers the experience between heterogeneous robots, enabling each robot to use the whole sum of the experience of the multi-robot team. The proposed approach uses correlation to combine individual neural networks that assess the traversability of individual robots. The presented method has been verified in a real-world deployment of multi-legged walking robots with different cost assessment policies.

Keywords: Transfer learning · Traversability assessment · Mobile robot

1 Introduction

The studied transfer learning is motivated by deployments of a heterogeneous team of multi-legged walking robots, each exploring and perceiving various terrain types. It is desirable to explore as quickly and efficiently as possible during terrain exploration. As the robots are deployed, they collect a large amount of information about the environment and experience the traversability cost of the traversed terrain. The collected traversability information can be encoded into a traversability cost model that can assess the cost; however, such a quality assessment is limited to the experience collected by the individual. A group of cooperating robots can improve their performance on a given task by sharing their experiences. The knowledge of each robot can be enhanced by sharing the

© Springer Nature Switzerland AG 2022
J. Mazal et al. (Eds.): MESAS 2021, LNCS 13207, pp. 333–344, 2022.
https://doi.org/10.1007/978-3-030-98260-7_21

obtained knowledge among the robots. Hence, it can enable the team to improve its overall performance. Motivated by groups of social animals learning experiences from one another [17], we aim to implement such transfer learning patterns into multi-robot systems.

Furthermore, missions such as exploration of unknown environments can be speeded up by parallelization of the exploration process using a large group of robots [23]. Thus, having multiple robots, we can explore transferring the collected knowledge between the robots. For homogeneous teams with a single robot type, the knowledge transfer that is called inductive transfer learning [14] can be utilized [21]. Such a transfer is possible only when all robots have the same morphology and sensor equipment; otherwise, the homogeneity of the team is lost. Changes in morphology or equipment can lead to variation in the terrain perception. However, changes of the identical robots can be caused by damage to the robot during the mission or hardware updates in later operational deployments. With inductive transfer learning, changed robots would not contribute to the shared knowledge nor benefit from it. In that perspective, heterogeneity seems to be natural, highlighting the importance of transfer learning in multi-robot heterogeneous team [13].

Fig. 1. The used hexapod walking robot in the experimental deployment of the proposed method in the Bull Rock Cave, where it builds elevation map and collects the dataset.

We propose a transfer learning method to share knowledge among heterogeneous robots to enhance their cost assessment capabilities. In the proposed approach, the robots transfer their individually learned models implemented as the Convolutional Neural Network (CNNs) regressor. For knowledge transfer between two different robots, correlation of predictions on terrains traversed by both robots is used to determine the relationship between the models, thus creating an augmented model. After the relation between models is learned, the robots are ready to exchange the traversal experience they have already collected. The proposed approach has been experimentally verified on data from the deployment of a real hexapod walking robot with adaptive locomotion control [5] in a natural cave system, see Fig. 1. Based on the achieved results, the proposed method allows two robots to share the knowledge and exploit the traversability cost models experienced by the other robot.

The paper is organized as follows. Section 2 summarizes related work with the emphasis on traversability assessment and transfer learning techniques deployed in robotics. The proposed transfer learning method is introduced in Sect. 3. In Sect. 4, we report on the experimental results of the proposed method using real hexapod walking robots. The paper is concluded in Sect. 5.

2   Related Work

Traversability assessment is studied in various fields such as planetary exploration [6,20], search and rescue missions [3], and agriculture, or off-road driving [8]. Two main classes of approaches can be identified in the literature: traversability classification and prediction of traversability cost as a continuous score. The simplest terrain classification can be a binary classifier to determine whether the terrain is traversable or not [10]. However, the authors of [6] report improved path planning results avoiding impassable terrain and also better-optimized paths using a continuous score. Therefore, in this paper, we follow the idea of traversability as a continuous score.

The traversability assessment can be based on proprioceptive and exteroceptive sensory signals, where the exteroceptive data processing approach can be further categorized into geometry- or appearance-based [15]. Nevertheless, hybrid methods might benefit from combined approaches. The rest of this section provides an overview of the most related traversability approaches to support our traversability assessment choices.

Proprioceptive traversability assessment uses information captured by sensors that measure the robot's internal properties during the robot's interaction with the terrain, e.g., speed, tilt, shakiness, energy, or vibration. Thus, the proprioceptive traversability assessment can estimate traversability only on currently traversed terrains. An example of traversability assessment based on the energy expenditure is reported in [12].

The exteroceptive, geometry-based approaches use range measurements such as LiDARs and RGB-D cameras to construct maps of the perceived environment. The maps are then used to examine terrain properties such as roughness, edges, slope, or features the robot might not be able to traverse. Obstacle extractions from the maps using filtering and clustering are presented in [16]. On the other hand, the visual appearance of the terrain can be studied using image-processing and classification of terrain types into categories with defined properties [1]. Methods using appearance and geometrical properties might suffer from wrongly classified terrains in cases where range sensing is not sufficient, e.g., unexpected covered hole [22]. Therefore, we have chosen hybrid approaches to leverage the advantages of the individual methods.

In addition to traversability assessment, transfer learning is assumed to improve the assessment by exploiting individual experiences of the particular robots in a team. Transfer learning can be defined as a machine learning approach to boost the knowledge in the target domain by the transfer from the source domain [24]. Transfer learning is already established technique in the fields such as text [7] and image [18] classification. In [24], the authors combine text and image classification using the matrix factorization method to enhance image classification by information extracted from their annotations, thus merging the two tasks.

Similar to text and image classification, robotics is a domain where labeled data are costly to obtain. Besides, it is relatively hard to train robots to adapt to the demands of various environments. The knowledge transfer is a way to benefit from deployments of multi-robot teams. The authors of [4] adopt transfer learning to reduce the learning time of the particle swarm optimization for faster optimization of robot's gaits (walking patterns). Transfer learning applied in the learning of humanoid robots to solve tasks by observing human behavior is described in [11]. The idea is to transfer knowledge about a human motion to the robot that is requested to perform a similar motion. In [19], learned navigation patterns around obstacles are transferred into new environments to enhance planning capabilities.

The aforementioned transfer learning approaches in the robotics domain provide supportive evidence of successfully deployed techniques. Therefore, we focus on deploying transfer learning among heterogeneous robots that might yield different traversability assessments [9].

3   Method

The proposed method for transferring knowledge from one robot to another is motivated to improve cost assessment capabilities by learning from one another. Two roles of the robots can be distinguished: a provider of new information called teacher T; and receiving robot denoted student S. Each i-th robot collects a dataset $D^i = \{(t_1^i, c_1^i), (t_2^i, c_2^i), \dots\}$ that consists of features (t_j^i, c_j^i) describing the perceived terrain $t_j^i \in \mathcal{T}$ and labels describing the cost of traversing a particular segment of the terrain $c_j^i \in C$. Hence, observations of the i-th robot are stored in the dataset D^i that represents the robot's experience with the environment.

In this case, the two robots are in the roles of teacher and student, respectively, and the experience with the same terrain, i.e., $t^S = t^T$, the observed costs might not necessarily be equal, $c^S \neq c^T$, because of different terrain perception. The proposed approach targets to extrapolate individual datasets on newly observed terrains. The extrapolation is realized through analyzing the relation between the cost assessment of the student and the teacher. The relation is then used to enrich the student's extrapolation capabilities by the teacher's cost assessment.

The remainder of this section describes the proposed method for transferring knowledge between the robots with heterogeneous terrain perceptions. First, the proposed cost assessment learning model is introduced in Sect. 3.1, together with the training procedure. The transfer learning framework is then introduced in Sect. 3.2.

3.1 Cost Assessment Learning Model

The individual robot's cost assessment model $M = (r, a)$ is trained using its dataset D. The model comprises the regressor $r : \mathcal{T} \to C$, which returns the cost estimation, and certainty evaluation $a : \mathcal{T} \to U$, where $u \in U = \mathbb{R}$ denotes the certainty of the model over the particular terrain segment $t \in \mathcal{T}$.

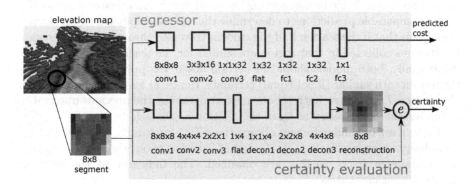

Fig. 2. The elevation map is segmented and sent into the model M. The model M is composed of a regressor (orange) providing the cost prediction r and autoencoder (olive) from which the certainty a is computed. The input of the model is an 8×8 segment of elevation map, which is processed by the neural networks. The depicted architecture of the neural networks shows the convolutional (conv), flattening (flat), deconvolution (decon), and fully connected (fc) layers with their dimensions. For certainty evaluation, the log of reconstruction error $\log(e(t; g))$ computation is indicated by the e-node. (Color figure online)

Both the functions r and a are implemented with separate neural networks, where the terrain t is represented as a set of elevation map segments, each encoded by $n \times n$ matrix of real values, $t \in \mathcal{T} = \mathbb{R}^{n \times n}$. Topological dependencies between the matrix values are the same as in images; therefore, convolutional layers are used similarly to image processing. Thus, for processing the elevation map segments, convolutional layers are added to the neural networks for the regressor r and certainty evaluator a. The learning architecture is depicted in Fig. 2.

The certainty evaluation a is trained indirectly with a convolutional autoencoder. The autoencoder $g : \mathcal{T} \to \mathcal{T}$ maps given segments t to reconstructed segments $g(t)$, where the reconstruction error $e(t; g) = ||t - g(t)||$ is minimized during the training. Trivially, the reconstruction error would be zero for all $t \in \mathbb{R}^{n \times n}$ if the map g is an identity function. However, due to the bottleneck architecture of the autoencoder, the map g cannot be an identity function; so, the segments have different reconstruction errors. Here, we assume the trained autoencoder has a low reconstruction error on segments presented in the dataset and a higher reconstruction error for other segments. The certainty of the model is thus represented by the log of the reconstruction error $a(t) = \log(e(t; g))$, where higher values correspond to the terrain segments that are dissimilar to segments the model has been trained on.

3.2   Transfer Learning Framework

The proposed framework uses trained models M_S and M_T where the student uses the teacher's knowledge by considering the relation κ. Similar features are used to obtain comparable predictions to determine the relationship between the models. We assume that if the teacher and the student traversed through the same region, the features collected in that region are similar for the teacher and the student. Additionally, both models should be certain about the previous observation of the terrain with similar (if not equal) certainty. Hence, at least one similar terrain observation T_{sim}, where both robots are certain about the previous observation of the terrain, is necessary to learn the relation κ successfully.

Using a set of similar terrain observations T_{sim} containing n samples of terrain observations, predictions of M_S and M_T models about the cost $C = (c_i)_{i=1}^n$ and certainties $U = (u_i)_{i=1}^n$ are obtained. The indicator of the certainty $A = (\alpha_i)_{i=1}^n$ is created as

$$\alpha_i = \begin{cases} 0 & \text{if } u_i^S < \theta \vee u_i^T < \theta \\ 1 & \text{otherwise} \end{cases}. \tag{1}$$

The individual indicator is zero for samples where one robot's certainty about the terrain observation is less than the empirically set threshold θ. The relation κ between student's and teacher's cost assessment models is determined by the average of the oriented differences between corresponding cost predictions $c \in C$

$$\kappa = \frac{1}{\sum_{i=1}^n \alpha_i} \sum_{i=1}^n \alpha_i (c_i^S - c_i^T), \tag{2}$$

where the certainty indicator α_i is used to remove samples where the robot model is not certain enough.

The obtained relation κ is used to enhance student's future predictions about the newly observed terrain cost c^p to facilitate better path planning decisions. With the next terrain observation t_{new}, both models M_S and M_T of the student and teacher, respectively, are used to predict the cost and certainty $(c, u) = M(t_{new})$. Then, the certainties are compared and the prediction with the higher certainty is selected. If the teacher's prediction is selected, the cost c^T is corrected by the relation between the models κ as

$$c^p = \begin{cases} c^S, & \text{if } u^S > u^T \\ c^T + \kappa, & \text{otherwise} \end{cases}. \tag{3}$$

The feasibility of the proposed cost models and transfer learning framework has been empirically validated using real datasets. The achieved results are reported in the following section.

4   Results

The proposed method has been verified in an experimental deployment using a real hexapod walking robot shown in Fig. 1. The robot is equipped with the Intel RealSense D435 RGB-D and T265 tracking cameras, and terrain's features are stored into an elevation map [2]. During the deployment, the robot collects datasets further used in the model learning, knowledge transfer, and evaluation of the learned traversability cost assessment models. In particular, the used dataset has been collected in the Bull Rock Cave, Czech Republic, and the proposed method has been evaluated as follows.

Various cost perceptions are simulated using different cost calculation methods instead of deploying heterogeneous robots. The student's costs are computed as an angular distance of the pitch and roll from the leveled position of the robot. On the other hand, the teacher's costs are computed as the robot's relative slowdown compared with the commanded velocity v_{cmd}, which characterizes the difficulty of the terrain as a resistance difference from regular walking. An individual terrain segment i holds the information about the robot's state changes between two consecutive feature collection places. The cost of the i-th segment is defined as the median over multiple traversed consecutive segments

$$c^T = \text{median}\{c_{sk}\}_{k=1}^n \tag{4}$$

for $c_{si} = v_{cmd}\Delta t_i / s_i$ with the segment duration Δt_i and length s_i.

The student and teacher models are trained as described in Sect. 3.1. The cost regressor is trained using 2000 epochs, and the autoencoder is trained for 100 epochs. It is presumed that the autoencoder is uncertain in the previously unobserved terrains. Therefore, the model benefits from overfitted autoencoder. The architecture of the regressor and autoencoder neural networks is as in Fig. 2, and for each layer, ReLU activation functions are utilized.

Datasets have been collected in three parts of the Bull Rock Cave with different traversability properties. The student's model is trained on data collected

(a) Chiffon   (b) Room   (c) Hall

Fig. 3. Different terrains of the Bull Rock Cave used in the evaluation of the proposed method.

in the *Chiffon* and *Hall* parts of the cave, while the teacher's dataset is collected in the *Room* and *Hall* parts. Terrains in both *Room* and *Hall* consist of similar leveled, packed surfaces. In *Chiffon*, the robot has experienced a slightly sandy surface, which makes the robot's movement marginally slower due to its legs sinking into the sand during motion. The visual appearances of the terrains are displayed in Fig. 3.

Since both the teacher's and the student's models are trained on datasets collected on the *Hall* terrain, both models should have certain predictions for the respective terrain, which is ideal for demonstration of the transfer learning. Hence, the *Hall* dataset is selected as the training dataset for the transfer relation κ with the uncertainty threshold set to $\theta = 3$ that has been found empirically. The relation between the models is determined to be $\kappa = -0.52$, indicating that the teacher's cost assessments are, on average, by 0.52 higher than the student's assessments. Therefore, 0.52 is subtracted whenever the student uses the teacher's prediction.

The *Room* dataset is selected as the testing environment for the scenario, where the teacher's knowledge enhances student's predictions. In this setup, the teacher should be able to make better cost predictions than the student because the teacher previously observed the terrain in the *Room*. The results presented in Fig. 4 show that during the evaluation phase, the teacher's cost assessments are used more often because the teacher is more certain about the terrain sample. In most cases, transfer learning improved the cost assessment. The values of the *Mean Square Error* (MSE) using the transfer learning are depicted in Table 1, where we can also observe the negative transfer.

Only the student is trained in *Chiffon*, and therefore, the student accumulated better knowledge about the *Chiffon* terrain. From Fig. 4, we can observe that the teacher's and student's predictions are used almost equally by the transfer learning component. The comparable prediction usage might be caused by the fact that the flat terrain in *Chiffon* is partially similar to the teacher's training domain of *Hall* and *Room*. However, teacher's cost assessments rarely improve the student's knowledge, thus resulting in a decrease in cost assessment capabilities using the transfer learning in this scenario, which is further shown in Table 1.

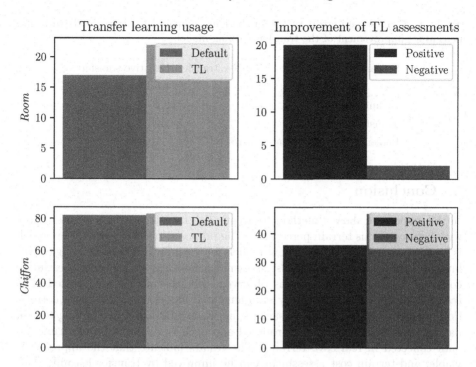

Fig. 4. Results of the transfer learning performed in *Room* (top row) and *Chiffon* (bottom row). The left column shows the amounts of usage of the transfer learning cost assessments versus the default estimations made by the student's model. The numbers of positive and negative improvements achieved by transfer learning assessments compared to the student's estimation are illustrated in the right column.

The sparse improvement of the cost assessment using the transfer learning is likely caused by the sandy surface being fairly similar to the packed surfaces in *Hall* and *Room*. Note that the real cost of moving over the sandy terrain is higher than on packed surfaces.

Discussion – The results show that the knowledge transfer from the teacher improved the student's ability to assess cost in the case of the terrain previously observed by the teacher. The positive transfer learning is represented by the transfer over the *Room* terrain. However, there can be confusion in the model selection during the transfer learning phase, which can be observed for the *Chiffon* scenario. Nevertheless, in *Chiffon*, both models performed similarly well, producing lower MSE than in *Room*. The negative transfer might be solved by making the transfer learning component more strict and enforcing stricter requirements on the teacher's estimation.

Table 1. Mean square errors of the predictions compared to the ground truth and percentage improvement of the transfer learning model.

Scenario	Positive transfer	Negative transfer
	Room	*Chiffon*
Default (MSE)	6.19	2.70
Transfer (MSE)	4.10	2.98
Percentage improvement	33.76%	−10.00%

5   Conclusion

In this paper, we show a method to accomplish transfer learning across robots with heterogeneous terrain perception. The proposed transfer learning approach is based on creating cost assessment models for the individual robots using convolutional neural networks. The cost assessment models are then used to estimate cost and uncertainty for terrains observed after the models are created. The transfer learning is addressed by an augmented model created using a correlation between the students and teachers models established on a training terrain dataset. The feasibility of the proposed approach is validated on experimental data collected in real cave terrains. The results indicate that the approach is viable, and terrain cost assessment can be improved by transfer learning. Different cost assessment models are used to simulate heterogeneous robots. In our future work, we plan to experimentally evaluate the method using robots with different morphology and sensory equipment. Besides, we also aim to deploy the learning method directly during the exploration task. It is expected to improve the individual robot's performance by avoiding hard to traverse areas experienced by the other robots.

Acknowledgment. The presented work has been supported by the Czech Science Foundation (GAČR) under research projects No. 19-20238S and 21-33041J. The support of the Defense Advanced Research Projects Agency under agreement No. HR00112190014 is gratefully acknowledged. We would also like to acknowledge the support of the speleologist branch organization ZO 6-01 for providing access to the Bull Rock Cave testing site.

References

1. Ball, D., et al.: Vision-based obstacle detection and navigation for an agricultural robot. J. Field Robot. **33**, 1107–1130 (2016). https://doi.org/10.1002/rob.21644
2. Bayer, J., Faigl, J.: Speeded up elevation map for exploration of large-scale subterranean environments. In: Mazal, J., Fagiolini, A., Vasik, P. (eds.) MESAS 2019. LNCS, vol. 11995, pp. 190–202. Springer, Cham (2020). https://doi.org/10.1007/978-3-030-43890-6_15
3. Cafaro, B., Gianni, M., Pirri, F., Ruiz, M., Sinha, A.: Terrain traversability in rescue environment. In: IEEE International Symposium on Safety, Security, and Rescue Robotics (SSRR), pp. 1–8 (2013). https://doi.org/10.1109/SSRR.2013.6719358

4. Degrave, J., Burm, M., Kindermans, P.J., Dambre, J., Wyffels, F.: Transfer learning of gaits on a quadrupedal robot. Adapt. Behav. **23**(2), 69–82 (2015). https://doi.org/10.1177/1059712314563620

5. Faigl, J., Čížek, P.: Adaptive locomotion control of hexapod walking robot for traversing rough terrains with position feedback only. Robot. Auton. Syst. **116**, 136–147 (2019). https://doi.org/10.1016/j.robot.2019.03.008

6. Gennery, D.B.: Traversability analysis and path planning for a planetary rover. Auton. Robot. **6**, 131–146 (1999). https://doi.org/10.1023/A:1008831426966

7. Howard, J., Ruder, S.: Universal language model fine-tuning for text classification. In: 56th Annual Meeting of the Association for Computational Linguistics (ACL), pp. 328–339 (2018). https://doi.org/10.18653/v1/P18-1031

8. Huertas, A., Matthies, L., Rankin, A.: Stereo-based tree traversability analysis for autonomous off-road navigation. In: IEEE Workshops on Applications of Computer Vision (WACV/MOTION 2005), pp. 210–217 (2005). https://doi.org/10.1109/ACVMOT.2005.111

9. Kolvenbach, H., Bellicoso, D., Jenelten, F., Wellhausen, L., Hutter, M.: Efficient gait selection for quadrupedal robots on the moon and mars. In: 14th International Symposium on Artificial Intelligence, Robotics and Automation in Space (i-SAIRAS) (2018). https://doi.org/10.3929/ethz-b-000261939

10. Langer, D., Rosenblatt, J.K., Hebert, M.: A behavior-based system for off-road navigation. IEEE Trans. Robot. Autom. **10**(6), 776–783 (1994). https://doi.org/10.1109/70.338532

11. Makondo, N., Hiratsuka, M., Rosman, B., Hasegawa, O.: A non-linear manifold alignment approach to robot learning from demonstrations. J. Robot. Mechatron. **30**(2), 265–281 (2018). https://doi.org/10.20965/JRM.2018.P0265

12. Martin, S., Corke, P.: Long-term exploration & tours for energy constrained robots with online proprioceptive traversability estimation. In: IEEE International Conference on Robotics and Automation (ICRA), pp. 5778–5785 (2014). https://doi.org/10.1109/ICRA.2014.6907708

13. Murphy, R.R., et al.: Use of remotely operated marine vehicles at Minamisanriku and Rikuzentakata Japan for disaster recovery. In: IEEE International Symposium on Safety, Security, and Rescue Robotics (SSRR), pp. 19–25 (2011). https://doi.org/10.1109/SSRR.2011.6106798

14. Pan, S.J., Yang, Q.: A survey on transfer learning. IEEE Trans. Knowl. Data Eng. **22**(10), 1345–1359 (2010). https://doi.org/10.1109/TKDE.2009.191

15. Papadakis, P.: Terrain traversability analysis methods for unmanned ground vehicles: a survey. Eng. Appl. Artif. Intell. **26**(4), 1373–1385 (2013). https://doi.org/10.1016/J.ENGAPPAI.2013.01.006

16. Peng, Y., Qu, D., Zhong, Y., Xie, S., Luo, J., Gu, J.: The obstacle detection and obstacle avoidance algorithm based on 2-D lidar. In: IEEE International Conference on Information and Automation, pp. 1648–1653 (2015). https://doi.org/10.1109/ICInfA.2015.7279550

17. Pongrácz, P., Miklósi, Á., Kubinyi, E., Gurobi, K., Topál, J., Csányi, V.: Social learning in dogs: the effect of a human demonstrator on the performance of dogs in a detour task. Anim. Behav. **62**(6), 1109–1117 (2001). https://doi.org/10.1006/ANBE.2001.1866

18. Quattoni, A., Collins, M., Darrell, T.: Transfer learning for image classification with sparse prototype representations. In: IEEE Conference on Computer Vision and Pattern Recognition (CVPR), pp. 1–8 (2008). https://doi.org/10.1109/CVPR.2008.4587637

19. Saha, O., Dasgupta, P., Woosley, B.: Real-time robot path planning from simple to complex obstacle patterns via transfer learning of options. Auton. Robot. **43**(8), 2071–2093 (2019). https://doi.org/10.1007/S10514-019-09852-5
20. Singh, S., et al.: Recent progress in local and global traversability for planetary rovers. In: IEEE International Conference on Robotics and Automation (ICRA), vol. 2, pp. 1194–1200 (2000). https://doi.org/10.1109/ROBOT.2000.844761
21. Stone, P., Veloso, M.: Multiagent systems: a survey from a machine learning perspective. Auton. Robot. **8**, 345–383 (2000). https://doi.org/10.1023/A:1008942012299
22. Tennakoon, E., Peynot, T., Roberts, J., Kottege, N.: Probe-before-step walking strategy for multi-legged robots on terrain with risk of collapse. In: IEEE International Conference on Robotics and Automation (ICRA), pp. 5530–5536 (2020). https://doi.org/10.1109/ICRA40945.2020.9197154
23. Tribelhorn, B., Dodds, Z.: Evaluating the Roomba: a low-cost, ubiquitous platform for robotics research and education. In: IEEE International Conference on Robotics and Automation (ICRA), pp. 1393–1399 (2007). https://doi.org/10.1109/ROBOT.2007.363179
24. Zhuang, F., et al.: A comprehensive survey on transfer learning. Proc. IEEE **109**(1), 43–76 (2021). https://doi.org/10.1109/JPROC.2020.3004555

SWORD RAS Project

Salvatore De Mattia[✉]

NATO Modelling and Simulation Centre of Excellence, Piazza R. Villoresi 1, 00143 Rome, RM, Italy
mscoe.cde04@smd.difesa.it

Abstract. The NATO Modelling & Simulation Centre of Excellence (M&S COE) has realized a synthetic environment, based on a federate simulation, where several RAS platforms have been modeled in terms of physical and attitudinal characteristics. In the concept/capability development field a project was broken out using SWORD constructive simulator of MASA Company, where the main idea was to adapt the SWORD powerful capabilities to RAS entities in military operations, building a specific operative vignette for a proof of concept.

The combination of these aspects was born through an idea to adapt SWORD main features to automatic processes in a military operation involving RAS systems. The main task was to create a versatile synthetic platform able to develop and experiment RAS functionalities from technical, operative and informative point of view, for concepts and capabilities development.

The SWORD RAS tool is able to accept as inputs parameters all the variables presented both in the physical and behavioral database. In particular, it is possible to study the output of the simulations in terms of dynamics of robotics platforms, sensor network used (technological aspects) and the autonomous and manual missions developed inside the robotic system according to rules of engagement (ROE), TTPs, missions and level of autonomy. Thanks to DAI (Direct Artificial Intelligence) of SWORD, these physical and attitudinal aspects are processed and correlated with external factors from technical and operative points of view.

The main output produced by SWORD RAS platform are Courses of Action (COA) for M multiple parallel simulation in order to statistically characterize results according to N parametric variables defined in the inputs and for several and disparate output variables.

The project's aim is to provide a mathematically quantification of stochastic results managed by simulator in order to better defined multi-factorial aspects connected to RAS deployment in the Future Operating Environment.

Further developments of SWORD RAS project consist in the implementation of a federated simulation, through a mapping process, with VRForces or VBS4, as a proof of concept, in order to have also a 3D visualization of the simulation. Furthermore, VBS4 gives also the virtual capability to the project, following LVC paradigm (Live-Virtual-Constructive). In conclusion, an architecture of a synthetic environment is proposed in the capability development field of RAS systems, involving, besides SWORD RAS tools, also expert simulated systems (e.g. Matlab/Simulink and/or CEMA simulator) in order to investigate specific area of interest (e.g. swarm logic and/or robotics counter-measures). Implementing a synthetic-real bridge, for Live capability of the architecture, a project is proposed finalized to an electronic board development, called LoA implementation board.

© Springer Nature Switzerland AG 2022
J. Mazal et al. (Eds.): MESAS 2021, LNCS 13207, pp. 345–356, 2022.
https://doi.org/10.1007/978-3-030-98260-7_22

Keywords: MASA SWORD · RAS · Concept development · Behavioral codes

1   Synthetic Platform for Concept and Capability Development

1.1   Main Idea

NATO M&S COE has realized a synthetic environment, based on a federate simulation, where several RAS platforms have been modeled in terms of physical and attitudinal characteristics. This architecture, called R2CD2 Evolution has demonstrated a new idea for concept and capability development, focusing on a different paradigm regarding Modelling and Simulation tools as a fundamental element for analyzing and studying the military robotic technology. As a result, through the generation and development of behavioral algorithms for simulated RAS entities, was demonstrated that with a cyclic diagram among simulated and real world, could be possible to train the automatic attitude of RAS platforms, allowing to characterize and to reduce uncertainty conditions in the platform's behavior in the future operative environment. This diagram was synthetized in Fig. 1, highlighting the idea that the synthetic world could be considered as an arena in order to experiment behavioral algorithms and physical parameters characterized each robotic systems. The basic idea is to translate the behavioral codes developed in the simulated mission in feasible algorithms in a real electronic board able to connect to an unmanned platform and to reproduce specific LoA (Level of Autonomy) according to external sensors and technical-operative aspects. In this field, the feedback elements are fundamental in order to enhance simulation fidelity with respect to real world operations, training RAS systems to respond to specific conditions of scenario, reducing progressively the error probability during a military mission (Fig. 1 – conceptual flow chart).

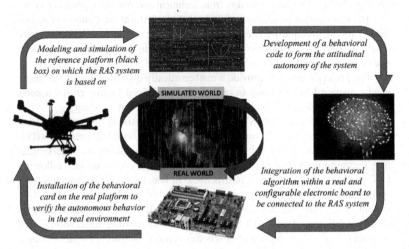

Fig. 1. Main phases for RAS capability development using Modelling and Simulation tools.

In this way, it is possible to develop new autonomous behaviors, translate it on an electronic board connected to a real platform and test technically and tactically the system attitude. The aforementioned attitude has to be correlated to specific LoA, type of mission of system and external factors of the scenario in terms of weather conditions, structure of terrain, TTPs, electromagnetic and cybernetic conditions.

The next step of the project was conducted on SWORD constructive simulator of MASA Company, where the main idea was to adapt the SWORD powerful capabilities to RAS entities in military operations, building a specific operative vignette for a proof of concept [4].

The main characteristics of SWORD simulator are following reported:

- Constructive aggregate simulator;
- Analysis tool which allows studying different Course of Action (COA) also for Wargaming purpose;
- Physical database where is possible to model the main structural and conceptual characteristics of synthetic entities;
- Direct Artificial Intelligence (DAI) that defines the logic algorithms inside the simulator calculations to accomplish orders on simulated entities in relation to specific boundary conditions of the scenario.

The main characteristics of RAS systems knowledge, developed by NATO M&S COE personnel thanks to synthetic environment designed, are following reported:

- Level of Autonomy and type of mission;
- Behavioral algorithms in LUA scripts.

Starting from these reciprocal characteristic, it is fundamental to understand the basic idea on which this project is based on, through a simple association between two different fields of application, trying to match them together to explore the relative correlated and joint advantages (Fig. 2).

The combination of these aspects was born through an idea to adapt SWORD main features to automatic processes in a military operation involving RAS systems. The main task was to create a versatile synthetic platform able to develop and experiment RAS functionalities from technical, operative and informative point of view, for concepts and capabilities development.

The proposed project can be analyzed as a black box with inputs and outputs, where the SWORD RAS platform performs calculations exploiting the internal logic functions implemented inside SWORD algorithm engine. For this task, the platform developed, although has been realized according a scenario consisting of a specific operative reconnaissance vignette, it was designed in order to be flexible and parametric.

The main inputs parameters on which this project is based on are the following:

- Physical database modelling;
- Behavioral database modelling;
- External conditions;
- Scenario, terrain and TTPs;
- Different LoA and RAS autonomous mission.

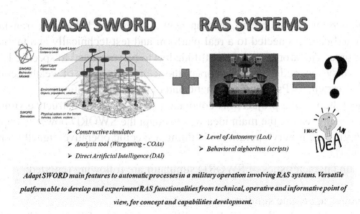

Adapt SWORD main features to automatic processes in a military operation involving RAS systems. Versatile platform able to develop and experiment RAS functionalities from technical, operative and informative point of view, for concept and capabilities development.

Fig. 2. Main idea of SWORD RAS project as a combined platform able to combine the RAS autonomous operations with the Direct Artificial Intelligence of MASA SWORD simulator.

The "black box analysis", reported in Fig. 3, represents a meaningful element to delve into the final aim of this project, which is broken out following specific directions in terms of versatility, mathematical-based and multi-factorial parametric analytics.

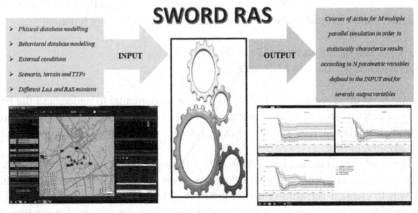

Fig. 3. "black box analysis" reported as a function of inputs and final output of the SWORD RAS project.

To sum up, the platform is able to accept as inputs parameters all the variables presented both in the physical and behavioral database, where is possible to study the output of the simulations in terms of dynamics of robotics platforms, sensor network used (technological aspects) and the autonomous and manual missions developed inside the robotic system. These aspects need to be associated with rules of engagement (ROE), TTPs, type of missions and level of autonomy. Thanks to DAI of SWORD, these physical and attitudinal aspects are processed and correlated with external factors from technical and operative points of view.

The main output produced by SWORD RAS platform are Courses of Action (COA) for M multiple parallel simulation in order to statistically characterize results according to N parametric variables defined in the inputs and for several and disparate output variables.

The methodology inside this project is explained through the parametric analysis of the simulated mission from multi-factorial points of view; the probabilistic results give a mathematical meaning for concept and capability development for military operations with RAS systems. The results represent a combination between the behavioral algorithms in terms of code typing and flow diagrams modelling autonomous operations and computer assisted simulation through engine calculator able to contextualize decisional aspects with the scenario's boundary conditions.

In conclusion, this project represents an external layer in the RAS concept and capability development field through modelling and simulation tools. The main idea broke out in this project consists in creating a platform able to characterize many parameters in a simulated mission involving robotic platforms in military operations. In particular, this project's aim is to provide a mathematically quantification of stochastic results managed by simulator in order to better defined multi-factorial aspects connected to RAS deployment in the Future Operating Environment. This platform could bring the following results:

- Behavioral algorithms development, validation and verification;
- Physical characterization of sensors, weapons and structural data;
- Definition of LoA, missions and TTPs.

As further developments of this project, could be important to federate this platform with virtual simulator (VBS4). This federation could be meaningful in order to allow:

- 3D visualization extending the graphical characteristic of SWORD (only 2D simulation);
- Human-robot interface development, according the parameters and concepts defined through SWORD RAS platform, in order to test a real operator performing a military mission with other human simulated entities and with robotic platforms. The interaction between human and robot has to be defined according to LoA, human-robot hierarchical chain and technical parameters of RAS platform (such as sensors and behavioral capabilities).

1.2 SWORD RAS Simulation

The platform data-setting scenario was developed implementing an urban environment based on a future (2035) mega-city model (Archaria 5x5 km synthetic terrain) [2].

Although the basic characteristic on which the SWORD RAS project was developed is the versatility, it is currently on a specific vignette. The simulation created in the MASA SWORD tool deals with a tactical action, where both blue and red forces have RAS entities modelled. In particular, the platoons modelled inside the simulator are:

- Blue platoon nr. 1 with mission of taking control of the critical area;
- Red platoon with mission of defend the critical area;
- Blue platoon nr. 2 with mission of patrolling an area different from the critical one.

These platoons, according to physical and doctrinal databases, are based on infantry platoons, with addition of RAS platforms. The robotic platforms present different autonomous behaviors, which were developed modifying LUA codes inside DAI (Direct Artificial Intelligence) engine of the simulator.

Focusing on RAS modelled entities, the main physical and behavioral characteristics created in this project are the following:

- RAS of blue platoon nr. 1 – Platform: UGV. Default behavior: follow human platoon and gather information through sensors. Reactive behavior at platoon enemy engagement: kinetic action on enemy with 12.7 mm machine gun. Sensors: Doppler radar, thermal sensor, 6× binoculars, camera and VHF radio.
- RAS of red platoon – Platform: UGV. Default behavior: follow human platoon and gather information through sensors. Reactive behavior at platoon enemy engagement: kinetic action on enemy with 12.7 mm machine gun and 14.5 mm machine gun. Sensors: Radar, Doppler radar, thermal sensor, 8× binoculars, camera, light intensifier and civilian radio.
- RAS of blue platoon nr. 2 – Platform: UGV and UAV. Default behavior: follow human platoon and gather information through sensors. Reactive behavior at platoon nr. 1 enemy engagement: UAV deployment with autonomous flight mission planner according to the position of discovered enemies; kinetic action on enemy with 12.7 mm machine gun with UGV; non kinetic action with UAV. Sensors of UGV: Doppler radar, thermal sensor, 6× binoculars, camera and VHF radio. Sensors of UAV: real time camera (for ISR mission) and portable on board jammer.

Furthermore, the simulation is connected to user-defined parameters for UAV deployment, where it is possible to configure the number of rounds for ISR part, the number of rounds for jamming part and the radius of the jamming area. Besides all the physical, behavioral and external characteristics of the scenario, the output COAs are correlated to the mentioned user-defined parameters, since they define:

- the amount of information gathered by UAV system and shared with incoming blue platoon for situational awareness enhancement;
- the effectiveness of non-kinetic attack on enemy platoon, in terms of persistence and performance, which inhibit the communication enemy network and the RAS platform operational capability.

The described scenario was completely tested using SWORD RAS project, allowing having a practical proof of concept for concept development application [8].

2   SWORD RAS Federated Simulation

As second phase in the development stage of the SWORD RAS project, a federated simulation has been implemented. In particular, the short-term aim of second phase implies a 3D visualization of simulated environment build in SWORD simulator. For this reason, the third part simulators do not play any entities but they are just used for graphical aspect.

The architecture for federated simulated involves, besides SWORD, also:

– VRForces;
– VBS4 (this gives also the virtual component to the whole architecture for further developments).

In the Fig. 4 the architecture is shown, in addition to technical considerations performed during the implementation phase, in order to summarize the current situation of the project.

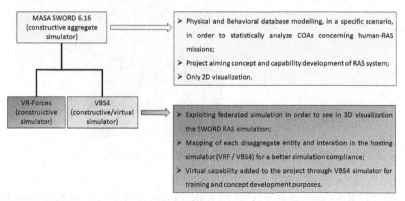

SWORD RAS – current situation

Fig. 4. SWORD RAS federated simulation and current situation analysis of the project.

The current situation of this project is fundamental in order to depict future considerations projected to project's enhancement.

– Align the scenario terrain in the federated simulation in terms of geographical and structural data (e.g. streets, buildings, etc.);
– Enhance mapping process performed in the federated simulation in order to reach the highest level of matching both in the 2D and in the 3D simulation, to better study some final results in a multi-perspective domain;
– Enhance aggregate to disaggregate management of simulated entities, in order to reach the maximum matching level in terms of logic algorithms used by SWORD aggregate simulator and managed by third part simulator through HLA publishing in disaggregate mode [1];

- Extend SWORD RAS project to further scenarios based on the versatility character-istic of the developed tool;
- Prepare an evaluation of the project's effectiveness in the Concept and Capability Development field;
- Prepare a comparison table between typical RAS wargaming and Computer-Assisted wargaming using this project;
- Develop a human-machine interface able to interconnect constructive (SWORD) and virtual part of the project (VBS4), in terms of human-RAS doctrine, physi-cal/behavioral model of RAS, type of mission and LoA.

An example of federated simulation is shown in Fig. 5, where the same situation is represented in different simulators, through HLA1516e protocol.

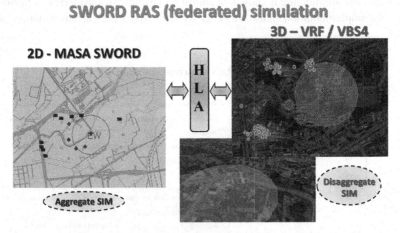

Fig. 5. SWORD RAS (federated) simulation example.

SWORD RAS project is a proof of concept able to exploit computational power of a simulator in order to develop concept and capability on RAS systems in military operations. The analysis tool allows plotting different COAs, in a Wargaming context, giving a mathematical support in terms of probability functions. Analysis process is a very important aspect of SWORD RAS project, because each outcome could be enhance and correlate to operative and technical elements in a multi-domain contextualization. In addition to COAs, for instance, the simulated reports shared between entities in the synthetic environment, represent a meaningful data set, which is useful to figure out the conceptual, and logical processes of the tactical scenario in a timeline view [6]. This aspect is also potentially compatible with C2SIM protocol, allowing the interoperability with Command and Control systems [7].

One of possible application for SWORD project could be express in a translation of physical and behavioral modelling projected on real world, mostly for capability devel-opment purposes. In particular, the physical database translation could affect sensors or weapon systems characterization of a real system, in a bidirectional way, in order to:

– evaluate technical feasibility for specific sensor or device that gives the desired COAs for the specific mission. After that evaluation, a more specific phase of technological implementation, modelling and simulation could be performed using proper tools (e.g. expert system such as Matlab/Simulink, CEMA domain simulators, etc.);
– carry out reverse engineering on specific physical characteristics of real RAS systems in order to create a model in the synthetic environment for verification and characterization through modelling and simulation tools.

Similarly, a behavioral translation could be performed, exploit the aforementioned bidirectional process defined for physical part, in order to verify RAS autonomous functionalities in real and simulated environment. For this aspect, it is important to study the RAS system architecture, highlighted on autonomous engine and physical interface with external environment, in order to understand how and where implement the behavioral translation from synthetic to real and vice versa, obtaining a high fidelity result. The translation process is very important in order to reduce the gap between synthetic and real environment, creating a matching for capability RAS development through SWORD RAS project from physical and behavioral point of view. This idea is described in Fig. 6 by a possible conceptual application.

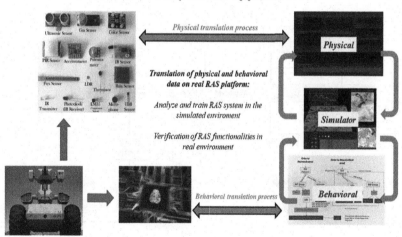

Fig. 6. SWORD RAS possible application for capability development.

Considering SWORD RAS project in a bigger architecture, exploiting Modelling and Simulation tools for concept/capability development of autonomous systems for military operations, it is possible to analyze the problem under two different main point of view. The SWORD RAS project is useful to define doctrinal, physical and behavioral characteristic of a RAS system for capability development (answer the question "WHAT?"), in order to fix the preliminary structure aimed to analyze disparate type of considerations. Consequently, through dedicated tools (expert systems) could be possible to delve deeply in specific matters reaching a greater knowledge of implementation (answer the question "HOW?"). The two described processes could be correlated through LVC (Live-Virtual-Constructive) paradigm for a comprehensive approach. In particular:

– Constructive: useful to develop concepts on RAS system in a specific tactical scenario involving both human and robotic entities;
– Virtual: useful to train real operator to accomplish specific missions involving robotic platform, also developing the human-robot interface through which the real soldier could interact with simulated robotic entities. Developing the human-robot interface has to be linked to sensors, mission, LoA and autonomous functions modelled for the synthetic RAS platform;
– Live: useful to analyze and verify the simulation results in terms of physical and behavioral aspects. Developing a synthetic-real bridge is a fundamental part of the architecture for LVC paradigm implementation and interact directly on real devices. The live part could be realized to implemented hardware or software in the loop and allows interactions between real systems and simulated entities, for example in a single platform or swarm context [5].

For further and future developments of the SWORD RAS project is presented in Fig. 7 an architecture adding some expert systems for specific modelling and simulation.

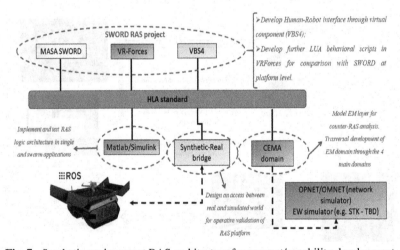

Fig. 7. Synthetic environment RAS architecture for concept/capability development.

In this case, as expert systems are inserted:

– Matlab/Simulink for implementation and testing of RAS logic architecture in single and swarm applications.
– CEMA domain, using for example an electronic warfare (EW) simulator, for modelling electromagnetic (EM) layer in a counter-RAS application [9]. The EW virtualization is a meaningful element studying RAS platforms because of transversal development of EM domain through the four main domains.

To sum up, the expert systems involvement part in the architecture and the LVC paradigm as a cornerstone of synthetic environment RAS, the proposed whole process is described to give a comprehensive approach to RAS concept/capability development. Afterwards concept development phase, the capability development starts, which, basically, could be focused on specific areas of interest. The following potential phases are not essentially sequential, but their implementation depends on the specific sector in which develop deeper.

- Working with Matlab/Simulink for modelling RAS architecture, in terms of sensors and behavioral algorithms to implement inside the platform also exploring swarm application;
- Modelling Electromagnetic layer in order to test CEMA activities. This task is useful in order to study and to define RAS counter-measures, through tool like Matlab/Simulink, EW simulator and/or OPNET Network simulator [3];
- Develop synthetic-real bridge through a script able to interface the real world with HLA protocol. In this way could be possible to perform a cyclic process among simulated and real world, building such as electronic board able to interface with robot's sensors and autonomous algorithms.

The reference flow diagram is based on the cyclic process between real and synthetic environment, exploiting modelling and simulation tools to train the platforms and verify the mission accomplishment in a real scenario with different external stimulus of boundaries conditions. The cyclic process could allow reducing gaps and error percentages based on stochastic results given by simulated environment.

3 Conclusion

In conclusion, a conceptual idea for a possible application of synthetic environment RAS has been analyzed, in order to concretize a possible project in a specific field. The idea of the project lies on an electronic board development/adaptation, called LoA implantation board, through which could be possible to set up a feasible synthetic-real bridge. The electronic board could allow realizing the cyclic process described before, receiving as inputs the main outcomes generated by synthetic environment RAS and interfacing with Robotic Operating System for managing the autonomous functionalities of the real platform. Based on this assumption, LoA implementation board could synthetize the behavioral algorithms developed, analyzed and tested in the synthetic environment, obtaining a versatile platform able to customize different autonomous functions, based on missions, platform, LoA and sensors, and to communicate through ROS interactions.

The basic idea is to transform an unmanned COTS system to a RAS one, thanks to LoA implementation board translation function, in order to extend the robotic and autonomous capability to simple and versatile commercial platforms. In addition, it is important to implement this project as swarm systems network, in order to improve the versatility of this idea, reaching an interoperability between different platforms. The interoperability part is useful to improve situational awareness and extend decision-making process during a military operation. The unmanned to RAS transformation using modelling and simulation tools is something of pioneering in the field of robotics, different from military and dedicated RAS systems development, acting as a connection point to automate and to customize the attitude of simple commercial unmanned platforms for tactical and strategical purposes.

The artificial intelligence engine of RAS system could be remotely virtualized in a central calculation machine, implementing the paradigm of Cloud Robotics. In this way, most of autonomous functions is placed in a fixed workstation able to communicate with each platform in the field in order to integrate the autonomous algorithm of the RAS system, sending specific commands generated for specific mission. Modelling and simulation tool could support also the artificial intelligence development, exploiting new technologies, such as quantum information, which could foster machine learning algorithms effectiveness, paving the way for the future RAS systems development.

References

1. IEEE: 1516–2010 -IEEE Standard for Modeling and Simulation (M&S) High Level Architecture. IEEE Standard Association (2010)
2. NATO ACT: NATO Urbanization Project. (2015). http://www.act.nato.int/activities/natourban isation-project, Accessed Oct 2019
3. Biagini, M., Corona, F.: Modelling and simulation architecture supporting NATO Counter unmanned autonomous system concept development. In: Hodicky, J. (ed.) MESAS 2016. LNCS, vol. 9991, pp. 118–127. Springer, Cham (2016)
4. Biagini, M., Corona, F., Casar, J.: Operational scenario modelling supporting unmanned autonomous systems concept development. In: Mazal, J. (ed.) MESAS 2017. LNCS, vol. 10756, pp. 253–267. Springer, Cham (2018)
5. Biagini, M., Corona, F.: M&S based robot swarms prototype. In: Mazal, J. (ed.) MESAS 2018. LNCS, vol. 11472, pp. 283–301. Springer, Cham (2019)
6. Corona, F., Biagini, M.: C2SIM operationalization extended to autonomous systems. In: Mazal, J., Fagiolini, A., Vasik, P. (eds.) MESAS 2019. LNCS, vol. 11995, pp. 389–408. Springer, Cham (2020). https://doi.org/10.1007/978-3-030-43890-6_32
7. NATO Collaboration Support Office: MSG 145 Operationalization of Standardized C2 Simulation Interoperability. (2019). https://www.sto.nato.int/Pages/activitieslisting.aspx, Accessed Oct 2019
8. Biagini, M., Scaccianoce, A., Corona, F., Forconi, S., Byrum, F., Fowler, O., Sidoran, J.L.: Modelling and Simulation supporting unmanned autonomous systems (UAxS) concept development and experimentation. In: Proceedings of Disruptive Technologies in Sensors and Sensor Systems (SPIE Defense and Security), Anaheim, CA, USA, vol. 10206 (2017)
9. NATO ACT CEI CAPDEV: Autonomous Systems Countermeasures. (2016). https://www.inn ovationhub-act.org/project/counter-measures, Accessed Oct 2019

Utilizing the Maneuver Control System CZ in the Course of Wargaming Modelling and Simulation

Jan Nohel[✉] ⓘ, Ludovít Hradský ⓘ, Zdeněk Flasar ⓘ, Pavel Zahradníček ⓘ, and Dana Kristalova ⓘ

University of Defence in Brno, Brno, Czech Republic
{jan.nohel,ludovit.hradsky,zdenek.flasar,pavel.zahradnicek,
dana.kristalova}@unob.cz

Abstract. This article describes the use of mathematical algorithm models and digital terrain and relief models in the process of planning the maneuver axes of military units and military vehicles using the Maneuver Control System CZ. The maneuver axes of units on the battlefield are calculated based on the impact of the surface and terrain relief, the weather, and on the influence of the deployment of the enemy and friendly units. Calculated maneuver axes may be applied to analyze the variants of action both of friendly and the enemy's military forces, which may be carried out in the form of a so-called war game as part of the Commander's decision-making process. For the needs of the war game, the MCS CZ can quickly calculate the possible maneuver of enemy units and can subsequently also calculate the reaction – the maneuver of friendly units, and finally the possible counter-reaction of the enemy. Research results of the war game, using MCS CZ calculations, are described in case study. This case study analyzed the activities that were the most likely and most dangerous variants of enemy action, as well as the possible reaction of friendly forces in engaging in defense. Calculations of the axes of maneuver and times were compared with the simulation in the software MASA SWORD. The calculated axes of maneuvers in both programs confirmed the tactical correctness of the deployment of the friendly combat security detachments.

Keywords: Terrain passability · Course of action · Decision-making process · Multiple maneuver model · Situation awareness

Gaining superiority over the enemy and meeting targets in a military operation is contingent on a quick understanding of the situation on the battlefield. In order to do this, pieces of information must be collected efficiently, analyzed quickly, and the evaluation results must be used in the performance of combat tasks. On such a basis, a qualified estimate of the enemy course of action (COA) can be compiled, followed by the timely and efficient deployment of friendly forces exactly where they are needed. No decision-making by commanders and staff can be haphazard. In NATO armies, at the tactical level of command and control, the so-called military decision-making process (MDMP) is used to give the decision-making process a logical structure, efficiency, and the ability

© Springer Nature Switzerland AG 2022
J. Mazal et al. (Eds.): MESAS 2021, LNCS 13207, pp. 357–373, 2022.
https://doi.org/10.1007/978-3-030-98260-7_23

to co-operate with all commanders and unit staffs in the operation. The MDMP is a cyclic planning methodology carried out to analyze and understand the situation and the task, to produce and contrast variants for friendly units' operations, and to issue an operational order to implement the selected COA of friendly forces [1, pp. 2–17].

A significant influence on the COA of friendly forces is one's estimation of the intention of the enemy, which is one of the most demanding parts of intelligence security in a military operation. It is affected by the considerable degree of uncertainty about the objects and targets of the enemy, and the associated means of achieving them successfully. An analysis of the situational factors, the enemy, and other threats needs to be carried out in order to make a qualified estimate of the position of these targets and the course of the main, secondary and supporting maneuver to implement them. In the event that the presumed enemy COA is dynamic, coupled with a unit maneuver in order to gain control over a particular area, then a key factor in the evaluation will be the overall passability of the area, with the least hazard risk. If the enemy's maneuvering is expected to achieve its targets as fast as possible, then the course of its COA can be estimated as the fastest passable and safe.

The war game, which is based on the outputs of the process of intelligence preparation of the battlefield (IPB), is applied in NATO armies to analyze and evaluate the COA of friendly forces, see [2, pp. 1–1, 3–1, 6–20]. One of them is an estimate of the enemy COA, assessed in terms of probability and hazard [2, pp. 1–4]. Based on a qualified estimate of the enemy COA, intelligence-gathering efforts can be more accurately targeted, as can preliminary measures be implemented to safeguard the activities of friendly forces, including avoiding an unexpected attack by the enemy. The result of the war game is then a simulation of a planned maneuver by friendly forces, along with the expected reaction of the enemy forces, supplemented by a counter-reaction by friendly forces. For all the considered COAs are assessed, e.g. the time of implementation, advantages and disadvantages, strengths and weaknesses, the element of surprise, combat effectiveness and the real ability to achieve the planned targets of friendly units as well as the anticipated targets of the enemy forces. The ability to coordinate the efforts of units on the battlefield is equally important. The war game is a structured process by which commanders and staff attempt to anticipate the development of an operation. It uses a method of action-reaction-counter reaction in assessing the development of each critical event in the considered COA, in the interaction of friendly forces and those of the enemy, which is applicable to all types of military operations [3, pp. B–21]. It can be performed manually, using graphically assessed terrain passability, military vehicles capability tables, and tactical activity models on a map. With this variant, assessing the tactical characteristics of the terrain and the environment, the threats, and the conduct of the war game can take – depending on the experience of the analysts, the scale and the predictability of the development of the operation – from 6–12 h. These days it is much more efficient to use digital land and relief models, mathematical algorithms, modelling, and computer simulations.

1 Literature Review

The MDMP is described in a number of US military publications, e.g. in [1, 3, 4]. Its use in dealing with crisis situations, with its problematic acquisition, understanding and use of information is further characterized, e.g. in [5]. The goal of the MDMP is always to resolve a situation as quickly and comprehensively as possible, or to plan the execution of a mandated task. The result can take the form of creating axes for maneuver of friendly units, which in combination with a high mobility for enemy units, can be time-consuming. Digital geographic data models and mathematical algorithms can be used to solve this problem in the 21st century digital battlefield environment.

The mathematical modelling of the maneuver axis is based on the information layers of the digital terrain and relief model, which are described in more detail in [6–8]. Creating these models takes considerable time. The automated assembly of digital models is currently performed using various sensors, algorithms and computer graphics, see [9–12]. Assessing the impact of soil, obstacles, microrelief and field cover on military vehicle mobility and decision-making is depicted in [13–15]. Modelling the combined impact of relief, microrelief, soils, vegetation, hydrology, built-up areas, and meteorological factors on the speed deceleration of off-road vehicles in the raster geographic data format is then specified in [16–18].

The maneuver of units in military operations is currently largely carried out by road, due to the limited passability of the wheeled fighting vehicles, the efficiency of redeployment at an acceptable risk, as well as efforts not to cause unnecessary damage to agricultural land. Various road passability parameters and assessments of maximum speed values that can be used during redeployment are evaluated in [19]. However, if the road network does not allow this, or the safety risk is disproportionately high, the maneuver must be taken through terrain off the roads. The issue of identifying the course of the shortest route between two points in a computerized environment of geographic information systems using algorithms is addressed in [20]. It is then possible to find the creation of axes for multiple maneuvers using different algorithms in [21]. Analyzing the COA of friendly units by applying software to conduct a war game during the MDMP is described in [22, 23]. In the current battlefield environment, there are likely to be large numbers of units, military vehicles, armed and criminal groups, as well as religious and non-profit organizations. Software using artificial intelligence (AI) and the expected activity of a large number of entities in the operating space can recommend improvements to the commander and staff in the course of action (COA) of friendly units. The possibilities of using AI in planning the axis of a war game maneuver in the computer environment can also be found in [24]. Research and experiments in using expert computer systems in the execution of war games to minimize prejudices and risks associated with decision making is described in [25–27]. The innovative contribution here is to use dynamic modelling and the simulation of new approaches in designing theoretical models of capability development that shape the core of expert systems. The purpose is to draw a qualitative comparison between the two proposed sets of capability requirements.

The Tactical Decision Support System (TDSS) was developed at the University of Defence of the Army of the Czech Republic in 2006, see [28]. The TDSS is designed to support the decision-making process of commanders and staffs at the tactical level of command and control, using mathematical models and raster representations of tactical-geographic data. Current military operations are typically conducted in a complex environment, where the activities of friendly units are influenced not just by the enemy, but also by the civilian environment, with lots of entities with different types of socio-cultural behavior. To conduct a war game in the complex environment of current military operations, it is advisable to use so-called multi-agent modelling, evaluated e.g. in [29].

The "optimal movement route model" was implemented in TDSS in 2015, see [30, 31], which combines the effects of situational factors in its calculations. The result of the mathematical algorithms is that the axis of the maneuver is optimized on the basis of predefined criteria for the maneuver's speed and safety. The method for rapidly forming the axes of the four fundamental tactical maneuvers of units is made possible by the Maneuver control system (MCS CZ), implemented in the TDSS in 2019, whose functionality is described in [32]. These are the maneuvers of frontal attack, envelopment, turning movement and attack by fire. The spatial expression of these maneuvers, then, provides guidance for the movement of units on the battlefield in order to take advantage of the enemy. Based on its structure, information inputs, mathematical algorithms and linkages, graphical outputs, time calculations and processing speed, the MCS CZ is suitable for use in war gaming.

2 Maneuver Control System CZ

The basis for calculating the MCS CZ maneuver axis – optimized in terms of time-constraints and safety – is the "optimal movement route model" created by the author, see [30]. The model combines the influences of different spheres of the battlefield through map algebra, mathematical linkages, and criterion assessment. Implemented in the TDSS, this model assesses the effects of surface and terrain relief, weather, enemy activity, and friendly forces on the battlefield, in the form of raster mathematical layers. The MCS CZ allows the calculation of maneuver axes of a group of elements coordinated with each other to achieve the same target. It can be used both to calculate the axes of maneuvers of friendly units as well as the supposed maneuvers of the enemy. The MCS CZ mathematical models of unit maneuvers and military vehicles use a raster format representing the digital land model data and a digital relief model from the space of operation.

The structure of the optimal movement route model as the basis for MCS CZ is made up of several matrix layers that represent individual groups of horizontal and vertical passability factors (motion intensity) of the space. These layers (factors) influence the choice of the axis of maneuver of units and military vehicles in terms of terrain passability and situation safety. Each raster cell contains a numerical value for the difficulty of surmounting it, derived from the current state of impacts of mission variables at a given position in space. These are represented by the horizontal factors of cost surface (CS1), weather (HF3), influence of enemy forces (HF4), supporting influence of friendly units (HF5) and the vertical challenge factor of the terrain relief maneuver (VF2), dependent on the criterion assessment of their occurrence characteristics, shown in Fig. 1 and described in [30, p. 557]. The criterion scoring metric varies for each layer, taking into account its character and composition. The basic data for calculating it are the cell dimensions, the average redeployment speed for the selected element on a given type of terrain surface that moves across the cell, and the magnitude of resistance of the factor under consideration.

The basis for the model calculations is CS1, characterizing the speed of maneuvers on individual surface types for the geographic conditions of Central Europe. The model generally considers larger bodies of water, such as ponds, dams or lakes impassable for conducting tactical maneuvers. They are surmountable for floating forces and resources, depending on their capabilities. Swamps and marshes are always rated impassable. The speed of maneuver on each type of surface may vary, depending on the technical characteristics of the vehicles, the experience of the operator and the readiness of the infantry units. The terrain elevation layer is formed by its relief, which enters the final combined cost surface (CCS) calculation with the vertical ground slope factor (VF2). Its definition can be defined as the degree of difficulty of moving on the terrain relief. In the layer weather (HF3) and its effects, only snow and rainfall are considered as blanket effects on ground passability, see [30, p. 559]. The enemy situation layer (HF4) assesses the passability of an area against the safety criteria for redeploying elements of friendly forces, depending on the deployment, armament, range and visibility of enemy units. The layer of the factor of influence of friendly military forces (HF5) expresses the degree of elimination of the influence of enemy activity, in the sense of support for surmounting (passability) the hazardous area of the task. It represents the space within the visibility and distance of conducting effective fire with friendly units not performing the calculated maneuver. The mathematical calculation of the combined influence of all the considered layers of influence factors on passability and speed of surmounting space is represented by the resulting matrix layer of CCS, see the equation below (1). The CCS calculation is based on the average repeatedly measured speed of a particular vehicle type on a specific ground surface. The result of this calculation is in the form of hundredths of a second. It is further modified by mathematical linkages and criteria evaluation of the influence of other situational factors such as terrain relief, weather, the enemy and the support of friendly forces. The value of the time of overcoming each matrix takes into account the combined influence of all factors of the situation, calculated using (1), which creates the CSS of the whole space. The shortest and safest path is then given by the sum of the matrix values with the smallest cumulative value. The CCS – as the result of the calculations of the model optimal axis of the maneuver – is then used to calculate the fastest safe axis of the maneuver in the TDSS using the modified Floyd-Warshall and Dijkstra's algorithm for finding the shortest route, see [33, 34].

$$CCS = \frac{CS_1}{VF_2 \cdot HF_3 \cdot \min(1, HF_4 \cdot HF_5)} \tag{1}$$

The MCS CZ offers calculations for 4 offensive maneuvers, i.e. frontal assault, attack by fire, envelopment and turning movement, as described in [32]. Frontal assault uses a search algorithm to find the shortest route on the basis of the CCS optimal movement route model. Its objective is the fastest safe axis of maneuver to the target object or enemy unit. The fire attack maneuver also uses a search algorithm for the shortest route on the base of the optimal axis CCS model. The axis target of its maneuver is not an object or position of an enemy unit, but the nearest edge of the visibility area to the target enemy at the distance of effective fire for the purpose of its firing destruction. The encirclement maneuver is based on the assumption of coordinated implementation with units acting frontally. To specify its course, it uses an invisible layer with a circular-shaped impassable space between the initial position of the friendly maneuver unit and that of the target enemy. Its radius is equal to the distance D, which is the distance between the two positions, no more than 1 km. This distance is adjustable and was chosen because of the average distance for the effective firing of units operating frontally and the time possibilities to surmount such a semicircular distance to the target. Specifications for the course of maneuver circumvention were also established on the assumption of its coordinated execution with units acting frontally. It is delineated by an invisible layer with an impassable space in the form of two heart-shaped semi-ellipsoids, formed on both sides between the starting point and the target point of the maneuver, at the distance D of no more than 1 km. The shape of the two semi-ellipsoids has the largest length equal to 1.3D and the largest width is 0.5D; see Fig. 1 – Maneuver models layer.

All four models of MCS CZ maneuvers can be used to calculate the so-called Group Unit Maneuver. A group maneuver is the collective maneuver of a set of units (companies, platoons, squads and smaller elements or a group of UGV) to act in a coordinated manner on the same target object. The individual axes of maneuvers are then calculated in a sequence, based on the subordination or organizational division of the individual elements, in three-person formations. Based on the size or type of tactical element (company, platoon, squad or UGV), it is then necessary to adjust the width of the restricted area, located on the sides of the axes of each entity's maneuvers. The defined restricted zone creates tactical spacing between the axes of the formation maneuvers. Its width can be variant-adjusted for each tactical element. The total width of the impassable zone, double-sided from the calculated axis of the maneuver, may take values such as 50 m for a UGV, 100 m for a squad or army tank, 200 m for a platoon and 400 m for a company. The value of the distance of this restricted area on the sides of the maneuver axis from the target enemy's position can be adjusted into the calculation depending on the distance requirement for maintaining tactical spacing when deployed in a pre- or post-combat formation.

The MCS CZ group maneuver model can be used both in calculating the assumed variant of the enemy unit maneuver and for the maneuver of friendly units. Using it, one can identify the likely axes of maneuvers of enemy units, the narrowed points with anticipated increased concentrations of units, and the areas deployed in the battle formation. Based on the results thus calculated, it is then possible to identify the target positions of the counter-maneuver reaction of friendly units. Applying the MCS CZ, it is then possible to calculate the axes of the counter-maneuver and its execution time. It is feasible to model different variants in the development of the situation on the battlefield by means of variable changes in the input characteristics of the MCS CZ, such as the deployment of enemy units and obstacles, or changes in the speed of redeployment across different types of surface.

The MCS CZ is primarily focused on solving tactical situations on the battlefield, in which it is necessary to make decisions quickly and under time pressure based on the analysis of most of important influences of the situation. The axes of the maneuvers are calculated as the fastest and safest, which corresponds exactly to the offensive maneuver of the enemy forces as well as the possible reaction of friendly forces. The time and safety of the maneuver is, therefore, a priority evaluation characteristic. In military practice, the MCS CZ can be used especially in the situations when enemy and friendly forces need to perform an offensive maneuver, retreat or move from the target position with

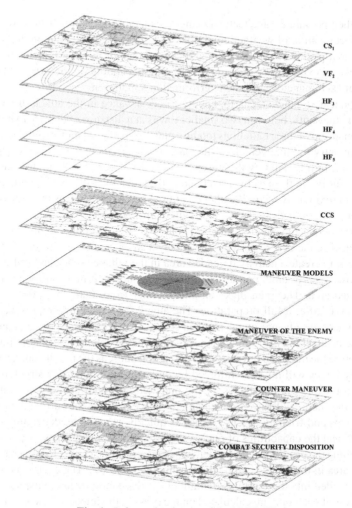

Fig. 1. Information layers of the MCS CZ

a minimal risk of attack by the enemy and in the shortest possible time. Some of them are described in [35–37]. The MASA SWORD simulator can be used to validate the maneuver axes and times calculated by the MCS CZ, as an alternative to the actual execution of maneuvers by units in the field.

3   Case Study: Modelling the Enemy Course of Action

In the following case studies, calculating the maneuver axes of both the attacking forces of the enemy and the reaction maneuvers of friendly forces using MCS CZ and the MASA SWORD simulator are compared. The purpose of this comparison will be to verify the ability of the MCS CZ to generate real axes of maneuvers and times to cover them that are usable in a war game. Digital land and relief models, including mathematical algorithms to assess their passability and tactical maneuver models, have already been stored in the system. Specification of weather conditions and the tactical situation on the battlefield, including unit positions and the tactical technical data of weapon systems in the MCS CZ can take an experienced operator 10–20 min, depending on the extent. Calculating the axes of each maneuver are then performed in a matter of seconds.

In a tactical situation scenario, enemy forces advance in the attack direction from the north to the south, between two smaller towns at a distance of approximately 20 km. They are made up of three battalions totaling nine companies of motorized infantry, using wheeled infantry fighting vehicles. The band of advancement is made up of undulating terrain relief, farmed land, with densely scattered villages. It is bounded naturally by woodland, further to the east by a river and to the south by hilly terrain, with an average width of approximately 6 km. Weather conditions were simulated as the previous 3 days without precipitation, clear, 20 °C, light wind 5 m.s^{-1}.

Halfway across the enemy's advance zone, our friendly units rapidly took up a line of defense, with the task of slowing the enemy's advance until reinforcements arrived. It is planned to create a heavily defended area in the village at the center of the line for this purpose. Enemy units preparing to advance southward must surmount this heavily defended area. They can make an attack either from marching, after a shortened preparation, or after a full preparation. The attack from marching represents a steady continuation of the advance on the planned course of attack, using speed and the element of surprise in attacking the rapidly engaged, unprepared defense of friendly forces. This option may be the most dangerous for the defending units, as the rapidly engaged defense of friendly forces will not be reinforced by engineer-built defensive structures, explosive and non-explosive barricades, or the prepared efforts of artillery and air support. However, the enemy will have to surmount the space between the current deployment of enemy troops and friendly defense forces as quickly as possible, depending on terrain passability and the capabilities of one's vehicles and weapon systems.

Modelling the progress of the maneuvers of enemy units seeking to breach the heavily defended area in the village can be done using the MCS CZ. The enemy's cooperating units are divided into three groups (battalions) of three motorized companies. The axes of maneuver of each unit are calculated on the basis of the impact of the surface and the relief of the terrain, the weather, and the actual deployment of enemy units and friendly forces in two variants – as tactical and fastest. The axes of tactical maneuver are shown in Fig. 2 in red and represent a maneuver in the tactical formations of the units with mutual security and control of space. The axes of fastest maneuver are shown in yellow in Fig. 2 and are mainly led using paved roads that allow maximum maneuver speeds to be reached.

The anticipated time constraints for the execution the enemy's maneuvers was calculated at 34:00–38:23 min for the Eastern Battalion Task Force, at 29:00–33:37 min for the Central Battalion Task Force and 32:51–36:11 min for the Western Task Force.

The speed variables are set to be the highest average for a tactical maneuver of a particular type of combat technique in a space where contact with enemy units is likely. The speed values take into account the nature of the specific terrain surface, the technical capabilities of the vehicle, the mutual coverage of the vehicles and control of the surroundings, in order to identify enemy forces in a timely manner and minimize the risks of destruction. Their values have been generated based on the author's practical experience in driving combat vehicles and commanding maneuvering units within military exercises, and are shown in Fig. 3 on the left. Their values may vary in variation depending on the field conditions, type and state of the technology, and the tactical situation on the battlefield. Larger bodies of water and streams are generally considered impassable in the MCS CZ due to the significant risks of loss of mobility and combat capability caused by the nature of the bottom, banks, depth and speed of the watercourse.

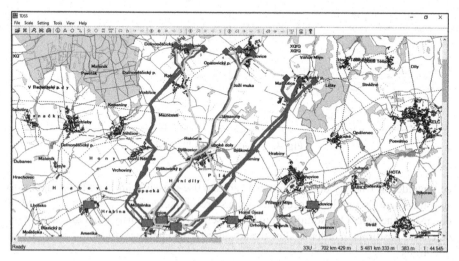

Fig. 2. Enemy courses of action [TDSS – adjusted] (Color figure online)

For the variant of the fastest enemy unit maneuver, the MCS CZ enemy approach times to friendly units in defense were calculated at 8:31–15:35 min for the Eastern Battalion Task Force, 6:31–11:52 min for the Central Battalion Task Force, and 7:04–11:25 min for the Western Task Force. Speed variables are set as the fastest possible speeds on the types of terrain considered. Their values were measured in experimental testing, described in [30], and are shown in Fig. 3 on the right.

In the event of a tactical approach by enemy companies into a heavily defended area of friendly forces, initial contact can be expected from the center of its formation in approximately 29 min from launch. It can therefore be assumed that a rapidly engaged defense of friendly forces has a minimum of 29 min to retake defensive positions or deploy roadblocks and combat security units.

In the event of the enemy's fastest possible approach into heavily defended space, as the most dangerous variant of enemy activity, first contact can be expected from the center of the enemy's formation in approximately 6:31 min. To do so, the enemy is likely to use two roads in particular, running through the center of space. Most of the enemy unit's maneuver is then likely to be surmounted by moving along the road in a flow, and deploying them off the roads will occur outside the perimeter of the defense area. The short approach time does not even allow for the basic deployment of the main weapons of friendly units, including any subsequent build-up of a defense fire system. Unless the

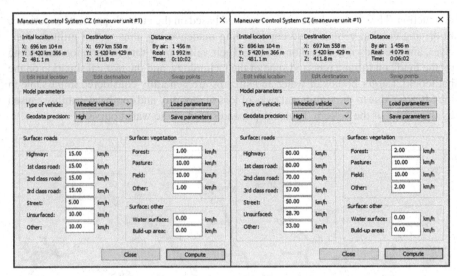

Fig. 3. Speed variables [TDSS]

commander of friendly defense forces is able to use massive air or artillery support to stop the enemy's rapid advance, their forces are unlikely to be ready for any kind of defense of the area. One solution could be to send combat security units into temporary defensive positions in order to slow the advance of the enemy and identify its approach directions.

4   Case Study: Modelling a Friendly Forces Counter Maneuver

Friendly forces have rapidly established a line of defense and are expecting an immediate attack from the enemy. However, they are not simply static, but seek to actively adapt the combat set-up structure for combat provision, due to an early warning about the enemy units attacking, forcing them to deploy prematurely. They can also take advantage of e.g. the setting up of roadblocks or the pre-planning of artillery and air support areas, see [38, 39]. Using mathematical calculations and axes modelling of the offensive maneuver of enemy units, one can estimate their probable course and length of time. In the event that friendly forces are still not in contact with the enemy, but are trying to act proactively and estimate the current position of its units, the MCS CZ allows you to calculate the fastest safe axis for the predicted maneuver of enemy units. Based on this, the commander is able to send smaller combat security units to the precise areas of anticipated concentration of the enemy's maneuver axes. The objective of these units is to rapidly engage in advantageous spaces and lines to slow the maneuver of the enemy units. Similarly, it is able to prepare and plan in time for the action of combat support units against the maneuver of the enemy, into the areas of anticipated concentration of enemy units.

As a variant for the activity of friendly forces 2 (COA2), combat security unit detachments will take up their planned positions at times: 13:12 min eastern unit, 14:02 min central unit, and 17:06 min western unit, in case they move out immediately after launching an attack by the enemy. The axes of maneuvers of the three combat security detachments were calculated at their defensive positions, at a direct distance of 2,697–3,132 m from the starting positions, graphically shown in dark blue in Fig. 4.

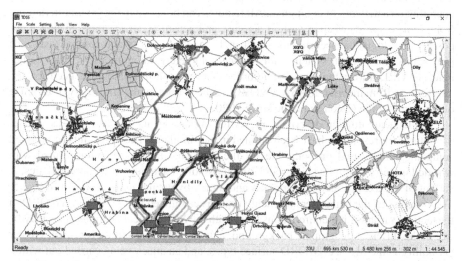

Fig. 4. Counter maneuver of friendly forces [TDSS – adjusted] (Color figure online)

In the case of a lack of time, when enemy forces have already taken over villages located to the north of the line of defense, friendly combat security units can be alternatively deployed to a position about half the original planned distance. In Fig. 4, these maneuver axes are graphically shown in light blue and identified as COA 1. Even with this option, rapidly engaged defensive positions are situated in spaces with assumed concentrations of the maneuver axes of enemy units. Defensive positions in these locations may be located in smaller woodland areas or behind a terrain barrier that provides cover and concealment. The combat security detachments are, based on geographic and meteorological characteristics of the terrain, able to take the COA1 defensive position at 7:06 min eastern unit, 5:52 min central unit and 6:38 min western unit. All three axes of maneuver of these units are identical to those of the COA2 maneuver, but they are only driven to a direct distance of 1,203–1,403 m from their starting positions. The advantage of COA1 is a faster execution over a shorter distance and the possibility to use the support of units at the strongpoint of friendly forces, including their cover when withdrawing under enemy pressure.

5   Comparing the MCS CZ Calculation and the MASA SWORD Simulation

MASA SWORD is a complex analytical tool designed for the automated constructive simulation of various tactical scenarios in user-defined field conditions powered by artificial intelligence. For simulation, it uses height, raster and vector geographic models of space as a base. It allows you to simulate a number of variable situations in highly real-istic environments, such as military conflicts, stabilization operations, terrorist threats, or natural disasters. It includes the server environment, a simulation client, a time sheet, scenario preparation tools, the ability to customize physical and decision models, sim-ulation evaluation tools, and stand-alone preparation, including a web environment for operating on a network of computers. All features can be set up in the simulator, ranging from vehicle speed, weapon system activity, and sensor accuracy, through unit com-position and logistical security, to a doctrinal model of activity and combat tasks, see [40]. Units and weapon systems can operate completely autonomously on the battle-field. SWORD allows them to give an order that they follow without the need for any interference from the user. MASA SWORD has been used by the University of Defence of the Army of the Czech Republic since 2020.

The tactical situation described above, the variants of its development, as well as the options for resolution have been defined in the simulator. The same units (motorized companies) were used on both sides of the simulated tactical situation, armed with wheeled infantry fighting vehicles and infantry soldiers with identical characteristics. A variant of the axes of tactical approach and attack by enemy units is shown in Fig. 5. The progress of the maneuvers of enemy units as calculated by the MASA SWORD in the tactical variant is not entirely consistent with the MCS CZ calculations. This may be due to the different geographic background, a divergent speed setting for much greater variability in the MASA SWORD surface types, and an attack being carried out within a defined spatial band.

In the case of calculating the fastest variant of the maneuver axes, in MASA SWORD these are for the most part led along two paved roads in the center of the space, identically evaluated by the MCS CZ. The initial direct distance between friendly forces' defensive positions and the enemy units was approximately 6,500 m. The calculated fastest time for the enemy to reach friendly units' defensive positions was 29 min in the MCS CZ, namely a battalion company in the middle of the enemy's offensive lineup. A possible reaction of friendly forces was to send in combat reinforcements in order to create the conditions for a defense buildup and early warning. Calculations of the axes of enemy maneuvers into the space of the captured strongpoint and the planned defensive positions of combat security with COA1 and COA2 are shown in Fig. 5 for comparison and supplemented by the execution times in Table 1.

Horizontally, the times shown in Table 1 are divided according to the target areas of the maneuvers, and vertically by which units are performing the maneuver. The units' maneuver times into specific areas are further divided according to the method and especially the speed of its execution into tactical and the fastest.

The comparison of the resulting maneuver times, calculated in the MCS CZ and MASA SWORD in Table 1, identifies insignificant differences. The times calculated by MASA SWORD are generally of a greater value than the analogous variant calculated in the MCS CZ. This may be due to different underlying geographic data and, for MASA SWORD, more detailed speed specifications on a larger number of terrain surface types and applying a doctrinal model for the combat activity of the enemy units. The greatest time variation was recorded in the variant of deploying friendly combat security

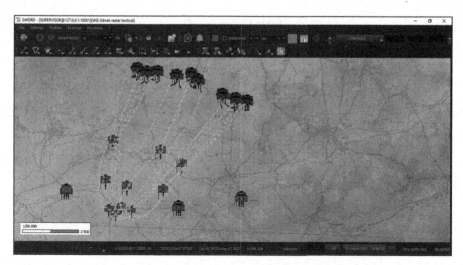

Fig. 5. Tactical variant of enemy attack [MASA SWORD]

Table 1. Maneuver of the first-line companies [MCS CZ, MASA SWORD]

	Method of implementation	Space of strongpoint			Combat security COA 1			Combat security COA 2		
		West	Center	East	West	Center	East	West	Center	East
Execution time of friendly forces' maneuver MCS CZ	Tactical	0	0	0	6:38	5:32	7:06	17:06	14:02	13:12
	Fastest	0	0	0	2:56	1:37	4:50	6:14	3:26	5:44
Execution time of friendly forces' maneuver MASA SWORD	Tactical	0	0	0	7:40	4:50	8:10	18:54	16:00	14:50
	Fastest	0	0	0	3:20	1:52	5:53	7:08	4:10	6:50
Execution time of enemy forces maneuver MCS CZ	Tactical	32:51	29:00	34:00	27:33	23:30	25:56	17:30	14:28	20:30
	Fastest	11:55	6:31	12:41	8:37	5:47	8:31	4:08	4:02	7:04
Execution time of enemy forces' maneuver MASA SWORD	Tactical	33:28	30:46	35:24	28:47	24:20	26:40	18:22	15:20	21:45

(*continued*)

Table 1. (*continued*)

	Method of implementation	Space of strongpoint			Combat security COA 1			Combat security COA 2		
		West	Center	East	West	Center	East	West	Center	East
	Fastest	12:26	7:10	13:20	9:16	6:05	9:07	5:02	4:50	8:21
Execution time of enemy forces' maneuver MASA SWORD while taking COA2 position by friendly forces detachments	Tactical	53:34	51:00	56:10	42:10	40:20	44:00	24:10	22:10	28:10
	Fastest	19:49	23:20	25:10	14:38	17:20	20:20	7:30	8:23	12:30

detachments in position with COA2 prior to the commence of the enemy attack, when a doctrinal model also enters the maneuver simulation.

6   Conclusion

In case studies, using the MCS CZ and MASA SWORD, the axes of the enemy units' offensive maneuver were calculated in the fastest and tactical variants of execution, advancing over rapidly engaged defenses of friendly forces. Using mathematical algorithmic models and digital data of territory and relief, it was possible to generate axes of unit maneuvers while dealing with time-constraint situations on the battlefield. Evaluating terrain passsability and calculating individual maneuver axes in the tactical situation took several seconds using the MCS CZ and MASA SWORD. They produced the fastest passable axis of maneuver to the target, calculated depending on speed parameters. The resulting COA can be modelled in a variable way, taking into account changing developments in the battlefield situation, information currently gathered about the enemy, or changes in the weather.

Based on the results of the war game conducted using the MCS CZ, a variant of the actions of friendly units can be chosen in a situation induced by the case study.

In the event that the enemy decides to proceed immediately in the direction of the attack, the first contact with its combat units can be expected, according to the MCS CZ, most likely in 29 min when implementing the tactical variant. However, when implementing the worst – i.e. fastest – variant, friendly units in defense can be attacked in as little as 6:31 min. The enemy forces outnumber friendly and have a high degree of mobility, and for that reason friendly units have rapidly taken up defense. For early warning, combat security units in the strength of three reinforced squads have been assigned to take up planned defensive positions in two COAs. In the case of the closer COA1, based on the MCS CZ calculations, defensive positions can be taken in time using tactical secured redeployment, without the risk of being attacked by the motorized companies of the enemy. However, in the case of applying COA2, it is necessary to use the fastest possible means of taking up combat security defensive positions in order to prevent the occupation of these spaces by the enemy. But even with the fastest means of engaging the COA2, the planned western defensive position may be overrun by the enemy two whole minutes prior to the arrival of friendly combat security unit, based on the MCS CZ calculations.

As a solution to this tactical situation, it is therefore advisable to take up the COA2 positions as fast as possible, provided there is constant monitoring of the area by means of aerial reconnaissance. In the case of identifying enemy presence in the COA2 position area, the combat security units would deploy or retreat to defensive positions on the COA1 under pressure.

The differences can be identified when comparing the calculations of axes and maneuver execution times using the MCS CZ and the MASA SWORD. Even so, the calculated axes of maneuvers in both programs confirmed the tactical correctness of the deployment of the combat security units in the COA1 and COA2. The implementation of individual MCS CZ maneuvers was mostly calculated faster than calculations done with the MASA SWORD simulator. Based on the results mentioned above, calculations of the MCS CZ can therefore be considered the COA under the so-called "worst case scenario" of the enemy forces maneuver. The worst-case scenario of the enemy COA for a battlefield situation is a very common commander's approach when planning and implementing the countermeasures of friendly units in military operations.

References

1. ADP 5-0 the operations process. Department of the Army, Washington, D.C., p. 105 (2019). PIN: 102805-000
2. ATP 2.01.3 Intelligence Preparation of the Battlefield. Department of the Army, Washington, DC, p. 228 (2019)
3. FM 5-0 the operations process. Department of the Army, Washington, D.C., p. 252 (2010). PIN: 082115-000
4. FM 6-0 commander and staff organization and operations. Department of the Army, Washington, DC, p. 394 (2014). PIN: 104216-001
5. Glarum, J., Adrianopoli, C.: The military decision-making process (2020). https://doi.org/10.1016/B978-0-12-815769-5.00002-6
6. Wang, M., Chang, J., Zhang, J.: A review of digital relief generation techniques. In: ICCET 2010 - 2010 International Conference on Computer Engineering and Technology, Proceedings, vol. 4 (2010). https://doi.org/10.1109/ICCET.2010.5485636
7. Hirt, C.: Digital terrain models. In: Grafarend, E. (ed.) Encyclopedia of Geodesy, pp. 1–6. Springer, Cham (2014). https://doi.org/10.1007/978-3-319-02370-0_31-1
8. Galin, E., et al.: A review of digital terrain modeling. Comput. Graph. Forum **38**(2), 553–577 (2019). https://doi.org/10.1111/cgf.13657
9. Forkuo, E.K.: Digital terrain modelling in a GIS environment. Int. Arch. Photogramm. Remote Sens. Spat. Inf. Sci. **37**(B2), 1023–1029 (2010)
10. Wasklewicz, T., Reavis, K., Staley, D.M., Oguchi, T.: Digital terrain modeling. In: Shroder, J., Bishop, M.P. (eds.) Treatise on Geomorphology. Academic Press, San Diego (2013). https://doi.org/10.1016/B978-0-12-374739-6.00048-8

11. Zhang, Y.-W., Wu, J., Ji, Z., Wei, M., Zhang, C.: Computer-assisted relief modelling: a comprehensive survey. Comput. Graph. Forum **38**, 521–534 (2019). https://doi.org/10.1111/cgf.13655

12. Alcaras, E., Falchi, U., Parente, C.: Digital terrain model generalization for multiscale use. Int. Rev. Civ. Eng. (IRECE). **11**, 52 (2020). https://doi.org/10.15866/irece.v11i2.17815

13. Mazal, J., et al.: Modelling of the microrelief impact to the cross-country movement. In: Proceedings of the 22nd International Conference on Harbor, Maritime and Multimodal Logistic Modeling & Simulation (HMS 2020). Virtual 2020, vol. 22, pp. 66–70 (2020). ISSN: 27240339. ISBN: 9788885741461. https://doi.org/10.46354/i3m.2020.hms.010

14. Rybanský, M.: Determination the ability of military vehicles to override vegetation. J. Terramech. **91**, 129–138 (2020). ISSN: 0022-4898. IF 2.043. https://doi.org/10.1016/j.jterra.2020.06.004

15. Rybanský, M.: Soil trafficability analysis. In: International Conference on Military Technology Proceeding, ICMT 2015, pp. 295–299. University of Defence, Brno (2015). ISBN: 978-80-7231-976-3. https://doi.org/10.1109/MILTECHS.2015.7153728

16. Rybansky, M., Hofmann, A., Hubacek, M., Kovarik, V., Talhofer, V.: Modelling of cross-country transport in raster format. Environ. Earth Sci. **74**(10), 7049–7058 (2015). https://doi.org/10.1007/s12665-015-4759-y

17. Rybanský, M., Hubáček, M., Hofmann, A., Kovařík, V., Talhofer, V.: The impact of terrain on cross-country mobility - geographic factors and their characteristics. In: Proceedings of the 18th International Conference of the ISTVS, Seoul, Korea, pp. 1–6. Seoul National University, Seoul (2014). ISBN: 978-1-942112-45-7

18. Dohnal, F., Hubáček, M., Šimková, K.: Detection of microrelief objects to impede the movement of vehicles in terrain. ISPRS Int. J. Geo-Inf. **8**(3), 1–16 (2019). ISSN: 2220-9964. IF 2.239. https://doi.org/10.3390/ijgi8030101

19. Bekesiene, S., Hubáček, M., Bureš, M.: Modelling possibilities of the vehicle movement on communication network for defense and crisis management. In: 7th International Conference on Military Technologies, ICMT 2019, p. 8870024. Institute of Electrical and Electronics Engineers Inc., Brno (2019). ISBN: 978-1-7281-4593-8. https://doi.org/10.1109/MILTECHS.2019.8870024

20. Fitro, A., Bachri, O.S., Purnomo, A.I., Frendianata, I.: Shortest path finding in geographical information systems using node combination and Dijkstra algorithm. Int. J. Mech. Eng. Technol. **9**, 755–760 (2018)

21. Zafar, A., Agrawal, K.K., Anil Kumar, W.C.: Analysis of multiple shortest path finding algorithm in novel gaming scenario. In: Singh, R., Choudhury, S., Gehlot, A. (eds.) Intelligent Communication, Control and Devices. AISC, vol. 624, pp. 1267–1274. Springer, Singapore (2018). https://doi.org/10.1007/978-981-10-5903-2_132

22. Schwartz, P.J., et al.: AI-enabled wargaming in the military decision making process. In: Proceedings SPIE Volume 11413, Artificial Intelligence and Machine Learning for Multi-Domain Operations Applications II, 114130H, 22 April 2020. https://doi.org/10.1117/12.2560494

23. Evensen, P.-I., Martinussen, S., Halsør, M., Bentsen, D.: Wargaming Evolved: Methodology and Best Practices for Simulation-Supported Wargaming (2019)

24. Bedi, P., Taneja, S.B., Satija, P., Jain, G., Pandey, A., Aggarwal, A.: Bot development for military wargaming simulation. In: Deka, G.C., Kaiwartya, O., Vashisth, P., Rathee, P. (eds.) ICACCT 2018. CCIS, vol. 899, pp. 347–360. Springer, Singapore (2018). https://doi.org/10.1007/978-981-13-2035-4_30

25. Hodický, J., et al.: Computer assisted wargame for military capability-based planning. Entropy **22**(8), 861 (2020). ISSN: 1099-4300. IF 2.494. https://doi.org/10.3390/e22080861

26. Hodický, J., Özkan, G., Özdemir, H., Stodola, P., Drozd, J., Buck, W.: Dynamic modeling for resilience measurement: NATO resilience decision support model. Appl. Sci. **10**(8), 2639 (2020). ISSN: 2076-3417. IF 2.679. https://doi.org/10.3390/app10082639

27. Hodicky, J., Prochazka, D., Prochazka, J.: Automation in experimentation with constructive simulation. In: Mazal, J. (ed.) MESAS 2018. LNCS, vol. 11472, pp. 566–576. Springer, Cham (2019). https://doi.org/10.1007/978-3-030-14984-0_42

28. Stodola, P., Mazal, J.: Tactical decision support system to aid commanders in their decision-making. In: Hodicky, J. (ed.) MESAS 2016. LNCS, vol. 9991, pp. 396–406. Springer, Cham (2016). https://doi.org/10.1007/978-3-319-47605-6_32

29. James, A., Hanratty, T.P., Tuttle, D.C., Coles, J.B.: Agent based modeling in tactical wargaming. In: Proceedings SPIE Volume 9851, Next-Generation Analyst IV, 985106, 12 May 2016. https://doi.org/10.1117/12.2230916

30. Nohel, J.: Possibilities of raster mathematical algorithmic models utilization as an information support of military decision making process. In: Mazal, J. (ed.) MESAS 2018. LNCS, vol. 11472, pp. 553–565. Springer, Cham (2019). https://doi.org/10.1007/978-3-030-14984-0_41

31. Nohel, J., Stodola, P., Flasar, Z.: Model of the Optimal Maneuver Route. IntechOpen, 3 April 2019. https://doi.org/10.5772/intechopen.85566. https://www.intechopen.com/online-first/model-of-the-optimal-maneuver-route

32. Nohel, J., Flasar, Z.: Maneuver control system CZ. In: Mazal, J., Fagiolini, A., Vasik, P. (eds.) MESAS 2019. LNCS, vol. 11995, pp. 379–388. Springer, Cham (2020). https://doi.org/10.1007/978-3-030-43890-6_31

33. Pradhan, A., Kumar, M.G.: Finding all-pairs shortest path for a large-scale transportation network using parallel Floyd-Warshall and parallel Dijkstra algorithms. J. Comput. Civ. Eng. **27**(3), 263–273 (2013). https://doi.org/10.1061/(ASCE)CP.1943-5487.0000220

34. Stodola, P., Mazal, J.: Planning algorithm and its modifications for tactical decision support systems. Int. J. Math. Comput. Simul. **6**(1), 99–106 (2012). ISSN: 1998-0159. http://www.naun.org/journals/mcs/17-474.pdf

35. Nohel, J., Stodola, P., Flasar, Z.: Combat UGV support of company task force operations. In: Mazal, J., Fagiolini, A., Vasik, P., Turi, M. (eds.) MESAS 2020. LNCS, vol. 12619, pp. 29–42. Springer, Cham (2021). https://doi.org/10.1007/978-3-030-70740-8_3

36. Nohel, J., Zahradníček, P., Flasar, Z., Rak, L.: Possibilities of modelling the coordinated maneuver of units in difficult terrain conditions. In: 2021 Communication and Information Technologies Conference Proceedings, vol. 1, pp. 113–117. The Armed Forces Academy of General Milan Rastislav Štefánik, Liptovský Mikuláš (2021). ISBN: 978-1-6654-2880-4. https://doi.org/10.1109/KIT52904.2021.9583742

37. Nohel, J., Zahradníček, P., Flasar, Z., Stodola, P.: Modelling the manoeuvres of ground reconnaissance elements in urban areas. In: 2021 Communication and Information Technologies Conference Proceedings, vol. 1, pp. 118–123. The Armed Forces Academy of General Milan Rastislav Štefánik, Liptovský Mikuláš (2021). ISBN: 978-1-6654-2880-4. https://doi.org/10.1109/KIT52904.2021.9583749

38. Ivan, J., Šilinger, K., Potužák, L.: Target acquisition systems - suitability assessment based on joint fires observer mission criteria determination, pp. 407–414 (2018). https://doi.org/10.5220/0006835204070414

39. Šustr, M., Potužák, L., Blaha, M., Ivan, J.: Střelba s jed notným dopadem střel a možnosti využití v Armádě České republiky. Vojenské rozhledy **29**(4), 084–093 (2020). ISSN: 1210-3292 (print), 2336-2995 (on-line). https://doi.org/10.3849/1210-3292. www.vojenskerozhledy.cz

40. MASA SWORD: Our constructive simulation with AI controlled units. MASA: Smarter simulations (2019). https://masasim.com/wp-content/uploads/2019/09/sword-uk-recto-2019.pdf. Accessed 27 Jan 2021

Future Challenges of Advanced M&S Technology

Enhancing Requirements Completeness of Automated Driving System in Concept Phase

Ahmad Abbadi$^{(\boxtimes)}$ⓘ and Vaclav Prenosilⓘ

Faculty of Informatics, Masaryk University, Brno, Czech Republic
Ahmad.abbadi@mail.com, prenosil@muni.cz

Abstract. During the product lifecycle, a change in the requirements may involve expensive consequences on the system development. Nevertheless, having a decent understanding of the system and documenting the correct requirements from different perspectives help to minimize the changes caused by missing functionalities, therefore, reduce the development cost.

Developing comprehensive understanding of the system in the concept phase promotes the completeness and reduces the requirements' changes. That helps also to create a good model of the system and develop simulation and test cases, which reveal the bugs and the design issues early. This work focus on thinking strategy to answer the research questions, "How to analyze the Automated Driving System (ADS) to improve the requirements' completeness".

ADS is a complex system that works in non-deterministic environment, in addition, it is a safety related system, meaning that, any malfunction during the operation can cause a harm to people or properties.

Three systematic methods were investigated. First, identify the gaps of the stakeholders. It uses product life cycle to identify a list of internal and external stakeholders and then identify their expected needs. The second systematic process utilizes holistic thinking perspectives method to build a broad understanding of ADS and its neighbor systems. This process tries to direct the system definition using external, internal, progressive, quantitative and scientific perspectives. The last method deals with safety requirements identification tools.

Keywords: Requirement engineering · System engineering · Holistic thinking · Automated driving system

1 Introduction

The uncertainty and incomplete requirements lead to project failing. On the other hand, a good and complete understanding of a problem helps to formulate the problem correctly, and to write requirements that drive the development of the product [1–3].

ⓒ Springer Nature Switzerland AG 2022
J. Mazal et al. (Eds.): MESAS 2021, LNCS 13207, pp. 377–396, 2022.
https://doi.org/10.1007/978-3-030-98260-7_24

Having a good overview of the system and of required functionalities help to create a suitable model for the system. Additionally, when a simulation for a complex system is needed, the integrity between the outputs of every submodule and the output of the system, as whole, can be tested. The completeness of high-level requirements, in this case, helps to define the correct boundary between the submodules to be modular, scalable and iterable in the development. Moreover, a good completeness level of the high-level requirements reveal the hidden factors that affect the simulation process and the simulated output. For example, the restrictions that are created because of safety or because of compatibility with other systems. Furthermore, identifying and isolating the expensive and technically complex sub modules, help to reduce the cost of re-modeling of the system and adapting the interfaces in the simulations, because those modules most likely will be changed when an optimization of the system is needed.

In new emerging systems, the real needs, the expected behavior, and the expected quality may not be obvious neither to the development teams nor to the customers. This is the case of ADS especially in highly automated levels. This situation leads to change of the requirements during the project life cycle. It is different when comparing ADS to small projects, adapting already existing solutions, or comparing it to non-safety related systems. Because in the small project, small development teams can manage the requirements' changes. The teams have a holistic view of the project and the effect of the changes in the schedule and the cost. In [3] the authors present the effects of changing the requirement in relation to the project size, where the smaller and shorter projects can tolerate the requirement changes more than the bigger projects.

When comparing the ADS development, from scratch, to adapting already existing solutions, the task in adapting existing project will be estimating the applicability of the new changes, including the estimation of the effects on the cost, schedule and quality. In existing solutions, the models, the behaviors, and the general needs are most likely matured, and the high-level requirements have a good completeness level. The author in [4] states that the completeness of the highest-level requirements is the most difficult problem, while the completeness of the lower level of requirements can be achieved by allocation and traceability. The lower level completeness is discussed in Sect. 2.1. Although current ADS are composed of a set of existing and new features developed over time, ADS in Level 3 and beyond [5] do not have a matured example that can be adapted to estimate the change consequent.

In the safety related systems, any fundamental change can cause enormous deviation in the cost or schedule, because it is not acceptable to compromising the safety quality. Which requires enough tests to proof that the safety level is acceptable. In ADS case, any change in the requirement may require expensive efforts to clarify the consequence and to ensure the quality i.e. by simulation multiple thousands of kilometers of driving, drive on closed area, and drive on public roads. For this reason, having complete safety requirements can save the project resources.

The incomplete and uncertain definition of high-level requirements is a known issue in the software domain. Iterative methods of development are proposed to mitigate this problem. They try to minimize the release patches and get feedback for a narrow scope of the solution as fast as possible, in this case, missing functions, wrong definition of a problem, or a wrong understanding of the behavior can be corrected to match what is really needed in the final product. However, even in the iterative methods, the major product functionalities should be clearly understood in the concept phase [6,7]. In ADS case, it should also follow automotive rules, which add more restrictions when using agile methods. Automotive domain has many rules to ensure the quality, and they are originated mostly from waterfall development method [8]. Over the time, those practices evolve to include some degree of agility in the development. An example is the usage of agile methods with ASPICE (automotive software performance improvement and capability determination). The authors in [9] present some potentials and challenges of using ASPICE with agile development methods.

In addition to the improvement of the development methods, there are many tries to improve the management methods of the big projects e.g. by centralizing and modeling the system's knowledge using MBSE (model based system engineering) [10,11]. The aim is to reduce the overheads caused by the requirement-based development, while keeping a good quality of the product.

This paper uses the terms "requirement" and "need" with different meaning, the "need" is used to express a narrative statement that represents what is required from the stakeholder point of view, but it does not follow best practice of writing a "requirement" [14]. While a requirement is a broken down of a need to respect the best practice of writing requirements, mainly uniqueness, singularity, testability and other quality [15].

The main goal of this work is to present systematic thinking strategies to enhance the completeness and present some tools to answer the research question "How to analyze ADS systematically to enhance the requirements' completeness". First, a method to identify the stockholders and organize their needs are used. Then, the holistic thinking method is utilized to generate a more complete understanding of the system and enhance the completeness of the stakeholders list. the holistic thinking method provides systemic and systematic views of the system, from external, internal, progressive, quantitative and scientific perspectives. In each perspective, a set of data collection tools can be used e.g. brainstorming, interview, mind maps, or documentation review, etc. [12,13]. Finally, a risk analysis tools are described as systematic methods to generate more requirements that are necessary for safety related systems.

This paper is organized as follows. Section 2 introduces the requirements and requirements' levels in the automotive field, where the completeness in the lower levels of the requirements is discussed. Then, in Sect. 3, a systematic method for stakeholder identification is discussed. Section 4 presents the usage of system perspectives as sources to identify the high-level requirements. The Sects. 3 and 4 present a methods to enhance the completeness of the requirements in higher level. Whereas the Sect. 5 provides insights into a systematic approach for safety requirements identification from different safety views. Finally, Sect. 6 summarizes the main points.

2   Requirements in Automotive Field

Requirements engineering is an essential part of the product development. It establishes a foundation for a successful project, and ensuring that the requested product is delivered as desire.

The requirements are a set of the product description, capabilities description, conditions, or product relevant information that must be met to ensure that a solution meets the stakeholders' needs. ISO/IEEE system and software standard defines the requirement as "Statement which translates or expresses a need and its associated constraints and conditions" [16].

The requirements are used as a language for intercommunication between different teams and departments in different abstraction level. The clear and unambiguous requirements ensure that the problem is defined and the development teams know what functionalities they will deliver and in which quality.

The requirements can be divided based on their abstraction level. Ranging from high-level requirements to detailed technical requirements. An example of the high-level requirements can serve the business needs, goals, policies, business rules, regulations, or stakeholders' needs, while low-level requirements represent the use of a specific hardware part, use an algorithm in a software domain, or use a specific material in mechanic domains. The requirements can be also viewed based on their content. The main two categories are the functional and the non-functional requirements, where the functional requirements define what the system shall do to satisfy the user and the business needs. Whereas the non-functional requirement defines how well the system's quality is [17].

2.1   Requirements Levels and Requirements Presentation

This section presents how the requirements are broken into levels and the semantic relations between the levels.

The high-level requirements represent the purpose of the system for the main stakeholders and other regulatory body. They ensure that the organization's goals are met, and the product complies with the regulations. On the other hand, the needs of other stakeholders, e.g. customers, beneficiaries and users, describe what is needed from the product, in which quality, and in which context.

As an example of high-level requirements is the intent of the traffic law, e.g. "The road user shall not act in a way that harm or endanger others". This high-level policy can be translated into a set of rules, e.g. "A person operating a vehicle moving behind another vehicle must, as a rule, keep a sufficient distance from that other vehicle" (German traffic low [19], Sect. 4.1).

Stakeholders' requirements can be represented using textual statements of what is expected of the system, they also could be written as use cases or user scenarios. A use case or a use scenario is not a requirement, but it is a structure helps to elicit requirements. The use cases and use scenario help to understand how the users will interact with the intended system [17].

Figure 1 shows an example of the relation between an organization's goals down to the technical requirements. The dashed links hide other levels that may be required based on project complexity.

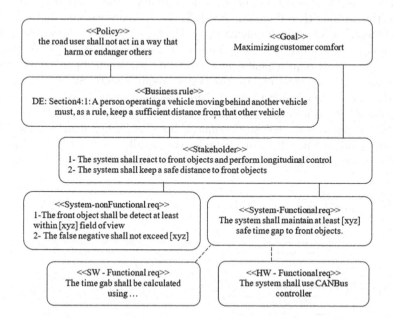

Fig. 1. Example of breaking down the high-level goal and policy into technical level. The dashed links hide other levels.

When the high-level requirements are elicited, they are broken down into more detailed and testable requirements, which describes the system as a black box. One of the most essential characteristic of system requirements is, they deal with what the system should do, not how.

The main types of system requirements are the functional and non-functional requirements. Functional requirements focus on what system does and what are the system capabilities in a measurable manner. They define in details the behavior when the system interacts with users and other systems, in addition, they define the degradation behavior e.g. when the system has an error. The non-functional requirements include quality requirements that define how good the system is when performing its tasks, e.g. performance, reliability, security, maintainability and safety constraints. The non-functional requirements also include other requirements related to the capabilities of resources, schedule, cost, operation environment, documentation, logistics, and diagnosis [20].

The requirements completeness means identifying the needed requirements and representing them in a correct quality. Every requirement of the product should have a reason to exist. Introducing unnecessary requirements make the product more complicated, in addition, the quality, time and cost factors will be affected negatively.

Organizing the requirements in levels makes the development process easier and less risky. The lower levels inherit their necessity form the upper levels. In the requirements engineering, the traceability mechanisms between the requirements in different levels plays an important role to manage the complexity [21].

In the automotive domain, ASPICE forces bidirectional links between requirements' levels, which ensure the consistency between the levels and the consistency in each level [22]. An example of the traceability can serve the relations of the system requirements, where software requirements satisfy the system requirements. Test cases validate the system requirements. System architecture allocates the system requirements. Figure 2 shows the recommended ASPICE's requirements levels and the traceability between them.

ASPICE focuses on software domain, however, other domains also follow the same guidance usually. For example, the middle part of Fig. 2, domain or subsystem level, is adapted to fit hardware, mechanical, or other domains.

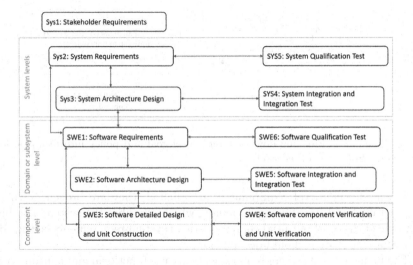

Fig. 2. Requirements levels and traceability in the Automotive-SPICE

The main principle to ensure the completeness and the consistency using the traceability is, in every level, the requirement should describe what is expected at this level, then a solution is selected, then its detailed requirements are created in the lower level. The traceability and allocation provide some proof of the completeness. In each level, each requirement should be unique and address one aspect only. It should also be achievable, if a requirement is not, then the system will not be able to deliver the stakeholders' needs. The traceability to the upper and lower requirements' levels shall be maintained to identify the gaps in the completeness, and ensure that the overall set of the requirements cover all needs [18]. A requirement should be independent from other requirements in the same level. They should not have any statement that is overlapped, conflict, or restate the same need or part of it as other requirements. Every requirement must be unambiguous and understandable in one way only. Ambiguous requirements can lead to project delays, over budget, or having the wrong functionality or behavior [18]. In order to guarantee good quality, and satisfy the above mentioned constraints, there are many proposals for writing a good requirement [15–17,23].

Practically, when a project is started, the detailed information not always available, specially, in the case of innovative products or products with unclear customer needs [24]. In this case, it is better to start with prototypes or system models and progressively writes suitable requirements or start with initial requirements that contain undermined aspects, and then, improving their quality during the development iterations. In all cases, the high-level requirements should be clear, then, the details can be filled out at the suitable time.

The completeness in the lower levels can be proofed to some degree using techniques like suitable abstraction level of the requirements, having good requirements quality and establish traceability between the levels. What missing is whether the requirement cover all aspects of the product or whether the high-level requirements are complete. The next two sections discuss those points. They present methods to enhance the completeness in the higher level requirements, i.e. enhancing the completeness of the system stakeholders who define the high-level needs, and having a wide overview of the system when defining the needs.

3   Stakeholders Identification

During the concept phase of the product life cycle, the stakeholders are identified with the aim of gathering every stakeholder's expectations. Those needs are grouped, prioritized and then written as requirements or modeled to drive the development of the product in later stages [25].

In [26] the authors cite multiple definition of the stakeholders, including, The people and organizations affected by the application.

Multiple researchers tried to define a list of the stakeholders, or classify them in order to manage the list easily. For example, Cotterell and Hughes cited by [26] define three categories of the stakeholders, internal to the project team, internal to the organization but external to the project team, and external to both the project team and the organization. Newman and Lamming cited by [26] suggest a division into those who will use the system directly or indirectly, and those who will be involved in developing the system. Macaulay, cited by [26] identifies four categories, those responsible for design and development, those with a financial interest, those responsible for the introduction and maintenance and those who have an interest in its use. The authors in [26] propose an approach that focuses on interactions between stakeholders, they classify them as 'baseline' including users, developers, legislators and decision makers, and satellite stakeholders who interact with the baseline which communicating, reading, and searching for information. Other classifications and identification methods, of empirical studies concerning the usefulness of stakeholder identification methods in requirements software, are reviewed in [13].

This work uses a systematic method to enhance the completeness by identifying a list of possible stakeholders based on the product life cycle phases, and then elicit the stakeholders' needs. Having a successful product means that the product satisfy all expectations from all stakeholders, and having a list of all stakeholders can be a starting point to proof good completeness level. Carson et al. [4], express this meaning by stating the following definition "A set of requirements is complete if and only if all stakeholders approve the set of requirements".

The product life cycle stages can be adapted based on the domain. Figure 3, shows example of life cycle models [27] for system, software and hardware domains.

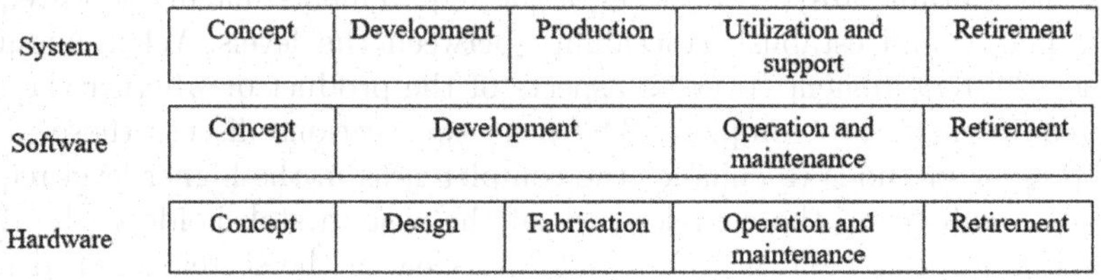

System	Concept	Development	Production	Utilization and support	Retirement

Software	Concept	Development		Operation and maintenance	Retirement

Hardware	Concept	Design	Fabrication	Operation and maintenance	Retirement

Fig. 3. Example of system, software and hardware products' life cycle

In every phase, the internal and external stakeholders are listed, and then prioritized. During the concept phase, the potential internal stakeholders include the managers and people of the research department. Those stakeholders initiate many non-functional requirements e.g. cost, schedule, and compliance with organization strategy and processes. External stakeholders include system acquirer, standardization body, regulators, and certification authorities. Their needs mostly non-functional e.g. product quality, delivery schedule, support, political compliance, environmental compliance, and technical compliance.

During the development phase, the main internal stakeholders are the developers and the engineers, their needs summarized as a clear process of the development, teams' boundaries, development chain, tasks management, work safety, and rewards. Also in this phase, the suppliers, as external stakeholders, need a clear requirements, and delivery process.

In the production phase, the stakeholders list includes internal producers, quality team, and validation and verification teams. External stakeholders include supplier, and external assessors for standards compliance. The needs of those stakeholders are mainly about the test equipment, test strategy, documentation and reporting. The organization management and customer may have additional cost and waste requirements.

In the utilization and support phases, the internal and external stakeholders list includes the end users, supporters, trainers, maintainers. Those groups of the stakeholders drive most of the functional requirement. It is very important to identify most or all stakeholders in this stage, because the changes in other phases are usually limited and slow, however, the requirements and the expectation, of the stakeholder in utilization phase, may change quickly during the product development. That can happen either because of system acquires when they change their strategy, or because of wrong identification of the real needs, e.g. having wishes or nice to have features defined as essential needs.

The completeness of identifying those stakeholders plays two important rules, first, having a clear and complete set of requirements that drive the developments and reduce the changes. Second, helping to manage the changes of the stakeholders' needs during the product development. While the first point dominates all requirements elicitation process in many projects, the second point is, most of the time, forgotten. Managing the stakeholders' needs in general is important because even if the needs in one point of the time is complete, if a stakeholder change the focus or the strategy, then, this completeness will be obsolete. This topic is discussed in some researches, for example, in [28], the authors present a method of continuously managing the stakeholders and changes in requirements throughout the life cycle.

Lastly, in retirement phase, the stakeholder list includes disposers and certification body. Their needs could be environmental compliance or compatibility reports.

In highway assist example, during the utilization phase, the driver is the main stakeholder. Additionally from steering system regulation UNECE R79 view [29], The HMI teams can be considered as stakeholders candidate, based on the statement "When the system is in standby mode (i.e. Ready to intervene), an optical signal shall be provided to the driver".

The generated list of stakeholders can be huge in the first place, and each of the stakeholders generates multiple requirements. However, this list can be optimized to have more useful and abstract representation of the stakeholder. One method is to identify the primary and secondary stakeholders and establish priority list. Additionally, all stakeholders that have a lower priority should be marked as a project risk, where those risks are reviewed during the progress of the project to ensure that no fundamental new requirements comes from those stakeholders that may have expensive consequences.

The systematic approaches help to enhance the completeness, nevertheless, relay on them solely do not guarantee a complete list of the stakeholder, because the systematic process helps to direct the thinking, but the identification of a stakeholder in every direction is a subjective process.

The holistic perspective method is presented later to enhance the completeness when elicit the stakeholders' requirements, and the same method can be used as an additional source for the stakeholders list.

3.1   Stakeholders Requirements Elicitation

Having a generic list, of potential stakeholders and their needs, helps to identify the gaps. The stakeholders' needs can be captured by applying some tools before and during development of the system. For examples, before the development the needs can be collected using strategy documents, technical documents, prototypes reports, related system documents, interfaces, pictures, trade studies, brainstorming sessions with stakeholders and interviews. Other resources that help to perceive stakeholders' needs during the concept phase of development include modeling and simulation reports, prototypes feedbacks, verification and validation feedbacks, reviews feedbacks, quality processes, quality analysis i.e. fault tree analysis, failure modes and effects analysis, and hazard and safety analysis [25, 30].

Stakeholders may express their needs in ambiguous and vague statements. They may also express their desires and wishes instead of the real needs. Based on the experience, the questions language, when extracting the needs, play an important role to find the real and the correct problem. For example, to elicit the needs of people that interact with the systems, a question like "What is the purpose of the system?" can be used, because, when considering the purpose of a system, people think what they want to be able to do with the system, rather than how the system do it. Therefore, the answers, which starts with "I want to be able to ... " most likely, will represent a direct stakeholder need or wish. Asking for the priority of the needs help to distinguish the real needs from the wishes.

Alternative technique is to ask "why" questions multiple times, until the answers match one of the high-level goals or reach a point where no clear answer can be provided. That helps to define the correct problem, especially when dealing with technical people who have pre-assumed solutions, or a people have difficulty to express their needs clearly.

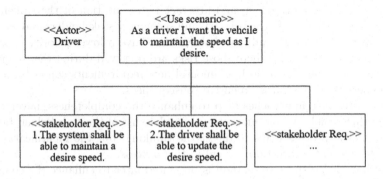

Fig. 4. Example of eliciting requirements from a use scenario

Additional technique represents the need by use cases, user stories or use-scenarios [6,31]. Those representations help to model what people do or want to be able to do. Those techniques are very close to the people way of thinking. Use-scenarios are used typically as part of the Agile development process, to provide a simple and clear description of what a user does as part of the role they perform [18]. The scenario can be used as a basis for discussing the required capabilities and extract requirements statements [17]. The use-scenario or the use cases are by themselves, not requirements, but they can be used to elicit the requirements. Figure 4 shows an example of eliciting requirements from a use scenario.

The stakeholder' needs can be documented using textual requirements, models or diagrams in addition to use scenarios and use cases. Every representation has its strength and weakness. However, the goal is to promote the understanding of the system and define the correct problem to be solved.

4   Enhance Completeness Using Holistic Perspectives

When a system is required, it is important to have a good understanding of the problem that the system tries to solve. Because, not documenting the correct problem statement leads to having incorrect requirements, and in consequence, solving the wrong problem.

Additionally, having a good understanding of the system, and finding the correct and the complete list of needs, help to minimize the changes caused by missing some functionalities, and help to minimize the changing loops between the development's levels. Even when using iterative development, introducing new requirements, that change the architecture dramatically or introducing a need for new system components, may cause expensive ripple changes.

Looking to the system from multiple views systemically and systematically, enhance the system understanding, which means, thinking about the system as part of a bigger system and think of it as a whole to understand its context and its internal aspects. The need for systematic techniques is more obvious in a complex system, where the amount of information cannot be handled or organized easily.

Holistic thinking perspectives method presented in [32]. It consists of multiple perspectives, mainly, external (big picture and operational), internal (functional, structure), progressive (generic, continuum, and temporal) and other (quantitative and scientific), as shown in Fig. 5. This work utilizes holistic thinking perspectives method to elicit the need and fill the gaps of the stakeholder list. This method can be considered as thinking strategy, and it can be applied on any information collection tools, e.g. interviews, workshops, brainstorming, prototyping, modeling, and user stories [1,12,13,33]. When looking at a system from a specific perspective, the question words {who, when, where, what, why, and how} can be used to explore the information.

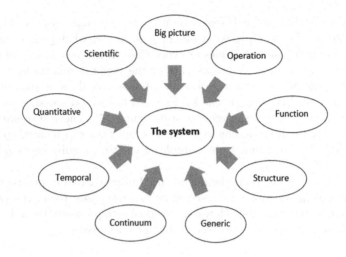

Fig. 5. Holistic Thinking perspectives

4.1 Big Picture

Big picture perspective is looking for the context, the information about the mission, and the goal. Additionally the big picture deals with information to understand the competitors' and the partners' states.

ADS from this perspective is a system in a vehicle (the where question), it communicates with neighbor systems like braking, steering, propulsion, lighting, and other HMIs. The main purpose of the system is to help the drivers by taking over the driving tasks. It lowers the fatal accident rate (What is the purpose, what is the problem to be solved). The external stakeholders are the carmakers as a system acquirer, the driver as an end user, the management body of the producer and the acquirers, and governmental body that legalize the system and certify its safety (Who).

The needs' statements out of this perspective can be formulated as, 1 - The ADS shall communicate with braking, steering, propulsion, lighting, and HMIs. 2 - The ADS shall take over the driving tasks. 3 - The ADS shall follow the safety standards as required by the producer, the acquirers, and the government, and so on. Those needs are refined later and written as a set of requirements.

4.2 Operational Perspective

Operational perspective describes the system as a black box. This perspective looks for the inputs, the outputs, the behavior, the operational environments and the operational conditions. The main users of ADS are the driver, the carmaker, or its representative e.g. workshops (who operate), where the driver can activate or deactivate the system with a suitable interface. The workshop repair, maintain, administrate the parameter, and update the system (what users can do).

The system performs driving tasks on highways, where no pedestrian, animals, traffic lights are expected. Additionally, this system is not intended to work in a construction zone. The system is expected to work with other vehicular means. It works also on curves, multi lane roads. (Where to operate, what are the operational scenarios). The inputs of the system are the system control signals, surrounding environment status, and the vehicle conditions. The outputs of the system are the braking, accelerating, steering, and other HMI signals.

ADS should be able to operate on environmental condition as set by carmaker for a specific country. Regarding this point, the operation design domain (ODD) has a special importance in ADS because many safety-related analyses based on the definition of ODD, and there are systematic techniques to define the ODD as presented in [34].

4.3 Functional Perspective

Functional perspective looks to what functions the system provides to perform the missions in the operating conditions. For example, The ADS taking over the driving tasks by controlling the longitudinal and lateral movements. That includes maintaining the speed in cruising mode, adapt the speed based on frontal object, perform steering to keep the vehicle in a lane (What activities performs the system mission).

The system shall provide a means to the driver to activate, deactivate or change the parameters of the system (Who). In addition, the system shall work in ODD and ask the driver to take over the control if this condition not fulfilled (where, when).

Regarding the functionalities identification, in ADS context, there is a systematic process to identify the needed functionality from an external point of view, it is based on ODD and expected object on the operational scenarios, it is called object and event detection and response (OEDR) [34].

4.4 Structural Perspective

Structural perspective looks how the system is organized from logical and physical views. Questions like (What parts the system has and where system elements are located logically and physically) can be used to explore the information from this perspective. ADS's functions can be classified as 1 - input receivers, 2 - the processing functions 3 - Output processors that send the signals to other systems to perform the required actions. In ADS, there is a need also to special kind of functions between the receivers and the processors. Their function is to perceive the situation and process the input data to model how the scene elements interacting with each other. ADS, in most cases, follows sense-understand-plan-act framework.

One important aspect in structural perspective is to decouple the system structure needs from the solutions and build an abstract view. For example, a statement like "the system shall use a radar sensor to perceive the objects in the environment" can be rewritten as "the system shall have sensors to perceive the environment". Initially, this requirement may be ambiguous, but challenging this statement from other perspectives e.g. continuum perspective as described later, will provide more understanding of the need. To build knowledge that is more abstract the following questions can be raised "what part can be replaced, what is the interfaces between the elements". The answers to those questions help to reveal other possible solutions i.e. the sensor can be a physical component attached to the vehicle, or it can be remote data providers. In addition, the communication can be wired or wireless. Using this methodology of challenging the needs, identified from a structural perspective, helps to define the real needs regardless of the solution.

4.5   Generic Perspective

Generic perspective is looking for how the system similar to other systems, and how the knowledge from other domains can be reused. For example, asking a question like "What this system reminds you" can trigger a thinking process to explore the other similar functions. e.g., ADS has some similarity to the airplane cursing controller. Airplane uses data logger to record the failure of cruise controller, ADS can use that also (What function this system inherit from similar system). Then a high-level requirement can be, ADS shall record failure information, or ADS shall be able to retrieve recorded failure, etc.

4.6   Continuum Perspective

Continuum perspective looks at the differences between the system of interest and other similar systems. It deals also, with what can go wrong and with extreme situation or assumptions. For example, 'why ADS is different from airplane system', the answers can be, the uncertainty in the ADS operation environment is higher, and the operating condition is different e.g. speed and temperature. The answers add a need to define the ranges of sensors, the environment definition, and the alternatives to perceive the surrounding information and getting the weather information.

Another question that help in this perspective is what-if questions, e.g. what happen if an elements fail or provide unexpected high values or unexpected low values. This question helps to set up requirements for the general degradation behavior i.e. ADS shall detect if inputs has abnormal values and perform a degradation function.

Other questions, like what is the most expensive part of the system and how to reduce it, can help to formulate a proper uniqueness level of the requirements, because this thinking helps to identify the parts that have a higher chance to be replaced in the near or far development future.

4.7   Temporal Perspective

Temporal perspective is looking to the system through the time and sequence view. It deals with dynamic behavior of the system, e.g. what is the conditions before the system goes in operation, what the sequence needed to change the system, who will interact with the system and what is the sequence. From this perspective, a sequence of action and states of the system can be derived, in addition to set of requirements about the limitation before, during and after the operation.

Another view in this course is looking to the system during its life cycle, for example, the effects of a sensor's ageing factor on the behavior.

4.8 Quantitative Perspective

The quantitative perspective looks for the numbers. It gives insights of the non-functional requirements. For example, asking questions like, What are the limits, what are the allowable faults per time, delay limits, cost limits, physical limits (shape, inference, moisture, and jerk, etc.), and sensing range limits, etc. Some of those values may not be available in the concept phase of the project e.g. the sensing range and time delay, however recording statements about them will set a placeholder for further stages of development where those values are calculated based on the functionalities.

4.9 Scientific Perspective

Scientific perspective deals with hypothesis, assumptions, alternatives, feasibility, and missing concepts, etc. The inputs of this perspective, most of the time, come from the above perspectives. For example, from the big picture, the scientific perspective tries to prove the statement "ADS lowers the fatal accident rate". Another example "ADS is a system in a vehicle" looking into this entry from continuum view, trigger a question like what is the alternative, i.e. having a remote system distributes functions between the vehicle and the infrastructures. The scientific view to this statement try to give insights, evaluations and evidences to challenge the alternatives. Another example, on missing concepts, can serve "the available sensor's range is not enough for the functions". Making a scientific study helps to find and evaluate solutions. For example, challenge the statement from continuum and generic perspectives, set up a benchmark for identifying the weaknesses and strength, suggest changes or alternatives, and then evaluate the new proposals.

Those nine perspectives help to explore the system' needs from different views. One important point when discussing different system aspects is the narrative language. It is not easy to be abstract all the time. As humans, most of the time, we use examples, or formulate the problem bounded with a solution, which is easier for understanding and easier for discussion. However, after recording all the ideas, a necessary step should be performed to ensure that the real problem is formulated and it is expressed in a functional language not a solution language. Every statement should be challenged if it contains a solution or an embedded solution, then the solution shall be replaced by its equivalent concept that is abstract enough. To identify a suitable concept, asking "why" questions multiple times, can help to get the real problem. e.g. "the radar shall detect cars in front of the ego-car". Challenging this statement to be a solution free, a question like why a radar, why the front of the car, why the ego-car? What the functions of those 'names', and can they be replaced by other 'names'. This way of thinking leads to formulating the statement, as "a perception module shall report objects in front of ego-vehicle". Where the aim of the "radar" is to perceive the things, and this can be replaced by other sensor or set of sensors. In the same way any frontal objects on the road, including other cars, trucks, buses, motors, or even a debris, can be abstracted to replace the "cars" word by general word like "objects".

5   Requirement Elicitation from Safety Perspective

In automotive, the safety related systems, as ADS, shall comply with a set of safety standards. Every standard has specific processes to remove or reduce the hazards that can cause a harm. The differences between the standards come from the type of the causes. e.g. functional safety (FuSa) standard, ISO 26262, deals with the hazards caused by elements' failure [35]. ISO 21448, the safety of the intended functionality (SOTIF), deals with the hazards arising from the insufficiency of the functionalities or the unintended misuse [36]. The cybersecurity standards [5,37] consider the threatens and deals with intended misuse of the system.

In concept phase, there is a pattern that can be recognized and utilize when applying those standards. First, the safety related functions of the system are identified, in addition to the corresponding hazards and risks. Then, the risks are evaluated based on relevant standard criteria, where the risks are classified and abstracted to generate high-level safety goals. Then, the safety concepts are derived from the safety goals, with the aim to eliminate or mitigate the risks. After that, the technical safety requirements are elicited from the safety concepts. The standards, then, define a general procedure for verifying and validating the system. When a hazard exceeds the safety limits, the safety process should be repeated by introducing a new safety concept to reduce or eliminate the risk and then write a requirement to verify that the risk is within acceptable level.

The safety concepts may require updating of the existing functions, introduce new functions, or even remove the function that has this risk. In all cases, introducing new requirements that change the architecture dramatically or introducing a need for new system components may result in expensive ripple changes, because the safety assessment, technical safety requirements, and safety mechanism are built on top of established architecture and well defined safety related functions [35,38].

Safety related systems development is different from the normal systems development, by the additional safety assessment processes. Those processes require manual human decisions e.g. hazard and risk assessment and safety mechanism. Adding or deleting a safety related function needs repeating the safety assessments for all related functions and components.

A good level of completeness, in this case, helps to reduce the changes. It can be viewed from two perspectives. First, the completeness of the high-level safety related functions, and this issue is similar to identifying the high-level need. Second, The completeness of identifying the hazard in every function, because detecting a new risk will trigger the expensive safety assessment process again. For this reason many tools are proposed to find the risk systematically, including hazard and operability method (HAZOP) [39], fault tree analysis (FTA) [40], failure mode and effects analysis (FMEA) [41], and System-Theoretic Process Analysis (STPA) [42].

An example in context of ADS, the function "adapting speed" performs accelerating and decelerating to adapt the speed on the highway based on front vehicle. The hazards can be initiated if the function did not decelerate when needed, not brake enough, brake too much, brake too late, intermittent braking, or braking without request. Out of those possible hazards, a safety goal can be written as, "the item shall prevent unintended braking".

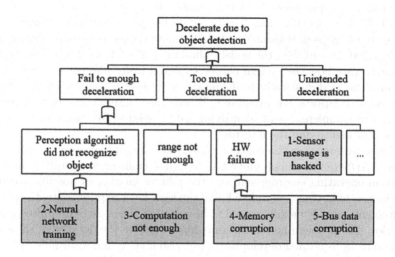

Fig. 6. FTA of deceleration during speed adaptation

Figure 6 shows the fault tree analysis of deceleration function during speed adaptation. Multiple type of hazards are identified, "1 - Sensor message is hacked", this is an intended misuse of the system, and it is related to cybersecurity domain. The safety concept can be "The Sensor message shall be protected from changes". The causes of 2 and 3 are related to SOTIF because they are not faults. Rather, they are insufficient functionalities that lead to a risk. The safety concept to mitigate those hazards can be written as "The computation power shall be enough for calculation. The false negative shall be less than x for y kilometers". The false negative means the system not reporting an existing object, which can be an indicator for the neural network detection performance. The last type of causes, on 4 and 5, are FuSa related because they are a failure that lead to risks. The general safety concept can be, "error detection on memory data shall be applied", and "error detection on communication bus data shall be applied".

Each of those high-level safety concepts should be further decomposed into requirements that are more technical according to the corresponding standards. Additionally, effects analysis shall be performed on other requirements in order to check if there is any inconsistency.

6   Summary

This work presents some systematic processes in an automated driving domain that help to enhance the completeness and minimize the changes during the development.

First, the requirements in the automotive domain and their levels is presented. In addition to how tractability, abstraction level and quality contribute to enhance the completeness in the lower requirement levels. Second, the completeness in higher requirement level is presented by utilizing a systematic method for stakeholders identification. It is based on the product life cycle, and it helps to identify the gaps in stakeholder definition and ensure that most of their needs are identified. In addition, the holistic thinking perspectives method is employed to demonstrate how the system can be explored from different perspective systemically and systematically to define the high-level requirements. Finally, the safety-related aspects are presented. Where the completeness in this regards, include the completeness of identifying safety related functions and the completeness of identifying the hazards.

This work focus on the concept phase, however, in development phase and beyond, ADS development embeds additional challenges, resulting from the uncertain operating environment and the infinite number of possible scenarios. Even in case of having a complete list of the high-level requirements, the detailed requirements cannot be guaranteed to be complete, because of unlimited relative scenarios. The problem here is not only how to derive the detailed of the system completely, as in normal development, but also how to find a suitable method to validate the proposed functionalities. For this reason, ADS projects are usually started by defining the high-level functions, in parallel, defining how to validate those functionalities, then, the details of the functionalities are finalized. Some researches try to define the possible states of possible actors in ADS context and then generate set of scenarios automatically to validate the performance of the system [43,44]. Again, systematic methods are needed to manage the huge amount of scenarios and the validation results, with the aim of enhancing the completeness.

References

1. Palomares, C., Franch, X., Quer, C., Chatzipetrou, P., López, L., Gorschek, T.: The state-of-practice in requirements elicitation: an extended interview study at 12 companies. Requirements Eng. 26(2), 273–299 (2021). https://doi.org/10.1007/s00766-020-00345-x

2. Chatzipetrou, P., Unterkalmsteiner, M., Gorschek, T.: Requirements' characteristics: how do they impact on project budget in a systems engineering context? In: 45th Euromicro Conference on Software Engineering and Advanced Applications (SEAA) (2019). https://doi.org/10.1109/SEAA.2019.00048
3. Kulk, G.P., Verhoef, C.: Quantifying requirements volatility effects. Sci. Comput. Program. (2008). https://doi.org/10.1016/j.scico.2008.04.003
4. Carson, R.S., et al.: Requirements completeness. In: INCOSE International Symposium, p. 14 (2004). https://doi.org/10.1002/j.2334-5837.2004.tb00546.x
5. SAE J3016 Taxonomy and Definitions for Terms Related to Driving Automation Systems for On-Road Motor Vehicles (2018)
6. Cohn, M.: User Stories Applied: For Agile Software Development. Addison-Wesley, Boston (2004)
7. Cleland-Huang, J., Vierhauser, M.: Discovering, analyzing, and managing safety stories in agile projects. In: 2018 IEEE 26th International Requirements Engineering Conference (RE) (2018) https://doi.org/10.1109/RE.2018.00034
8. Royce, W.: Managing the development of large software system. In: Proceedings of the 9th International Conference on Software Engineering (1987)
9. Heimicke, J., Spadinger, M., Li, X., Albers, A.: Potentials and challenges in the harmonization of approaches for agile product development and automotive SPICE (2020)
10. Hölldobler, K., Michael, J., Ringert, J.O., Rumpe, B., Wortmann, A.: Innovations in model-based software and systems engineering. J. Object Technol. (2019). https://doi.org/10.5381/jot.2019.18.1.r1
11. Wheatcraft, L., Ryan, M., Svensson, C.: Integrated data as the foundation of systems engineering. In: INCOSE International Symposium, vol. 27 (2017). https://doi.org/10.1002/j.2334-5837.2017.00438.x
12. Zowghi, D., Coulin, C.: Requirements elicitation: a survey of techniques, approaches, and tools. In: Aurum, A., Wohlin, C. (eds.) Engineering and Managing Software Requirements, pp. 19–46. Springer, Heidelberg (2005). https://doi.org/10.1007/3-540-28244-0_2
13. Pacheco, C., García, I., Reyes, M.: Requirements elicitation techniques: a systematic literature review based on the maturity of the techniques. IET Softw. **12**, 365–378 (2018). https://doi.org/10.1049/iet-sen.2017.0144
14. Génova, G., Fuentes, J.M., Llorens, J., Hurtado, O., Moreno, V.: A framework to measure and improve the quality of textual requirements. Requir. Eng. **18**, 25–41 (2013). https://doi.org/10.1007/s00766-011-0134-z
15. Alexandrovich, A., Igorevich, K.: INCOSE Guide for Writing Requirements. Translation experience, adaptation perspectives. In: CEUR Workshop Proceedings (2019)
16. ISO/IEC/IEEE 29148-2018 Systems and software engineering - Life cycle processes - Requirements engineering (2018)
17. Dick, J., Hull, E., Jackson, K.: Requirements Engineering. Springer, Cham (2017). https://doi.org/10.1007/978-3-319-61073-3
18. Sparx Systems: Requirement Models, technical report (2019)
19. Straßenverkehrs-Ordnung, StVO (german road traffic regulations). www.stvo.de/strassenverkehrsordnung. Accessed 04 May 2021
20. Glinz, M.: On non-functional requirements. In: 15th IEEE International Requirements Engineering Conference (RE 2007) (2007). https://doi.org/10.1109/RE.2007.45
21. ASPICE 3.1 GUIDE (2020). https://knuevenermackert.com/wp-content/uploads/2020/02/KM-ASPICE-Guide-3rd-022020-002.pdf

22. INCOSE - Requirements and architecture within modelling context. Presented at the (2016)
23. Mavin, A., Wilkinson, P., Harwood, A., Novak, M.: Easy approach to requirements syntax (EARS). In: 2009 17th IEEE International Requirements Engineering Conference (2009). https://doi.org/10.1109/RE.2009.9
24. Spork, G.: Efficient requirements management considering automotive standards: best practice sharing of mechatronic engineering within MAGNA groups (2011)
25. INCOSE: Guide to the Systems Engineering Body of Knowledge (SEBoK), version 2.2 (2019)
26. Sharp, H., Finkelstein, A., Galal, G.: Stakeholder identification in the requirements engineering process (1999). https://doi.org/10.1109/DEXA.1999.795198
27. ISO/IEC/IEEE 24748–1 Systems and software engineering - Life cycle management (2018). https://doi.org/10.1109/IEEESTD.2018.8526560
28. Nomura, N., Aoyama, M., Kikushima, Y.: A continuous stakeholder management method throughout the system life cycle and its evaluation. In: 2015 IEEE 39th Annual Computer Software and Applications Conference (2015). https://doi.org/10.1109/COMPSAC.2015.111
29. UN-Regulation: R79 v4 - Uniform provisions concerning the approval of vehicles with regard to steering equipment.pdf (2018)
30. NASA: NASA Systems Engineering Handbook (2007)
31. Lucassen, G., Dalpiaz, F., van der Werf, J.M.E.M., Brinkkemper, S.: Improving agile requirements: the quality user story framework and tool. Requirements Eng. **21**(3), 383–403 (2016). https://doi.org/10.1007/s00766-016-0250-x
32. Kasser, J.E.: Holistic thinking: creating innovative solutions to complex problems (2015)
33. Nuseibeh, B., Easterbrook, S.: Requirements engineering: a roadmap. In: Proceedings of the conference on The future of Software engineering (2000). https://doi.org/10.1145/336512.336523
34. NHSTA: A Framework for Automated Driving System Testable Cases and Scenarios. National Highway Traffic Safety Administration (2018)
35. ISO2626-3-2018: ISO2626-3 Road vehicles-Functional safety-Part 3: Concept phase (2018)
36. ISO-PAS_21448-2019 SOTIF Road vehicles - Safety of the intended functionality (2018)
37. SAE-J3061-2016: SAE J3061 Cybersecurity Guidebook for Cyber-Physical Vehicle Systems (2016)
38. ISO2626-4-2018: ISO2626-4 Road vehicles-Functional safety-Part 4: Product development at the system level (2018)
39. BS: IEC61882 HAZOP guide (2001)
40. Robert Bosch GmbH: FTA fault-tree-analysis (2015)
41. Ford: FMEA Handbook (2011)
42. Nancy Leveson, JOHN THOMAS: STPA handbook (2018)
43. Pegasus Method. https://www.pegasusprojekt.de/en/pegasus-method. Accessed 13 July 2020
44. Bock, F., Sippl, C., Heinzz, A., Lauerz, C., German, R.: Advantageous usage of textual domain-specific languages for scenario-driven development of automated driving functions. In: 2019 IEEE International Systems Conference (SysCon) (2019). https://doi.org/10.1109/SYSCON.2019.8836912

Operationalizing a Medical Intelligence Platform for Humanitarian Security in Protracted Crises

Walter David[1]([✉]), Michelle King-Okoye[1,2], Gianluca Sensidoni[3],
Alessandro Capone[3], Irene Mugambwa[4], Stanislava Kraynova[5],
and Beatriz Garmendia-Doval[6]

[1] Ronin Institute, Montclair, NJ 07043, USA
{walter.david,michelle.king-okoye}@roninstitute.org
[2] University of Edinburgh, Edinburgh EH8 9YL, UK
Michelle.King-Okoye@ed.ac.uk
[3] Expert.ai, 41123 Modena, Italy
{gsensidoni,acapone}@expert.ai
[4] International Organization for Migration, 1211 Geneva, Switzerland
imugambwa@iom.int
[5] Crisis Management and Disaster Response Centre of Excellence, 1606 Sofia, Bulgaria
stanislava.kraynova@cmdrcoe.org
[6] Masa Group SA, Group, 75002 Paris, France
beatriz.garmendia-doval@masagroup.net

Abstract. Most of present humanitarian crises are protracted in nature and their average duration has increased. Climate change, environmental degradation, armed conflicts, terrorism, and migration are producing exponentially growing needs to whom humanitarian organizations are struggling to respond. Novel infectious diseases such as COVID-19 add complexity to protracted crises. Planning to respond to current and future medical threats should integrate terrorist risk assessment, to safeguard population and reduce risks to aid workers. Technologies such as Artificial Intelligence (AI), and Modelling and Simulation (M&S) can play a crucial role. The present research has included the conduct of the United Nations HNPW 2021 session on AI and Medical Intelligence and an exercise on a real scenario. Focusing on medical and terror threats in North East Nigeria operating environment, authors have successfully deployed and tested the Expert.ai Medical Intelligence Platform (MIP) jointly with the MASA SYNERGY constructive simulation, with the aim to improve situational awareness to support decision-making in the context of a humanitarian operation.

Keywords: Protracted crises · Humanitarian security · Medical intelligence · Artificial intelligence · Machine learning · Modelling & simulation · Natural language understanding · Cognitive algorithms · Digital disease detection · Deep semantic analysis · Social media · Situational awareness · Decision-making

© Springer Nature Switzerland AG 2022
J. Mazal et al. (Eds.): MESAS 2021, LNCS 13207, pp. 397–416, 2022.
https://doi.org/10.1007/978-3-030-98260-7_25

1 Introduction

The average length of a humanitarian crisis involving an UN-coordinated response has increased from 5.2 years in 2014 to more than 9 years, due to the impact of climate [1, 2]. Humanitarian organizations are already struggling to meet exponentially growing needs.

Over the next two decades we can expect increasing desertification, flooding, droughts, large scale cyclones, heat waves, mega fires, extreme weather. The disruption of agricultural livelihoods and food systems, and its cascading economic consequences [3] put dramatic pressure to populations [4–6].

Climate change, with already reported effects on crises' location, frequency, scale, and complexity [3], links with migration and presents an unprecedented challenge [7]. In 2020, disasters caused the displacement of about 30.7 million people of all 40.5 displacements recorded [4]. Indirectly, climate increases the risk of conflict by exacerbating already existing social and economic factors.

Countries enduring armed conflict lack the ability to adapt; in fact, of the 20 most climate vulnerable countries, 12 are mired in conflict [8, 9]. Populations, in particular minorities, living in conflict zones are among the most vulnerable.

Infectious diseases such as the COVID-19 challenge humanitarian actors and it is unlikely to be the last health emergency globally.

1.1 Research Approach

Aim and Objectives

The COVID-19 pandemic teaches us to break the panic-then-forget cycle, respond to present threats and plan to address future threats. Integrating risk assessment in humanitarian operations is crucial to safeguard people and minimize risks to aid workers.

This research aims to present a novel approach to humanitarian security, by deploying AI and modelling and simulation (M&S) in order to improve the situational awareness to support decision-making. The objectives are to find a consensus among humanitarian organizations on the deployment of AI and test a medical intelligence system on a real operating environment.

Methodology

Authors combine different methods, in order to produce a richer and more comprehensive understanding.

The first part of the research has included qualitative methods such as literature review, interviews, and brainstorming conducted to generate creative and diverse ideas.

A workshop has been organized at the United Nations HNPW 2021 Conference with experts from World Health Organization (WHO), International Organization for Migration (IOM), UN Office for the Coordination of Humanitarian Affairs (OCHA), the Crisis Management and Disaster Response Centre of Excellence (CMDR COE), academia, medical intelligence and simulation industry. A Strengths, Weakness, Opportunities, and Threats (SWOT) analysis seminar which is an effective strategic planning analytical tool has been conducted.

Following preliminary observations and the seminar's outcomes, an exercise has been conducted, and the Expert.ai Medical Intelligence Platform (MIP) has been deployed in Nigeria.

Based on the availability of Natural Language Understanding (NLU) software, capable to understand data in the form of sentences using text or speech, authors have hypothesized that AI supported intelligence can improve situational awareness to support better decision-making in humanitarian operations.

Finally, authors have used MASA SYNERGY constructive simulation system to build a mathematical model containing key parameters of the real physical model.

1.2 Use Case - Northeast Nigeria Operational Scenario

Field updates and humanitarian security requirements have been provided by IOM, including variables such as disease and security. The basin of Lake Chad, whose waters have shrunk from 26,000 sq. km to 1,500 sq. km in just 50 years, represents an outstanding example of the present environment degradation trend [10].

The operational context is characterized by huge humanitarian needs and decreasing international aid funding. A complex crisis, driven by climate, poverty, disruption of food production and distribution, and the 11 years long extensive insurgency fought by non-state armed groups (NSAG) such as Boko Haram and the Islamic State in West Africa Province (ISWAP) in Nigeria, Niger, Chad and Cameroun, is affecting some 17 million people [4, 11].

More than 37,500 people have been killed from 2011 to 2020 [12]. 1.7 million Internally Displaced Persons (IDP), mainly women and children, live in fragile shelters [4] of 284 overcrowded camps, informal settings and more than 1,000 locations [2], often lacking basics such as food, clean drinking water, shelter and sanitation [13–16].

The security situation is unpredictable, with 896 incidents (kidnappings, attacks, roadblocks, and bombings) recorded in the region between January and April 2019. Since December 2020, 8 aid workers have been abducted and a significant number have been killed, prompting organizations to reduce their outreach activities [14].

Access is much limited by violence and there are limitations to communication and electricity networks, security related restrictions, and roadblocks. The rainy season further complicates movements, with affected people cut off, due to flooding and waterlogged roads [15, 16]. In addition to the already registered increase of diseases (malaria, cholera, and hepatitis E [17]), the onset of the COVID-19 has exacerbated the vulnerability of IDPs. A SWOT analysis seminar has provided the opportunity to analyze (Fig. 1) and identify solutions to improve the approach to humanitarian security.

Sharing of experiences and new ideas about medical/healthcare intelligence have helped in identifying synergies between the solutions of Expert.ai and those presented by the WHO Public Health Intelligence team.

Among the most promising tools to increase global health security and improve situational awareness are those utilising AI methods and components, based on their inherent forecasting, early anomaly detection, alerting, and advanced analytics capabilities to provide actionable intelligence to reach the Information Superiority.

The key to maximising the performance of analysis tools is data and there is no shortage of it. The key to maximising the use of available data is to analyse it in the

Fig. 1. Outcomes from the SWOT analysis seminar.

right spatiotemporal context and a key to having the data available is a commitment by governments, public and private sector, for open sharing of information and knowledge.

2   AI Supported Medical Intelligence

Big data analysis facilitates disease surveillance [18] as key indicators can be generated for effective relevant data clustering, going to discover weak signals, providing information for further reasoning.

Event-based Internet bio surveillance systems search, and gather contents from sources in the surface, deep and dark web. Data harvested from social media, news aggregates, tweets, websites pages, call transcripts, survey responses, slangs, and dialects can be analyzed, for instance, for:

- monitoring behavioural/social risks by identifying emotions (social impact),
- creating graphical views useful to improve the situational awareness,
- identifying trends and patterns related to current diseases.

Such systems discover insights referred to specific problems on a part of the population during a dedicated temporal reference in a location [18], identify and extract details to assess risk, to apply prediction to estimate the likelihood of a future outcome based on trends and patterns.

2.1   Expert.ai Medical Intelligence Platform (MIP)

Embedding ethical principles is a critical requirement to safeguard against bias in AI systems [19] but *explainability* is still an issue for AI systems based on pure Machine

Learning (ML)/Deep Learning (DL) algorithms and neural networks whose training processes can result in parameter values difficult to correlate with results [20].

Such issue has been mitigated by a hybrid approach in the Expert.ai Medical Intelligence Platform (MIP) (Fig. 2) eXplainable and trustworthy AI (XAI) technology. MIP's engine is the COGITO deep semantic Natural Language Processing (NLP) that is based on a rich, "knowledge graph" representation of the real world where concepts are defined and connected to one other by semantic relationships.

MIP utilises ontology-powered methods based on the International Press Telecommunications Council (IPTC) Taxonomy, the Medical Subject Heading (MESH) and the SNOMED CT clinical vocabulary, readable by computers, well-known healthcare knowledge base.

Exploiting both rule-based and ML algorithms, raw unstructured data [21] is loaded into a warehouse, aligned, and combined with existing structured data [22] for extracting information and developing predictive and reasoning models that can be run against all the relevant features [23]. Third party software, such as automatic translation and SpeechToText engines, allows the transcription from audio/video contents.

MIP users are free to choose the AI technique, on a case-by-case base, by creating semantic rules or by applying ML/DL algorithms. Such flexibility contributes to ensure responsibility and accountability for AI outcomes.

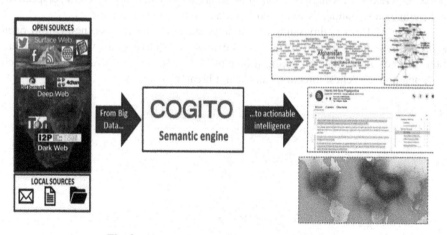

Fig. 2. The Medical Intelligence Platform (MIP).

AI supported MEDINT supports strategic decision-making by improving the ability to:

- *Understand:* semantically indexing content, categorizing it based on relevant domain taxonomies, identifying events and entities.
- *Augment*: enabling end-users to process more information, faster, anticipating threats and acquiring knowledge and insight from unstructured data.
- *Automate*: supporting all phases of the intelligence cycle, from source acquisition to deep analysis and dissemination.

- *Monitor in real-time the target:* receiving proactive alerts triggered by relevant news from validated sources; specific dashboards provide views of target information.

2.2 Use Case. Mapping COVID-19 and Borno-Related News

A contingency plan helps an organisation to avoid the risk of losses, reduces the risk of uncertainty, ensure its ability to continue operating after a disaster. Scenario planning involves the development of specific scenarios, which are then used as a basis for developing a response plan.

A MIP scenario has been designed and an use case has been developed, focusing on medical and terror threats in the State of Borno, in North East Nigeria, in particular concerning the IDPs population, and deployed health and aid personnel.

Data available in social media platforms can be collected by use of log-data [24], web scraping or application programming interfaces (API). Authors have made use of the Twitter API that provides researchers with access to non-public internet environments (requiring authentication through login and password) as the collection of data runs through the back-end of the social media service [25].

The MIP GeoMap module integrates Geographic Information Systems (GIS) features (Gmap, Esri, etc.) for advanced geospatial analysis. Using graphical widgets, it has been possible to refine the search by submitting smart queries by using different perspectives to leverage on semantically extracted metadata on the area of interest.

During MIP tests, thousands of online documents (tweets, rss, news from open sources) have been filtered to retrieve the most relevant ones in which users mention "Borno" AND the "breathing problems" concept, including related words such as "pneumonia", "chest infection" and "can't breathe" (Fig. 3).

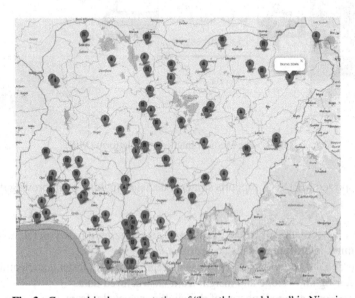

Fig. 3. Geographical representation of "breathing problems" in Nigeria.

Results have been analyzed by using domain-specific taxonomies that contain, arranged in a hierarchically mode, a set of topics related to a specific domain. It has been possible to identify the number of documents related to a specific topic and further refine the search query adding a new constraint.

2.3 Health, Sentiment and Emotions Picture

Information truthfulness and reliability are critical to inform medical threats and safe-guard general public's security [26]. A challenge, in applying a reasoning on the social and psychological impact of a pandemic, is understanding emotions but also the writing style about specific verbal groups, jargons, dialects, morphology on expressing opinions, what is being done in stylometric analysis.

In the deep understanding of the information dynamics that lead decision-makers to trust or not on a reasoning and to consequently adopt it on specific missions, it is crucial to have real news and to be aware about fake ones [27–29]. This has been achieved by measuring sites for consistency, spread, meaning underlying language, logic versus emotional contexts of data, comparing credible media sources and understanding social bots versus human web interactions.

During tests on Borno, it has been possible to reach super-specialized documents dealing with emotions related to COVID-19 symptoms and monitor, in *real-time*, public opinion, sentiment and more than 70 different citizens' emotions (e.g.: fear, anxiety, stress, hope, etc.) and their impact (Fig. 4).

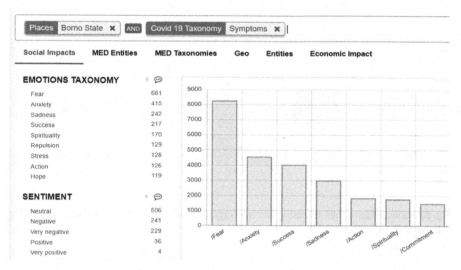

Fig. 4. Social and emotional impact related to COVID-19 symptoms in Borno.

More different situations have been highlighted such as, an overview of the public order issues connected to the pandemic or to terrorist attacks in a selected geographic area.

Finally, it has been possible to create a dynamic graphical widget and generate statistics to visualize a multi-level overview of the infectious diseases and of the terrorist organizations active in the region (Fig. 5).

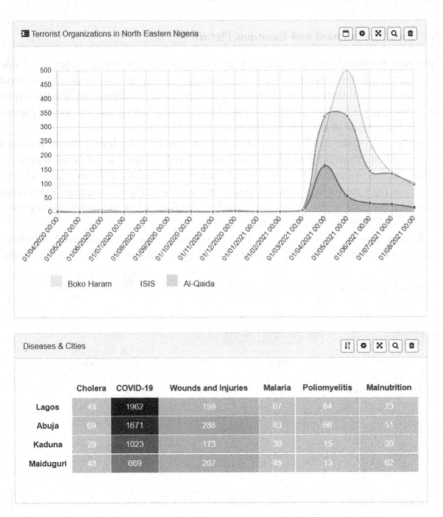

Fig. 5. Public order events related to terrorist organizations, and number of diseases linked to cities in MIP.

3   SYNERGY Simulation for Humanitarian Operations

In this section, authors propose a novel approach, exploiting MIP intelligence in the SYNERGY constructive simulation software suite developed by MASA to simulate human behaviors in the face of disasters.

SYNERGY contains the same advanced AI technology of the SWORD, able to simulate the behaviour of thousands of artificial intelligent agents in a realistic way and used for staff training and scenario analysis to support decision-making. SWORD has already been used for the elaboration of epidemiological results computed by the Spatiotemporal Epidemiological Modeler (STEM) [30].

SYNERGY models the behaviour of the population both in normal and emergency situations. It simulates human and material damages, resulting from natural or human-made disasters. By allowing to generate complete and realistic crisis scenarios, the system supports decision-makers, such as law enforcement, critical infrastructures managers or local authorities, both for training purposes and for testing contingency plans, and find areas of improvement.

MASA systems' open interfaces allow the interoperation with third party simulations or data providing systems. For example, SYNERGY was used in the city of Blumenau (Brazil) for training and testing their contingency plan when facing a crisis scenario of city floods after heavy rains.

Blumenau suffers regularly (80 times in the last 140 years) episodes of heavy rains leading to the two main rivers flowing through the town to overflow and cause major crisis. Data available on Blumenau's city records was used to create a realistic scenario: cartographic and topographic data provided by Blumenau's Defesa Civil were easily imported into SYNERGY to recreate the maps printed and made available to the crisis cell; the data used to simulate the flooding was taken from the city records of previous real flooding episodes; the steps on the contingency plan were made available to the user inside the system so the plan could be validated over a real scenario at the same time that it was being simulated.

SYNERGY simulated not only the flooding but also the movement of the population affected by the flood: the opening of the shelters defined in the contingency plan, the movement of the population either leaving the city or, when that was no longer an option, moving towards their designated shelter, the blockage of different roads, not just by the water but by landslides caused by the heavy rains, etc.

This was just an example of the type of data that can be imported into SYNERGY to create more realistic scenarios. If data is available representing other types of disasters like forest fires [31], chemical spillage, spread of a disease, the data can be introduced into SYNERGY to better simulate real conditions. Of course, when data is not available, plausible data can be created manually by the user.

Although SYNERGY is used mostly for training and analysis, it has been used to support operations, to simulate several scenarios related with COVID-19 to analyse the way the coronavirus expands through a population and to help planning the delivery of the vaccines, the logistics involved, the number of vehicles needed, the different transport options through difficult access terrain and other features.

Using external tools as data providers, SYNERGY has been able to import data from flooding, fires, epidemic spread, chemical spillage, terrorist attacks, and other man-made or natural disasters.

4   Linking MIP Intelligence with SYNERGY Simulation

Considering humanitarian operating theatres, the closer to the centre of the crisis, the more information is needed, in more detail and in *real-time*; the faster the flow of data and the more tolerance for error and delay [32].

In this research, authors have investigated how to exploit data discovered and analysed by a strategic MEDINT tool in order to improve the situational awareness related to the challenges in the specific North East Nigeria theatre, with a particular focus on:

- medical threats (current and future diseases),
- terroristic threats against humanitarian workers and population, including IDPs.

The scope of the article is not to explore all possible uses cases of AI supported MEDINT but to understand if its deployment can help in supporting planning and the conduct of a humanitarian operation, in particular, when it is coupled with a constructive simulation tool such as SYNERGY.

A use case exercise has been built with SYNERGY, by modelling convoys to perform medical supply in restricted access area.

During operation planning, MIP is used for what the military call "Intelligence Preparation of Battlefield (IPB), an analytic framework for collecting, organizing and processing information to provide timely, accurate, and relevant intelligence during the military decision-making process. Medical Intelligence is applied to the theatre of operations [33].

When a new scenario is created in SYNERGY, the terrain, weather, and other environmental conditions can be defined. This is a first step where the information discovered through MIP can be used to create a more realistic scenario.

In this exercise, SYNERGY has been used initially for analyzing locations and size of IDP camps, medical and humanitarian facilities, as from IOM requirements. MIP was used to upgrade this initial information with as much detail as possible.

The analysis normally performed with M&S systems is based on a static scenario. In the exercise, in the initial scenario, the data provided by UN agencies and government, related to IDP camps have been included. Such official information has been supplemented through a search by MIP, focused to detect the presence of informal camps. Discovered informal settlements have been introduced into SYNERG, including information such as their name, location, maximum capacity, and current occupancy.

This has made possible to analyse the logistics requirements to support IDPs camps: food, water, shelter, and again, the relevant security issues. When the scenario is simulated, this information is transformed into logistic actions dealing with supplies and medical treatments.

In general, the static information can be enriched through the information discovered by MIP, including actions happening at the same time of the scenario preparation. For example, a sudden influx of people in a camp due to the flooding of a nearby village; or a COVID-19 cluster happening in one of the camps.

By exploiting this type of information, the initial scenario can be modified to match as close as possible the current circumstances in the defined area. With the given examples, we can modify the amount of people assigned to a camp and modify the initial status of the people in the camp to represent the cluster.

A scenario for Borno has been created in SYNERGY (Fig. 6). A humanitarian logistic unit is deployed to deliver vaccines and essential medicines to selected IDP camps. The needs of the camps have been determined so that it has been identified the requirement for a distribution hub to be established at the Regional Hospital. Five remote IDP camps have been selected to receive 40,000 doses of vaccines each, to be transported from the distribution hub.

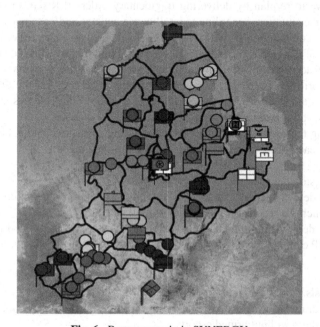

Fig. 6. Borno scenario in SYNERGY.

However, in operations planning, plans need to be modified as new information is received about changes in the situation on the ground.

Here, it has been hypothesized that combining SYNERGY and MIP helps to modify the scenario and to support the development of workable solutions, evaluate, compare, and select the best course of action (COA). This hypothesis has been tested through the simulation and analysis of a scenario and the COAs defined for it.

Like in the military decision-making process, at least two "Enemy COAs" have been considered, concerning the Non-State Armed Group (NSAG) who control the territory or large areas of it:

• the "most dangerous COA": the NSAG controlling the area decides to stop every effort from the international community to conduct a vaccine campaign by killing all health and humanitarian operators and destroying their assets;

- the "most probable COA": the NSAG will disrupt the activities of humanitarian organizations by seizing convoys cargo, and eventually kidnapping workers, but will not attack all convoys in order not to prompt a quick reaction of military forces.

The plan has been prepared to confront the most probable COA but taking in consideration that a contingency plan (not the focus of the present research) is needed in case organized armed groups adopt the "most dangerous COA".

However, even after the adoption of a COA, and conducting the current operation, we often have to re-plan by delivering fragmentary orders (FRAGO) to adapt to the evolution of the situation, according to intelligence updates and/or to react to incidents.

Here, the combination of MIP and SYNERGY plays a major role, allowing the modification of the scenario as soon as news concerning that area is received. Planning can take in to account the latest information as soon as they are available.

The scenario has been tested assuming different itineraries for the distribution of vaccines, injected incidents such as small arms attacks to the convoys and how they affect the distribution of vaccine.

According to the general security situation in the area, three own COA have been selected. For each of them, the following assumptions are made:

- COA 1: to stop the vaccination campaign;
- COA 2: to deliver vaccine using trucks; this course of action is the reference scenario against which the results of a second COA will be compared;
- COA 3: to deploy a mix of trucks and cargo drones to reach the most difficult access remote IDP camps.

COA Analysis

The logistic units ensuring the distribution of resources to IDP communities are made up of aid workers without armed escorts.

For both COA 2 and 3, ten executions of the simulation were performed. In the first COA the vaccines are sent by regular convoys, and we assume there are no armed attacks. This is used as the baseline. Even there, sometimes not all vaccines arrive in the give time due to random breakdowns on the way.

The simulation use of random number generation introduces variations in the outcome of each execution. The average values of monitored variables have been computed for each COA. Unlike COA 2 (where no action is taken), in COA 3 it is decided to use drones to deliver the vaccines in the conflictive areas.

Authors believe that information feed from the AI in COA 2 could lead to a recommendation to use COA 3 especially if the situation is critical for any medical threat, yet the terroristic threat combined with impassable roads due to flooding.

Once a COA has been selected, it is possible to connect to MIP and add the information received (in an automatic way in the future). The analysis is reiterated by applying constructive simulation on the modified scenario and develop new COAs, compare them and select the best.

The evolution of the situation on the ground is reflected in changes to the simulation scenario, either by taking into account the MIP updated data, or even (in case no action-able intelligence is provided) by using SYNERGY to calculate the threat evolution, taking into account environment variables, terrain and other factors.

In fact, in the rain season, many roads are impassable or difficult to use. For example, the "Arabic Camp" is on one of those difficult roads. As for "Rann Camp", it is usually impossible to use the road in rain season due to the overflow of the river connecting Rann Camp and Arabic Camp. Therefore, in real life, sometimes cargo is taken through Cameroun.

During tests, when MIP has delivered a terror alert, it has been simulated as a NSAG entity. MIP intelligence data, discovered and analysed in *real-time*, related to terror threats have been imported into SYNERGY to create disaster objects that simulate terror-related events/incidents. Figure 7 shows the Borno scenario in SYNERGY.

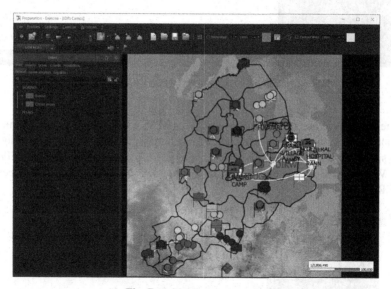

Fig. 7. Initial Borno scenario.

Once the scenario has been built, SYNERGY has made it possible to analyse several itineraries as means to deliver medical resources from the Regional Hospital vaccine hub to the IDP camps (Fig. 8).

When using the simulation for this type of use case, it becomes important to receive and process any news affecting that area, potentially impacting the analysis being done, i.e., a terrorist attack, a blocked road, a landslide, etc., so that the analysis can be reiterated taking this new information into account.

For the exercise, different searches have been created in MIP, so that new information appearing on the news and social media, related to the state of Borno, and to critical situations (e.g. guerrilla activities and terrorist armed attacks) have been received and transformed into alerts.

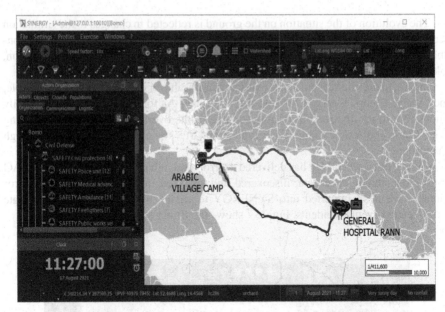

Fig. 8. Investigating different itineraries to reach a camp.

The alerts have been sent automatically to a MASA e-mail address, containing the "semantically highlighted" of each content matching with the Boolean query associated to the specific alert. Subsequently, it has been possible to modify the SYNERGY scenario. When a MIP alert is delivered, a terrorist attack in a close-by area is reflected in SYNERGY (Fig. 9):

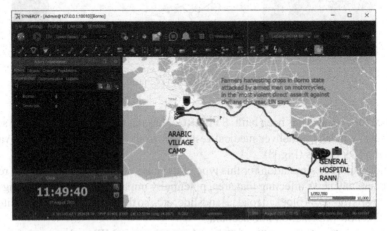

Fig. 9. New alert is received in SYNERGY from MIP intelligence update about an attack.

MIP alerts contain the coordinates of the armed attack incident; they are used to mark the point over the map.

When the terrain is deemed too dangerous for land vehicles, it is decided to use DJI Matrice 300 industrial drones repurposed as cargo drones as the delivery option to cover the last 25 km [34] (Fig. 10). The use of cargo drones is generally discouraged as the organized armed groups are also using them for attacks on camps, unless they have been cleared by the military authority and humanitarian community.

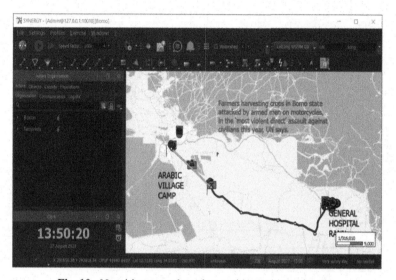

Fig. 10. New itinerary selected to avoid a problematic area.

The following graphs show the result of several simulations of the COAs described above, while planning the delivery of vaccines to the IDPs. Each camp has been assigned 40,000 vaccines. The initial idea is to use trucks for the delivery. The first graph shows the result of 10 runs of the simulation where the vaccine delivery has been simulated assuming no problems in the roads, just checking for the best roads to make the delivery. Even assuming no problems in the roads, in one of the runs, two camps did not receive the vaccines due to technical problems.

In the following graphs, the number of vaccines reaching the target IDPs camps is plotted (Fig. 11).

In the second graph a terrorist ambush has been simulated in one of the roads used be the trucks delivering the vaccines. On average, two of the IDP camps do not receive the vaccines as two convoys are destroyed on the way (Fig. 12). The attack was simulated in the area where the MIP alarm was received as shown in Fig. 9.

The next graph (Fig. 13) shows the outcomes from the use of cargo drones to deliver the vaccines to the two IDP camps for which trucks have to traverse that dangerous terrain. As indicated before, drones need clearance by the military and international organizations. Because of this, they take longer to be ready but are safer to use.

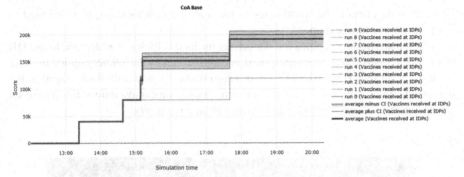

Fig. 11. No attacks. Number of vaccines reaching target IDP camps.

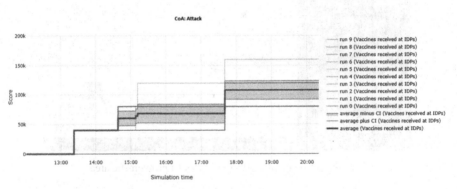

Fig. 12. Terrorist ambush. Number of vaccines reaching IDPs camps.

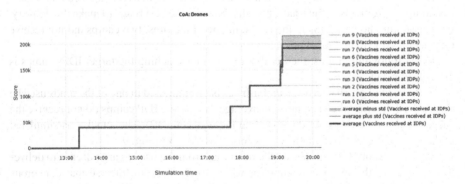

Fig. 13. Use of cargo drones for vaccines delivery to IDPs camps.

Discussion

MIP, supported by the COGITO advanced semantic engine, have proved the ability to acquire data from different sources (libraries, news, tweets, etc.) incorporating multiple terms and phrases associated with selected concepts, providing actionable insight, and identifying risks and opportunities.

In the case of medical threats, it has rapidly visualized results associated to "breathing problems", and produced interactive graphs, tabs, and facet.

In this study, some assumptions were made, restricting the search to documents (tweets, rss, news, etc.) available from open sources, in English language. MIP has been tested in Nigeria for a period of two months and evaluated by IOM subject matter experts (SME). The feedback from hands on practice highlighted:

- the flexibility of the platform, easy to use, manipulate and explore;
- the *intelligence* of predicting emotions based on algorithms; SMEs could view/access multiple information on similar topics which they strongly believe had already been filtered to avoid fake news; and
- the overall quality of the pulled information.

IOM field Staff requirements on AI medical intelligence include tailored demonstrations of AI tools capabilities for deeper insight on the prospects of operationalizing AI tools for humanitarian security risk management, even looking at camp-based solutions (at sub-tactical level) where they can easily search details of incidents at camp level.

In the exercise, MIP intelligence has proved to enhance the analytic power of SYNERGY simulation. Events captured in *real/near-real time* have been used to update the simulation scenario; finally, the result of changes has been observed.

It has been found that the same could not be done by a time-consuming web search; it is the deep linguistic analysis performed by MIP that allows the feeding of real news as they happen into SYNERGY, facilitating the development of updated COAs.

5 Conclusions and Way Ahead

The present research has addressed, from a novel perspective, the challenge of security issues in protracted humanitarian crises that often are complicated by medical threats. Thinking out of the box and investing in innovative technologies such as AI, ML and M&S are crucial for integrating a risk assessment management in the context of humanitarian security concerning operations.

In crises, it is crucial to have an avant-garde information-seeking and-smart-analysis capability revealing where data is transformed into information. Faster detection of prodromal signals of potential threats, such as disease outbreak or terroristic attacks, is critical to surveillance and control.

Intelligence platforms are always linked to strategic contexts and activities. However, they can be used at the operational and tactical level, where customers are provided with just fewer required features, focused on specific and super dedicated sources.

A Medical Intelligence Platform customized to be deployed in humanitarian operations, could generate key indicators for effective relevant data clustering going to discover even weak signals. Such capability can help to forecast, predict, anticipate action, finally improving the situational awareness available to decision-makers.

Further research is critical to support IOM requirements on AI medical intelligence including tailored demonstrations of AI tools capabilities for deeper insight on the prospects of operationalizing AI tools for humanitarian security risk management. This should be focused on the creation of new opportunities for collaboration.

AI supported medical intelligence can be coupled with simulation analysis to improve situational awareness to support decision-making in an operation.

Modelling & Simulation (M&S) and text analytics technologies can be combined:

- to exploit the NLU output as an added value to M&S, going to consider an optimal life cycle between the two technologies;
- to enable easier reasoning and deduction by inserting in the simulation workflow multiple language structured/unstructured contents from a huge number of sensors;
- to enhance the intelligence processes so to reach the *Information Superiority* crucial for decision-makers to select the optimal COA.

Authors have also started to investigate the possibility to make intelligence information available directly inside SYNERGY. MIP can analyse a document and extract key elements and geographic locations from the text. Such information could be delivered as JSON files to be processed automatically and to be visualized in SYNERGY.

An evolution of MIP could provide new ways to automatize the transfer of actionable intelligence to SYNERGY simulation. For instance, after receiving information about a blocked road due to a landslide, instead of just a marker over the terrain, an obstacle can be built over that road, representing the blockage. Once it is represented as an obstacle, SYNERGY will take this information into account when calculating routes and will avoid using that road.

As we improve such interaction, we can obtain the automatic representation of the news feed over the terrain as it happens. That way, in an exercise supported by SYNERGY, the supervisor can receive the MIP intelligence directly in the application, and decide to modify the current scenario to take into account the evolution of the situation on the ground.

References

1. United Nations OCHA: World Humanitarian Data and Trends (2018). https://www.un-ili brary.org/content/periodicals/24118419. Accessed 23 May 2021
2. United Nations OCHA: Global Humanitarian Overview (2021). https://reliefweb.int/sites/rel iefweb.int/files/resources/GHO2021_EN.pdf. Accessed 23 May 2021
3. Food and Agriculture Organization of the United Nations – FAO: Agriculture on the proving grounds. Damage and loss (2021). http://www.fao.org/home/digital-reports/disasters-in-agr iculture/en/?utm_source=twitter&utm_medium=social+media&utm_campai. Accessed 23 May 2021
4. IDMC Coronavirus crisis: internal displacement. https://www.internal-displacement.org/cri ses/coronavirus. Accessed 10 June 2021
5. ND-GAIN Index Country Rankings. https://gain.nd.edu/our-work/country-index/rankings/. Accessed 10 June 2021

6. International Committee of Red Cross – ICRC: When rain turns to dust: understanding and responding to the combined impact of armed conflicts and the climate and environment crisis on people's lives (2020). https://reliefweb.int/report/world/when-rain-turns-dust-und erstanding-and-responding-combined-impact-armed-conflicts. Accessed 23 July 2021
7. International Organization for Migration - IOM: Migration, Environment and Climate Change: assessing the evidence, Geneva (2009). https://publications.iom.int/system/files/pdf/migration_and_environment.pdf. Accessed 15 July 2021
8. Abel, G., Brottrager, M., Crespo Cuaresma, J., Muttarak, R.: Climate, conflict and forced migration. Glob. Environ. Chang. **54**, 239–249 (2019)
9. IPPC: Global Warming of 1.5 °C Special Report Ipcc.ch (2021). https://www.ipcc.ch/sr15/. Accessed 12 June 2021
10. Salkida, A.: Africa's vanishing Lake Chad (2021). https://www.un.org/africarenewal/mag azine/april-2012/africa's-vanishing-lake-chad. Accessed 12 June 2021
11. Institute for Security Studies: "Islamic States Determined Expansion into Lake Chad Basin" (2020). https://issafrica.org/iss-today/islamic-states-determined-expansion-into-lake-chad-basin. Accessed 23 Sept 2021
12. International Organization for Migration – IOM: World Migration Report 2020: Chapter 9 Human Mobility and Adaptation to Environmental Change. https://publications.iom.int/books/world-migration-report-2020-chapter-9. Accessed 15 July 2021
13. NATO: NATO Strategic Foresight Analysis Report, Regional Perspectives Report on North Africa and the Sahel (2018). https://www.act.nato.int/futures-work. Accessed 12 June 2021
14. United Nations OCHA: Nigeria Humanitarian Response Plan (2021). www.humanitar ianresponse.info/en/document/nigeria-2021-humanitarian-response-plan-summary, https://www.humanitarianresponse.info/en/operations/nigeria/document/nigeria-2021-humanitar ian-needs-overview. Accessed 12 June 2021
15. Armed Conflict Location and Event Data Project – ACLED. https://acleddata.com/#/das hboard. Accessed 21 May 2021
16. International Organization for Migration – IOM: DTM Round 32/33 Reports Nigeria, DTM human mobility in the context of environmental and climate change. Assessing current and recommended practices for analysis within DTM (2020). https://environmentalmigration. iom.int/sites/environmentalmigration/files/Human%20Mobility%20in%20the%20context% 20of%20Environmental%20and%20Climate%20Change%20DTM-MECC%20%28002% 29.pdf. Accessed 12 July 2021
17. Medicins Sans Frontiers – MSF: Lake Chad Crisis (2021). https://www.msf.org/lake-chad-crisis-depth. Accessed 09 Sept 2021
18. United Nations OCHA: Nigeria Situation Report (2021). https://reports.unocha.org/en/cou ntry/nigeria/. Accessed 22 May 2021
19. Liang, H., Zhu, J.J.H.: Big data, collection of (social media, harvesting). In: The International Encyclopedia of Communication Research Methods. Wiley, Hoboken (2017)
20. European Commission, Directorate-General for Communications Networks, Content and Technology, Ethics guidelines for trustworthy AI, Publications Office (2019). https://data. europa.eu/doi/10.2759/177365. Accessed 14 June 2021
21. Zhang, D., Yin, C., Zeng, J., et al.: Combining structured and unstructured data for predictive models: a deep learning approach. BMC Med. Inform. Decis. Making **20**, 280 (2020)
22. Kandikonda, S.: Artificial Intelligence and Data Science in Healthcare (2020). https://www. thedatasteps.com/post/data-science-in-healthcare. Accessed 08 June 2021
23. Expert Systems, Expert System Achieves a World First by Enhancing Different AI Capabilities Accelerating Human-Like Language Comprehension for Text Analytics (2021). https:// www.prnewswire.com/news-releases/expert-system-achieves-a-world-first-by-enhancing-different-ai-capabilities-accelerating-human-like-language-comprehension-for-text-analyt ics-300647843.html. Accessed 05 Jan 2021

24. David, W., King-Okoye, M., Capone, A., Sensidoni, G., Piovan, S.E.: Harvesting social media with artificial intelligence for medical threats mapping and analytics. In: Proceedings of International Cartographic Association, vol. 4, p. 24 (2021)
25. Lomborg, S., Bechmann, A.: Using APIs for data collection on social media. Inf. Soc. **30**(4), 256–265 (2014)
26. Twitter (2020). https://s22.q4cdn.com/826641620/files/doc_financials/2019/q1/Q1-2019-Slide-Presentation.pdf. Accessed 14 Jan 2021
27. The Royal Society: Explainable AI: the basics policy briefing (2019). https://royalsoci ety.org/-/media/policy/projects/explainable-ai/AI-and-interpretability-policy-briefing.pdf. Accessed 15 June 2021
28. Ahmad, I., Yousaf, M., Yousaf, S., Ahmad, M.O.: Fake news detection using machine learning ensemble methods. Complex. **2020**, 11 p. (2020). Article ID 8885861. https://doi.org/10.1155/2020/8885861
29. Aldwairi, M., Alwahedi, A.: Detecting fake news in social media networks. Procedia Comput. Sci. **141**, 215–222 (2018)
30. David, W., Baldassi, F., Piovan, S.E., Hubervic, A., Corre, E.: Combining epidemiological and constructive simulations for robotics and autonomous systems supporting logistic supply in infectious diseases affected areas. In: Mazal, J., Fagiolini, A., Vasik, P., Turi, M. (eds.) MESAS 2020. LNCS, vol. 12619, pp. 86–107. Springer, Cham (2021). https://doi.org/10.1007/978-3-030-70740-8_6
31. David, W., et al.: Giving life to the map can save more lives. Wildfire scenario with interoperable simulations. In: Advances in Cartography and GIScience of the ICA, vol. 1, p. 3 (2021)
32. Goodness, C.: Data in the time of crisis (2021). https://unite.un.org/blog/data-time-crisis. Accessed 13 July 2021
33. U.S. Department of the Army. Medical Intelligence in a Theater of Operations: Field Manual 8-10-8 (1989). https://digitalcommons.unl.edu/cgi/viewcontent.cgi?article=1087&context=dodmilintel. Accessed 13 Jan 2021
34. WeRobotics. https://blog.werobotics.org/2021/01/12/teaming-up-with-pfizer-on-new-cargo-drone-project/. Accessed 07 July 2021

New Ways for Ground-Based Air Defence Personnel Training Using Simulation Technologies

Jan Farlik[1]([✉]) [iD], Lukas Gacho[1] [iD], and Vlastimil Hujer[2] [iD]

[1] University of Defence, Kounicova 65, 66210 Brno, Czech Republic
{jan.farlik,lukas.gacho}@unob.cz
[2] Joint Operations Command, Armed Forces, Prague, Czech Republic
vlastimil.hujer@unob.cz

Abstract. Armed forces around the world have more and more sophisticated weapon systems, sensors and C2 systems. The training on these systems is very demanding and exploitation of simulation technologies is a key requirement for successful personnel preparation. The area of ground-based air defence (GBAD) is not different. However, the special needs for GBAD training require also special simulation technology. This paper presents key attributes of such a simulation technology and brings one of possible solution (concept) for effective and cheap GBAD personnel training and education.

Keywords: Air defence · Modelling and simulation · Ground based air defence

1 Introduction

The last two decades have brought tremendous progress in the development of simulation technologies. Thanks to more powerful hardware and more sophisticated software, it is now possible to implement solutions that could not be solved at all a few years ago. Special attention in this area is represented by military technologies, or technologies that enable the training of military personnel. A special category of users in this respect is ground-based air defense personnel. This category has special requirements for training, resulting from the uniqueness of the use of ground-based air defense technologies. The specificity is mainly the fact that it is a ground element, which is mostly interested in airspace and related aspects of direct and procedural management. Thus, ground-based air defense units must not only strictly adhere to the current division of airspace, but also respond to a ground situation that could endanger them. The second category of interest is the operation of their own weapon systems, which personnel must know flawlessly and be able to control. Together, it is a complex set of tasks, whose training and coaching on simulation technologies requires a high degree of authenticity, not only visual, but especially procedural. This article aims to summarize and present the existing knowledge of the Department of Air Defense (University of Defense) in the field of modeling and simulation for the purpose of training ground-based air defense personnel.

© Springer Nature Switzerland AG 2022
J. Mazal et al. (Eds.): MESAS 2021, LNCS 13207, pp. 417–423, 2022.
https://doi.org/10.1007/978-3-030-98260-7_26

2   Development of Air Defense Simulation Technologies and the Needs for Future Functionality

Simulation technologies have been available in a figurative sense since the 1960s. Since the beginning of the existence of more complex air defense missile systems, simple simulation algorithms have been implemented in the weapon systems themselves. Most of them were simple algorithms enabling simulation of missile launches and guidance to a simulated or real target in training mode. This ensured the basic requirements for the training of personnel on real systems, where the simulation technologies were not evolved yet. However, this functionality only allowed the practice of shooting itself. It was not yet possible to train staff and unit commanders more comprehensively in more complex scenarios, including other aspects of combat activities, such as synchronized activities in the airspace with its own air force, training in larger combat groups (clusters) and the like.

Over the last ten years, however, due to the advent of powerful computer technology, there have been opportunities to expand the originally limited training opportunities on simulators and trainers. Thanks to the graphical performance, the battlefield can be simulated more or less visually realistic. Thanks to the computing power, it is now possible to simulate for example missile movements in real or nearly real time based on real properties such as missile engine thrust, maneuverability, destructive effect of combat charge, missile guidance method etc. However, all these possibilities hide one aspect, which in many cases prevents the use of simulation technology. Companies and programming teams that are able to program these processes do not have knowledge of real systems. In order to create a simulation environment for the training of ground-based air defense personnel, it is necessary to merge a team of programmers with a team of experts in the field of air defense. Otherwise, created products will be inaccurate and will not have adequate training value. Paradoxically, this can lead to a situation where the resulting product causes bad habits and knowledge, which can then cause the failure of the real combat mission in real combat environment. In the development of simulation technologies by a commercial entity, close coordination with the so-called Subject Matter Experts is necessary, i.e. experts with deep knowledge of the matter.

3   Aspects of Advanced Air Defense Simulators and Requirements for Modern Simulators for Air Force Training

Current military air force simulators are usually not very robust. Very often these are so-called "single role" simulators, such as flight simulators and trainers, simulators for unmanned aerial vehicle (UAV) operators, fire simulators, etc. These simulators are designed to train a narrow range of users from military operational personnel and are unable to provide a wider range of tasks. However, current trends show [1, 2] that it is more advantageous in terms of training benefits to train a more comprehensive range of personnel together and thus enable coordination between different types of troops during simulated missions. In the case of ground forces, there is a wide range of more complex simulators such as Presagis, OneSAF, VBS3, etc. However, for the ground-based air defense there are no such options. Usually, simulations for ground-based air defense are solved separately, where the emphasis is mainly on trainers, without wider possibilities of involvement in the simulation of joint or more complex operations.

The simulator, designed for ground-based air defense training should have the ability to implement a wide range of tasks. This would provide a more realistic simulation of both command and control centers and tactical units. A comprehensive simulation environment for ground-based air defense training should include the following entities:

1. Air Force Command and Control Centers, in which at least the following could be simulated:

   a. Positive and procedural control of subordinate forces and resources.
   b. Allocation of simulated weapons (air and ground).
   c. Sensor models and their fusion.

2. Air Force (airborne):

   a. Basic models of Wing Operation Centers (WOC) in terms of procedural management.
   b. Models of aircraft (airplanes, UAVs, helicopters, etc.)
   c. Tactical procedures and techniques used in the preparation phase and then for combat operations [3, 4].

3. Ground-based air defense forces

   a. Basic models of ground-based air defense operations centers (SAMOC - Surface to Air Missiles Operation Center) in terms of procedural control, e.g. [5].
   b. Models of ground-based air defense weapon systems (models of ground-based air defense batteries, or just stand-alone vehicles), e.g. [6].
   c. Tactical procedures and techniques used in the preparation phase and then for combat operations.

4. Sensors

   a. Modeling of detection capabilities of radars and other sensors (e.g., air observer equipment, etc.).

5. The enemy

   a. Modeling the capabilities and actions of enemy forces and resources.

6. Environment

   a. Modeling of the battlefield from the environment point of view, weather, terrain, etc., e.g. [7].

However, the requirements for modeling and simulation are, from the point of view of complex simulation of air force activities, highly dependent on the user, because each of them (num item 1 to 4) needs different information inputs in real operation. For a realistic simulation, aircraft pilots need the most reliable modeling of the view from the cockpit and models of instruments in the cockpit, as well as a credible visualization of enemy assets however this is not the case for GBAD units. Personnel or service of ground-based air defense means have different requirements. Next, the staff of operating centers or radar sensors have also completely different requirements in the form of yields of radar information and information important mainly for the decision-making process.

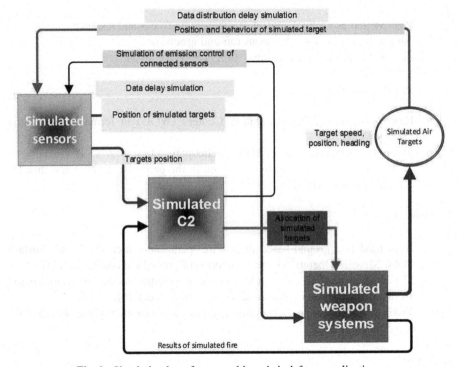

Fig. 1. Simulation loop for ground-based air defense applications.

From the point of view of requirements for a modern simulation environment, it is necessary to simulate not only the process of firing against air targets, but also the flow of information from sensors, commands from superior command and control elements and distribution of reports from subordinate simulated elements to superiors (see Fig. 1). Such an environment then contains the following functionalities:

1. Simulation of information processing

   a. Fusion of information from simulated sensors.
   b. Distribution of this information to relevant subordinate entities (firing element models, etc.).

   c. Registration of information from subordinate entities (status of entities, e.g. combat-ready/damaged/destroyed).

   d. Registration and distribution of information from superior entities (if simulated) to subordinate entities.

   e. Registration of hostile or potentially hostile simulated targets.

2. Decision-making process

   a. Analysis of the simulated air situation

      (1) Prediction and evaluation of the position of simulated own and enemy air objects, e.g. [8].

   b. Registration of statuses of subordinate simulated entities (ground-based air defense weapon systems models).

   c. Status change of the subordinate simulated entities.

   d. Allocation of weapon systems.

      (1) Commands to fire simulated aircraft entities and deconfliction with assignment of targets to ground-based air defense entities.

      (2) Commands to fire to simulated ground-based air defense entities.

   e. Calculations for procedural control

      (1) Calculations of the range of firing units depending on the position, speed and maneuver of the target.

      (2) Calculations of the intersection of air target routes with the areas where the firing units operate and the algorithm for allocation the target to the unit in case two or more units have a target within its range.

3. Monitoring and saving of the state of combat and evaluation of simulated combat.

   a. Calculation of simulated missile consumption.

   b. Simulation of the delay in calculating the ammunition replenishment time.

Another "challenge" for future ground-based air defense simulators is the prediction of target trajectories. In current command and control systems, as well as simulators, the prediction of the target position is generated based on the calculation of the straight flight trajectory, which in cases of air combat or ground-based air defense is out of the question, because the target will try to maneuver and get out of range of weapon systems. Correct prediction of the trajectory of targets (whether in real systems or in simulators) should have a more realistic prediction, e.g., in the form of the application of Apollo circles. A similar prediction would thus take into account the current course of the enemy target and the position of the defended object with the calculation of the probability of change of the target course based on a possible maneuver.

However, the most important function of the ground-based air defense simulator is the procedurally correct allocation of targets to the relevant subordinate weapon systems during the simulated combat activity. In the simulation environment, the target allocation process must be linked to criteria and conditions that allow the target to be optimally allocated:

1. Evaluation of whether the target has already been allocated to another simulated weapon system. This evaluation must be performed periodically after certain time intervals, usually given by the information recovery interval.
2. Assessment of the availability of a sufficient number of simulated aircraft at a distance to the target to allow approach, combat and return to the nearest or designated landing position. The distance to the target must be constantly recalculated in the simulation environment.
3. Evaluation of whether a sufficient number of simulated activated means (units, batteries) of the ground-based air defense is available.
4. Assessment of whether the fire density of the available simulated ground-based air defense systems is greater than the enemy density. If not, the simulated ground-based air defense grouping is overwhelmed and the command and control algorithms must reallocate available aircraft resources to cover the reduced ground-based air defense capabilities.

The whole target distribution algorithm in the simulation is conditioned by activated/inactivated simulated radar sensors. The algorithm is executed only if the target is within range of the simulated radar station. In the case of an activated radar network, it is necessary to perform an evaluation for a positive display of targets to the operator. This evaluation would be performed based on the presence of targets in the detection area of at least one simulated combat-ready radar. This simulation would then be performed either in the presence of a DMR (digital model of relief) and evaluating the radar visibility depending on the relative position of each target and each radar. The change in the displayed data should also be influenced by simulating the recovery time of the radar information.

4   Conclusion

The above described concept of extending existing simulation technologies with advanced elements that will allow not only realistic, but especially comprehensive training of ground-based air defense personnel was tested in the research activities of the Department of Air Defense, University of Defense (PROKVES 2016–2020 project). To support these extensions, several experimental software was created, e.g. [9, 10], which confirmed the correctness of the arguments defining the requirements for advanced simulation technologies of ground-based air defense. The list of attributes and properties of the simulator in the article contains the most important characteristics that a ground-based air defense simulator should have in order to be used for modeling and simulation of combat operations involving multiple air force entities. Thus, not only more than one ground air defense unit (two or more - the so-called clusters), but also coordination with air elements of the air force, such as aircraft, helicopters, etc.

References

1. Exploring new command and control concepts and capabilities. Rev. 2002. North Atlantic Treaty Organization, Research. Neuilly-sur-Seine Cedex, France (2007). ISBN 978-928-3700-708
2. Stanton, N., Baber, C., Harris, D.: Modelling Command and Control, Event Analysis of Systemic Teamwork, 1st edn. Ashgate, Burlington (2008)
3. Stefek, A., Casar, J., Stary, V.: Flight route generator for simulation-supported wargaming. In: Maga, D., Hajek, J. (eds.) Proceedings of the 2020 19th International Conference on Mechatronics–Mechatronika (ME). Institute of Electrical and Electronics Engineers Inc., Prague (2020)
4. Biagini, M., Corona, F., Casar, J.: Operational scenario modelling supporting unmanned autonomous systems concept development. In: Mazal, J. (ed.) MESAS 2017. LNCS, vol. 10756, pp. 253–267. Springer, Cham (2018). https://doi.org/10.1007/978-3-319-76072-8_18
5. Stary, V., Farlik, J.: Aspects of air defence units C2 system modelling and simulation. In: Mazal, J., Vasik, P. (eds.) MESAS 2019. LNCS (LNAI and LNB), vol. 11995, pp. 351–360. Springer, Cham (2020). https://doi.org/10.1007/978-3-030-43890-6_28
6. Stary, V., Gacho, L.: Modelling and simulation of missile guidance in WEBOTS simulator environment. In: Maga, D., Hajek, J. (eds.) Proceedings of the 2020 19th International Conference on Mechatronics - Mechatronika. Institute of Electrical and Electronics Engineers Inc., Prague (2020)
7. Mazal, J., et al.: Modelling of the microrelief impact to the cross country movement. In: Bottani, E., Bruzzone, G. (eds.) Proceedings of the 22nd International Conference on Harbor, Maritime and Multimodal Logistic Modeling & Simulation, HMS, vol. 22, pp. 66–70 (2020)
8. Farlik, J., Hujer, V., Kratky, M.: Utilization of modeling and simulation in the design of air defense. In: Mazal, J., Fagiolini, A., Vasik, P. (eds.) Modelling and Simulation for Autonomous Systems. Lecture Notes in Computer Science, vol. 11995, pp. 291–298. Springer, Cham (2020). https://doi.org/10.1007/978-3-030-43890-6_23
9. Farlik, J.: GBAD configurator. Software (2016)
10. Farlik, J.: 2D and 3D visualization of missile effective coverage zones. Software (2016)

Building Trust in Autonomous Systems: Opportunities for Modelling and Simulation

Thomas Mansfield[1](✉), Pilar Caamaño[1], Sasha Blue Godfrey[1], Arnau Carrera[1],
Alberto Tremori[1], Girish Nandakumar[2], Kevin Moberly[2], Jeremiah Cronin[2],
and Serge Da Deppo[3]

[1] NATO STO Centre for Maritime Research and Experimentation, Viale San Bartolomeo 400,
La Spezia, Italy
{thomas.mansfield,pilar.caamano,sasha.godfrey,arnau.carrera,
alberto.tremori}@cmre.nato.int
[2] Old Dominion University, 5115 Hampton Blvd, Norfolk, VA, USA
{gnand002,kmoberly,jcronin}@odu.edu
[3] NATO ACT Innovation Hub, 4111 Monarch Way, Norfolk, VA, USA
serge.dadeppo@innovationhub-act.org

Abstract. Advances in artificial intelligence and robotics development are providing the technical abilities that will allow autonomous systems to perform complex tasks in uncertain situations. Despite these technical advances, a lack of human trust leads to inefficient system deployment, increases supervision workload and fails to remove humans from harm's way. Conversely, excessive trust in autonomous systems may lead to increased risks and potentially catastrophic mission failure. In response to this challenge, trusted autonomy is the emerging scientific field aiming at establishing the foundations and framework for developing trusted autonomous systems.

This paper investigates the use of modelling and simulation (M&S) to advance research into trusted autonomy. The work focuses on a comprehensive M&S-based synthetic environment to monitor operator inputs and provide outputs in a series of interactive, end-user driven events designed to better understand trust and autonomous systems.

As part of this analysis, a suite of prototype model-based planning, simulation and analysis tools have been designed, developed and tested in the first of a series of distributed interactive events. In each of these events, the applied M&S methodologies were assessed for their ability to answer the question; what are the key mechanisms that affect trust in autonomous systems?

The potential shown by M&S throughout this work paves the way for a wide range of future applications that can be used to better understand trust in autonomous systems and remove a key barrier to their wide-spread adoption in the future of defense.

Keywords: Modelling and simulation · Trusted autonomy · Model based systems engineering · Future of defense

© Springer Nature Switzerland AG 2022
J. Mazal et al. (Eds.): MESAS 2021, LNCS 13207, pp. 424–439, 2022.
https://doi.org/10.1007/978-3-030-98260-7_27

1   An Introduction to Trust and Autonomous Systems

Despite continued advances in artificial intelligence and robotics development that provide the technical abilities that will allow autonomous systems to perform complex tasks in uncertain situations, incorrect levels of trust are a key barrier preventing autonomous systems from fully achieving their potential. A lack of trust leads to inefficient system deployment, increases supervision workload and, under certain conditions, could lead to potentially catastrophic mission failure.

In response to this challenge, trusted autonomy is an emerging scientific field aiming at establishing the foundations and framework for developing trusted autonomous systems. One area of research within this field is working to identify the mechanisms that affect human trust in autonomous systems. The first line of research in this field is defining standard methods to measure changes in human trust while training, operating and making decisions based data obtained from of autonomous systems. A second research track is working to identify methods that allow humans to better understand autonomous system behaviors and the operation of emerging technologies in order to encourage the correct level of trust to be obtained. Despite these efforts, challenges stem from the rapid pace of system capability and complexity. As autonomous systems rapidly evolve, the opportunities for humans to interact with systems and manually understand data sets generated by their use is reduced. Further, the accelerating complexity of emerging technologies and sophistication of autonomous system decision making is placing pressure on the pace of progress in the field of trusted autonomy.

Building upon the recent successes of using model-based methodologies in the communication of emerging technologies and systems [1, 2], the focus of this paper is the design and test of the first iteration of a model based framework that, coupled with existing wargame approaches, promise to provide a powerful technology analysis capability. Utilizing standardized modelling and simulation (M&S) approaches, the framework provides two layers of support; the first allows users to interact with emerging technologies and to understand the behavior of autonomous systems. Secondly, the model-based framework supports the collection of actionable data to reveal changes to the player's level of trust throughout system operation.

The remainder of the paper provides a review of related work in Sect. 2 and a description of the framework's architectural design in Sect. 3 before discussing initial test results in Sect. 4 and highlighting the main conclusions in Sect. 5.

2   Related Work: Trusted Autonomy and Measuring Trust

2.1   What is Trust?

In the field of Trusted Autonomy [3], a single definition of trust is emerging and beginning to be widely adopted. This definition of trust is:

> **"Trust is the attitude that an agent will help achieve an individual's goals in a situation characterized by uncertainty and vulnerability"** [4, 5]

Within this definition there are two important elements that must be present in any experiment investigating the concept of trust:

1. Uncertainty [6] – The problem is not simple or trivial, there must be an element within the scenario that the trustor does not fully control. However, the trustor must have an inherent understanding of the trustee's ability to achieve goals.
2. Vulnerability [6, 7] – The presence of vulnerability implies the trustor will be less likely to relinquish control in the presence of low trust. The need for there to be a 'loss' to the trustor is vital in order to test, understand and measure system trust. Without it, trust is not important and players will just play with interesting things.

While a single, common definition of trust is emerging in current literature, the field of trusted autonomy has not yet established a single, agreed and effective measure of trust. The primary challenge with the measurement of trust is that it is an abstract, human opinion that is difficult to record in a robust or reliable manner. A popular way to address this shortcoming in the field of trusted autonomy is to instead measure the reliance that a user has on a system, inferring trust based on the reliance the human has on the system under test [8]. In the process of inferring trust from reliance, other factors that contribute to reliance, such as perceived risk and self-confidence must also be considered [9, 10]. A typical causal flow, highlighting some of the key interrelations among human, machine and environmental factors in the context of user trust in depicted in Fig. 1.

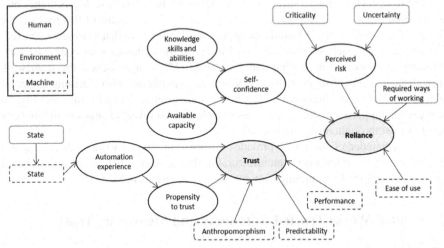

Fig. 1. A causal flow diagram summarizing the interactions that effect trust and reliance adapted from [8]

Based on this rationale, the investigation reported in this paper will focus on the measuring of autonomous system reliance and control the perceived risk and user self-confidence, allowing trust to be inferred.

2.2 How Can Reliance Be Measured?

While the measure of system reliance can be used to infer trust, measuring it still requires the consideration of challenging human aspects. Typical measures of reliance break down into three main methods; subjective surveys, measuring psychophysiological human characteristics and indirect assessment of behaviors influenced by trust.

Subjective surveys require humans to consciously report their level of trust. Difficulties exist with this measurement as not all humans can accurately characterize or understand their current trust attitudes or may not be willing to report their true attitude [11]. However, basic subjective surveys are often easy to implement in many military wargame settings and provide a source of actionable data.

To overcome the problems with subjective trust measurement, many researchers have tried to create objective measures of trust. One form of objective trust measures is to associate trust with different psychophysiological human characteristics, such as electroencephalography (EEG) [12] or galvanic skin response (GSR) [13]. Unfortunately, these measures need to be subsequently calibrated against some subjective measure of trust to ensure they provide a meaningful measure of trust. Additionally, the ability to measure psychophysiological human characteristics is significantly diminished due to practical barriers such as the common availability of technology and the extensive training required to operate it reliably.

Finally, a third form of trust measurement involves indirect assessment by measuring behaviors influenced by trust. This includes the ability to monitor and assess behaviors via the user's interaction with software [14]. This approach may be of particular value in projects based on the use of modelling and simulation methodologies which may result in the creation of interactive and immersive software applications.

2.3 When Can Reliance Be Measured?

Most experiments, games or events that focus on automation may be broken into three distinct sections; before, during and after the application of automation [5]. At each stage of the event, any direct or indirect measure of trust may be applied. Further, it is assumed that throughout this work the user will iteratively cycle through repeated stages, as shown in Fig. 2.

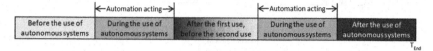

Fig. 2. Non-operational time may be before or after an automated event.

One clear example of an iterative approach [7], allowed players under pressure (money or a timer) to escape from a maze with the assistance of an autonomous, robotic guide. In the first iteration, the autonomous guide exhibited behaviors associated with the technology under test. The key aspects were also explained to the subjects using videos, images, or text and providing information that allowed the participant to evaluate the risk associated with choosing to follow the robot. In a second iteration, the players were then asked to play a second time, with the option of not using the robotic guide. This selection was used to infer trust for each of the technologies under test.

Monitoring when the human takes action to activate the automation, as well as instances where the user does not take action, provide a good indication of the reliance the user has in the system [15]. It may also be noted that the underlying trust once the mission is complete is manifested differently when the human has authority to activate the automation rather than when the machine activates automation as the positive or negative response to their action is likely to strongly affect both the self-confidence and perceived risk of subsequent interactions with the autonomous system.

Once an automation has begun to act, the authority to turn an automation off will influence the way trust behaviors are expressed. When an automation is running, if the human can override the machine, an act of reliance will constitute inaction (i.e. he or she will choose to not turn it off). In this case, an act of non-reliance will look like turning the automation off. Once the automation has finished, another type of authority that could be monitored is the human decision to repeat or re-run the activity taken by the autonomous system. Authority for the human to repeat or re-run only pertains once the automation has completed its action.

2.4 NATO Investigation Frameworks

In order to investigate future technologies and their impact on future military missions, NATO typically utilizes a wargame framework known as the Disruptive Technology Assessment Game (DTAG) [16]. The DTAG flow, shown in Fig. 3, allows opposing blue and red teams to build upon a starting vignette, allowing likely confrontations to be played and understood.

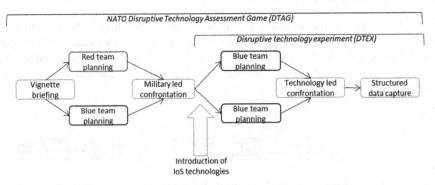

Fig. 3. DTAG framework, adapted from [16] (Color figure online)

Once this baseline has been set, new and emerging technologies are added into the wargame in the form of Ideas of Systems (IoSs). These ideas of systems briefly describe the key technologies, operational benefits and drawbacks of future technologies. Once understood by the game players, competing blue teams try to identify the best combined use of the technologies to achieve the vignette aims. Initially based on the DTAG model,

DTEX improves it and expands its range of application, through computer assisted processes, and distributed participation. The Disruptive Technology Experiment (DTEX) approach was created by the NATO Innovation Hub in Norfolk, Virginia, USA in partnership with Old Dominion University's Institute for Innovation and Entrepreneurship (IIE) and the NATO STO's Center for Maritime Research and Experimentation (CMRE) to reduce the time required to evaluate technologies, increase the level of input into the evaluation by leveraging virtual/distance methods, and employ synthetic environments (SENs), in addition to subject matter experts, to quickly provide outcomes based upon participants choices of ideas of systems. The objective was to create an approach that would allow the Innovation Hub to quickly and regularly test new ideas, concepts, and solutions sourced through its open innovation efforts such as the NATO Innovation Challenge.

3   Methodology: A DTEX Framework Based on Modelling and Simulation Approaches

This paper investigates the design and test of a framework that, based on the use of modelling and simulation methodologies, supports geographically distributed DTEX event which immerse the player into a scenario and collect data via subjective surveys and the indirect assessment of their decisions and software interactions. In this paper, the framework will be utilized to extract information about the level of trust users have during a scenario that relies on the successful operation of autonomous systems.

The design of the DTEX framework uses a model based approach to progress the design in three parallel and complementing streams: a gameplay stream, a human stream and a technology stream. Key attributes within each stream are listed in Fig. 4.

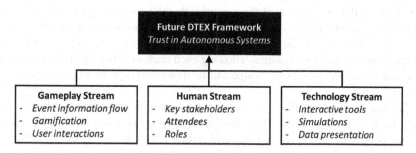

Fig. 4. Three areas of investigation

3.1   Gameplay Stream

A key principle of NATO's DTEX is to use a wargame to structure the interaction between competing blue teams. Within each team, gamification techniques provide the structure, rules and motivation to identify innovative solutions to challenging scenarios and record the discussion points and rationale behind the decisions made. With a focus on investigating trust in autonomous systems, a series of key gameplay design decisions have been made and implemented. These design decisions focus largely on the flow of information provided to the players during the event along with managing their interactions with the supporting game tools.

Before the event, the pre-existing bias each player has with respect to their trust in autonomous systems is recorded via a subjective survey. The results of this survey are kept for later analysis following the completion of the gameplay activities. A description of the event objectives are then shared following this survey, allowing the event players to understand the topics to be addressed for the first time.

The first significant and tangible set of information provided to the players is a description of the scenario. The scenario describes how autonomous systems are currently used to survey a harbour following reports of suspicious behaviour by a terrorist organisation. The scenario also contains several motivating aspects for the timely and reliable completion of the mission with economic consequences of the harbour being closed coupled with a second concern about disrupting military peace keeping operations. Model based approaches were utilised to communicate the scenario in two specific ways; first the overall scenario was communicated in a view that forms part of the NATO Architectural Framework (NAF) guidance [17]. Secondly, specific technologies and interactions within the scenario need to be communicated within focus areas, such as operating behaviour and limits of autonomous underwater vehicles. The main advantages of using a model based approach to communicate the scenario content is the ability to quickly and effectively brief players with a varied range of skills and previous experiences with the key scenario elements in an intuitive and understandable way. Further, the approach provides the tools required to monitor player interactions at this stage of the event, highlighting areas in which users may have limited or differing experience (Fig. 5).

Following the explanation of the scenario, the players move into a series of three iterating confrontation stages. Implementing the iterative autonomy approach described in Sect. 2.3 of this paper, the approach provides multiple opportunities for the players to engage the autonomous systems, allowing each team member's self-confidence and perceived risk of the system to change and be monitored depending on the success of pervious events.

The first execution of the scenario is contained within a Baseline Confrontation in which none of the IoS technologies are incorporated. The main motivations for this baseline confrontation are to reinforce and further illustrate the key events of the scenario and allow the players to better understand the current operation and limitations of autonomous systems while beginning to build momentum and an understanding of the DTEX event stages. At the end of the baseline confrontation, the outputs of the autonomous systems are presented and the players are asked if they trust the results enough to reopen the harbour to traffic. This final question again, clearly asking if the user trusts the system enough to open the harbour to traffic, provides an opportunity to record the player's trust in autonomous systems and allow a comparison with the biases recorded at the start of the event. The structure of this baseline confrontation also provides the event controllers the potential to investigate ability to normalise the

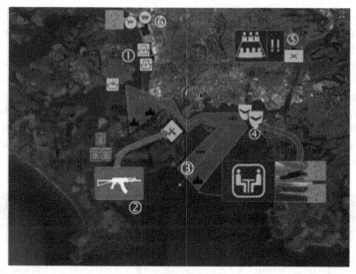

Fig. 5. A NAF scenario overview. Each symbol and number shown in the map signifies a briefing focus area.

level of trust the teams have in autonomous systems; if the players show a higher than average level of trust in the systems, the results shown here could show the autonomous systems performing poorly, potentially lowering their trust before the subsequent confrontations. Conversely, if a team has been shown to be sceptical of autonomous system performance and low levels of trust are shown, the results of very successful autonomous system operation could be communicated prior to subsequent runs of the scenario.

Following the completion of the baseline confrontation, the first of two IoS technology-assisted confrontations is played. The technology assisted confrontations first require the players to review and discuss a wide range of potential future technologies that may assist them in the execution of the scenario and the analysis of the autonomous system outputs. A key consideration in the design of the DTEX framework is in supporting and recording the discussion within each blue team in an attempt to identify the anticipated benefits and drawbacks of each of the IoS technologies. Due to the fact that in this first event, the teams will be presented with over 60 separate IoS technologies, the DTEX framework has been designed to support their analysis in two stages. The first stage involves a quick sort of the technologies focusses on the initial reaction following the review of a small set of supporting information including the technology name, method of operation and key benefits. The IoS card technologies that the teams consider suitable for further investigation are added to a technology shortlist. This shortlist then forms the basis for the following activity in the confrontation where the teams are asked to discuss and combine technologies, resulting in a final three technologies that will be applied to a re-run of the scenario. Each run provides an augmented data set on which the teams can decide whether to open the harbour for traffic. This decision point, where the players are presented with an updated set of autonomous system outputs and asked to make a decision, again provides an opportunity to collect information on their level of system trust with a further subjective survey. The change in trust implied by the survey results provides a data set that may allow the effectiveness of the technologies used to be assessed, a key output from this activity.

A second technology based confrontation allows the player's learning to be applied, repeating all of the technology shortlisting and review actions form the previous spiral to be executed, providing an opportunity for each team's experiences and learning to be discussed during the selection of an updated technology combination. The DTEX event ends with a recap of the key activities and a request to support any further, offline exchanges of information in the coming days. An overview of the complete, model-based gameplay stream designed for the event is provided in Fig. 6.

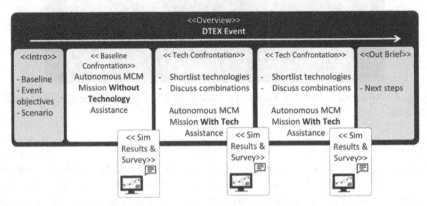

Fig. 6. Overview of the designed DTEX gameplay stream

3.2 Technology Stream

To support the identified DTEX event gameplay stream, a range of modelling based technologies have been employed that allow it to be executed effectively and efficiently in an online, distributed environment. The complete event is designed to be executed on a range of commercially available video-conferencing (VTC) platforms, the key requirements of which are ease of access to allow all participants to join and video recording, allowing subsequent analysis of the team's discussion and software interaction.

Model based methodologies, applied at a technical level, allowed the design of an interactive and intuitive NAF-based dashboard. This web-based dashboard provides a range of user specific views, via the login options in Fig. 7, that guide the user through the complete DTEX event. Figure 8 provides one example of the ability to communicate key event structures, where an interactive and linked NAF view of the game flow is presented.

Fig. 7. A screenshot of the NAF-based dashboard allowing customised login

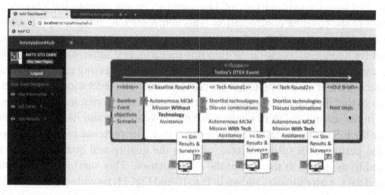

Fig. 8. A screenshot providing players with an interactive and navigable overview of the DTEX event

The availability of this prototypical dashboard allows all of the subsequent technology stream developments to be linked and utilized by the players. Ordered by game flow, the first role of the technology is to support the elicitation of information from the teams in a series of subjective surveys. These surveys utilized online questionnaire provider platforms [18] to create intuitive and interactive surveys that could be linked to the relevant sections of the NAF gameplay flow in the dashboard. The results of each survey can be saved and stored for analysis after the event.

A range of M&S technologies have been applied to support the articulation of the event scenario. With an example provided in Fig. 9 where the operation and limitations of an autonomous underwater vehicle is being investigated, multi-media visualizations have been created to provide intuitive information concerning all of the key scenario technologies and events.

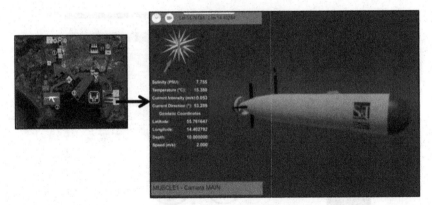

Fig. 9. Linked information allowing users to investigate specific scenario aspects

The multi-media explanations of key scenario events prepare the players for the execution of the baseline confrontation. This confrontation, along with the subsequent iterations, utilizes a complete federation of simulators to execute the scenario without the need for human intervention. Within this scenario, key autonomous system aspects such as asset motion, sensor performance and on-board processing capabilities, run as if the mission were being completed in the scenario area. The federation logs results of the scenario so that they can be displayed after confrontation, as required by the gameplay stream. In the second and third confrontation, the simulation and its related results also consider the IoS technologies selected by the teams. The presence of these technologies affect the performance of the system, altering the quality and quantity of the results presented to the teams at the end of the confrontation. This varying set of simulation generated data, with examples shown in Fig. 10 and Fig. 11, provide a key input that will alter the team's response to the final question, asking if they would reopen the harbor, and allow an estimate of the ability for the technology to shape trust to be made.

In addition to providing a framework around the M&S federation that runs the scenario and generates technology dependent results, the NAF–based dashboard also provides a number of tools that support the shortlisting and analysis of the IoS technologies provided to each team. To support shortlisting, a user interface has been developed that provides players with an overview of key IoS information and three options; to discard, shortlist or potentially consider its inclusion in the confrontation. An example of this functionality is contained in Fig. 12. Once complete, a second stage in the shortlisting process allows the players to review and modify their shortlisted cards, with the option to discard further technologies if desired or required. Once a suitable shortlist has been established, the NAF-based dashboard stimulates group discussion by providing the players with the capability to graphically drag, drop and sort the cards into a series of editable categories. A screenshot of this functionality, shown directly after the completion of card shortlisting, is provided in Fig. 13.

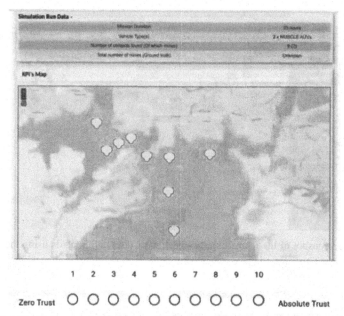

Fig. 10. Basic simulation outputs before the application of IoS technologies

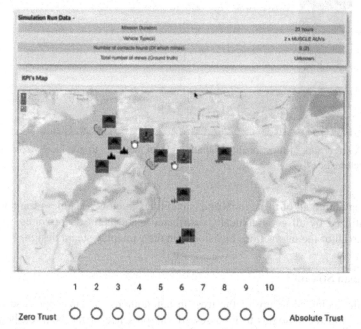

Fig. 11. Simulation outputs augmented with additional information following the application of IoS technologies

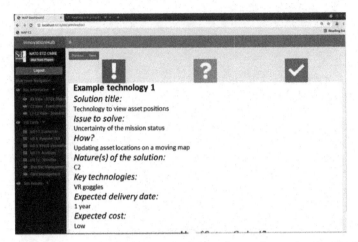

Fig. 12. A screenshot of the teams rapidly selecting or discounting cards using the '!', '?' and '✓' buttons

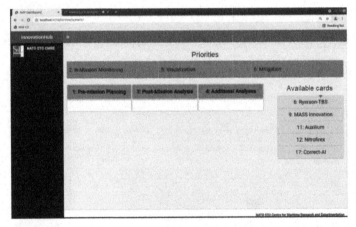

Fig. 13. A screenshot of functionality that allows team members to graphically sort, group and organise selected cards

By combining all of these tools and capabilities into one, seamless framework that can be utilized by all geographically distributed team members provides a technical solution to all of the interactions envisioned in the gameplay steam design.

3.3 Human Stream

Despite the technical nature of the trust in autonomous systems topic and the large number of IoS technologies to be assessed, the presence of the NAF-dashboard and model based approaches utilised in its design increase the ability to include players from communities outside maritime autonomous systems, providing access to other skills and experiences. This allows the test of the platform, and the acquisition of valuable data on the technologies that are likely to affect trust in autonomous systems, to be executed without further detailed consideration of the human stream. Each DTEX event should contain blue team players that are external to the design activities described in this document and should be facilitated by team members that can assist with time-management and ensure the correct use of the prototype tools. In addition, observers in both M&S and gameplay development activities may be present during the DTEX event to drive further improvements to the approach.

4   Event Execution Results and Discussion

The distributed, model-based DTEX framework discussed in this paper was tested in a NATO DTEX prototype trial event in June 2021. The event, advertised with the flyer shown in Fig. 14, allowed the complete set of tools linked within the NAF-based dashboard to be utilized to guide a single blue-team consisting of two players through the process. While limitations were present in the form of reduced IoS technology coverage, lower simulation fidelity and few blue team players, observers could comment on all event stages. Further, the blue team players were provided with the opportunity to comment on the success of the event.

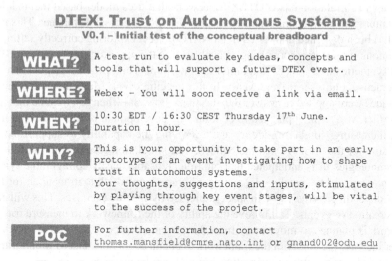

Fig. 14. Invitation to the first test of the DTEX framework prototype

Feedback from the observers and players identified that the structure of the event was clear and all participants could contribute where required. One particular success in this area stemmed from the fact that the players in the blue team had vastly differing levels of experience of autonomous systems prior to the event, with one player expert on their use and another a relative newcomer to the subject. The use of the NAF-based dashboard and all of the linked capabilities provided a common reference point that allowed both players to contribute their opinions in search for the optimal selection of technologies.

This task is likely to have been very difficult in the absence of such a clear and intuitive toolset providing a common reference point for both players' opinions. Conversely, one potential shortcoming of the designed architecture was observed in the collecting of information in the subjective surveys. One survey was provided per team and consensus was required to reply with a single, integer on a scale of trust. Due in part to the differing backgrounds of the blue team players, achieving consensus was not always possible. While the discussion this prompted was useful in the assessment of the technologies, there may be a future opportunity to allow the players to respond separately or to input their perceived level of trust onto a non-numeric scale.

5   Conclusion and Future Activities

The work reported in this paper marks a first step in exploring the ability of emerging and potentially disruptive technologies to shape human trust in autonomous systems in the context of maritime military deployments. The final objectives of the work are to identify the emerging technologies that most effectively set the correct level of human trust in autonomous systems, allowing the conceptual models (mechanisms) that each solution uses to set the correct level of trust to be identified and understood.

In pursuit of these objectives, this paper presents the development of a future distributed and simulation based DTEX framework that uses model-based methodologies to communicate complex information in an intuitive and accessible manner. This framework has been designed and tested in an NATO project exploring correctly setting trust in autonomous systems, supporting an analysis into the opportunities for a future M&S-based synthetic environment to monitor operator inputs and provide outputs in a series of interactive, end-user driven events. In this specific event, M&S methodologies and techniques were applied to answer the questions; how and when can trust be measured?

Further work is now planned to continue the development of this prototype framework, increasing simulation fidelity and improving the capability to support the communication of the scenario and the collection of simulation results with the full set of IoS technologies. It is anticipated that multiple, reusable, easily configurable synthetic environments will be developed, creating a library from which to draw upon for future DTEX depending on the problem space addressed by specific exercises. This will enable rapid creation of events. To improve the ability of the framework to measure trust, further work is planned to move away from the current reliance of subjective surveys and further increase the ability of the framework to obtain actionable data by monitoring and analyzing player's software interactions throughout the wargame.

All three streams of the model-based framework identified in this paper are intended to remain as a persistent capability, allowing M&S methodologies to support the analysis of a wide range of emerging technologies in a across a broad spectrum of domains, disciplines and activities.

Acknowledgements. The work reported in this paper has been funded by NATO Allied Command Transformation (ACT) Innovation Hub.

References

1. de Rosa, F., Mansfield, T., Jousselme, A.-L., Tremori, A.: Modelling key performance indicators for improved performance assessment in persistent maritime surveillance projects. In: Ahram, T.Z., Karwowski, W., Kalra, J. (eds.) AHFE 2021. LNNS, vol. 271, pp. 295–303. Springer, Cham (2021). https://doi.org/10.1007/978-3-030-80624-8_37
2. Harper, A., Navonil, M., Yearworth, M.: Facets of trust in simulation studies. Eur. J. Oper. Res. **189**, 197–213 (2021)
3. Abbass, H.A., Scholz, J., Reid, D.J. (eds.): Foundations of Trusted Autonomy. SSDC, vol. 117. Springer, Cham (2018). https://doi.org/10.1007/978-3-319-64816-3
4. Lee, J.D., See, K.A.: Trust in automation: designing for appropriate reliance. Hum. Factors J. Hum. Factors Ergon. Soc. **461**(1), 50–80 (2004)
5. Bindewald, J.M., Rusnock, C.F., Miller, M.E.: Measuring human trust behavior in human-machine teams. In: Cassenti, D.N. (ed.) AHFE 2017. AISC, vol. 591, pp. 47–58. Springer, Cham (2018). https://doi.org/10.1007/978-3-319-60591-3_5
6. Evans, A.M., Revelle, W.: Survey and behavioral measurements of interpersonal trust. J. Res. Pers. **46**(2), 1585–1593 (2008)
7. Robinette, P., Wagner, A.R., Howard, A.M.: Investigating human-robot trust in emergency scenarios: methodological lessons learned. In: Mittu, R., Sofge, D., Wagner, A., Lawless, W.F. (eds.) Robust Intelligence and Trust in Autonomous Systems, pp. 143–166. Springer, Boston (2016). https://doi.org/10.1007/978-1-4899-7668-0_8
8. Rusnock, C.F., Miller, M.E., Bindewald, J.M.: Framework for trust in human-automation teams. In: Industrial and Systems Engineering Conference, Pittsburg, USA (2017)
9. Johnson, N., Patron, P., Lane, D.: The importance of trust between operator and AUV: crossing the human/computer language barrier. In: OCEANS 2007, Aberdeen, UK (2007)
10. Wu, X., Stuck, R.E., Rekleitis, I., Beer, J.M.: Towards a framework for human factors in underwater robotics. Proc. Hum. Factors Ergon. Soc. Ann. Meet. **59**(1), 1115–1119 (2016)
11. OECD: Measuring trust. In: OECD Guidelines on Measuring Trust, Paris, France, pp. 115–154. OECD (2017)
12. Hu, W.-L., Akash, K., Jain, N., Reid, T.: Real-time sensing of trust in human-machine interactions. Cyber-Phys. Hum.-Syst. **49**(32), 48–53 (2016)
13. Khawaji, A., Chen, F., Zhou, J., Marcus, N.: Using galvanic skin response (GSR) to measure trust and cognitive load in the text-chat environment. In: 33rd Annual ACM Conference Extended Abstracts on Human Factors in Computing Systems, Seoul (2015)
14. Andre, H., Sihombing, P.P., Sfenrianto S., Wang, G.: Measuring consumer trust in online booking application. In: ICITISEE (2019)
15. Feigh, K.M., Dorneich, M.C., Hayes, C.C.: Toward a characterization of adaptive systems: a framework for researchers and system designers. Hum. Factors J. Hum. Factors Ergon. Soc **56**(4), 1008–1024 (2012)
16. NATO Allied Command Transformation: Disruptive Technology Assessment Game Handbook. NATO ACT, Norfolk, USA
17. North Atlantic Treaty Organization (NATO): NATO Architecture Framework. Architecture Capability Team Consultation. Command and Control Board, Brussels, Belgium (2018)
18. Google: Google - Create effortless forms. Google. https://www.google.com/forms/about/. Accessed 21 July 2021

Autonomous Systems and Cyberspace: Opportunities for the Armed Forces

Jakub Fučík[1](✉) [iD], Libor Frank[2] [iD], and Richard Stojar[2] [iD]

[1] Cyber and Information Warfare Command, Czech Armed Forces, Kounicova 65, 602 00 Brno, Czech Republic
jakub.fucik@unob.cz

[2] Centre for Security and Military Strategic Studies, University of Defence, Tučkova 23, 602 00 Brno, Czech Republic
{libor.frank,richard.stojar}@unob.cz

Abstract. The development of autonomous systems represents an opportunity for states to enhance their military power and change their status in the system of international relations, probably with the presumptions of the new Revolution in Military Affairs (RMA). Characteristics like flexible deployment, efficiency, adaptability, rapid upgradability, and/or the capacity to absorb losses that crewed systems cannot constitute potential game-changing technology in terms of military capabilities. The usage of the autonomous systems may vary from support functions (such as intelligence gathering, reconnaissance, or transportation) to coercive ones (autonomous weapon systems) in different operational domains. However, all types share a common characteristic – their interconnection with information technologies and cyberspace. Features like the complex structure, absence of specific borders, and diminished role of distance clearly distinguish this domain from land, sea, air, and even space and change the logic of how the security is provided there. Autonomous cyber systems (ACS) that operate in and through this domain represent unique military systems even among autonomous systems themselves. This paper aims to examine the interconnection between ACS and cyberspace and identify opportunities for a state's military power represented by armed forces. This relation addresses not only the dependency of the ACS on cyberspace and cybersecurity but also possibilities to enhance the cyber capabilities of the state to promote its national interests. The paper will focus on opportunities of the relation between ACS and cyberspace for capabilities of the NATO member states' armed forces. The Main Capability Areas (MCA) analytical framework will be used to analyze these aspects. Orientation on Prepare; Project; Engage; Sustain; Consult, Command and Control; Protect; Inform areas will identify opportunities to develop the full spectrum of potential capabilities and highlight main aspects of the transformation of the armed forces.

Keywords: Autonomous systems · Armed forces · Cyberspace · Main capability areas

© Springer Nature Switzerland AG 2022
J. Mazal et al. (Eds.): MESAS 2021, LNCS 13207, pp. 440–451, 2022.
https://doi.org/10.1007/978-3-030-98260-7_28

1   Introduction

Nowadays, the armed forces of more than eighty countries worldwide are using remotely controlled Unmanned Systems (UxS) for reconnaissance, survey, or monitoring purposes [1]. The number of states which employ armed UxS is also gradually growing. We can assume that this general trend, i.e., the growing number of states which operate UxS of various categories, will only intensify in all physical military domains (land, sea, air, space). Compared to remotely controlled systems, the Autonomous Systems (AxS) require no or only minimal human operator involvement [2]. Individual systems should be able not only to obtain information about the environment but also to process (evaluate) this information and take appropriate decisions on their own. The motivation to establish those systems is directly based on their increased effectiveness in combat. Similar to remotely controlled systems, the idea of minimizing the human losses on the part of the operator's armed forces plays a key role [3]. Moreover, AxS enable to reduce (or altogether remove) the cognitive load of their operators. The development of AxS represents an opportunity for states to enhance their military power and change their status in the system of international relations. All mentioned features fundamentally distinguish AxS from "traditional" conventional weapon systems. They create new ways of warfare and even make the current one obsolete. This status establishes the new Revolution in Military Affairs (RMA) presumptions. All types of AxS share a common characteristic – their interconnection with information technologies and cyberspace. Features like the complex structure, specific borders, and diminished role of distance clearly distinguish this domain from land, sea, air, and even space and change the logic of how the security is provided there.

Autonomous cyber systems (ACS), which operate in and through this domain, represent unique military systems even among autonomous systems themselves from this point of view. This paper aims to examine the interconnection between ACS and cyberspace and identify opportunities for the state and its military power represented by armed forces. This relation addresses not only the dependency of the ACS on cyberspace and cybersecurity but also possibilities to enhance the cyber capabilities of the state to promote its national interests. The paper will focus on opportunities of the relation between ACS and cyberspace for capabilities of the NATO member states' armed forces. The findings within this paper were verified through expert meetings and workshops with the participation of members of the Ministry of Defence of the Czech Republic, the Czech Armed Forces, and representatives of the security community of the Czech Republic. Data gathering and evaluation on these events were carried out through the combination of structured interviews, questionnaires, and nominal group technique methods. Analyses of the opportunities are in this paper divided into several chapters. Section 2 discusses trends in the development of cyberspace which is mainly related to the transformation of the security and operational environment as well as to elements of military power. This chapter also provides necessary information for evaluating opportunities (see below). Section 3 provides general characteristics of the AxS and pinpoints on specificities of the ACS. Section 4 defines the theoretical framework of the paper – Revolution in Military Affairs and Main Capability Areas – which serves to identify and evaluate the opportunities for armed forces. Finally, Sect. 5 discusses opportunities arising from the connection between ACS and cyberspace in the set-theoretical framework. Through this analysis, the paper should contribute to the discussion about the possible enhancement of NATO member states' military capabilities through ACS.

2 Cyberspace

Strengthening the strategic importance of cyberspace is directly linked to the development of information technologies and their use in virtually all areas of human life. Information globalization enables any actor (state and non-state) to have almost instantaneous and unrestricted access to a vast amount of data and their subsequent processing and use for their own needs. In this sense, the information became a strategic commodity usable both for shaping the position in this dimension and the functioning of the real environment. From the perspective of state and non-state actors, ensuring permanent and secure access to this domain is a prerequisite for the effective fulfillment of their interests. In this sense, the so-called cyber-attacks or malicious cyber activities - for example, in the form of the ability to deny an opponent's access to this domain and degrade his ability to exploit his systems or deliver harmful effects to our - represent essential tools for achieving the set goals [4], which are generally characterized not only by a very favorable utility ratio (investment/profits from the discussed activities) but also reduced ability to attribute such attacks and low probability of retaliation from the harmed entity.

The development of the Internet of Things (IoT) is gradually evolving into the Internet of Everything (IoE), which not only facilitates more effective use of the comprehensive information links (e.g., to ensure monitoring and decision-making in real-time) but also deepens the overall dependency on the stable and efficient operation of this space, resulting in user vulnerability. Building and developing 5G information networks take the discussed issue to a qualitatively higher level, both in terms of opportunities (faster data processing and bandwidth expansion) and potential threats (more vectors through which adversaries can attack). Ensuring security to particularly critical information infrastructure must consider this trend. This presumption is especially valid if we look at the potential abuse of devices within the so-called botnets to conduct targeted attacks against the information systems of both relevant state and non-state actors (for example, attacks on the availability of Telegram services in Asia in 2019 [5]).

Simultaneously intensifying the interconnection of humanity within this area increases the number of networks created and used on the distributive principle, i.e., without the existence of a central control or management "node". One such approach is the so-called "blockchain" technology, which is used by current cryptocurrencies and is gradually being introduced in other areas (e.g., banking or data and supply chain management and sharing) [6, 7]. The final form of this trend is an increase in the importance of the so-called "deep web", or in a narrower sense with the security connotations of "dark web" and "darknet" [8].

Notably, the dark web/darknet is directly linked to illegal activities across all areas (from illegal information gathering to trafficking in arms, addictive substances, or people). In addition to organized crime, similar means/possibilities are used, for example, by terrorist organizations and, in principle, by the states themselves. The consequence of this development is a further weakening of state power in the ability to control and regulate the activities and actors concerned and intervene against them as needed. This situation causes conflict between the protection of national interests on the one hand and the utility of such networks on the other hand. The interdependence of all areas of human society with cyberspace further develops a dependence on the availability of information. The digitalization of state administration and the transfer of links between the citizen and the state into this domain (e.g., in the form of electronic identity cards or elections) directly reflects this phenomenon, which, however, also brings new forms of vulnerability (e.g., the issue of manipulation with electoral systems). From this perspective, the Internet allows increased transparency in almost all activities in the real environment. Social media, such as Facebook, Instagram, Twitter, YouTube, or TikTok, allow almost constant monitoring and keeping track of the activities of individuals. At the same time, it serves as an ideal tool and platform for conducting information operations by both state and non-state actors. Therefore, the capacity to monitor these networks or their providers can be considered as an essential prerequisite for controlling and influencing public opinion in general. On the other hand, this aspect helps defend against a potential adversary's meaningful activities effectively. Building the independent Russian "Internet" RuNet (successfully tested in 2019 [9]), or increasing the effectiveness of the so-called "Great Chinese Firewall" [10] in this respect, combines the two characteristics discussed above.

Fundamental importance (not only) for this domain will be the full implementation of quantum (computational) technologies, which by their very nature surpass the performance of individual computing systems. Consequently, new possibilities such as processing and storing extensive data (Big Data), calculating corresponding threats/opportunities for current encryption tools and procedures, i.e., protecting data and information itself, are associated with this. In January 2019, the first "commercial quantum" computer (IBM Q System One) was introduced [11]. Quantum technologies have the potential to enhance the capabilities of AxS and their performance. Significantly ACS could benefit from mentioned big data processing and the possibility of preparing and running more complex scenarios and prediction models or algorithms [12]. On the other hand, these are still the first explanatory steps, and ensuring the widespread use of this technology is still a matter of long-term research and development.

3   Autonomous Systems

The development of AxS is mainly related to the development of information technology, where many factors enable the creation of systems capable of performing enormous amounts of calculations per second. Along with the advancement of robotics and mechatronics, it is now possible to create robotic systems with capabilities that fell into the science fiction category a decade ago. Individual systems should be able not only to obtain information about the environment but also to process (evaluate) this information and take appropriate decisions on their own [13]. By general definition, autonomy is the ability of an entity to make conscious, unforced decisions. To describe nuances among AxS, it is necessary to express the level of autonomy concerning the environment and the role of the human operator. For military purposes, there are several definitions of the level of control or the level of independence of the system on the human operator. Scharre, for example, distinguishes among semi-autonomous (Man-In-The-Loop), autonomous with supervisor (Man-On-The-Loop), and fully autonomous without supervisor (Man-Out-Of-The-Loop) systems [14].

The motivation to field these systems as soon as possible directly results from the increased demand for higher combat efficiency. As in the case of remote-controlled devices, the idea of minimizing casualties of own or friendly armed forces and non-combatants is represented [15]. Systems based on AI/machine learning elements more effectively suppress and eliminate human beings' physical and psychological limitations (including the need for sleep and the effects of fatigue or stress).

On the other hand, there are unanswered severe ethical and legal questions, e.g., the degree of autonomy that should these systems enjoy, and whether, at least from an ethical point of view, a decision to use force against human being can be taken purely by AxS [16]. This aspect is increasingly being discussed throughout the professional community [17] and is becoming a motivation for efforts to establish and enforce the arms control regime at the international level (e.g., under the auspices of the UN) [18]. On the other hand, it is necessary to point out that, following historical examples (e.g., cluster munitions or anti-personnel mines), the probability of achieving an overall ban across all states and enforcing it is somewhat unrealistic.

We can identify some aspects of these technologies on Guardium vehicles that can operate in a semi-autonomous mode [19]. Similarly, these elements are used in long-range missions of unmanned drones, where the human operator takes control of the UAV in the target area or surpasses air defense systems (e.g., the Phalanx point defense system) [20].

Furthermore, we can identify the considerable potential of the ACS based on AI/machine learning elements. The virtual identity of these systems ensures that they are most suitable to adapt to the regularities of cyberspace and fulfill tasks like the collection, evaluation, and processing of data and information in general. Their development and performance bring new possibilities, for example, for the detailed analysis of a large number of documents, visuals, or audio records. Consequently, related to this is the ability to imitate such data accurately, make copies, or even give them brand new features (such as a virtual person) almost indistinguishable from reality/originals (so-called deepfakes). Inherit connection of the ACS and cyberspace also enable further exploitation of this domain and related trends (see Sect. 2).

4   Theoretical Framework

4.1   Revolution in Military Affairs

For the purpose of this paper, we define RMA as the process and condition of revolutionary changes in the nature or method of warfare based on the external manifestations (actions) which employ the threat of force or the use of force to achieve political aims [21]. The "revolutionary" then refers to the radical nature of these changes, which, concerning the original system and its elements, must occur abruptly de facto preserving just a minimum similarity (e.g., in features by which the system is identified). Therefore, we cannot speak of a progressive (gradual) transition and the establishment of new elements into the existing framework and its evolutionary transition. Regarding the military dimension of this revolution, we can use the modified characteristics defined by Jeffrey R. Cooper, who speaks about: "… discontinuous increase in military capability and effectiveness" [22].

Relevant changes in the method of warfare are founded on a technological level with the introduction and use of advanced weapons and information systems (e.g., precision-guided munitions - PGM, unmanned aircraft, and remote sensing devices/sensors). From this point of view, the character of AxS (see Sect. 3) represent technology which disruptiveness is based on new opportunities (as well as threats) how-to, for example, deliver harmful effect and thus the transformation of ways of war [23].

The character of ACS further correlates with changes in the doctrinal dimension. They are represented by the establishment of the concepts of so-called System of Systems (SoS) and Network Centric Warfare (NCW). The first concept is based on two fundamental elements - information and integration (cooperation). The prerequisite amalgamates particular systems and components, such as command, control, computers, communications, and information (C4I), into one coherent functional framework [24]. This structure should provide situational awareness on the battlefield in real-time for all relevant components of the armed forces.

The second concept is associated with the very existence and use of communication links among the units on the battlefield and their integration into the mentioned framework. Their interdependence allows them to maximize utilization of their combat skills and, on the other hand, compensate for weaknesses (e.g., through almost excellent fire support, information about the enemy's intentions). Full use of this potential is connected, e.g., to the implementation of the so-called "swarming" tactic, which in itself implies synchronized and highly flexible combat deployment of a large number of small clusters (military units) [25].

In practical terms, the army, which fully applies both concepts, is allowed to interfere (invade) the opponent accurately at his most vulnerable areas to prevent his possible attempts to initiate counterattacks or enact countermeasures and therefore wholly take over the combat initiative and paralyze the opponent.

4.2 Main Capability Areas (MCA)

The character of the armed forces could be analyzed in many ways and approaches. To ensure that conclusions of this paper would be valid in terms of NATO's documents and planning processes (especially NATO Defence Planning Process), NATO's analytical framework will be used. This framework covers the spectrum of tasks and phases in military operations that armed forces should fulfill successfully and prepare their capabilities for them. The concept of Main Capability Areas covers all this necessary development. These capability areas demonstrate a complex approach to tools of the military power of (NATO) states with the ambition to identify the ideal composition of forces that would maximize the potential of each state and diminish the threats and risks. The analytical framework identifies seven areas – *Prepare; Project; Engage; Consult, Command and Control (C3); Sustain; Protect; Inform* [26]. Interconnection between cyberspace and ACS in terms of RMA will be in this paper analyzed in all these areas to identify possible opportunities.

Prepare subsumes capabilities to establish, prepare and sustain adequate presence at the right time, including the ability to build up forces, through appropriate and graduated readiness, to meet any requirements, keeping sufficient flexibility to adapt to possible changes in the strategic environment. *Project* represents capabilities to conduct strategic deployment of the armed forces in support of any NATO and national mission. *Engage* is characterized by performing the tasks that contribute directly to achieving mission goals within the context of collective defense, crisis management, and cooperative security. It includes all capabilities required to defeat adversaries as well as accomplish the goals of other non-combat missions. *C3* are capabilities of commanders to exercise authority over the full spectrum of assigned and attached forces in the accomplishment of the mission. Including, for example, the capability to communicate and coordinate with other actors who are present or involved in the operational area and effective information exchange with the political and military leadership. *Sustain's* capabilities serve to plan and execute the timely support and sustainment of forces, including essential military infrastructure, transportation, military engineering support, contracting, supply/maintenance/services management, basing support, and health and medical support. *Protect* represents capabilities to minimize through a common multinational and holistic approach of Force Protection the vulnerability of personnel, facilities, materiel and activities to any threat and in all situations, to include towards the effects of WMD, while ensuring the Allies freedom of action and contributing to mission success. Finally, *Inform* subsumes capabilities to establish and maintain the situational awareness and level of knowledge required to allow commanders at all levels to make timely and informed decisions [27].

5   Opportunities for the Armed Forces

5.1   Prepare

Interconnection between cyberspace and ACS gains two fundamental features in *Prepare* capabilities. The first one subsumes the possibility to train and prepare all armed forces branches through ACS. The second one targets how to train/prepare AxS in other domains to enhance our armed forces. In the first case, ACS exploits trends such as virtual reality development discussed in Sect. 2. Cyberspace represents here a domain that enables the existence of such training programs and could be used as a training field for military personnel. Comply with the use of enhanced or virtual reality, ACS creates an opportunity to bring simulation as close as possible to the real world and scenarios. For example, soldiers would be able to employ even lethal munition against an (artificial) adversary without endangering another human person who would, in traditional scenarios, take this role. Simultaneously, ACS connection with cyberspace and related computing tools enables preparation, running, and evaluating much more complex and comprehensive (real-time) scenarios than human observers could implement. These features are even highlighted in the case of training scenarios in cyberspace. Even today, the insufficient technological development of power sources and other related scientific fields cannot limit the ACS in this domain. Practical application try to address, for example, Neural 3D holography project [28].

The second-mentioned feature – how to train physical AxS – cyberspace represents an ideal platform for effective and quality learning. Direct connection of the ACS to this domain enables the development of required traits, skills, and knowledge faster and more effectively than would be possible through classical educational methods. Of course, nowadays, this process is mainly represented by machine or deep learning. However, we can presume that further development of cloud and edge computing. Related analytical tools could move this feature beyond this concept and, through new levels of complexity, evolve it into the possibilities of almost full-fledged AI. The critical challenge in this process is to ensure that these systems would learn what we want and need and diminish the possibility of "mislead and failed" cases.

5.2 Project

The almost non-existent influence of distance in cyberspace is further highlighted by the computing and processing power of the ACS. This feature enables the precise and simultaneous projection of much higher numbers of units and other armed forces components than is possible by traditional means and processes. Of course, outside cyberspace, "conventional" transportation components are valid and mainly establish projection capabilities. On the other hand, even in these domains, ACS could minimally enhance, for example, air or naval transports through synchronization and control of their components to, for example, reduce fuel consumption, or ensure efficient use of cargo space. In cyberspace, ACS can project power almost independently. Virtual domain diminishes any physical limitation, and AI can fully employ all advantages over human operators. The impact of establishing new capabilities or enhancing existing ones in this area is incredibly revolutionary if we discuss small military powers. The combination of ACS and cyberspace enables them to gain global military power projection, which would be almost impossible under different conditions.

5.3 Engage

Practically, even nowadays, elements of ACS could be deployed as full-fledged agents. All kinds of cyber operations are manageable by ACS. It does not matter if the task is to steal information about the adversary's forces, disrupt his chain of command by DDoS attacks, or lead a disinformation campaign through its communication channels and even other (social) media. Contrary to human operators, ACS profit from their direct connection with cyberspace. They are part of it. Basically, and with a bit of imagination, we can consider cyberspace something like the natural environment for ACS. This feature ensures that these systems are best suited for operations in it. From this point of view, each discussed characteristic of the ACS provides enhancement and new possibilities for breaking through the adversary's defense or fulfilling any other tasks. This presumption is highlighted by other mentioned tools such as quantum computing which would provide new levels of performance. For example, dedicated ACS could establish not only a botnet of "zombie" systems but a network of autonomous agents that will have the advantage of the hive mind [29]. Their operations will be much faster, precise, and less detectable than nowadays. The same advantages we can identify in information operations that influence or deceive the adversary. ACS has made the process of creating convincing fake videos much easier and faster [30]. So-called deepfakes impersonate ACS to perpetrate various information campaigns, including phishing attacks. Fake politicians or commanders could deliver our messages with unprecedented credibility and reliability to the targeted audience. All these aspects enhance social engineering tools and potentially sow distrust into the capability to distinguish the real world from the virtual one.

5.4 C3

C3 capabilities and the role of ACS are deeply connected with the concept of Network Centric Warfare and System of Systems described above. Emphasis on complex interconnection among all elements of armed forces and network of functional links increases demand on communication systems and data gathering and analysis. This is another case where so-called Big Data is produced, processed, and converted into the information necessary for all elements of armed forces according to their needs. ACS could fulfill such tasks and provide a robust network even with (near) real-time functions. Simultaneously, ACS provides enhanced support to the decision-making process. At tactical, operational, or strategical levels, such support establishes ideal conditions for evaluation all possibilities, choosing the best one, and implementing it. In these terms, ACS also provides enhanced opportunities for modeling and simulations. Access to processed and evaluated Big Data creates a unique database for different M&S tools [31]. ACS ensures that this database will be up to date and employ mentioned tools in real-time, and constantly provide necessary information to command structure and even make well-timed and precise decisions in a given situation.

5.5 Sustain

The usability of ACS to enhance sustain capabilities can be identified through the whole logistic and support system. Some possibilities, such as reducing fuel consumption or efficient use of transportation capacities, were discussed earlier. However, ACS provides and ensures complex support to armed forces in this area. Through cyberspace and physical sensors, ACS could process data from all components in real-time. They can provide information about their status and needs, evaluate them and issue orders to relevant agents. For example, autonomous medical systems can monitor the health of the soldiers and, in the case of degradation, send proper help even the soldier realizes that something is wrong.

Similarly, an autonomous logistic system can provide timely supplies or repairs. Simultaneously, ACS is not limited to these tasks. They could prepare predictive models for future (logistical) conditions updated constantly through data processing and evaluation. Application in so-called Deep Logic Networks [32] effectively enables the allocation of available resources and diminishes the possibility of shortage or unavailability of services. Based on the level of autonomy, ACS could also set and maintain the whole structure of sustain capabilities and decide about employing them.

5.6 Protect

Protect capabilities in cyberspace are critical parts of credible and effective exploitation of the ACS and other information-based systems in general. In this case, cybersecurity became the vital interest of all actors. The character of ACS is inseparably connected with information technologies and cyberspace itself. This nature implicitly creates vulnerabilities of the ACS, especially in terms of malicious cyber activities which could target these systems. It does not depend on if we discuss cyber sabotage or cyber espionage. Practically every system could be compromised and used against our interests. Also, the same threats are valid for every domain where AxS would be deployed. Essentially, adversaries could hijack an autonomous plane like an AI agent in cyberspace. This possibility should be considered at every level of decision-making (from tactical to strategic) and developing safety- and countermeasures reflecting the share of dependency on these systems.

On the other hand, ACS provides new possibilities to ensure credible security and defense (not only) in cyberspace. Contrary to human operators, AI can analyze, process, and even control vast data flows. Practically, this increases the probability of detecting any malicious cyber activity targeting our (information) systems and provides new tools to counter it. The same logic could be applied to possibilities of deepfakes identification and bolstering resilience against adversaries' information operations [33]. Simultaneously, ACS enhances capabilities to identify such attacks in the preparation phase and provide early warning and enough time to related authorities to react. However, such conditions are related to the interoperability of deployed systems. Without it, AI would not develop its full potential throughout the whole protected structure and its elements.

5.7 Inform

ACS' role in this area is deeply enrooted in possibilities from *C3* capabilities. These elements manage to provide and control the whole network of sensors in every domain and follow-on communication among all components. Moreover, all intelligence disciplines (OSINT, IMINT, TECHINT, even HUMINT) are dependent on the capability to gather and analyze required information and provide it to stakeholders. Transmission and processing of all data in cyberspace help gain (near) real-time connection, which pace and throughput depend more on human limitations than on technical ones. ACS could also ensure that threat of the information overwhelming will be diminished. Permanent presence and information monitoring through these systems enable to filter off all data much more preciously and with a proactive approach to providing them to relevant subjects.

6 Conclusion

Autonomous systems and their interconnection with cyberspace represent opportunities and challenges for the NATO member states' armed forces. Opportunities can be identified in all Main Capability Areas - Prepare; Project; Engage; Sustain; Consult, Command and Control; Protect; Inform. Specific features of cyberspace combined with ACS characteristics like suppression and elimination of the human beings physical and psychological limitations (including the need for sleep and the effects of fatigue or stress) provide new possibilities related to information basis of every (military) operation and network of NATO member states' armed forces components. The development of ACS is also connected with enhanced exploitation of concepts System-of-Systems and Network Centric Warfare in terms of Revolution in Military Affairs. In *Prepare,* opportunities arise not only in training and preparation of all branches of armed forces through ACS but also ACS themselves. *Project, C3, Sustain* and *Inform* areas could benefit from modeling and simulation capabilities and new ways of data processing. *Engage* area is influenced by the inherited relation between ACS and cyberspace. AI agents represent the most effective tool to fulfill set tasks in this domain. ACS could bolster the cybersecurity and defense of Allies' systems and counter malicious cyber activities in the Protect area.

On the other hand, the main challenges come from the discussed dependency of ACS on cyberspace and information technologies. This connection creates potential vulnerabilities that adversaries could exploit. Especially in *Protect* area is crucial to develop and implement safety- and countermeasures on every level of the military structure, which could at least mitigate them. There is also a need to ensure interoperability of deployed systems, not only with the allies within NATO but also internally, through generations of weapons and other hardware classes. Interoperability and mutual compatibility strengthen the resilience of the entire structure (robustness and substitutability) and increase the efficiency of individual elements.

References

1. Gettinger, D.: Drone Databook Update: March 2020. Center for the Study of the Drone at Bard College (2020). https://dronecenter.bard.edu/projects/drone-proliferation/drone-databook-update-march-2020/
2. Hodicky, J., Prochazka, D.: Challenges in the implementation of autonomous systems into the battlefield. In: 6th International Conference on Military Technologies, ICMT 2017, Brno, pp. 743–744 (2017)
3. Stojar, R., Frank, L.: Changes in armed forces and their significance for the regular armed forces. In: The 18th International Conference. The Knowledge-Based Organization: Conference Proceedings 1 - Management and Military Sciences, vol. 1, pp. 142–145 (2012)
4. NATO: Strategic Foresight Analysis Report (2017)
5. Shieber, J.: Telegram faces DDoS attack in China…again (2019). https://tcrn.ch/2ZJgfbc
6. Karl, A.: Blockchain Technology for Cloud Storage: This Looks Like Future. Tech Genix (2018). https://1url.cz/yM4zC
7. Kelly, J.: Top Banks and R3 Build Blockchain-Based Payments System. Reuters (2017). https://1url.cz/vM4zs
8. Sui, D., Caverlee, J., Rudesill, D.: The Deep Web and Darknet: A Look Inside the Internet's Massive Black Box. Wilson Center (2015). https://www.wilsoncenter.org/sites/default/files/media/documents/publication/deep_web_report_october_2015.pdf
9. Wakefield, J.: Russia 'successfully tests' its unplugged internet. BBC (2019). https://bbc.in/2X7vyst
10. Wyciślik-Wilson, M.: It is getting harder than ever for VPNs to break through the Great Firewall of China. Beta News (2019). https://bit.ly/3gr4v32
11. Russell, J.: IBM Quantum Update: Q System One Launch, New Collaborators, and QC Center Plans. HPC wire (2019). https://1url.cz/kM4K2
12. Swayne, M.: Four Ways Quantum Computing Will Change Artificial Intelligence Forever. The Quantum Daily (2020). https://thequantumdaily.com/2020/01/23/four-ways-quantum-computing-will-change-artificial-intelligence-forever/

13. Hodicky, J.: Modelling and simulation in the autonomous systems' domain – current status and way ahead. In: Hodicky, J. (ed.) MESAS 2015. LNCS, vol. 9055, pp. 17–23. Springer, Cham (2015). https://doi.org/10.1007/978-3-319-22383-4_2

14. Scharre, P.: Autonomy, "Killer Robots," and human control in the use of force – Part I. Just Security 7(1) (2014). https://www.justsecurity.org/12708/autonomy-killer-robots-human-con trol-force-part/

15. Stojar, R.: Bezpilotní prostředky a problematika jejich nasazení v soudobých konfliktech (The Unmanned Aerial Vehicles and Issues Connected with Their Use in Contemporary Conflicts). Obrana a strategie 16(2), 5–18 (2016)

16. Fučík, J., Frank, L., Stojar, R.: Legality and legitimacy of the autonomous weapon systems. In: Mazal, J., Fagiolini, A., Vasik, P. (eds.) MESAS 2019. LNCS, vol. 11995, pp. 409–416. Springer, Cham (2020). https://doi.org/10.1007/978-3-030-43890-6_33

17. Autonomous Weapons: An Open Letter from AI & Robotics Researchers (2017). https://goo.gl/X2N6CA

18. Gill, A.S.: The role of the united nations in addressing emerging technologies in the area of lethal autonomous weapons systems. UN Chronicle 50(3&4), 15–17 (2018). https://1url.cz/3zZjq

19. Army-Technology.Com: AvantGuard Unmanned Ground Combat Vehicle, Israel (2016). https://goo.gl/knZqWb

20. Raytheon: Phalanx Close-in Weapon System: Last Line of Defense for Air, Land and Sea (n.d.). https://goo.gl/Ky3RD1

21. Gray, C.S.: Strategy for Chaos - Revolution in Military Affairs and the Evidence of History. Frank Cass, London (2005)

22. Cooper, J.R.: Another View of the Revolution in Military Affairs. Strategic Studies Institute (1994). https://ssi.armywarcollege.edu/another-view-of-the-revolution-in-military-affairs/

23. Worcester, M.: Autonomous Warfare – A Revolution in Military Affairs. ISPSW, Berlin (2015). https://www.files.ethz.ch/isn/190160/340_Worcester.pdf

24. Owens, W.A., Offley, E.: Lifting the Fog of War. Farrar Straus Giroux, New York (2000)

25. Alberts, D.S., Gartska, J.J., Stein, F.P.: Network Centric Warfare: Developing and Leveraging Information Superiority. CCRP, Washington D.C. (1999)

26. NATO: MC 400/3, MC Guidance for Military Implementation of Alliance Strategy (2012)

27. NATO: C3 Taxonomy Baseline 3.1 (2019)

28. Choi, S., et al.: Neural 3D holography: learning accurate wave propagation models for 3D holographic virtual and augmented reality displays. ACM Trans. Graph. 40(6) (2021). https://www.computationalimaging.org/wp-content/uploads/2021/08/NeuralHolography3D.pdf

29. Ciancaglini, V., et al.: Malicious Uses and Abuses of Artificial Intelligence, p. 35 (2020). https://documents.trendmicro.com/assets/white_papers/wp-malicious-uses-and-abu ses-of-artificial-intelligence.pdf

30. Lyu, S.: Deepfakes and the New AI-Generated Fake Media Creation-Detection Arms Race. Scientific American (2020). https://www.scientificamerican.com/article/detecting-dee pfakes1/

31. Hodicky, J., Prochazka, D.: Modelling and simulation paradigms to support autonomous system operationalization. In: Mazal, J., Fagiolini, A., Vasik, P. (eds.) MESAS 2019. LNCS, vol. 11995, pp. 361–371. Springer, Cham (2020). https://doi.org/10.1007/978-3-030-43890-6_29

32. Pandey, S.: Opportunities to use artificial intelligence in Army logistics (2019). https://www.army.mil/article/216389/opportunities_to_use_artificial_intelligence_in_army_logistics

33. Divišová, V., et al.: "The whole is greater than the sum of the parts" towards developing a multidimensional concept of armed forces' resilience towards hybrid interference. Obrana Strat. 21(1), 8–15 (2021). https://doi.org/10.3849/1802-7199.21.2021.02.003-020

Current State and Design Recommendations of Exoskeletons of Lower Limbs in Military Applications

Lydie Leova[1]([✉]), Slavka Cubanova[1], Patrik Kutilek[1], Petr Volf[1], Jan Hejda[1], Jan Hybl[1], Petr Stastny[2], Michal Vagner[2], and Vaclav Krivanek[3]

[1] Faculty of Biomedical Engineering, Czech Technical University in Prague, Sitna Sq., 3105, Kladno, Czech Republic
{leovalyd,slavka.cubanova,kutilek,petr.volf,jan.hejda, jan.hybl}@fbmi.cvut.cz

[2] Faculty of Physical Education and Sport, Charles University, José Martího 31, Prague, Czech Republic
{stastny,vagner}@ftvs.cuni.cz

[3] Faculty of Military Technology, University of Defence, Kounicova 65, Brno, Czech Republic
vaclav.krivanek@unob.cz

Abstract. With the development of new wearable technologies, the use of exoskeletons is gradually gaining ground in the world's advanced militaries. The aim of this paper is to explore the current status of exoskeletons and describe the requirements of exoskeletons for military use towards the specifics of use. The most important movement of a soldier in the field is walking, and therefore the analysis of the current state will focus primarily on lower limb exoskeletons and their subsystems. The paper will compare active and passive lower limb exoskeletons and their use in military practice. Subsequently, the requirements for individual subsystems of the mechatronic system of the exoskeleton will be specified. Sensor subsystems, actuator subsystems, control subsystems and man-machine interface will be described. Recommendations for further analysis and use of exoskeletons in the military can thus contribute not only to increased performance but also to the safety of the users themselves - individual soldiers.

Keywords: Exoskeleton · Lower limbs · Military

1 Introduction

Soldiers have to perform many physically demanding activities on a daily. These tasks can lead to increased neuromuscular fatigue and decreased mobility, which can ultimately lead to acute or chronic injuries. Whether acute or chronic, injuries sustained by soldiers can cause significant financial costs and loss of time to the armed forces. Therefore, militaries around the world have an eminent interest in keeping their soldiers healthy and protected from injury.

J. Mazal et al. (Eds.): MESAS 2021, LNCS 13207, pp. 452–463, 2022.
https://doi.org/10.1007/978-3-030-98260-7_29

Although injuries can occur during combat, non-combat injuries are more common and can arise for a variety of reasons. For example, approximately 75% of U.S. Army recruits sustain a non-combat musculoskeletal (MSK) injury during basic or advanced training. During such exercises, and depending on the specific tasks and requirements, personal protective equipment and a backpack are often worn, all of which can total approximately 20–70 kg, which in some cases must be carried for many miles over many days. As a result, this amount of additional weight carried can lead to unusual movement patterns and increased stress on joints, which can increase the likelihood of non-combat MSK injuries.

These injuries are often related to movement during high-speed changes of direction during sprinting, jumping, and landing, including the transition between these movements. According to a technical report from the U.S. Army Defense Technical Information Center, 97% of injuries are due to mechanical energy transfer, while 86% and 69% are due to MSK and cumulative microtrauma injuries, respectively. This increased incidence of injuries is related to the lack of physical training of soldiers and the overall amount of weight soldiers carry and wear while marching, running, and jumping, with most injuries occurring in the lower extremities and lower back. Soldiers' physical fitness must therefore take these loads and risk factors into account. For the above reasons, military experts consider the use of the exoskeleton to be a promising means of addressing these issues. The aim of this article is to describe existing technologies and to highlight those promising for military service.

2   An Overview of Current Types of Lower Limb Exoskeletons in Military Applications

2.1   Methodology

The aim of this article was not to conduct a systematic literature review with precise search criteria but to propose a thematic review focused on military applications based on recent articles and those already known to the authors.

Two databases were used for the literature search - IEEE Xplore (Institute of Electrical and Electronics Engineers and Institution of Engineering and Technology) and ScienceDirect (Elsevier). Combinations of keywords such as an exoskeleton, military, kinematics were used as cardinal search terms. Preference was given to publications from 2010, however, in some cases, this range was extended. In addition to the database searches, a manual search of a reference list of selected articles and the authors' institutional databases was conducted.

To identify sources that provide more explicit information on lower limb exoskeletons in military use, we further reviewed searches focused on lower limb exoskeletons [1–4]. However, none of the articles focused on the lower limb exoskeleton in military applications.

2.2   Types of Lower Limb Exoskeletons in Military Applications

In general, exoskeletons can be divided into two categories – medical and non-medical [5]. Medical exoskeletons are out of the scope of the paper/and not more discussed in the article. Non-medical exoskeletons belong to the category of personal care robots, which are designed for various applications to assist healthy individuals in demanding activities [5]. Non-medical exoskeletons also include military exoskeletons, which were primarily developed to enhance the soldier's abilities so that the weight of the transported cargo is reduced to the lowest possible weight to reduce the metabolic cost of transportation [5, 6]. Lower limb exoskeletons based on joint motion (hips, knees, ankles) are classified into several types. Actuators can drive only one joint, a combination of joints, or all 3 joints, [7]. The source of actuator movement is distinguished as active and passive. Active exoskeletons use an energy source to activate the actuator and can be electric, pneumatic, or hydraulic [7, 8]. Passive exoskeletons, on the other hand, have no energy source. This type uses kinematic forces, for example, by using springs and shock absorbers (see Fig. 1).

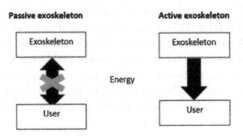

Fig. 1. Types of exoskeletons.

2.3 Passive Exoskeletons

With this type of exoskeleton, no external power supply is needed, which is quite an advantage as it works for a long time. Through mechanical levers supported by biological mechanics and human movement and weight-bearing mechanisms, the exoskeleton moves the soldier's load to a better position [9]. This reduces the metabolic cost of transport, improves endurance, and reduces soldier fatigue. In recent years, some countries have made significant progress in the development of passive exoskeletons [9]. For example, a Canadian company, Bionic Power, has developed a new prototype passive exoskeleton called the Power Walk. The principle is that a gearbox converts the rotational speed of the knee to a higher speed for efficient power generation, and a generator then converts the mechanical energy produced into electrical energy. The energy conversion circuit then converts the electricity to recharge, for example, Li-ion batteries. This development can eliminate the carrying of heavy batteries, especially when the user is on a mission, and reduce the logistics tail and the unit's reliance on field resupply and extend the duration and effectiveness of the mission [9]. Lockheed Martin of the United States has developed a prototype passive exoskeleton called FORTIS that sends the weight of the load directly into the ground, significantly reducing the weight of the soldier's equipment. The latest passive exoskeleton we have featured is a prototype called the Marine Mojo from the British company Twenty Knots Plus (20KTS+). Unlike the previous prototypes, this exoskeleton has a different specialized use. For the coastguard, police, and government wildlife agencies to patrol bays and rivers, as this prototype can absorb shock and vibration from standing on fast-moving small watercraft. This exoskeleton provides relief from muscle fatigue, especially in the knee area, which reduces the likelihood of injury and increases crew alertness. Not only exoskeleton development is under progress but also the rules to define them for special forces. For example, during the years 2017 and 2019 there was a NATO group to define conditions to deploy exoskeletons for support of explosive ordnance disposal (see Fig. 2).

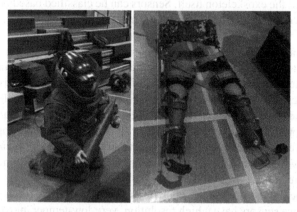

Fig. 2. Military application usage for support of explosive ordnance disposal [10].

2.4 Active Exoskeletons

In this type of exoskeleton, electrical, hydraulic and/or pneumatic actuators are required to power the human joints in place of muscles. Approximately two decades ago, a research group at Karolinska University in Berkley developed the prototype Berkeley Lower Extremity Exoskeleton, abbreviated BLEEX, intending to help soldiers carry heavy loads during missions [11]. The exoskeleton consists of two powered anthropomorphic legs, a propulsion unit, a frame, and a backpack on which the load can be placed. The exoskeleton can carry up to 100 kg of cargo and move at a speed of 4.68 km/h with a cargo weight of 34 kg. More recent prototypes include an active exoskeleton called ONYX from Lockheed Martin. Its control system, through sensors placed on the feet, knees and hips, can detect the nature of the terrain and to some extent predict the user's next move, providing effective assistance regardless of whether they are carrying a load. The Hercule V3 system from French company RB3D has Support points at the joints of the feet and legs and can carry a load of 100 kg. The battery allows the load to be carried 20 km at a movement speed of 4 km/h. The South Korean company LIG NEX1 has developed a prototype called LEXO with hydraulic propulsion that increases the endurance and strength of the user. The exoskeleton is foldable and can be easily transported in a hardened container. The system can support a maximum load of 90 kg with a battery life of 4.5 h.

2.5 Types of Sensors and Actuators of Lower Limb Exoskeletons in Military Applications

Sensors. Many types of sensors can be used for evaluation purposes in the case of lower limb exoskeletons. These sensors can sense the activity of the exoskeleton itself or the user's biosignals, e.g., in the form of lower limb electromyography (EMG) [12]. The electrodes are placed on the musculus tensor fasciae latae, musculus rectus femoris, musculus biceps femoris and musculus gastrocnemius [12]. This section focuses only on the sensors of the exoskeleton itself. Sensors can be classified based on the measured quantity or based on the physical nature of the measurement. In this case, the first option will be used.

Inductive angular rotation sensors are non-contact and are mainly used in places exposed to high vibrations or shocks or in aggressive environments like industrial applications, army, automotive, vehicles and ships. Measurement of the angle using the change of the magnetic field is allowed by one or more Hall effect chips implemented in the sensor [13]. The sensors have a body made of metal and fully hermetically sealed electronics with protection class IP68 in industry or IP69 deep underwater applications. The sensors measure magnetically different domains around the circumference of the shaft or in versions without shaft, the sensors measure the position of one or more permanent magnets mounted on the measured object.

Optoelectric angle sensors have code disks whose tracks are scanned numerically. [2, 14, 15]. These sensors have a high resolution, very low-temperature dependence.

In industry, automatization is used optical absolute angle encoders like the ECA4410 series from Heidenhain GmbH. The sensors can be used in single-sensor systems for applications with SIL2 category control (as per EN 61 508) and power level "d" (as per EN ISO 13 849). Reliable position transfer is based on two independently generated absolute position values and error bits, which are then provided for safe control. The sensor uses serial data transfer via bidirectional interface EnDat 2.2. A similar industry alternative is ATOM Renishaw's plc. miniature non-contact optical incremental linear and rotary position measuring sensors with several interfaces, like Ti, Ri and compact ACi PCBs. In own made machines is possible to use low-cost optic sensors like HEDS-9100 from Avago Technologies. The sensor module, pair of sensors consist of a lensed (LED) source and a detector IC enclosed in a small C-shaped plastic package and code wheel disk. Due to a highly collimated light source and unique photodetector array, these modules are extremely tolerant to mounting misalignment. Sensors have standard resolutions between 96CPR and 512CPR and they are available for use with a HEDS-5120 code wheel or equivalent. The output signal of single and multi-turn sensors can be analogue current (in range 1 to 100 mA) or voltage (in range 3.3 to 24 V DC) or digital with CAN, SPI, I2C or RS-487 wires communication. A special feature of the multi-turn sensor is the signal platforms at the beginning and end of the range, which prevent a sudden change in signal. For example, the HSM22M magnetic rotary encoder detects angles up to 3 600° (10 turns) with absolute reliability. Metal plain bearings offer a service life of at least 20 million shaft revolutions. A permanent shaft control speed of up to 400 rpm is permissible.

Resistance rotation and position sensors use precision rotary single-turn or multi-turn resistance potentiometers [2, 16] or diaphragm potentiometers. They are suitable for lower speeds and have a limited-service life of tens of thousands of rotations. They are absolute sensors with a range of one to 10 rotations.

The relative position of the lower limb segments is further allowed by inertial measurement units (IMUs) combining an accelerometer, gyroscope and usually a magnetometer [17].

For foot-ground contact assessment, force-sensitive resistor sensors placed under the foot can be used [18, 19]. Other approaches are special hermetically sealed Micro-Electromechanical Systems (MEMS) pressure transducers [20] or optoelectric sensors using an optical distance sensor to measure the relative distance between two aluminium plates, which is dependent on the applied force [21]. Exemplary control systems and methods for controlling a prosthetic or orthotic device by proximity sensors, load sensors, accelerometers, tactile sensors, pressure sensors, and others are shown in the next picture inspired by a European patent EP2434992B1 and US 2006/0135883 patent. How the patents show, the future of reliable measurement and control is in independent sensors cooperating as cognitive network units (Fig. 3).

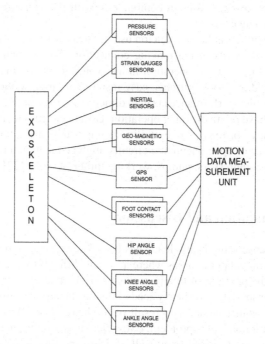

Fig. 3. Types of movement sensors in lower limb military active exoskeletons.

Another alternative to measuring the angle are encoders that use the measurement of the angular position of the rotary shaft, most often by the optical method [2, 14, 15]. Simple potentiometers can be used to measure the angle of the lower limb segments [2, 16]. Measurement of the angle using the change of the magnetic field is allowed by Hall effect sensors [13].

Torque or torsion sensors again use several technologies. Stress-optic effects are based on deformation of polymer optical fibre (POF) [22], torsion sensors based on measuring deformation by strain gauges of components such as shafts and beams by the action of a resistive force against the force of the actuator [23], magnetic-inductive torque sensors base on emits an alternating magnetic field, which penetrates the surface of the object and the mechanical forces acting on the object influence the alternating magnetic field and its properties and these changes are detected by the sensor. The last indirect possibility is measuring the input power of actuators and changes in actuators movement characteristics, like hold torque of DC motors or working pressure at the pistons.

Actuators. The purely mechanically driven exoskeleton is designed to use only mechanical elements (springs, cables, joints, etc.) without additional actuators [21]. The energy expended by the user in initiating the motion is used to complete it or compensate for the motion failed.

Direct Current (DC) servo motors [24, 25] and DC brushless motors (BLDC) [26] and stepper motors in some cases in combination with planetary gearheads [26], are most often used as actuators.

The pneumatic or hydraulic exoskeleton uses a linear piston on an articulated rotary joint between the two parts of the moving connection. The advantage of both types of technologies is the safety of operation in an explosive environment (no sparking), smoothness of speed and force settings, easy installation, long service life. Hydraulic drives achieve very high performance at small dimensions, have high control and regulation accuracy and have relatively quiet operation. Disadvantages are air compressibility (uneven movement of pistons), susceptibility to impurities in the transmission media and noise in the case of open pneumatic circuits, high acquisition and operating costs, size and weight of compressors, pumps, working fluid reservoirs [27].

3 Reccommendation and Discussion

The problem of transporting heavy loads has been known since antiquity. Particularly in terrains such as mountainous areas where access is difficult for mechanical transport systems, causing the problem of timely provision of the battlefield and logistical support. Thus, the transportation of heavy loads falls on the soldier himself, which leads to fatigue and subsequent injury as a result of the heavy load. Wearable exoskeleton systems can be a solution to this problem. Exoskeletons are multifunctional, mobile and user-friendly. They are designed specifically for transporting heavy loads in difficult terrain. The soldier becomes more efficient in the field because he is more logistically independent and can march faster and longer. Among the current military exoskeletons for lower limbs developed worldwide, the BLEEX (100 kg), Hercule V3 (100 kg), Lexo (90 kg) prototypes can be used for this purpose, which is designed to carry very heavy loads.

Soldiers deployed at high altitudes face inaccessible terrain, often having to climb mountain peaks. Rough terrain can cause several health problems due to physical overload. The solution is the use of lightweight exoskeletons that support the soldier during movement, improve performance, reduce fatigue, and prevent injuries. The soldier is thus able to cover longer distances in inaccessible terrain and much faster. For these purposes, the ONYX, FORTIS, Powerwalk prototypes can be used from the current military exoskeletons for lower limbs developed in the world. These exoskeletons are designed to assist the soldier in movement, reduce tension in the leg muscles, increase strength and endurance.

It can be assumed that in the future exoskeletons will become a standard part of every soldier's equipment. In general, exoskeletons should allow soldiers deployed on missions to lighten their load, enable them to march faster, reduce fatigue and increase their combat capability. When used for routine activities outside mission deployments, exoskeletons should increase work efficiency, reduce the risks associated with work activities and improve work performance. There is also a main difference to the medical exoskeleton. Because soldiers wear their equipment for several days, it must fit the solder body perfectly. If not, soldiers will refuse this part of their outfit. This generates pressure for exoskeleton price and delivery time.

Research on wearable exoskeletons has developed significantly in the last two decades. This is evidenced by countless studies, articles, and research that develops prototypes of military exoskeletons and tests them in practice. As noted, exoskeletons are used to improve endurance and reduce fatigue, increase walking ability over uneven terrain, reduce muscle strain, and carry heavy loads with minimal metabolic cost, thereby improving endurance and reducing fatigue in soldiers. The goal of a soldier's successful completion of a task is to increase combat fitness, which is closely related to the speed of movement in difficult terrain with heavy loads. However, there is a contradiction between the weight and speed of walking of soldiers. The passive exoskeleton weighs about 8 kg. When solder is without any equipment one feels it as a useless load. For eight kilograms and more of gear soldiers perceive exoskeleton as a benefit.

The load of the army in high mountain terrain should be less than 20 kg, which is commonly exceeded in practice as the demands on soldiers are increasingly higher. In real wars, a soldier has to carry even more than 60 kg [6]. With the development of military exoskeletons, this contradiction seems to have been resolved. However, the weight of the exoskeleton itself is still high, which can be uncomfortable for the user when wearing it. The main factor affecting the resulting weight of the exoskeleton is the material.

To make the exoskeleton mobile, most of them use independent power supply through batteries. Due to the high power, a problem arises during operation. To extend the operating time, the energy efficiency of exoskeletons needs to be increased, which, given current battery technology, means a high battery weight, which in turn affects the overall weight of the exoskeleton. However, there is a prototype Powerwalk from BionicPower that generates electricity from natural motion. The principle is that a gearbox converts the rotational speed of the knee to a higher speed to efficiently produce energy, and a generator then converts the mechanical energy produced into electrical energy. The energy conversion circuit then converts this electricity to recharge, for example, Li-ion batteries. In addition, ergonomic considerations are also important, in particular the human-skeleton interface. The exoskeleton is attached to the user at several points by straps.

Due to the anatomical differences of each of us, it is very difficult to design a universal clamping system. The interface should be adapted to the contours and anatomical needs of the individual. The dimensions of the exoskeleton should therefore be adjustable to suit different users. Special attention should also be paid to the safety of the user himself. The user is attached to the exoskeleton by straps. His movement is restricted. In the event of a malfunction or other problem, the exoskeleton should have an emergency shutdown and quick release from the suit. Also, the mechanical structure of the exoskeleton should be designed to limit the range of motion of each joint. Other safety factors should also be considered, specifically the protection of the battery and other measurement systems in, particularly challenging conditions.

In addition to the basic frame of the exoskeleton, equally important is the sensor part which allows the measurement of basic parameters. These parameters include, for example, the relative position of body segments, pressure distribution, etc. The information from these sensors can be used for analysis or feedback control of the exoskeleton, usually using DC stepper motors as actuators. In the future, we can expect the development of haptic technologies that would provide feedback on, for example, the type of surface under the foot. This information is often limited by the exoskeleton itself and can be a source of unsteadiness.

Electric actuators are commonly used in industrial manipulators, they are available in a myriad of types from hundreds of manufacturers. Why aren't biomedical skeletons widely available and used? One answer is weight. Industrial manipulators with outputs that can be used as alternative locomotives for humans are unacceptably heavy, reaching hundreds of kilograms to units of tons. Weight reduction is not a required feature of manipulators in the industry and therefore the price increases with each weight saving. For example, a worm gearbox (1 kg at 38 N) can be up to 5 times heavier than a harmonic gearbox (1 kg at 170 N) but up to 10 times cheaper. The value of lightweight parts and the price of control electronics are perfectly visible on the commercially available Atlas and Spot robots from Boston Dynamics. Table 1 provides an overview of the actuators and their characteristics depending on the type.

Table 1. Summary of the actuators with their benefits and disadvantages.

Actuators	Action member	Benefits	Disadvantages
Electric	DC servo motor, BLDC	Easy motor control, high power, easy connection of electrical monitoring devices	Lower load capacity, more complex construction design
Mechanic	Spring, cable	Easy installation, low weight, small dimensions	It is not possible to connect electrical devices, it is not self-moving
Pneumatic	Single-acting or double-acting air cylinder	Safety, smooth speed adjustment, easy installation	Uneven movement of pistons, noise, susceptible to medium quality
Hydraulic	Single-acting or double-acting liquid cylinder	High performance at a small size, precise control and regulation, quiet operation	High weight, high acquisition and operating costs

4   Conclusion

In this paper, we have provided an overview of lower limb exoskeletons, focusing on exoskeletons in military applications. We summarized the types and requirements for exoskeletons. Based on collected and compared available information, we can conclude that exoskeletons can be widely used in the military based on the above recommendations. We discussed key points considering exoskeleton characteristics for use in military applications. However, this paper is not only a summary of the current state but also outlines recommendations for further analyses and exoskeleton use in the army.

Acknowledgement. This work was supported by Development of wearable sensor systems and passive body protection in assistive technologies SGS21/081/OHK4/1T/17 project (CTU in Prague).

References

1. Viteckova, S., Kutilek, P., Jirina, M.: Wearable lower limb robotics: a review. Biocybern. Biomed. Eng. **33**(2), 96–105 (2013)
2. Dollar, M., Herr, H.: Lower extremity exoskeletons and active orthoses: challenges and state-of-the-art. IEEE Trans. Robot. **24**, 144–158 (2008)
3. Yan, T., Cempini, M., Oddo, C., Vitiello, N.: Review of assistive strategies in powered lower-limb orthoses and exoskeletons. Robot. Auton. Syst. **64**, 120–136 (2015)
4. Rupal, B.S., Singla, A., Virk, G.S.: Lower limb exoskeletons: a brief review. In: Conference on Mechanical Engineering and Technology (COMET-2016), pp 130–140. IIT (BHU), Varanasi (2016)

5. Rupal, B.S., Singh, A., et al: Lower-limb exoskeletons: research trends and regulatory guidelines in medical and non-medical applications. Int. J. Adv. Robot. Syst. **14**, 1729881417743554 (2017)
6. Jiang, W., et al.: Overview of lower extremity exoskeleton technology. In: OP Conference Series: Earth and Environmental Science, vol. 714. IOP Publishing (2021)
7. Pamungkas, D., et al.: Overview: types of lower limb exoskeletons. Electronics **8**, 1283 (2019)
8. Yeem, S., et al.: Technical analysis of exoskeleton robot. World J. Eng. Technol. **7**, 68–79 (2019)
9. Jia-Yong, Z., et al.: A preliminary study of the military applications and future of individual exoskeletons. In: Journal of Physics: Conference Series, vol. 1507, IOP Publishing (2020)
10. Ministry of Defense of the Slovak Republic Homepage. https://www.mosr.sk/44340-en/exo skeleton-2019-na-trencianskom-ostrove/. Accessed 19 Jan 2022
11. Chen, B., et al.: Recent developments and challenges of lower extremity exoskeletons. J. Orthop. **5**, 26–37 (2016)
12. Gordleeva, S.Y., et al: Real-time EEG–EMG human–machine interface-based control system for a lower-limb exoskeleton (2020)
13. Lu, R., et al.: Development and learning control of a human limb with a rehabilitation exoskeleton. IEEE Trans. Ind. Electron. **61**, 3776–3785 (2013)
14. Ma, W., Zhang, X., Yin, G: Design on intelligent perception system for lower limb rehabilitation exoskeleton robot. In: 13th International Conference on Ubiquitous Robots and Ambient Intelligence (URAI), pp. 587–592 (2016)
15. Long, Y., Du, Z.J., Wang, W., Dong, W.: Development of a wearable exoskeleton rehabilitation system based on hybrid control mode. Int. J. Adv. Robot. Syst. **13**(5), 1–10 (2016)
16. Liao, Y., Zhou, Z., Wang, Q.: BioKEX: a bionic knee exoskeleton with proxy-based sliding mode control. In: IEEE International Conference on Industrial Technology (ICIT), pp. 125–130 (2015)
17. Beravs, T., Reberšek, P., Novak, D., Podobnik, J., Munih, M.: Development and validation of a wearable inertial measurement system for use with lower limb exoskeletons. In: 11th IEEE-RAS International Conference on Humanoid Robots, pp. 212–217 (2011)
18. Kim, J.H., et al.: Design of a knee exoskeleton using foot pressure and knee torque sensors. Int. J. Adv. Robot. Syst. **12**(8), 112 (2015)
19. Cha, D., Kang, D., Oh, S.N., Kim, K.I., Kim, K.S., Kim, S.: Faster detection of step initiation for the unmanned technology research center exoskeleton (UTRCEXO) with insole-type force sensing resistor (FSR). In: 10th International Conference on Ubiquitous Robots and Ambient Intelligence (URAI), pp. 298–300 (2015)
20. Wheeler, J., et al.: In-sole mems pressure sensing for a lowerextremity exoskeleton. In: The First IEEE/RAS-EMBS International Conference on Biomedical Robotics and Biomechatronics, BioRob 2006, pp. 31–34 (2006)
21. Park, J., Kim, S.J., Na, Y., Kim, J.: Custom optoelectronic force sensor based ground reaction force (GRF) measurement system for providing absolute force. In: 13th International Conference on Ubiquitous Robots and Ambient Intelligence (URAI), pp. 75–77 (2016)
22. Leal-Junior, A.G., et al.: Polymer optical fiber for angle and torque measurements of a series elastic actuator's spring. J. Lightwave Technol. **36**, 1698–1705 (2018)
23. Saccares, L., Brygo, A., Sarakoglou, I., Tsagarakis, T.: A novel human effort estimation method for knee assistive exoskeletons. In: International Conference on Rehabilitation Robotics (ICORR), pp. 1266–1272 (2017)
24. Zawawi, M.Z.F.B.M., Elamvazuthi, I., Aziz, A.B.A., Daud, S.A.: Comparison of PID and fuzzy logic controller for DC servo motor in the development of lower extremity exoskeleton for rehabilitation. In: 3rd International Symposium in Robotics and Manufacturing Automation, ROMA, pp. 1–6 (2016)

25. Zhang, S., Wang, C., Hu, Y., Liu, D., Zhang, T., Wu, X.: Rehabilitation with lower limb exoskeleton robot joint load adaptive server control. In: International Conference on Information and Automation (ICIA), pp. 13–18 (2016)
26. Aliman, N., Ramli, R., Haris, S.M.: Design and development of lower limb exoskeletons: a survey. Robot. Auton. Syst. **95**, 102–116 (2017)
27. Sergeyev, A., Alaraje, N., Seidel, C., Carlson, Z., Breda, B.: Design of a pneumatically powered wearable exoskeleton with biomimetic support and actuation. In: Aerospace Conference (2013)

Modelling and Simulation Support to Medical Treatment Chain in Role 1

Dalibor Procházka[1][(✉)] [iD], Jan Hodický[2] [iD], Milan Krejčík[1] [iD], and Aleš Tesař[1] [iD]

[1] University of Defence, Brno, Czech Republic
dalibor.prochazka@unob.cz
[2] HQ SACT JFD Modelling, Simulation and Training Technologies, Norfolk, USA

Abstract. Role 1 medical treatment at the battalion level command focuses on the provision of the primary health care being the very first physicians and higher medical equipment intervention to casualties' treatment. Role 1 has the paramount importance in casualties' reduction in the current operations representing a complex system. The successful treatment depends on casualties' transportation from battlefield to Role 1, where lifesaving medical intervention takes place, and on processes in Role 1. It depends on resources available (transport vehicles, medical and supporting staff, materiel) and proper resources allocation, i.e. decision process. The paper deals with the task of transportation from the battlefield to Role 1 replicated by multi-agent modelling paradigm, and the medical treatment processes in Role 1 represented by process modelling approach. Initial data on casualties is taken from Computer Assisted Exercise (CAX) exercises and the outputs of the model are compared to decision process during the CAX. Based on simulation outputs, the Role 1 structure is proposed and decision risks are discussed.

Keywords: Casualty medical treatment · Role 1 modelling · Agent based simulation · Simulation experiment

1 Introduction

Casualty reduction is in addition to military objectives the main objective of any planned mission. There are two ways of casualty reduction. The first approach is to avoid casualties by means of higher operational effectiveness and better combatant protective equipment. The second approach is higher efficiency of medical treatment that reflects a case when casualty already happened and it includes medical treatment and organizational measures to save life and health of injured combatant. The paper deals with organization of transport and medical treatment of casualties in operations at the battalion level. Medical procedures are beyond the scope of the paper. The research question is an organization of the Role 1 (how many physicians, military ambulances and other necessary staff are needed ensuring effective first medical treatment and transportation, while operational environment and transport time taking into account. Another aspect considered in a decision process is prioritization in case transport requirements or lifesaving surgeries prevail available resources, i.e. ambulances, physicians or other staff.

© Springer Nature Switzerland AG 2022
J. Mazal et al. (Eds.): MESAS 2021, LNCS 13207, pp. 464–477, 2022.
https://doi.org/10.1007/978-3-030-98260-7_30

Founded in military history, NATO Nations military medical treatment is classified in five categories [1]. Each category defines an essential treatment capability it is able to provide. Medical treatment categories are referred to as Roles 0–4. The medical treatment capability differs from the first response carried out by every soldier, defined as Role 0, up to the full hospital treatment capability in Role 4.

Casualties occur on the battlefield. They are given immediate care by a Combat Life Saver (CLS) and they are transported into Role 0 assembly areas under company commander command, while waiting for transport to Role 1. Role 1 is responsible for transport from Role 0 to Role 1 and for primary health care which means first aid, triage (primary triage takes place in Role 0), resuscitation, stabilization and immediate intervention if necessary.

Role 1 is usually a part of a battalion or an equivalent unit. Assessing Role 1's capability is not an easy task. The assessment can be done comparing requirements for transport and medical treatment to real capability of the unit. There are problems with data for real requirements due to the fact that the Role 1 capability should respond to a regular force-on-force conflict and actual data for such a conflict are not available. Therefore, it requires using casualty estimations for the purpose.

There are two approaches to casualty estimation. The first one is based on statisti-cal analysis on historical data, see in [2–4] and [5], the second approach use modelling and simulation of combat operation. Peng [6] demonstrates a simulation approach to casualty estimates employing the agent-based modelling and simulation paradigm. The relevance of the approaches is discussed in [7].

2   Role 1 Organization and Tasks

2.1   Role 1

The Role 1 medical support is integral or allocated to a small unit, and it includes the capabilities for providing first aid, immediate lifesaving measures, and triage [8].

Role 1 is responsible for:

- transport casualties from Role 0 to Role 1,
- initial triaging in Role 1 (contaminated, seriously wounded – P1, moderately wounded – P2, slightly wounded – P3),
- medical treatment,
- preparation for transport to Role 2, a request for transport and prioritization of casualties,
- medical documentation of casualties.

To cope with these tasks Role 1 consist of two physicians, five members of supporting medical staff and it is equipped with two ambulances capable of transport up to four casualties (Czech Army Role 1 structure).

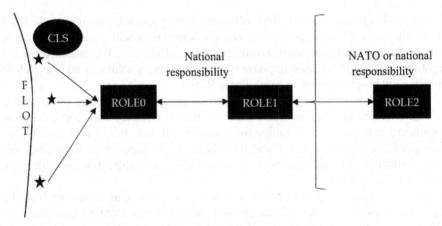

Fig. 1. Healthcare overview up to Role 2

Figure 1 [7] describes a simplified view of the healthcare treatment in the area of operation up to the Role 2 functions. This structure, its variants and the actual primary operational environment in which it is executed creates a complex system.

2.2 Initial Problem Formulation

This study starting point was a preparation of a new internal regulation for Role 1, based on new knowledge of the tactics of conducting contemporary management of combat activities and new standards of medical security. The outdated standards and principles of using the medical service in the field are based on experience and knowledge from past conflicts and the technologies, techniques and materials used.

The goal is to assess current Role 1 capabilities and to define and verify its new medical capabilities, principles, staff and equipment in providing medical assistance during the battalion combat activities in terms of setting time limits - Clinical Timeline (CTL): 10 min for first aid, 1 h for life-saving treatment and 2 h to reach Role 2.

The change in doctrinal documents necessitated a reassessment of existing procedures and thus a change in the existing regulation. Actual arrangements do not reflect doctrinal changes in Role 1, particularly:

- transport means,
- field medical equipment and materiel,
- command and control system,
- personnel,
- medical treatment time norms,
- Standing Operational Procedures (SOPs),
- tactical norms.

The first task was to assess the current arrangement of the medical treatment chain up to Role 1. To reflect the requirement, modified outputs from two battalion Computer Assisted Exercises (CAX) were used. The CAX in general use constructive simulations.

i.e. simulations involving simulated people operating simulated systems. Real people - operators - stimulate (make inputs) to such simulations. Combat simulation provided data for damaged or destroyed vehicles (time and locations), but to generate injury data for crews and other combatants the platform kill categories (mobility kill, firepower kill, catastrophic kill) were used. Based on these events casualties, characterized by time of occurrence, coordinates of the event, and the category (P1 – P3) were generated. The exact procedure is described in [7] and the resulting data are taken as given in this paper.

Problem formulation:

1. Validate the actual structure and capability of Role 1 to fulfil requirements, namely provide required transportation and medical treatment in given time, i.e. **from injury to Role 1 within one hour and to prepare casualties for transport to Role 2 within two hours.**
2. Propose recommended Role 1 structure that would satisfy the previous requirement in various tactical situations.

3   Methodology

3.1   CAX Experiment

The first problem was to assess the current Role 1 capabilities. The possible way is to set an experiment. From all different types of experiments, field experiments are those that best emulate the conditions that war fighters will likely face in combat. In such case, military units and their equipment reflect the reality. However, field experiments have many limitations. They are very costly, so usually the scale of the experiment must be reduced. Moreover, if a new capability is experimented, the environment of the field experiment is not applicable, because of non-existing prototypes or a simplified version of prototype do not replicate the future capability with an enough level of realism; therefore, it reduces the credibility of the experiment. Repetition of trials enabling reduction of uncertainty within the experiment is almost impossible because of cost of a single run of the experiment. The field experiment is very difficult to control and there is very limited ability to detect changes and isolating the true causes.

Modelling and simulations (M&S) can overcome the previously mentioned drawbacks of the field experiment. M&S can complement and extend a field experiment by integrating non-existing components or by augmenting the complexity with simulated events from pre-scripted scenario. M&S can increase the number of repetition of trials within the experiment or event support more than one design of experiment and therefore to maximize the invested money into the experiment venues. M&S helps with communication among all stakeholders of the experiment. The use of simulation and mainly 3D realistic visualization support the credibility of the results of the experiment. Moreover, interactive simulation helps decision makers developing improved understanding of a system.

On the other side, there are limitation of M&S use in support of an experiment. The credibility of results of the experiment supported by M&S is equal to the level of realism and rigor of models and the simulations involved. Simulation results needs to be analysed in statistical manners in order to process them and lately to interpret them.

Results credibility heavily depends on the skills of the designer of experiment and his knowledge on the M&S assets to be used during the experiment. The development of a simulation for a complex system replication is very expensive, a simulation execution can be costly because of operators, and computational time needs.

For the reasons given above, the first problem was attempted to solve by means of Computer Assisted Exercise (CAX) focused directly on Role 0 and Role 1 activities. There were used two scenarios from previous CAXs performed by Czech 7th mechanized brigade, namely the 73th battalion attack operation and the 71st battalion defence operation. Extended approach to bring more scenarios into the experimentation is following the methodology of Alternative Futures [9]. Data from these exercises were used to generate casualties and another CAX focusing on Role 0 and Role 1 activities was performed.

The following limitations took place:

- Two ambulances (BMP-2/AMB) were used to pick up casualties from battlefield and transport them to Role 0.
- Two ambulances (Land Rover/AMB) were used for transport from Role 0 to Role 1.
- Manoeuvre of Role 1 was not considered.
- Transfer to Role 2 was outside the scope of the CAX.
- No activities concerning dead or captured casualties were considered and simulated.
- Combat activities did not limit Role 1 activities (transport or treatment).
- Time for the casualty treatment was estimated (15 min for P1, 10 min for P2, 5 min for P3).
- The location of Role 1 was close to battlefield. Travel from Role 1 to Role 0 and vice versa took 7–11 min.

Generated casualties were transported by means of companies to Role 0. Transport means were individually operated by operators. Upon requests for transport ambulances were sent to Role 0, casualties were loaded and transported to Role 1 and procedures in Role 1 were played by the Role 1 staff – unloading, triage, treatment and corresponding time for each casualty was recorded. The resulting data for casualties' treatment are given in Fig. 2 for the attack scenario and in Fig. 3 for the defence scenario. Each bar represents one casualty and it represents total time from casualty event to time when treatment in Role 1 is finished and the request for transport to Role 2 is issued. Important stages are highlighted by different colours.

An initial assessment of these figures reveals that the actual operation of Role 1 does not satisfy the clinical timeline requirements given earlier. The average time from time of the injury to arrival to Role 0 in the attack scenario are P1: 2:04:32, P2: 2:38:06 and P3: 3:09:12, in the defence scenario P1: 1:59:11, P2: 1:20:49, P3: 2:41:40. These data reveal that for these scenarios it is impossible to fulfil requirements stated in the problem formulation, particularly to treat casualties in required time and to prepare casualties for transport to Role 2 within 2 h from time of the injury.

It is not a critical issue in case of categories P2 and P3, which treatment can be postponed, but for P1 category it is a serious problem, because life and health saving actions must be taken as soon as possible. Another observation is that the interval from the injury to transport to Role 0 takes big amount of total time, but it depends on battalion and companies organization and it is out of scope of the stated problem.

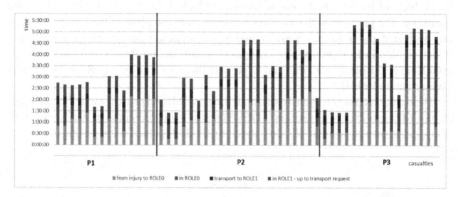

Fig. 2. CAX results - attack

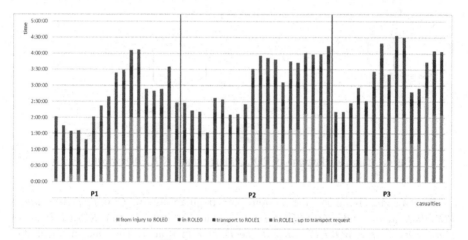

Fig. 3. CAX results – defence

The CAX as an experimental platform for the purpose of comparing variants is very limited. The main problems are:

- Experimentation in real time. Each trial of an experiment requires about 6 h to run.
- Manning requirement on supporting staff – operators, role players etc.
- Problematic repetitions caused by human-in-the-loop and the decision making process.
- Statistic meaningfulness of data – due to problems given above only a few trials can be performed.

This leads to an idea to create a simulation model focused on Role 1 processes and activities where the outputs of the CAX were used as data stimulating the model.

3.2 Agent Based Model of Casualty Treatment

The motivation to model and simulate the Role 1 processes and activities was driven by the problem of finding good enough structure of Role 1 that would ensure the required capability. The problem with casualties transport from the place of the injury to Role 0 was neglected and in further consideration it was supposed that the casualty will reach Role 0 in 30 min from injury, that changes requirements on Role 1 to transport injury from Role 0 to Role 1 in 30 min and to prepare them for transport to Role 2 in 90 min from arrival to Role 0.

The model is based on agent based modelling (ABM) and process modelling methodologies (discrete events simulation – DEVS) and it is implemented in AnyLogic modelling environment, which supports these methods together with system dynamics [10].

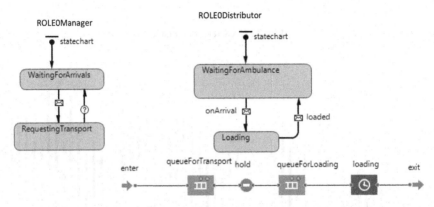

Fig. 4. Statechart diagrams of ROLE0Manager and ROLE0Distributor agents

The agents living in the system are populations of *Casualty, Ambulance, ROLE0Manager* and *ROLE0Distributor* agents in Role 0 and *ROLE1Manager, ROLE1Distributor, ROLE1Treatment* and *Doctor* in Role 1. The other medical staff and ambulance drivers are not modelled, as well as the activities in battlefield - first aid and transport to Role 0, which are not responsibility of Role 1.

Using the data provided by the CAX exercises, a casualty appears in Role 0 in given time, the *ROLE1Manager* agent sends a *transportRequest* message to the *ROLE1Manager* agent which manages requests and makes decision on sending ambulances. There can be several assembly areas in Role 1, depending on tactical situation and they can vary in time (usually one for each company). The *ROLE0Distributor* agent manages the casualties, i.e. prioritizes them based on the injury category and time and controls loading in an ambulance. The statechart diagrams for these agents are in Fig. 4 The agent *ROLE0Distributor* combines agent based model represented by the statechart and the process model presented by blocks passed by *Casualty* agents.

The scheme of the Casualty agent is in Fig. 5 on the left. The yellow boxes are agent's states and the arrows mean transition from one state to another. The transitions are driven by time, by a logical condition (symbol ⓐ), by a message from another agent (symbol ✉) or by arrival to destination (symbol ⊞). The transition from *InField* state to *InROLE0*, transition to *Loaded* state is caused by a message from the *Ambulance* agent as well as the next transition to *MovingToROLE1* state. The activities within *Unloading* and *ReadyForTransfer* are driven by the *ROLE1Treatment* agent. The statechart of the *Ambulance* agent is in Fig. 5 – on the right.

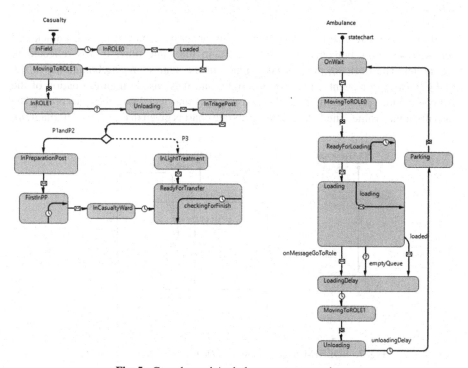

Fig. 5. Casualty and Ambulance agent state charts

The activities manipulating casualties in Role 1 are modelled by combination of *Casualty* and *Doctor* agents interactions and a process model represented by the *ROLE1Treatment agent.* The activities in Role 1 are in Fig. 6 and they reflect the activities performed in the CAX with the same time required for unloading casualties, triage and treatment. The physicians represented by the *Doctor* agents have different tasks. The first one triages casualties and the second one treats casualties in the Casualty Ward. There is a limit, only one bed is available for surgeries in case of two physicians (the default arrangement), thus in case of three physicians the capacity of casualty ward is limited by two beds etc. In the model the first physician works on triage and the second or the third physician treats casualties in the casualty ward.

Distances between Role 1 and Role 0 and speed of ambulances play important role in the whole medical treatment chain. The model uses only route time as a parameter, because it influences the Role 1 capability. To make a decision on the Role 1 location, a terrain, distances and a real vehicle speed must be taken into consideration [11], nevertheless these aspects allow estimating travel time.

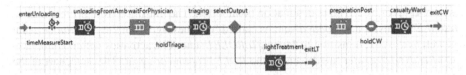

Fig. 6. Processes in Role 1 – *ROLE1Treatment* agent

In the CAX exercise four Role 0 assembly areas were played in both (attack and defence) scenarios but only two were active in the same time. It was reflected in the model. The decision process takes place in the state *OrderingAmbulance* of the *ROLE1Distributor* agent, Fig. 7, on the right, and it is based on an evaluation of the utility function, processing all uncovered requests and weighting them according the casualty injury value. The *ROLE1Manager* state chart is given in Fig. 7, the left diagram.

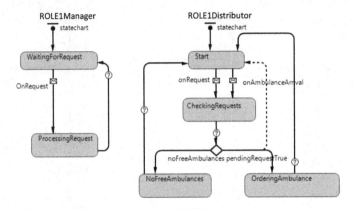

Fig. 7. ROLE1Manager and *ROLE1Distributor* agent state charts

Model validation was performed comparing activities in Role 0 and Role 1 described by stakeholder and corresponding agents and process blocks in the model. Verification of the model included comparison of the model outputs to outputs produced by the CAX experiment.

4 Experiment Setup and Simulation Results

4.1 Experiment Setup

The simulation model was run to support the assessment of a set of scenarios. As input data outputs from two scenarios played in the CAX exercise were used. The time of a casualty arrival to the ROLE0 was taken and then the casualty was processed as described above. The input independent variables for the scenarios were travel time from ROLE1 to ROLE2 used by the *Ambulance* agent (*timeToROLE0* = 6, 10, 15, 20 min), number of ambulances (*nAmbulances* = 2, 3, 4) and number of physicians in ROLE1 (*nDoctors* = 2, 3). The travel time *timeToROLE0* was the same for all pickup points in ROLE0.

Table 1. Experiment results – Attack (A) and Defence (D) scenarios.

Time to ROLE0	Number of ambulances	Number of physicians	From Injury to ready for transfer P1		From Injury to ready for transfer P2		From Injury to ready for transfer P3		From ROLE0 to ready for transfer P1		From ROLE0 to ready for transfer P2		From ROLE0 to ready for transfer P3	
			A	D	A	D	A	D	A	D	A	D	A	D
6	2	2	204	177	377	330	274	240	90	73	223	208	88	82
6	2	3	159	159	183	183	240	240	54	54	62	62	82	82
6	3	2	204	177	377	330	274	234	90	73	223	208	88	76
6	3	3	171	158	213	183	274	234	57	54	60	62	88	76
6	4	2	204	177	377	330	274	234	90	73	223	208	88	76
6	4	3	171	158	213	183	274	234	57	54	60	62	88	76
10	2	2	210	192	383	324	270	378	96	88	229	202	85	220
10	2	3	177	175	217	225	270	378	63	71	64	104	85	220
10	3	2	210	184	383	334	270	244	96	80	229	212	85	86
10	3	3	177	166	217	188	270	244	63	62	64	67	85	86
10	4	2	210	184	383	334	270	244	96	80	229	212	85	86
10	4	3	177	166	217	190	270	244	63	62	64	68	85	86
15	2	2	231	206	374	325	374	403	117	102	220	203	189	245
15	2	3	206	184	257	255	374	403	92	80	103	133	189	245
15	3	2	212	207	380	324	280	262	98	102	226	202	95	104
15	3	3	184	184	229	227	280	262	70	80	75	105	95	104
15	4	2	210	212	383	318	270	258	96	108	229	197	85	99
15	4	3	180	186	222	215	280	258	66	82	68	94	95	99
20	2	2	235	216	400	336	490	496	121	112	246	214	305	338
20	2	3	211	197	343	293	490	496	97	92	189	172	305	338
20	3	2	224	212	382	334	308	314	110	108	229	212	123	156
20	3	3	198	192	243	239	308	314	84	88	90	118	123	156
20	4	2	219	215	388	331	288	268	105	110	234	209	102	110
20	4	3	188	194	229	229	288	268	74	90	75	108	102	110

4.2 Simulation Results

The results of simulation are given in Table 1, where the scenarios are grouped by time to Role 0 and A and D columns are the attack and the defence scenarios respectively. The results presented in the table are average values for each casualty category (P1, P2 and P3). The reason why some rows with different input parameters gives the same values in output parameters (e.g. 6-3-2 and 6-4-2) is that there is a redundancy in the second case and all 4 ambulances are never used in the same time. Due to deterministic character of the simulation the results are the same for both scenarios. The most important category is P1 because time to reach life saving treatment is critical. The P3 category does not play important role in our considerations due to the (light) injury character.

The cells where average time from arrival to Role 0 to readiness for transport to Role 2 for P1 and P2 equals or is less than 90 min are highlighted.

For the most optimistic scenario, *TimeToROLE0* = 6, combination of the number of ambulances from 2 to 4 fits the required capability provided by the Role 1 is staffed by 3 physicians. It is necessary to say that it is the most unrealistic scenario, because the low value of travel time requires Role 1 location close to the battlefield which increases risk of a forced manoeuvre or even being under fire due to enemy force activities.

The scenarios with *TimeToROLE0* = 10 have two combinations fulfilling the requirements (at least in average values), namely 3 or 4 ambulances and 3 physicians, the same combinations cover requirements for *TimeToROLE0* = 15.

The combinations of 3 or 4 ambulances and 3 physicians in case *TimeToROLE0* = 20 fulfil the requirements for the attack scenario but only requirements on the P1 category for the defence. The case of 4 ambulances gives better values.

The simulation results are given in Fig. 8 and Fig. 9 for attack and defence scenarios respectively, where the curves depict time plots of casualties in Role 0, casualties in Role 1, cumulative values of casualties treated in Role 1 by categories and total number of casualties.

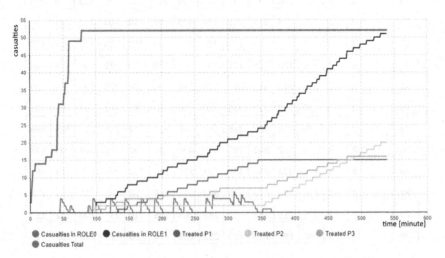

Fig. 8. Attack scenario, *TimeToROLE0* = 6, 4 ambulances, 3 physicians

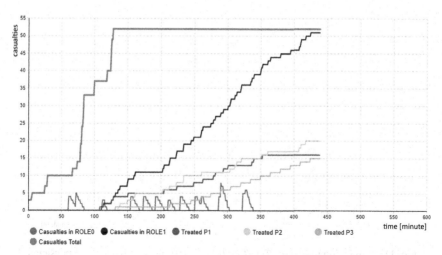

Fig. 9. Defence scenario, *TimeToROLE0* = 6, 4 ambulances, 3 physicians

To better understand an importance of the third physician, two defence scenarios are compared in Fig. 10 and Fig. 11. The values of the curve *Casualties in Preparation Post* are significant in case of 2 physicians because the limited treatment capability creates a queue in front of the casualty ward and it causes a problematic delay in the whole chain, while in case of 3 physicians the process goes more smoothly.

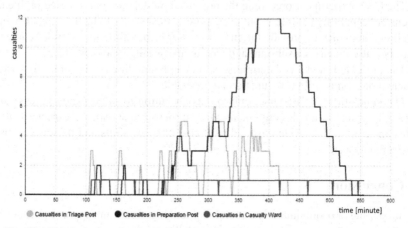

Fig. 10. Simulation results for *TimeToROLE0* = 15 min, 4 ambulances, 2 physicians

5 Discussion

The results of presented simulations can support decision on R1 structure, equipment and capabilities. The recommendation based on the presented cases are to reinforce the actual (Czech Army battalion structure) up to four ambulances and three physicians.

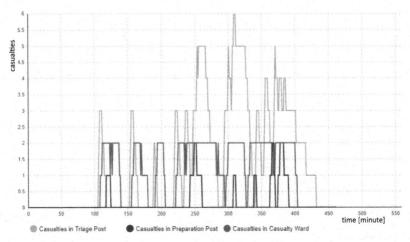

Fig. 11. Simulation results for *TimeToROLE0* = 15 min, 4 ambulances, 3 physicians

The third physician is crucial in all simulated scenarios, but it must be underlined that it is not only the physician as a person but it represents the capability to perform life saving treatment and it requires all necessary equipment including an operating bed and an assisting staff. These features and resources as a medical and supporting staff, materiel and equipment were not included in the model but must be taken into account in the decision making process.

The CAX experiment preceding the presented model creation revealed problems in the current capability of Role 1, but even more critical seems to be the capability of the battalion companies to provide the first aid and transportation to the Role 0 assembly areas, because it causes a substantial delay in the casualty treatment chain.

The presented model is deterministic, but can be modified to become stochastic, if sufficient data on distribution function are available.

The simulation is, of course, only one contribution to the decision process and it focuses on an operational requirement point of view. Other aspects, especially affordability in terms of life cycle cost and total cost of ownerships and personnel must be considered as well.

6   Conclusion

The modelling and simulation is a useful methodology that can support a decision process. The paper demonstrates agent based modelling and simulation approach to a problem of assessment the Role 1 capability to provide required transportation and medical treatment. The simulation results must be critically interpreted within the constraints of simplifications made in the modelling process. Any model is as good as data gathered to stimulate its inputs and to represent internal parameters of the model. The data used to stimulate the model were obtained from a combination of CAX exercises and an experiment using the same constructive simulation. The presented agent based model uses data as given but gathering more accurate data sets maybe by means of a live simulation exercise (a field exercise with simulated weapon effects) is a challenge.

References

1. NATO: AJP-4.10. Allied Joint Doctrine for Medical Support. NATO, Brussels, Belgium (2015)
2. Gottwald, S., Braun, D.: Bounded rational decision-making from elementary computations that reduce uncertainty. Entropy **21**(4) (2019). https://doi.org/10.3390/e21040375
3. Gawande, A.: Casualties of war – military care for the wounded from Iraq and Afghanistan. N. Engl. J. Med. **351**(24), 2471–2475 (2004). https://doi.org/10.1056/NEJMp048317
4. Travers, S., et al.: Five years of prolonged field care: prehospital challenges during recent French military operations. Transfusion **59**(S2), 1459–1466 (2019). https://doi.org/10.1111/trf.15262
5. Gerhardt, R.T., De Lorenzo, R.A., Oliver, J., Holcomb, J.B., Pfaff, J.A.: Out-of-hospital combat casualty care in the current war in Iraq. Ann. Emerg. Med. **53**(2), 169–174 (2009). https://doi.org/10.1016/j.annemergmed.2008.04.013
6. Peng, B., Zhang, W., Du, G., Xu, L.: A simulation research on casualty prediction based on system dynamics and agent-based modeling. Acad. J. Second Mil. Med. Univ. **39**, 510–514 (2018). https://doi.org/10.16781/j.0258-879x.2018.05.0510
7. Hodický, J., Procházka, D., Jersák, R., Stodola, P., Drozd, J.: Optimization of the casualties' treatment process: blended military experiment. Entropy **22**(6), 706 (2020). https://doi.org/10.3390/e22060706
8. NATO: NATO Logistics Handbook. NATO, Brussels, Belgium (2012)
9. Fučík, J., Melichar, J., Kolkus, J., Procházka, J.: Military technology evolution assessment under growing uncertainty and complexity: methodological framework for alternative futures. In: Proceedings of the 2017 International Conference on Military Technologies, pp. 682–689. Institute of Electrical and Electronics Engineers Inc., Piscataway, NJ 08854-4141, USA (2017). ISBN 978-1-5386-1988-9
10. Borshchev, A.: Multi-method modelling: anylogic. In: Brailsford, S., Churilov, L., Dangerfield, B. (eds.) Discrete-Event Simulation and System Dynamics for Management Decision Making, pp. 248–279. John Wiley (2014). https://doi.org/10.1002/9781118762745.ch12
11. Nohel, J.: Possibilities of raster mathematical algorithmic models utilization as an information support of military decision making process. In: Mazal, J. (ed.) MESAS 2018. LNCS, vol. 11472, pp. 553–565. Springer, Cham (2019). https://doi.org/10.1007/978-3-030-14984-0_41

Academic and Industrial Partnerships in the Research and Development of Hybrid Autonomous Systems: Challenges, Tools and Methods

Emma Barden[1]([⊠]) [iD], Michael Crosscombe[2] [iD], Kevin Galvin[1] [iD], Chris Harding[1], Angus Johnson[1], Tom Kent[2] [iD], Ben Pritchard[1] [iD], Arthur Richards[2] [iD], and Debora Zanatto[2] [iD]

[1] Thales Research, Technology & Innovation (RTI), Reading, RG2 6GF Palaiseau, France
emma.barden@uk.thalesgroup.com
[2] Faculty of Engineering, University of Bristol, Bristol BS8 1UB, UK

Abstract. Autonomous systems increasingly are integrated into larger, connected, and hybrid (Human-Machine) systems of systems, making them complex systems - which are hard to design and predicting emergent behaviour is difficult. These issues are faced increasingly across civil and military applications, both in the UK and NATO. A holistic approach is needed to fully quantify them. Working as a partnership between industry and academia has provided greater freedom to apply innovative technologies in the context of relevant use cases. This paper presents some tools and methods we have used in our research and development to support this approach and address the challenges of deploying autonomous systems in the future. We discuss the use of simulations and how they can support every step of the process, from academic experiments to digital twins; where the right level of fidelity is needed at different times to give maximum benefit. The use of a common simulation platform to align control design exploration with human factors research is discussed, enabling questions of human-machine teaming and trust. We highlight how foundational research on: architecture and modelling, network topology, decision making processes and human interactions impact on the overall development of a system. Included are our lessons identified from this partnership.

Keywords: Autonomous systems · Simulation · Human-machine team · Mission planning · Collaboration

1 Introduction

Thales, a global engineering company, has a worked closely with UK academia in research through collaborative projects, as part of a consortium, in a bi-lateral arrangement, or through sponsorship of PhD and EngD projects in emerging technologies. As the company's interest in Autonomy, Artificial Intelligence (AI) and Machine Learning (ML) has increased over the last decade, a number of initiatives between Thales

© Springer Nature Switzerland AG 2022
J. Mazal et al. (Eds.): MESAS 2021, LNCS 13207, pp. 478–493, 2022.
https://doi.org/10.1007/978-3-030-98260-7_31

and academia have been established. In the United Kingdom, two of these are with the universities of Bristol and Southampton, leading academic institutions in the fields of Autonomy, AI and ML. Although this paper will focus on the relationship with the University of Bristol (UoB) it will highlight where there is a crossover with some of the research with Southampton and other academic consortia that is being leveraged.

In 2017 Thales and UoB signed a Strategic Agreement and entered into a five year Engineering & Physical Sciences Research Council (EPSRC) Prosperity Partnership, a jointly funded project called the Thales – Bristol Partnership in Hybrid Autonomous Systems Engineering (T-B PHASE). We use "hybrid" in this context to include autonomous interaction with the ""open" (i.e. unlimited, uncontrolled, not lab) environment as well as with human interaction. Key objectives of this project are:

- To innovate new design principles and processes that integrate over the system life-cycle.
- To build new analysis and design tools that enable a complex system's interactions to be mapped, understood and bounded at design concept stage.
- To develop new whole-life-course monitoring approaches.
- To train/develop people with the skills required for leadership in systems engineering.
- To engage with live Thales use cases in Hybrid Low-Level Flight, Hybrid Rail Systems, and Hybrid Search & Rescue (SAR).
- To implement a programme of impact and integration activities that engage stakeholders, policy makers and the public.

In pursuit of these objectives, the programme has assembled a team of researchers from. Both UoB and Thales, and is carrying out a suite of component projects investigating both foundational topics and integrating questions including literature reviews and analysis of previous research programmes e.g. UK System Engineering and Integrated Systems for Defense and Semi-Autonomous Vehicles Defense Technology Centre (SEAS DTC). The remainder of the paper reports on the findings from foundational work as well as the tools and methods being applied to integrate them.

2 Autonomous Systems Foundational Research

Through academic links we have been able to grow several areas of foundational research. More specifically these are in the areas of architecture and modelling, network connectivity, tasking and Sense of Agency. Also we have a number of PhD candidates who are conducting research into relevant, albeit distinct, self-contained activities. This paper focuses on major postdoc-led work streams and additional projects and PhD studies are omitted for brevity, but can be found on the T-B Phase website [16]. These all involve the use of modelling and simulation and are now integrating aspects of their research into collaborative projects with the rest of the team.

This research has been mainly conducted in an academic setting, enabled by the partnership and with industrial inputs where needed.

2.1 Architecture and Modelling

Architecture and modelling is a key enabler to understanding both the functional and non-functional aspects of any system. It is essential that these are identified for hybrid autonomous systems so that they are then carried through to the design and realisation of those systems and support verification and validation. An architecture itself comprises of several artefacts that cover the conceptual, service, logical and physical specifications that are articulated in the form of a set of models. Use cases are key starting points and can be derived through several methods which include the development of ontology and associated taxonomies, in T-B PHASE the use cases selected are of particular interest to Thales (e.g. SAR).

Architecture frameworks provide a formalised approach to capture and query the architecture of a system. The NATO Mission Threads Methodology provides a mechanism for developing a Mission Thread and utilises the NATO Architecture Framework Version 4 (NAFv4) [1]. This methodology has been used in conjunction with the Simulation Interoperability Standards Organization (SISO) Guide to Scenario Development (GSD) published in 2018 [2].

One of the challenges, however, remains how Non-Functional Requirements are architected. These include importantly; Security, Safety and Human Factors and well as the '–illities' such as reliability, maintainability, interoperability, etc. This became an initial focus of the architecture research on the project as well as working to develop a framework using a use case based on SAR and how autonomous systems could be exploited.

2.2 Network Topology

For any autonomous system composed of two or more interacting agents, an implicit network (communications) topology can be identified which expresses the ability for pairs of agents to interact with one another. This network may be 'logical', defining which agents are allowed to interact, or the network may be 'physical' as determined by a combination of factors of the autonomous system, e.g. by the relative distance between the pair of agents and their respective communication radii. Many autonomous systems are also designed implicitly as a result of the formation of the system as required by its task.

In Fig. 1, we illustrate different network topologies that were studied in the context of a multi-agent system applied to the problem of consensus formation. This is a distributed problem whereby the system must reach a consensus about the state of its environment, i.e., given a set of propositions which describe features of the environment, which propositions are true/false?

Through simulation experiments we have demonstrated that overly constrained network topologies, such as the star topology, exhibit bottlenecks on the communication of the agents which cannot be overcome using traditional representations of the agents' beliefs. Such topologies are likely to occur in systems that are organised according to a command hierarchy, and so special considerations are needed. More generally, the results of our studies showed that total connectivity has a negative impact on the convergence dynamics of the agents during the consensus formation process. More accurate

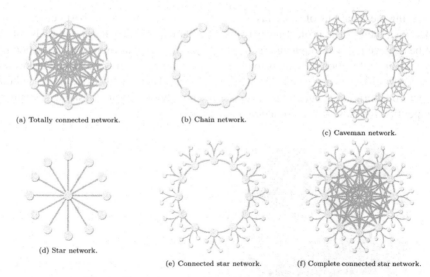

(a) Totally connected network.   (b) Chain network.

(c) Caveman network.

(d) Star network.

(e) Connected star network.   (f) Complete connected star network.

Fig. 1. Illustrations of various logical network topologies

learning of their environment is achievable with reduced levels of connectivity far below that of a totally connected network, even when the network is not especially structured but instead generated at random [3].

2.3 Decision Making

The term 'decision making' encompasses a rich and complex area of academic research. Widely used in game theory [4], economics [5], operational research [6], robotics and autonomous systems [7], it essentially asks: given our current understanding of the world, what should we do? Typically, this arises as an optimisation problem: given a number of input variables (i.e. sensor data, communications values, state information) how can we decide our control variables (i.e. direction to move, set a switch on/off, number and types of stocks to buy) in order to maximise our output (i.e. money, reward function, local/global utility function)?

When these decision-making processes are deployed in an 'autonomous' setting then the decision making is crucial to ensure that these decisions are executed in a timely, appropriate and safe manner. This often acts as a catch-all for making the system (whether vehicles, controllers, algorithms) 'do the right thing'. This can become increasingly complex when these systems contain multiple, cooperating/competing, and/or a mix of human or machine actors. When designing the decision-making system, it is therefore important to simulate, test and verify the impact that different choices have on the performance of the system; this includes ideas of resilience and robustness whilst trying to mitigate any undesirable emergent properties of the system. This has been one of the motivating factors in using our own simulation software outlined in Sect. 3.3, where a scenario containing Uncrewed Aerial Vehicles (UAVs) searching for tasks in the environment is presented. When a UAV "agent" finds a task, it adds it to its list and, using the locations of those tasks, decides on a route to take to best complete the tasks (visit them). The project investigates to what extent efficiencies of collaboration can be achieved only through emergence from distributed behaviour, without explicitly commanding allocations of agents to objects. For example, agents may share tasks they have seen with others in the hope that it improves overall performance. However, without proper collaboration agents, more agents knowing about the same tasks will result in competition for those tasks, producing inefficient behaviour.

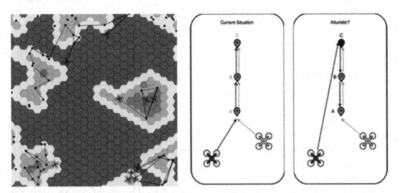

Fig. 2. TB-PHASE simulator (left) for tasking and routing for multiple agents. UAVs search an area for tasks (black dots) and optimally route (dashed lines) between them to complete all as quickly as possible. Using agent-intent modelling (right) we aim to remove potential conflicts.

If agents are able to share their locations with others, then agents can use that to predict intent and provide ways of being proactive. Figure 2(right) demonstrates this idea, where the blue agent sees that the other agent is closer to a task than it is. It predicts (but does not know) that the other agent might go for a task next, and so, instead of risking competition, it instead decides to reverse its route avoiding any potential conflict.

2.4 Human-Machine Interface and a Sense of Agency

The sense of control in psychology is referred to as the Sense of Agency. The Sense of Agency can be defined as one's prediction and perception of the effect of one's actions [8]. More specifically, it is "the subjective awareness of initiating, executing, and controlling one's actions in the world" [9].

In practical terms, the Sense of Agency (and control) arises when the human decides an action, causes a reaction and is aware of the consequence of the action. Considering Sense of Agency as one of the behaviour fundamentals, and investigating its role in the Human-Machine Interface (HMI), could in turn answer the specific issues that the HMI field is facing. Overall, investigations into the Sense of Agency and control perception have not made significant progress. No research on the role of both automation and workload on the Sense of Agency has been found; too little automation and the benefits of using the system may not be realised; too much, and the human supervisor may fail when malfunctions occur. Therefore, what constitutes a good 'level' of automation and workload that still means we feel in control? In considering the level of autonomy of the system, and the amount of workload, one would hypothesise the existence of an 'optimal situation' in which the user's sense of control is at its best. This would probably be at a middle stage of both machine autonomy and workload. Beyond that stage, and perhaps even before, control perception might decrease.

Results have shown that automation and mental workload are interconnected in playing a key role in influencing the Sense of Agency [10]. Both automation and mental workload have a degrading effect on the user's Sense of Agency. More importantly, results showed the presence of a residual Sense of Agency for the hybrid condition (namely the system warning). A combination of warning from the computer and low workload can keep the user in control and improve Sense of Agency. This has important implications for the design of future systems, for which mental workload should be cautiously balanced.

The present research demonstrates the possibility of integrating individual and machine actions while maintaining the individual's Sense of Agency.

3 Tools and Methods: Use of Simulation

The research strands discussed on Sect. 2 are united by their dependence on numerical simulation, not just for their scientific studies but for their connection to the motivating use cases. Meeting the needs of these diverse investigations demands an innovative approach to simulation. Moreover, in the light of the digital transformation, a simulation is not just a research tool but an engineering asset to accompany a system through its lifetime. For these reasons, simulation and visualization are cross-cutting activities of pivotal importance to T-B PHASE.

These kinds of tools can be used very differently between industrial and academic settings. The approach and availability will vary with factors such as the size of the project, stage of development, cost of software and equipment and the experience of users. This can make sharing information and recreating work between partners difficult.

3.1 Fidelity

As computing power and data storage improve dramatically there is a natural desire to take advantage of all the methods available, leading to simulation capabilities with more realism than ever before.

High fidelity simulations offer a great new opportunity to reduce the reliance on physical testing. However, this has associated risks; a great deal of investment of time and effort is needed to represent the systems within the simulation and this leads to large dependencies on accurate models of sensors, platforms, and the environment at a stage when that may not be well defined. There is also the need for specialist computing equipment that is not as easy to use (for example high performance Graphics Processing Units (GPUs), high resolution screens, headsets, and custom dashboards).

Conversely, low fidelity simulations can offer quick insights into the use of the system and any emergent behaviour but will be less realistic. Investing in such simulations early in the project can provide insight to basic questions of team composition or sensor selections before significant investment is made.

Careful consideration is therefore needed to pick the level of fidelity appropriate for the stage of the system's lifecycle and the needs of the system itself. The next sections discuss three different approaches to simulation projects which are at different levels of fidelity.

3.2   T-B PHASE Simulator

Initial investigations used bespoke simulations, but to promote collaboration and use-case engagement the project has decided to develop a common Agent-Based Modelling (ABM) simulator framework.

The framework provides a set of means and associated functions which allow an instance of a simulator to be built in a shared environment. The aims of this were to allow sharing of algorithms and methods across the research team as well as having a consistent environment for the development of scenarios.

There are many established platforms for ABM, in most languages. For ease of accessibility and familiarity, Python and the Mesa [11] modular Python framework for modelling and analysis have been selected. Visualisation modules that can be displayed via a server with a JavaScript interface simplify information dissemination.

We adapted this framework to facilitate the scenarios and behaviours we were investigating. This meant the adding of certain agent types, such as uninhabited air and water based vehicles, and environment functionality such as linear zone divisions and areas defined by polygons. All of these features are intended to be modular and adding to a growing code base that can be used by the whole team. By approaching the simulator in this way, it has allowed us to continually incorporate current, as well as combine previous, work in new ways.

As mentioned, the simulator framework developed is modular, consisting of server, model and agent files shown in Table 1. This allows developers to select functionality that has already been implemented and provides the rest of the team access to new developments.

Table 1. T-B PHASE simulator structure

Simulator		
Server File	Model File	Agent File(s)
• Visualisations	• Communications Network	• Movement rules
• Model parameters	• "World" rules	• Communication rules
• Maps and canvases	• Agent interactions	• Individual beliefs
• GUI elements	• Schedules for agents	

The basic features that have been implemented to complement those already within the framework have enabled us to support our foundational research as well as include them in our industrial use cases. Some of these features are outlined in Table 2:

Table 2. T-B-Phase simulator examples of useful features

	Environment elements such as barriers or areas
	Network based communications and visualisations
	Human control of elements and feedback (forms and logs) for human trials

It is the culture of academic research to conduct studies in a minimalist way, constructing the simplest possible experiment and then delivering a paper study of the results. Industrial impact often demands a different approach, involving more interactive demonstration and more reliant on the relevance of the application context. By embedding an agreed set of use cases into simulations, made as simple as possible, we have developed a tool that can enable scientific study and quick conversion to impact and uptake. This has provided new avenues for exploration, such as the combining of disparate concepts in one environment as well as overcoming data sharing constraints. It has also fostered more effective collaboration between the industry and academic team which will continue into the next phase of the project. For this we will use the simulator to explore the problem space and subtleties of design realisation that would typically be defined much later in the design cycle, namely in a User Requirements Document and Systems Requirement document.

3.3 TANDEM

TANDEM (Thales AutoNomy Domain Extensible Map) is a Thales tool developed to aid our work with academia. TANDEM is a standalone, lightweight web server, which has a SocketIO API, and the ability to extend core functionality, either via the API, through a client, or by writing a plugin to extend functionality on the server. Most aspects of the interface can be configured via a text file, to allow multiple use cases.

TANDEM was created as a lightweight, free-of-licensing tool, which allowed Thales to rapidly create visual demonstrators with our partners. In addition, the simulators which were then in use within Thales, required GPUs and restrictive (sometimes expensive)

terms of use. TANDEM enables close working with Third Parties without complex licencing and other overheads.

To maximise the number of use cases TANDEM could be deployed against, adaptation was a key requirement in its design. The resulting program can be run on hardware as simple as a Raspberry Pi with, or without, a network connection and is distributed as a standalone binary file which can be simply operated on Windows, Linux or MacOS. The core functionality is provided by TANDEM, with clients providing the specific business logic for a given use case.

These features provide some useful advantages when working with academia as it can be distributed and operated on existing hardware, without needing to build any libraries, and ensures the same setup on all machines for all parties involved. Additionally the interface allows multiple languages to communicate with TANDEM, and without tying developers to a particular language.

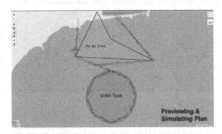

Fig. 3. TANDEM preview and simulating plan interface

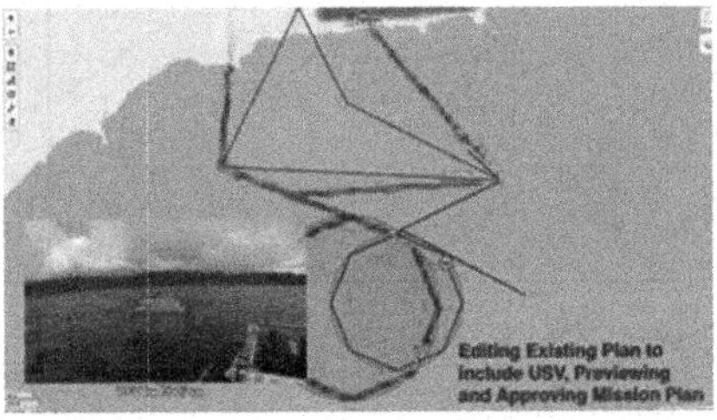

Fig. 4. TANDEM editing of existing plan

The Thales – University of Southampton Intelligent Mission Management System (IMMS) utilised TANDEM, as the mission interface, with clients connecting to a MAVLINK based UAV, and a Robot Operating System (ROS) based Uncrewed Surface Vehicle (USV) and an Uncrewed Underwater Vehicle (UUV) which communicated via Iridium in a successful live sea trial. High level tasks were set on the TANDEM Interface, and a planner client was triggered. This used the available data to construct asset level plans which could then be previewed on TANDEM (an example in Fig. 3). The operator could then verify the plan, and see deviations from allowed behaviour. Once the plan was approved, it was sent out to the assets to be acted on, and live asset monitoring was provided by TANDEM (example shown in Fig. 4). After the trial the data could also be replayed through TANDEM as an after action review tool.

Development on this tool is continuing with integration with the T-B PHASE simulator to provide for mission planning components, and incorporation with other Thales tools, for further experimentation opportunities.

3.4 Digital Twins for Maritime Test Environment

In other academic partnerships Thales has used a Digital Twin (DT) to support a number of applications ranging from rapid prototyping, training, and equivalence for virtual integration, verification, validation, qualification, acceptance and certification, monitoring and predictive optimisation of process, maintenance and control. A DT is a digital representation, often at physics level, of a system and its environment in all relevant aspects.

The DT can be used to simulate the process of observation/supervision of the physical entity, to predict future states of the physical entity based on data collected from the physical entity and possibly historical measurements and/or modelled behaviour. They can also be used to optimise the performance of the physical entity again based on data collected from the physical entity and possibly historical measurements and/or modelled behaviour by sending optimised data into the physical system.

To increase the use of synthetic environments in test and evaluation activities, the use of real world data to improve and tune the fidelity and accuracy of mathematical and physics models is essential. Using captured real world data and comparing with the tuned synthetic environment provides equivalence arguments that prove the safe use of the environments.

A DT was used to great effect in the MIMRee project (Maritime Inspection Maintenance and Repair in Extreme Environments) [12], discussed in Sect. 5.1, which required a safe and secure approach to the design, development, test and integration of cutting-edge maritime autonomy technologies. Many of the challenges presented by the MIMRee project can be resolved by the development and utilisation of two interoperable DTs: For the MIMRee project, Thales developed an environment DT of Plymouth Sound, the autonomous systems and key infrastructure. The DT utilises multiple data feeds to create an up-to-date virtual representation of the maritime area and is used in the first instance for supervising autonomous vessels to aid with mission planning, execution and debrief (strategy and planning). The DT accelerated the development, test, evaluation, and certification (assurance) of future autonomous systems operating in the maritime environment and any supporting infrastructure as part of an integrated system of systems, expediting the introduction of technologies for Net-Zero ambitions. The DT, in a Bayesian approach, will continually calibrate and optimise the system performance (performance optimisation) with collected sensor data to increase fidelity and situational awareness, opening up opportunities that require higher fidelity and safe and secure environments for training and testing.

In T-B Phase the level of fidelity that a DT provides was not appropriate but the lessons identified in developing such projects as MIMRee with academic partners has been applied.

4 Human Factors

Human factors are important in the development of autonomous systems, but perhaps not as prevalent in research as the algorithms, and is a growing area of research. The focus of Human-Machine Teaming (HMT) has traditionally focussed on:

- Optimising human performance – reducing the user's workload, costs, and errors while increasing precision and improving continual feedback [13]
- Improving acceptance, trust, and cooperation [14]
- Improving transparency and shared awareness [15].

From a psychological point of view, these challenges address predicting human behaviour when interacting with a system. However, the approaches used so far in the literature to investigate them is lacking a precise direction. The main focus has been, in fact, in analysing the variables that could affect the human behaviour in HMI from a design-specific point of view.

On the T-B PHASE project there is a mix of specialities, including psychology and human factors, enabling the inclusion of these considerations. Industry has often suffered from having the human elements considered too late in a product's development, or even disregarded completely. Our focus has been on the Sense of Agency (discussed in Sect. 2.4) and how this can help us understand a user's feeling of control.

4.1 Who is in Control?

Human-machine teaming strongly depends on cooperation and trust. These two components have been studied in different scenarios, although no specific framing around how they work in HMI design has been found to date. One of the key aspects that has not been fully considered is the human perception of control during the interaction with the system. Cooperation addresses the issue of reaching a common decision/goal with the system, whilst trust incorporates the feeling of commitment to reach that goal. This, in turn, implies that the human needs to be intentionally aware of their decision and the potential consequences of that decision. Hence, the human perception of control needs to be investigated in the context of HMT.

The question of human control perception arose several decades ago. To quote Baron [13] "Perhaps the major human factors' concern of pilots about the introduction of automation is that, in some circumstances, operations with such aids may leave the critical question, who is in control now, the human or the machine? At the extreme, some pilots argue that automation reduces the status of the human to a 'button pusher'."

Although the cockpit is just one example, this suits an enormous number of scenarios that involve cooperation with AI and machines. Overall, when the system fails, the users' supervision capabilities seem dramatically helpless in assessing the situation and determining the appropriate solution. This is mostly due to the absence of the system state awareness by the human.

4.2 Human Factors and Simulation

Future work will be focused on integrating the present findings in a simulated environment to investigate the Sense of Agency in a more applied context. The aim is to replicate and extend previous fundamental research in a scenario that can suit an HMI setting. Collision avoidance, loss of communication and different failure modes could be implemented to test their effect on the operator's Sense of Agency. This would also include different degrees of automation intervention. The hypothesis behind these studies would be that a higher Sense of Agency should be associated with greater performance, although it might be possible to find a hybrid condition where the intervention of the system does not undermine completely the user's Sense of Agency. These studies would deliver new information about the operator's mental and cognitive state under different type of tasks, such that new system design guidelines could be delivered.

Overall, the aim of this research is to embed Human Factors' features on the T-B PHASE simulator (Sect. 3.2), such that a dynamic environment for HMI can be created. This would take the shape of an interface that can address several HMI issues. From a psychology perspective, the benefit of using a common interface for different tasks would reduce the encounter of confounding effects and deliver more reliable results. From the engineering perspective, this research could offer the possibility of gathering useful information about the operator suitable for improving both the knowledge about HMI and the simulator platform itself.

5 Use Cases

Outside T-B PHASE Thales has been involved with many other projects with academia and other innovation organisations. The use cases that support two of these projects have proved useful in informing the approach to academic partnerships. In this section, we explore two use cases that highlight some of these key learnings applied in practice that were used to support T-B PHASE and vice versa where our outputs were used to inform other projects.

5.1 MIMRee

Thales collaborated in the Innovate UK project MIMRee, from the Robotics and AI in Extreme Environments call in 2018. Here we proved the concept, through a series of demonstrations, of how to utilise maritime autonomous systems to complete inspections and repair of offshore wind turbines without humans needing to be present in this dangerous environment (shown in Fig. 5).

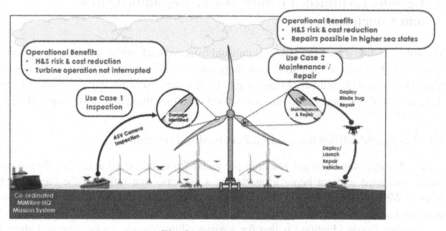

Fig. 5. MIMRee use case

The large and diverse consortium of Wootzano, Thales, Perceptual Robotics, Plant Integrity, Bladebug, The Welding Institute, Offshore Renewable Energy Catapult

(OREC), University of Bristol, The Victoria University of Manchester, Royal Holloway University, and Royal College of Art assembled their unique capabilities to address this challenge.

No single participant could have covered the full breadth of the problem space alone. Through close end-user engagement, facilitated by OREC, the project was able to define a new concept of operation and prove the concept of a new, coherent capability that could solve these challenges.

The lessons identified in MIMRee including the use of TANDEM and the concept Digital Twins have been instrumental the work in T-B PHASE, particularly the need for an easily deployable shared environment for all organisations involved.

5.2 Autonomous Last Mile Resupply

Although not directly linked to T-B PHASE, Thales led the development of the new Concept of Operations (ConOps) for Project Workhorse for the UK Defence and Security Accelerator (DASA) Autonomous Last Mile Resupply (ALMRS) challenge. In this project, the current, manned and unmanned, operations were identified and expanded with near and long-term development of operational practice. Working closely with the customer and our partners on that project meant we were able to apply the approaches defined from T-B PHASE, particularly the Engineering for Trust processes and the guidance for fostering trust in Human-Machine Teams.

The project pioneered new ways of working with autonomous systems for UK military and these new concepts are still being validated and experimented with today, such as recent trials with the Royal Marines and further work in project THESEUS (Framework Agreement for experimentation tasking in the field of Autonomous Logistics).

6 Lessons Identified, Further Work, Recommendations and Conclusion

The approach taken by T-B PHASE is unusual from an academic point of view, allowing researchers to combine their work and apply them to the applications that would not be available without the industrial input. We hope that the next phase will accelerate this as these new work packages are collaborative and introduce these real world requirements.

6.1 Lessons Identified from Initial Partnership Engagement

Although there was excellent research produced from individual activities, they were often conducted without being grounded in a use case that reflected the business of Thales. We have found having clear common themes, particularly in the form of use cases and scenarios help to overcome this.

Another lesson identified is that for a partnership to work, co-location and shared infrastructure is critical. We have found this to be the best way of ensuring that all participants work to a common vision while also facilitating knowledge exchange. Because Covid-19 meant that we could not be physically co-located for an extended period of time we were still able to work virtually because of the shared infrastructure.

A key lesson has been the use of the simulator framework (described in Sect. 3.2) which is being used to combine strategies and algorithms in a shared environment. This, in particular, has enabled creative sharing of ideas and knowledge while also allowing for industrial best practices to be explored. Alternative tools such as TANDEM (Sect. 3.3) have been used alongside other Thales tools, and to demonstrate to stakeholders and customers. Being able to use these open tools for collaboration between other industrial and academic partners has already proved to be of great benefit and is proving to be a corner stone in our T-B PHASE work streams.

6.2 Further Work

The first three years of the project were foundational research activities which although individually furthering our knowledge of autonomous systems it was clear that we needed to integrate the capabilities that have been developed in that research. As the T-B PHASE project moves forward, this research will be used to continue activities targeted towards realistic business use cases, combining research themes towards a cohesive simulator toolkit that can aid the Systems Engineer in informed early stage lifecycle decision making. Of particular interest is to ensure the Hybrid (Human and Machine) nature of autonomous teams is addressed with the inclusions of human factors exploration in technical demonstrations and experiments that will additionally lead to design guidelines. As part of this work, three sub projects have currently been identified to integrate the research strands that have been analysed in the project. These are focusing on the concept of empowered agency in autonomous systems, human swarm control training strategies and including failure modes and business specific use cases.

6.3 Recommendations

Co-location is critical to the success of a partnership and needs to be established from the onset including shared infrastructure. We have found that having a shared physical space with regular interactions between all parties helpful, but this can also be achieved virtually.

We have found relevant use cases need to be established to support the overall vision and goals of the research.

Working processes (for example filing and naming conventions, reporting timetables and templates) need to be agreed and implemented to the satisfaction of all partners.

Simulation plays a major role in all autonomous research but it needs to be cost effective and address the questions being asked. The level of fidelity needs to reflect the aims of the research. What we have found is that the simulation tools that are most useful in a partnership with academia are of a much lower fidelity that are usually used within the business to validate products. To support this the simulation environment again needs to be supported by a shared infrastructure that is accessible by all parties.

6.4 Conclusion

Industrial partnership has allowed for us to bring in experts and specialists, in specific domains or technologies (such as the maritime environment and use of communications and sensors) to share knowledge within the team and add credibility to the use cases. Reciprocally academic partners have been able to offer us creative thinking and perspectives on the cutting edge of their respective fields that are not always easily accessible in industry which can be more product focused and may be limited by processes.

So far we have found significant benefits from working in this way as it gives industrial partners greater access to the fundamental academic research, and therefore more able to quickly identify opportunities to imbed new ideas into future design and development of its products. There have also been improvements in collaboration and the breadth of research that has been able to be included, especially from new areas such as the psychological aspects for HMT or research into areas that we may not have considered applicable previously. Industrial partners have been able to share knowledge of working practices and use cases to enable researchers to produce more relevant outputs. Overall we have found this a very beneficial approach especially through a reliance on tools and simulations that are accessible to everyone.

Acknowledgments. This material was funded and delivered in partnership between the Thales Group and the University of Bristol, and with the support of the UK Engineering and Physical Sciences Research Council Grant Award EP/R004757/1 entitled 'Thales-Bristol Partnership in Hybrid Autonomous Systems Engineering (T-B PHASE)'. We would like to thank Professor Jan Noyes (UoB) for her support as well as the rest of the team.

We would also like to honour the memory of Angus Johnson who we sadly lost during the preparation of this paper, he was the driving force for the T-B Phase project and is greatly missed.

References

1. NAFv4: NATO ARCHITECTURE FRAMEWORK, Version 4, Architecture Capability Team, Consultation, Command & Control Board
2. SISO-GUIDE-006:2018: Guideline on scenario development for simulation environments
3. Crosscombe, M., Lawry, J.: The Impact of Network Connectivity on Collective Learning. Presented at the International Symposium on Distributed Autonomous Robotic Systems 2021 (2021)
4. Sanfey, A.G.: Social decision-making: insights from game theory and neuroscience. Science **318**(5850), 598–602 (2007)
5. Simon, H.A.: Theories of decision-making in economics and behavioural science. I:n Surveys of Economic Theory, pp. 1–28. Palgrave Macmillan, London (1966)
6. Adelson, R.M., Norman, J.M.: Operational research and decision-making. J. Oper. Res. Soc. **20**(4), 399–413 (1969)
7. LaValle, S.M.: Planning Algorithms. Cambridge university press, Cambridge (2006)
8. Haggard, P., Chambon, V.: Sense of agency. Curr. Biol. **22**(10), R390–R392 (2012)
9. Jeannerod, M.: The mechanism of self-recognition in humans. Behav. Brain Res. **142**(1–2), 1–15 (2003)
10. Zanatto, D., Chattington, M., Noyes, J.: Sense of agency in human-machine interaction. In: Ayaz, H., Asgher, U., Paletta, L. (eds.) AHFE 2021. LNNS, vol. 259, pp. 353–360. Springer, Cham (2021). https://doi.org/10.1007/978-3-030-80285-1_41
11. Masad, D., Kazil, J.: MESA: an agent-based modeling framework. In: 14th PYTHON in Science Conference, pp. 53–60. Citeseer (2015)

12. Multi-Platform Inspection Maintenance & Repair In Extreme Environments homepage. https://www.mimreesystem.co.uk/. Accessed 12 Dec 2021
13. Baron, S.: Pilot control. In: Human factors in aviation, pp. 347–385. Elsevier (1988)
14. Sarter, N.B., Woods, D.D., Billings, C.E., et al.: Automation surprises. Handb. Hum. Factors Ergon. **2**, 1926–1943 (1997)
15. Shively, R.J., Lachter, J., Brandt, S.L., Matessa, M., Battiste, V., Johnson, W.W.: Why human-autonomy teaming? In: Baldwin, C. (eds.) Advances in Neuroergonomics and Cognitive Engineering. AHFE 2017. AISC, vol. 586, pp. 3–11. Springer, Cham (2018). https://doi.org/10.1007/978-3-319-60642-2_1
16. Thales Bristol Partnership in Hybrid Autonomous Systems Engineering home page. https://www.bristol.ac.uk/engineering/research/t-bphase/. Accessed 21 Dec 2021

Emerging Technologies and Space

Antonio Carlo[1]([✉]) and Lucille Roux[2]

[1] Tallinn University of Technology, Tallinn, Estonia
ancarl@taltech.ee
[2] University of Paris-Saclay, Paris, France

Abstract. The fourth industrial revolution and breakthroughs in emerging technologies, namely in Big Data, Artificial Intelligence (AI) and Quantum Computing, have the potential to change the nature of space activities, ranging from space exploration, space tourism, to space activities. The practical applications of AI are manifold and include the automation of repetitive tasks in the manufacturing of satellites and spacecrafts, image recognition of satellite data and provision of assistance to astronauts in task and behaviour management. While emerging technologies offer new opportunities, the side effects of these promises rest on the negative issues associated. Ensuring the accessibility of space datasets, while also ensuring the identification and mitigation of cybersecurity implications of AI-powered devices and quantum computers, remains essential. These are some of the few challenges that scientists, design engineers, and decision-makers will have to face in both the short and long-term.

Based on the above, this paper will discuss the possibilities presented by emerging technologies, their associated risks in the space sector, and finally the potential mitigating measures addressing the degree of coordination that is necessary to ensure that space activities remain peaceful and beneficial for all humankind.

Keywords: EDTs · Artificial intelligence · Quantum computing

Abbreviations

AI   Artificial Intelligence
EDTs   Emerging Disruptive Technologies
GNSS   Global Navigation Satellite Systems
JAXA   Japanese Aerospace Exploration Agency
METOC   Meteorological
NASA   National Aeronautics and Space Administration
PNT   Position, Navigation and Timing
QT   Quantum Technologies
SatCom   Satellites of Communications

© Springer Nature Switzerland AG 2022
J. Mazal et al. (Eds.): MESAS 2021, LNCS 13207, pp. 494–508, 2022.
https://doi.org/10.1007/978-3-030-98260-7_32

1   Introduction

The recent fourth industrial revolution has allowed for significant scientific and technological progress, resulting in the blossoming of new technologies that have deeply impacted the civilian and military landscape in the cyber and space domains. Since the launch of the first artificial satellite, these two domains have been intertwined and have gradually become two faces of the same coin as one is essential to the other. The development of new technologies along with the access to relevant data have further influenced this interconnection, strengthening their need for defence and resilience. New technologies have driven the democratisation of the two domains, resulting in the increase of new activities in these fields by both public and private actors. Consequently, the close interconnection between the cyber and space domains has resulted in the inevitable development of Emerging Disruptive Technologies (EDTs) threatening critical infrastructures. For the purpose of this article, the concept of EDTs refers to, but not limited to, technologies ranging from Artificial Intelligence (AI) and autonomous systems, to hypersonic weapons, quantum computing, big data and biotechnological weapons. In this context, "emerging" does not correspond to the novelty of the technology but the novelty of their use and the threats associated with it. Additionally, the notion of "disruptive technologies" takes into consideration the issues (economic, legal, political, and strategic) raised by new technologies. As a result, the access to an increasing amount of data has led to a shift in the use of new technologies [40]. This article aims to provide an assessment and an analysis of the interrelations between outer space and emerging cyber technologies from a legal, strategic and policy standpoint.

2   Interrelations Between Cyber and Space

2.1   Opportunities and Application of EDTs

Applications of Artificial Intelligence in Space

AI is a technological innovation leveraging machines and computers to reproduce the human mind's cognitive capabilities for problem-solving and decision-making [39]. It is based on the assimilation by an algorithm of information and its performance of tasks that are specific to human intelligence (recognition, contextualisation, analysis, and problem solving). AI is often used as an umbrella term encompassing machine learning, natural language processing (NLP) computer vision, deep learning and cognitive computing. AI, falls into different categories: weak, strong and super AI. The 'weak AI', also called 'narrow AI', refers to a type of algorithm that specialises in a single task and is already widely used in virtual assistants, facial recognition or even self-driving cars. This kind of AI "repeats similar codes that were predefined by their makers and classifies them accordingly" [34]. On the other hand, 'strong' AI refers to systems that can replicate human intelligence and abilities - i.e. an AI that can learn and understand and potentially carry out tasks like a human being. Finally, "super" AI performs functions better than a human. This kind of AI is the long-term objective that today's scientists are trying to work towards.

While the concept of AI is not new and was theorised by Alan Turing in the beginning of the twentieth century [1], its application to the space sector has recently become exponential. Space assets generate an important amount of data amounting to thousands of terabytes of data per day and corresponding to a volume of petabytes of data[1] for 365 days [2]. The benefit of using data is not linked to the amount and capacity of collecting a large amount of data but rather to the ability to convert data into useful information.

For example, AI offers new opportunities to respond efficiently to the challenges of detection, identification, and mapping of either terrestrial or Martian phenomena. Earth analysis is illustrated by the AI4Copernicus [3] platform, developed at the European Union level, that facilitates the storage of and access to Copernicus data. In addition, MIT and the Qatar Computing Research Institute developed an AI model to tag road features based on satellite images with the objective of improving GPS navigation in places with limited map data [4]. Additional examples include the Mars rovers AI-based software that is used to improve data transmission by removing human scheduling errors and the ESA's Rapid Action Earth Observation dashboard which is capable of processing satellite data and extract economic indicators. However, at this stage, all these technologies are soft-AI based with the possibility to further develop and become more autonomous in the future. Currently, AI provides the analysis and recommendations to the human counterpart who, however, still retains the final decision.

As such, machine learning, understood as the performance of statistical research with regard to a computational process, has been increasingly applied to Earth Observation (geo-science, remote sensing) [5] and has been developed for Martian terrain mapping. On the latter, NASA has developed the AI4Mars online tool to train an AI algorithm to automatically read the landscape and support "Curiosity" [6] in its mission. This technique has been used to identify patterns and warn about changes based on the prediction of properties which are usually static over the observational time period. As spatial and temporal interrelations are not maximised, the scientific community has been looking forward to the development of deep learning, based on artificial neural networks creating hierarchical architecture thinking.

The combination between space and AI presents numerous opportunities in the military field for Intelligence, Surveillance and Reconnaissance (ISR); Meteorological (METOC); Position, Navigation and Timing (PNT); environmental monitoring, early warning systems (EWS), as well as for secure satellite telecommunications (SATCOM). The US Department of Defence (DoD) has partnered with the private sector to develop new solutions combining and leveraging space systems and AI tools. For instance, L3Harris Technologies supports the US DoD "to develop AI and machine learning (AI/ML) systems to help reduce the amount of time it takes to decipher usable intelligence from increasing amounts of data collected from space and airborne assets" [7]. Moreover, AI could improve Space Situational Awareness by assessing and numbering space debris, identifying Near Earth Objects (NEO), alerting to the probability of collisions, alerting to missile launches, and mapping Allies' and non-Allies' space assets in orbit. In this context, NATO and Luxembourg signed a Joint Statement to

---

[1] 100 terabytes of data collected in a year by commercial companies.

develop a Strategic Space Situational Awareness System (3SAS) at the 14 June 2021 Brussels Summit [8]. With regards to NATO's developments on EDTs, the 3SAS will likely incorporate AI systems.

Notwithstanding, the use of new technologies comes with risks. AI is based on training data sets, which could lead to falsification of satellite images, particularly for geospatial data [9] It creates new avenues for threats and enables a variety of attack methods. Such issues call for the development of a system of geographic fact-checking along with the development of security requirements and standards specific to deep learning algorithms used for space applications. This logic has been gaining traction in Europe, with the ongoing work of the European Union agency for cybersecurity on certification schemes for cybersecurity in the AI field [10]. However, the prospective certification requirements are not binding [11], unless otherwise stated, particularly when a critical infrastructure is targeted. Under the draft review of the Directive Network and Information Systems [12], space ground infrastructures are designated as critical infrastructures, so established standards could become compulsory for space stations' infrastructures. Yet, what remains unclear is how standards could be applied to in-orbit assets.

Space and Quantum Technologies

Our society is dependent on space technologies for daily life. The reliance on these technologies is crucial. The issues faced by the increasing reliance on space technologies have led authorities to invest more in promising emerging technologies, such as quantum technologies (QT).

QT works by using quantum mechanics' principles such as the physics of subatomic particles, quantum entanglement and quantum superposition. QT can be divided into several areas: quantum communication; quantum simulation; quantum computing and quantum metrology.

Taking the pace, the USA established in 2018 the "National Quantum Initiative Act" [13]. Meanwhile, Google developed its 53-qubit Sycamore quantum supercomputer based on electrons and superconductors. On the other hand, China developed a computer able to process 66 qubits, based on optical circuits and photons [14]. A significant milestone towards paving the way for EU strategic autonomy has been made by the European Commission QT flagship [15] which "is due to support the work of hundreds of quantum researchers over 10 years, with an expected budget of €1 billion from the EU" in the field of QT. The flagship aims to fund projects based on four core application areas: quantum computing, quantum simulation, quantum communication, and quantum metrology and sensing. Moreover, NATO has already started to deal with QT as a key component of the NATO2030 initiative [16] to foster the Alliance's technological edge. Consequently, the Alliance has been working on cryptography with Malta [17] and protecting cyberspace with QT [18]. Meanwhile, China established the world's first integrated quantum communication network, combining over 700 optical fibers on the ground with two ground-to-satellite links for quantum key distribution [49]. The remarkable examples explained above describe the concomitant importance of the use of QT for space applications and the exponential growth of patent filings in Research and Development since 2001 and 2021 [64].

Table 1. Evolution of patents filing between 2001–2020 [63]

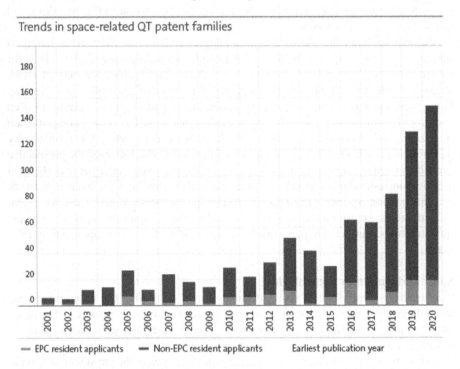

Trends in space-related QT patent families

— EPC resident applicants   — Non-EPC resident applicants   Earliest publication year

QT are based on quantum effects depending on the interaction and structure of particles. Major Space faring nations, including European countries have started to invest in QT. QT applied to space are particularly used in time and frequency transfer, secure communications, fundamental physics as well as Earth observation sensing (Table 2).

- QT for time and frequency transfer applications are based on Cold Atom Clocks providing precise frequency standards. Boosting time and frequency transfer performance may be used in the defence and security sectors to enhance cross-domain applications.
- QT for secure communications rest on Quantum Key Distribution (QKD) enabling two peers to build a key through a dialogue taking place on public channels (quantum channels and network link). Current fibre-based quantum key distribution prevents long-distance key distribution for encrypting communication links. Securing communications with QKD could protect data transmitted against cyber threats. Applying QT to space systems creates new opportunities to protect space assets. This is the objective of ESA led ARTES 4.0 programme and SAGA mission along with the EU Commission's driven initiative EuroQCI. In the long-term, these projects will support the European Strategic Compass towards resilience and EU strategic autonomy. Also, these projects set the basis for the deployment of 5G/6G networks and ensure the integrity, availability, and confidentiality of data. However, QKP raises issues. There are no indications in the current state-of-the-art "that key negotiation schemes

Table 2. Quantum Technologies in Space, White Paper, ESA [19]

QT for	Opportunities	Illustrations
Time and Frequency Transfer	Enhancing timekeeping and geolocation. QT increases fundamental frequency from the radio frequency domain (with accuracies of 1e-16) in the optical field (with accuracies of 1e-5) Enabling synthetic aperture optical astronomy; geodesy	-Improvements for GNSS services [20]; -ESA ACES [21] (atomic clock based on optical links)
Earth observation	Development of quantum sensors based on atom interferometry (more sensitive and precise tools) to monitor Earth resources and improve prediction of Earth events. The objective is to palliate the long-term drifts and noise at low frequency experienced by classical instruments	-ESA studies on the use of quantum technologies for Next generation gravity mission concepts [22]
Secure communications	Using asymmetric encryption through Quantum Key Distribution	-ESA ARTES 4.0 programme [23]; -ESA SAGA (Security and cryptoGrAphic) mission [24]; -EuroQCI initiative [25]; -Chinese satellite "MICIUS" [26]; -GALILEO (OS-NMA) [27]
Fundamental physics	Pushing the boundaries of the terrestrial understanding of general relativity, quantum mechanics, cosmology	-Ideas can be suggested on the ESA's Open Space innovation Platform [28]; -NASA's paper [29] on the ISS research environment; -JAXA new strategic L-class mission [30] for sky surveys of cosmic microwave background polarization

are robust against any adversary with unknown capacities". Also, QKD comes with implementation constraints, notably with the difficulties to implement QKD protocols [50]. Moreover, the transmission of sensitive data can be hindered as QKD devices may have weaknesses (software vulnerabilities or leakage of secrets by electromagnetic radiation). In addition, QKD have a point-to-point nature and dependence on physical characteristics of the channels used. Therefore, any large-scale deployment is complex and expensive as it would require the use of satellites [51].

- QT for Earth observation and Fundamental Physics use Cold Atom Interferometers. Tools are more precise and sensitive.

EDTs Opportunities to Enhance Space as a Cybersecurity Enabler

The view that new trends in Space, such as the use of EDTs, only result in more vulnerabilities, is however one-sided. Space is also a cybersecurity enabler. Galileo, one of the EU flagship programmes offers authenticated service. An open service has been developed to provide a Navigation Messages Authentication (OS-NMA) enabling the detection of spoofing attacks and thus increasing the reliability and trustworthiness of position and timing information. In addition, the Galileo Commercial Service increases robustness based on authentication services relying on signal encryption [31].

Furthermore, the growing development of optical communication satellites increases low latency data encryption and expands the capacity for secure data transmission [32]. These issues will rest on the mechanisms developed to secure ground-based stations, possibly prone to cyber operations that could impact space-based assets. Therefore, optical communication, such as the optical inter-satellite links of the European data relay system (EDRS) and ESA's prospective High Throughput Optical Network (HydRON), have been developed. The use of QT offers new opportunities however, it is associated with risks (Sect. 2.2).

2.2 Risks Associated to EDTs and Mitigation Measures

Context and Current Developments on the Risks Associated with Space and Cyber
The increasing use of cyber capabilities by space operators has resulted in the inevitable development of new threats. Nowadays, technologies such as communication infrastructures as well as navigation, positioning and timing depend on space assets through international transmission and connection. With many space operators performing a digital transformation, more and more satellites are controlled by digitised systems, carry digital payloads and use digital links and cyber capabilities to gather, store and transmit information. Hence, space assets are not only vulnerable to physical threats, but also to those in cyberspace.

The diversification of both public and private actors conducting space activities has sparked issues. The race to launching and deploying national satellites serves both security and defence purposes. This race is also strongly linked to the national impetus to deploy new capabilities to reach hegemony in the space field and remain independent. Space activities support a diverse set of other activities on the ground as space assets have global coverage to connect remote areas, monitor large zones all over the world, and offer accurate timing and positioning for many essential activities. While space assets are becoming more valuable, some actors are developing new types of devices with the intention of neutralising or temporarily invalidating systems and opposing capabilities in space. While neutralisation can be achieved through various ways, these new technologies allow cyber assets to inflict a decisive hit on the space sector. Satellites can be attacked directly through the space asset and indirectly via supporting assets like ground bases. In the event of such malicious activities, cyber and space issues are, in most cases, addressed by national laws and policies. Yet, both cyber and space normative systems are also dealt with by general public law, as well as customary international law and non-binding legal principles. In this context, and due to the evolution of cyber and space activities, developing laws and policies that would match the challenge of tackling all the issues stemming from these cyber and space activities is a vain attempt. Hence, an interpretation of general principles may apply to disputes relating to space activities using cyber capabilities. As technology continues to evolve, so do the opportunities and challenges it poses. The increasing dependence on technology in our daily life exposes us to a whole set of risks related to security and defence.

EDTs tend to be considered cutting-edge and represent potential opportunities to the Information and Communication Technology (ICT) sector. The range of EDTs is extremely broad and has significant overlaps: While data, AI, autonomy, space, and hypersonic are considered to be "disruptive"; developments in quantum, biotechnology, and materials are seen as "emergent" since they still require more time to mature. The latter are expected to mature in a timeframe of 20 years [33]. For instance, the NATO Defence Ministers Meeting held in October 2019, identified eight EDTs in the areas of data, quantum, AI, autonomy, space, hypersonic, biotechnology, and materials.

Depending on its category, AI can be used in the cyber and space field for both defensive and offensive purposes. AI has been taken into consideration by many national strategies due to its potential implication in the military field. Today, 36 nations take into consideration AI strategies in a military context. The importance attributed to AI by these countries can also be seen in the allocated military budgets. At the beginning of 2021, the Pentagon announced that it will be dedicating US $ 841 million to increase the use of AI in its military services [35]. Similarly, the UK has announced an increase in military spending amounting to US $ 21.8 billion over a period of four years which is meant to "cement the UK's position as the largest defence spender in Europe and the second largest in NATO" [36]. The funds are expected to be used for cutting-edge technologies, namely, space defence, cyber-offence and AI [37].

A major risk that the security and defence space sector might face is the lack of defence funding among NATO's member states. The size of defence budgets has direct effect on research and development (R&D), as well as the implementation of new technologies. [45] Often technologies are developed with a dual use factor in mind as they respond both to the military and commercial market. [48] A reduction in investment of either the military or commercial sectors in emerging technologies, could result in the stagnation of both research and development. An example is the winter AI, that refers to the decline in interest and funding in AI technologies that happened in the 80s. [47].

In the coming years, defence spending is predicted to fall as part of the adverse effects of the COVID-19 pandemic. [45] Then, the remaining question lies in whether and how the health crisis will impact the military and defence sectors along with international relations in a global context of an armament race.

Another major risk is the security supply chain. New technologies are highly dependent on computational hardware, which are indispensable to run these algorithms, that require ever greater computational performance and more complex manufacturing. Most of the manufacturing and the mining of rare material essential for key components of technology takes place outside of the US and the EU. South America, Africa and Asia are the main suppliers of these materials which make the supply chain result fragile and susceptible to international conflicts in these regions. [46] The commercial tensions between China and the USA in 2018 impacted export control regulations [52] and incidentally, concerned cybersecurity solutions.

In addition, supply chain might impact enormously on the development and use of new technologies, for instance it has also been targeted by malicious actors such as in the case of "solarwinds". The entire US security system was compromised due to a security breach to one of its suppliers. Another case can be non-malicious, during the COVID-19 pandemics, the extraction of raw material diminished leading to a decrease in production and lack of resources. As cascade effects, it impacted European companies. Supply chain is essential, and a strong security need to be ensured. This can be achieved through framework, introduction of international recognised standards, and certifications of the vendors. However, it should be also emphasised that the bulk of these problems is only solvable when there is close cooperation between the subject actors. [45] Security is essential in all its forms (physical, personnel and Communication Information System) and it doesn't have to be perceived as a technological inhibitor while as a technological insurer.

Another risk related to the use of EDTs lies within the conditions associated with data sharing, and governance. Both EDTs and particularly AI systems require a large amount of data to better perform analysis. It raises issues in terms of protection of personal data and the requirements associated in terms of transfer of personal data: which type of data would be used to train AI systems for Space? Would it be personal data? How will these data be protected? Where data should be located (location of the datacentre)?

- The **protection of personal data** worldwide generates debate and issues, especially while Law Enforcement Authorities aim to access data to facilitate judicial procedures. For instance, the Patriot Act [53] and the Cloud act [54] imply that in some cases the level of encryption over the internet is lower to let LEA access data. However, it is balanced with the right for privacy. It is the purpose of EU the GDPR [55], setting the basis for a general framework and the proposal for a regulation of AI systems in the EU that would correspond to the *lex specialis*.
- On **the transfer of personal data**, the CJEU [56] recently highlighted the fundamental right to privacy in the context of the transfer of personal data to third countries and invalidated the Privacy Shield[2]. [57] This decision questions then about the transfer of personal data in the context for Cloud services (and the ongoing work on the Cloud certification). [58]
- The **location of datacentres** is vital, especially when it relates to space data and intellectual property rights. The Cybersecurity legislation of the people's Republic of China [59] obliges companies that operate in the Republic's territory to stock their data in China. Similarly, the Russian legislation on the protection of personal data requires from companies collecting and processing data of Russian citizens to stock their data in Russia. However, it will be highly difficult to distinguish data from Russian citizens from those others.

Certainly, each EDT poses different risks. Therefore, it is important to estimate the risk associated with it. This can be estimated via the parameter portrayed below, which represents the impact and the likelihood that an event might happen. By crossing these two parameters, a matrix can be created through which five categories of risk can be identified: 'extreme', 'major', 'moderate', 'minor' and 'incidental'.

---

[2] A framework guaranteeing that the level of protection granted to personal data in the U.S. is essentially equivalent to that guaranteed within the EU, in line with the judgment.

Table 3. Risk - heat map

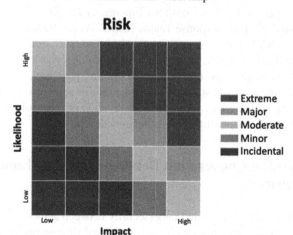

The "heat map" matrix allows for a two-dimensional identification of the potential risk impact. This allows, to evaluate the risk and its acceptance that users are willing to accept based on their risk appetite.

EDTs present opportunities in terms of offensive and defensive capabilities, enabling a better monitoring of data flow and easing decision-making process. However, it has led to the rise of new threats.

EDTs and the Rise of New Threats in the Space Field

Given the ultra-hazardous nature of outer space and the growing dependence of society on space, the familiar threats concept might extend beyond security to resilience. The utilisation of EDTs such as AI and QT in the space sector allow space actors to have a wider and more strategic view of the domain. However, these technologies can be used maliciously by different actors for monetary gain, political influence, espionage, etc. The use of AI and QT in cyber activities allow both the attacker as the defender to utilise an enormous computing power that would elaborate data in a faster and more precise way.

The development of EDTs led to the rise of new threats in the space field. These threats can be classified as kinetic and non-kinetic. Kinetic threats are those that attempt to strike directly or detonate a weapon near a satellite or other space assets. Non-kinetic threats involve weapons that have physical effects on space assets without any physical interference, namely electronic and cyber warfare. While cyberattacks are not a new threat in the space field, the number of cyber actors targeting space assets is rising. Cyber actors usually fall into one of the following categories [38]: nation state actors, private economic actors, hacktivists/natural persons, and international entities. These actors can either be instigating the attack, responsible for the attack, the victim or collateral victim of the attack.

Hostile cyber actors are continuously trying to break into close and highly secure systems. To counter these issues, space systems' security and defence need to be constantly updated, secured, and monitored. Many governments, companies, and international organisations have created ad hoc Computer Emergency Response Teams (CERTs) and Computer Security Incident Response Teams (CSIRTs) coordinated by Security Operational Centres (SOCs) in order to pre-empt and, if necessary, confront possible cyber events. Moreover, cyber hygiene, awareness and preventive cyber security measures are essential to further mitigate new threats and risks to the space sector.

Moreover, certification and standards are essential for new as well for old technologies to use them in a secure way. With the identification of international standards and a certification authority, allows to have a unique methodology.

3   Insights on the Cooperation Between Actors to Enhance Cybersecurity

Cooperation in the field of space and cybersecurity between actors is essential to work towards threats anticipation, sharing best practices, and sharing information on a non-sensitive level. The objective is to set the basis for coordinated actions between actors without hindering sovereign aspects tied to space and the cyber segments. To this end, it is suggested that at the international level, entities should foster dialogue on setting up norms of behaviour, standards, and guidance on threats anticipation, vulnerabilities detection and response. These rules will not be binding however it will provide useful food for thought to support both the private and public sectors to protect themselves/their capabilities against cyberthreats.

Also, the importance of cooperation in cyberspace has been put forward at the European level. In 2017, the European Commission adopted the Blueprint recommendations [60] for rapid emergency responses for large-scale, cross borders cyber incidents and crises. Two years later, during the second Blue OLEx 2020 exercise [61] has underlined that there is a need to implement an intermediate level between the technical (CSIRTs Network) and the political ones in the EU Cyber crisis management framework. Then, the Cyber Crisis Liaison Organisation network (CyCLONe) for crisis management and increasing European solidarity between Member States was launched. Its legal basis will be the revised version of the NIS Directive, Additionally, the European Commission announced the creation of a European network of SOCs as a key component of the prospective so-called "CyberShield". [62] Therefore, cooperation and coordination are presented as facilitating the resilience of infrastructures.

A strong international collaboration is essential to strengthen defensive and security capabilities in new cutting-edge technologies. Best practice was given by the defensive collaboration between the EU and NATO. At the Warsaw Summit, in July 2016, the two international organisations outlined areas of cooperation considering common challenges, including countering EDTs (hybrid threats, enhancing resilience, defence capacity building, cyber defence, maritime security and exercises) [42]. The two organisations share over 70 common measures to advance security and international cooperation [43]. Cooperation in the cyber field is also discussed at the OECD level.

Moreover, a strong cooperation between private and public actors is also essential to establish and maintain a strong resilience and enable private entities to scale up. For instance, the EU established strong Private Public Partnerships (PPPs) that allow the two parties to share best practices and lessons learned in order to respond reactively to possible malicious activities [44]. Coordination between private and public entities would be strengthened with the prospective implementation of the new Cybersecurity Competence centre in the field of research, new technologies and industrial skills. Indeed, the private community is represented in the Centre.

4 Conclusion

In the last decade, scientific and technological progress has led to the maturing of new technologies that have deeply impacted both the civil and military domain. Due to the increasing number of critical situations, the role of EDTs cannot be underrated. Humanity needs the assistance of machines capable of providing intelligence through fast and reliable analysis of data. EDTs have the potential to change the nature of space activities and will play a crucial role in international cooperation. Technologies such as AI and QT are highly efficient and could offer new types of high-impact solutions, improve decision making process, as well as support civilian and military operations. In the long run, investing in the research and development of new technologies could result in valuable benefits. However, the use of EDTs does not come without risks, and some cyber related ones can be identified. Cyberspace corresponds to the environment in which communication over computer networks takes place. AI and QT have significant impacts in cyberspace and any disruption to their systems could create risks in the collection and processing of data.

The current trend at a European and international level moves towards establishing principles and standards to be followed to prevent, and/or protect space-based assets. However, these are not binding commitments.

Glossary

AI: Artificial intelligence is a technological innovation leveraging machines and computers to reproduce the human mind's cognitive capabilities for problem-solving and decision-making [39].

EDTs: The Emerging Disruptive Technologies concept corresponds to, but is not limited to, technologies ranging from artificial intelligence and autonomous systems, hypersonic weapons, quantum computing, to big data and biotechnological weapons.

- In this context, "emerging" does not refer to the novelty of the technology but the novelty of their use and the threats associated with it.
- The notion of "disruptive technologies" takes into consideration the issues (economic, legal, political, strategic) raised by new technologies. As a result, the access to an increasing amount of data has led to a shift in the use of new technologies [40].

Quantum Computing and Quantum Technologies: According to the National Institute of Standards and Technology of the US Department of Commerce (DOC), quantum computers manipulate quantum bits of information, or qubits whereas classical computers process either higher or low voltages (0 and 1). Qubits is based on the superposition mechanical phenomena. These qubits can be both a 1 and a 0 at the same time [41].

References

1. Britannica: Alan Turing and the beginning of AI (2021). https://www.britannica.com/techno logy/artificial-intelligence/Alan-Turing-and-the-beginning-of-AI
2. Mohney, D.: Terabytes from space: satellite imaging is filling data centers. Data Center Frontier (2020). https://datacenterfrontier.com/terabytes-from-space-satellite-imaging-is-filling-data-centers/
3. Community Research and Development Information Service: Enabling AI and Earth Observation innovation through integration of AI4EU with DIAS platforms. European Commission (2021). https://cordis.europa.eu/project/id/101016798
4. Matheson, R.: Using artificial intelligence to enrich digital maps, MIT News Office (2020)
5. Reichstein, M., Camps-Valls, G., Stevens B., Jung, M., Denzler, J., Carvalhais, N.: Deep Learning and process understanding for data-driven earth observation science (2019)
6. AI4MARS: Teaching Mars Rovers How to Classify Martian Terrain (2021). https://www.zoo niverse.org/projects/hiro-ono/ai4mars
7. L3Harris: GPS World Staff, L3Harris to help US DOD with Artificial intelligence and Machine Learning (2020)
8. NATO: NATO and Luxembourg boost Alliance Space Situational Awareness (2021). https://www.nato.int/cps/en/natohq/news_185365.htm
9. Bo, Z., Shaozheng, Z, Chunxue, X., Yifan, S., Chengbin, D.: Deep fake geography? When geospatial encounter artificial intelligence (2020)
10. European Union Agency for Cybersecurity: Cybersecurity Challenges in the Uptake of Artificial Intelligence in Autonomous Driving (2021). https://www.enisa.europa.eu/public ations/enisa-jrc-cybersecurity-challenges-in-the-uptake-of-artificial-intelligence-in-autono mous-driving/at_download/fullReport
11. REGULATION (EU) 2019/881 OF THE EUROPEAN PARLIAMENT AND OF THE COUNCIL of 17 April 2019 on ENISA (the European Union Agency for Cybersecurity) and on information and communications technology cybersecurity certification and repealing Regulation (EU) No 526/2013 (Cybersecurity Act), Art. 56
12. European Commission: Cybersecurity-Review of EU rules on the security of network and information systems (2021). https://ec.europa.eu/info/law/better-regulation/have-your-say/ initiatives/12475-Cybersecurity-review-of-EU-rules-on-the-security-of-network-and-inform ation-systems_en
13. US Congress: NATIONAL QUANTUM INITIATIVE ACT, PUBLIC LAW 115–368—21 December 2018. https://www.congress.gov/115/plaws/publ368/PLAW-115publ368.pdf
14. Solca, B.: China now leading the quantum computing race, Notebook Check (2021)
15. Quantum Technologies Flagship: Midterm Report of the Quantum Technologies Flagship. European Union (2020). https://digital-strategy.ec.europa.eu/en/policies/quantum-technolog ies-flagship
16. NATO: NATO 2030: United for a New Era; Analysis and Recommendations of the Reflection Group Appointed by the NATO Secretary General (2020). https://www.nato.int/nato_static_ fl2014/assets/pdf/2020/12/pdf/201201-Reflection-Group-Final-Report-Uni.pdf
17. NATO works on quantum cryptography with Malta, https://www.nato.int/cps/en/natohq/ news_165733.htm. Accessed 01 Oct 2021
18. Changing lives and the security landscape – how NATO and partner countries are cooperating on advanced technologies. https://www.nato.int/cps/en/natohq/news_184899.htm. Accessed 01 Oct 2021
19. European Space Agency: Quantum Technologies in Space: Policy White Paper (2019). https://www.cosmos.esa.int/documents/1866264/3219248/BassiA_QT_In_Space_-_ White_Paper.pdf/6f50e4bc-9fac-8f72-0ec0-f8e030adc499?t=1565184619333

20. Calderaro, L., et al.: Towards quantum communication from global navigation satellite system. Quant. Sci. Technol. (2019). https://doi.org/10.1088/2058-9565/aaefd4
21. European Space Agency: Satellite Advanced Global Architecture (2020). https://artes.esa.int/projects/saga-satellite-advanced-global-architecture
22. Haagmans, R., Siemes, C., Massotti, L., Carraz, O., Silvestrin, P.: ESA's next-generation gravity mission concepts. Rendiconti Lincei. Scienze Fisiche e Naturali **31**(1), 15–25 (2020). https://doi.org/10.1007/s12210-020-00875-0
23. European Space Agency: About Advanced Research in Telecommunications Systems. https://www.esa.int/Applications/Telecommunications_Integrated_Applications/ARTES/About_ARTES. Accessed 01 Oct 2021
24. European Space Agency: European quantum communications network takes shape. https://www.esa.int/Applications/Telecommunications_Integrated_Applications/European_quantum_communications_network_takes_shape. Accessed 01 Oct 2021
25. European Commission: The European Quantum Communication Infrastructure (EuroQCI) Initiative (2021). https://digital-strategy.ec.europa.eu/en/policies/european-quantum-communication-infrastructure-euroqci
26. Kwon, K.: China reaches new milestone in space-based quantum communications. Scientific American (2020). https://www.scientificamerican.com/article/china-reaches-new-milestone-in-space-based-quantum-communications/
27. European Space Agency: Galileo Open Service Navigation Message Authentication. ESA navipedia (2021). https://gssc.esa.int/navipedia/index.php/Galileo_Open_Service_Navigation_Message_Authentication
28. European Space Agency: Get involved in ESA's quest for quantum. https://www.esa.int/Enabling_Support/Preparing_for_the_Future/Discovery_and_Preparation/Get_involved_in_ESA_s_quest_for_quantum. Accessed 01 Oct 2021
29. National aeronautics and space administration: a researcher's guide to: fundamental physics (2015). https://www.nasa.gov/sites/default/files/atoms/files/np-2015-04-021-jsc_fundamental_physics-iss-mini-book-508.pdf
30. Hazumi, M., et al: LiteBIRD: JAXA's new strategic L-class mission for all-sky surveys of cosmic microwave background polarization (2021). https://doi.org/10.1117/12.2563050
31. Duquerroy L.: Cyber security and Space-based services, ESA Webinar (2019)
32. Wu, B., Shastri, B.J.: Secure communication in fiber-optic networks (2013)
33. NATO Science & Technology Organization: Science & Technology: Trends 2020-2040 (2020). https://www.nato.int/nato_static_fl2014/assets/pdf/2020/4/pdf/190422-ST_Tech_Trends_Report_2020-2040.pdf
34. Carlo, A.: Artificial Intelligence in the Defence Sector. In: Mazal, J., Fagiolini, A., Vasik, P., Turi, M. (eds.) MESAS 2020. LNCS, vol. 12619, pp. 269–278. Springer, Cham (2021). https://doi.org/10.1007/978-3-030-70740-8_17
35. Ratnam, G.: Pentagon to expand the role of AI in military efforts, folsom: government technology (2021). https://www.govtech.com/products/Pentagon-to-Expand-the-Role-of-AI-in-Military-Efforts.html
36. PM to announce largest military investment in 30 years. Prime Minister's Office, London. https://www.gov.uk/government/news/pm-to-announce-largest-military-investment-in-30-years. Accessed 01 Oct 2021
37. Frith H.: What Boris Johnson has planned for his £16bn military spending spree. The Week, London (2020). https://www.theweek.co.uk/108713/boris-johnson-16bn-military-spend
38. Wallace, P., Schroth, R.J., Delone, W.H.: Cybersecurity regulation and private litigation involving corporations and their directors and officers: a legal perspective. Kogod Cybersecurity Center, Kogod School of Business, American University (2015)

39. Mccarthy, J.: What is artificial intelligence? (2004). https://homes.di.unimi.it/borghese/Tea ching/AdvancedIntelligentSystems/Old/IntelligentSystems_2008_2009/Old/IntelligentSyst ems_2005_2006/Documents/Symbolic/04_McCarthy_whatisai.pdf
40. NATO: Emerging and disruptive technologies. https://www.nato.int/cps/en/natohq/topics_ 184303.htm. Accessed 01 Oct 2021
41. Williams, C.J.: American leadership in quantum technology (2017). https://www.nist.gov/ speech-testimony/american-leadership-quantum-technology
42. Carlo, A., Casamassima, F.: Securing outer space through cyber: risk and countermeasures. In: International Astronautical Congress (IAC), Dubai (2021)
43. Rajagopalan, R.P.: Electronic and cyber warfare in outer space, UNIDIR (2019)
44. Public-Private Partnerships: European Research Area - Learn, European Union. https://www. era-learn.eu. Accessed 01 Oct 2021
45. Sabatino, E., Marrone, A.: Emerging Disruptive Technologies: The Achilles' Heel for EU Strategic Autonomy. IAI commentaries (2021)
46. Guillaume, P.: The Rare Metals War. Scribe UK (2021)
47. Carlo, A., Perucica, N.: Artificial intelligence: walking the line between military deterrence and interstate cooperation, U.S. Strategic Command (2021)
48. Christine, E.H.: Artificial Intelligence at NATO: dynamic adoption, responsible use. NATO, Brussels (2021)
49. Yuao, C., Jianwei, P., Chengzhi, P.: An integrated space-to-ground quantum communications network over 4,600 kilometers, University of Science and Technology of China, Nature (2021)
50. Agence nationale de la sécurité des systèmes d'information: Should Quantum Key Distribution be Used for Secure Communications? (2021)
51. Huang, A., Navarrete, Á., Sun S., Chaiwongkhot, P., Curty, M., Makarov, V.: Laser seeding attack in quantum key distribution. J. Quant. Phys. (2019)
52. U.S. Department of Commerce: Commerce Department Further Restricts Huawei Access to U.S. Technology and Adds Another 38 Affiliates to the Entity List. https://2017-2021.com merce.gov/news/press-releases/2020/08/commerce-department-further-restricts-huawei-acc ess-us-technology-and.html
53. U.S. CongressH.R.3162 - Uniting and Strengthening America by Providing Appropriate Tools Required to Intercept and Obstruct Terrorism (USA PATRIOT ACT) Act of 2001. https:// www.congress.gov/bill/107th-congress/house-bill/3162
54. U.S. Congress: H.R.4943 - CLOUD Act. https://www.congress.gov/bill/115th-congress/ house-bill/4943
55. European Union: Regulation on the protection of natural persons with regard to the processing of personal data and on the free movement of such data, and repealing Directive 95/46/EC (General Data Protection Regulation)
56. European Parliament: The CJEU judgment in the Schrems II case. https://www.europarl.eur opa.eu/RegData/etudes/ATAG/2020/652073/EPRS_ATA(2020)652073_EN.pdf
57. U.S. Department of Commerce: Privacy Shield framework. https://www.privacyshield.gov. Accessed 01 Oct 2021
58. European Data Protection Board: Letter to ENISA regarding the European Cybersecurity Certification Scheme for Cloud Services (EUCS) (2021)
59. People's Republic of China: Cybersecurity Law of the People's Republic of China (2017)
60. European Commission: Recommendation on an EU mechanism for Preparedness and Management of Crises related to Migration (Migration Preparedness and Crisis Blueprint) (2020)
61. European Union Agency for Cybersecurity: Blue OLEx 2021: Testing the Response to Large Cyber Incidents (2020)
62. European Commission: European Cybersecurity Competence Network and Centre (2018)
63. European Patent Office: Quantum technologies on the rise in the Space sector
64. European Space Policy Institute: Quantum Technology in the Space Sector. Springer (2021)

Author Index

Printed in the United States
by Baker & Taylor Publisher Services